Applications of Functional Analysis and Operator Theory

This is Volume 146 in
MATHEMATICS IN SCIENCE AND ENGINEERING
A Series of Monographs and Textbooks
Edited by RICHARD BELLMAN, *University of Southern California*

The complete listing of books in this series is available from the Publisher upon request.

Applications of Functional Analysis and Operator Theory

V. HUTSON

and

J. S. PYM

University of Sheffield

1980

Academic Press

A Subsidiary of Harcourt Brace Jovanovich, Publishers

LONDON NEW YORK TORONTO
SYDNEY SAN FRANCISCO

ACADEMIC PRESS INC. (LONDON) LTD.
24/28 Oval Road,
London NW1 7DX

United States Edition published by
ACADEMIC PRESS INC.
111 Fifth Avenue
New York, New York 10003

British Library Cataloguing in Publication Data
Hutson, V
Applications of functional analysis and operator theory.—(Mathematics in science and engineering; no. 133).
1. Functional analysis
I. Title   II. Pym, J. S.   III. Series
515′.7   QA320   79-50300

ISBN 0-12-363260-9

Text set in 10/12pt Times, printed in Great Britain by Page Bros (Norwich) Ltd, Mile Cross Lane, Norwich.

Preface

It is now widely accepted that functional analysis is a tool of great power in the solution of mathematical problems arising from physical situations. Nonetheless, those whose interest is primarily in applications and who are without special training in pure mathematics face a formidable barrier in bringing the apparatus of functional analysis to bear on the questions on which it can make the largest impact, since the treatment of these problems, of which differential equations and nonlinear equations are important examples, involve technicalities which may appear extremely difficult at first sight. However, it is our conviction that the "pure" functional analysis required in a wide range of applications is not very extensive (for example, much can be done with standard Banach space theory without explicit study of topology *per se*) and that a list of techniques adequate for many purposes may be acquired by the non-specialist without excessive effort.

The first object of this book is to provide by a careful selection of material what we regard as essential abstract techniques for applications. In order to make this material as accessible as possible we give a far greater amount of explanation than is usually found in standard texts on functional analysis and we frequently illustrate the abstract theory in terms of concrete objects, sets of functions and so on, which will be familiar to the reader. With the aim of bringing out the core of the argument unencumbered by an overabundance of detail, the most general treatment possible is sometimes not attempted, and occasionally the reader is asked to accept a result without proof if the argument used is not essential to the main development.

Our second object is to show how the abstract theory works in practice. At first the problems treated are necessarily simple, but later the theory is exploited in tackling some important and relatively deep problems which occupy a central position in applications. It is clearly impossible to cover the numerous topics to which functional analysis has been applied, and we have chosen to concentrate on a basic area which might be described roughly as "the solution of equations". Thus we consider applications of the spectral theory of self-adjoint operators to ordinary differential equations, an introduction to linear elliptic partial differential equations, certain central areas of numerical analysis, and some of the principle parts of the theory of nonlinear equations. Indeed one of the most significant contributions of functional analysis is probably to the theory of nonlinear equations. Since this is a topic of great importance, and one to which much current research is directed, considerable weight has been placed on it in this book, and throughout, the nonlinear theory is developed as soon as the appropriate theoretical apparatus becomes available. Thus at an early stage the relatively elementary Contraction Mapping Principle (the Banach Fixed Point Theorem) and the Newton–Kantorovich method are introduced, and later the deep Leray–Schauder degree theory and its application to bifurcation theory are discussed.

It is intended that this book should provide the basis of a postgraduate course or M.Sc. course for applied mathematicians, theoretically inclined engineers or physicists, or should serve as an introduction for research workers who wish to acquaint themselves with some of the powerful techniques of functional analysis. The prerequisites are some familiarity with real variable theory and a little linear algebra. Much of the material in Chapters 1–5 would be familiar to anyone who has done an introductory course on functional analysis, but is included since we hope to interest readers not so equipped. The theory of Lebesgue integration presents something of a problem to the non-specialist, for the $\mathscr{L}_p$ spaces of which this is a prerequisite are essential in applications, but the theory itself is somewhat inaccessible. An outline of Lebesgue integration is presented in Chapter 2, but the reader who is unwilling to cope with its technicalities may be reassured: only a few facts (the main one is the "completeness" of $\mathscr{L}_p$) are essential in the sequel, and this chapter need be used only for reference.

The numbering system used is as follows: all theorems, lemmas, definitions and examples are numbered consecutively by chapter and section and are always referred to by their full numbers (thus Definition 3.5.7 is followed by Example 3.5.8); equations are numbered independently, again by chapter and section, and are quoted in brackets; Problem 3.15 refers to a problem at the end of Chapter 3; an asterisk indicates that a problem is on the difficult side.

A list of symbols is provided at the end of the book. It may be useful to emphasize one point in advance; script symbols ae used for vector spaces, $\mathscr{B}$ and $\mathscr{C}$ always being Banach spaces and $\mathscr{H}$ a Hilbert space.

November 1979

V. Hutson
J. S. Pym

Acknowledgements

The authors are grateful to Drs D. Burley, C. Outhwaite and P. Harley for advice in the preparation of this book. We owe a special debt to Dr J. W. Baker, Professor L. E. Fraenkel and Professor I. S. Sneddon who spent a great deal of time in assisting us.

Contents

Chapter 1

Banach Spaces

1.1 Introduction

One of the earliest successes of an "abstract" approach to practical problems was in the discussion of the linear equation $Lf = g$ where L is an $n \times n$ matrix and f and g are n-vectors. By regarding L as a linear mapping on an n-dimensional vector space reference to individual variables is avoided, and this leads to simplification and conceptual clarification. However, the equations which arise in applications are for the most part either differential or integral equations and cannot usually be reduced to the above finite dimensional form; it is one of the principal tasks of functional analysis to uncover the analogous simple algebraic structure in these more difficult situations.

In finite dimensions a considerable part of the theory may be developed without reference to the convergence of sequences of vectors; in contrast, in infinite dimensions the notion of convergence is fundamental. In order to define such a concept, the space must be provided with a measure of distance, and for most applications it is sufficient to adopt a definition which generalizes the Euclidean distance; this idea is made precise in the definition of a "norm" on a vector space. Because the approach is abstract, we find that a given infinite dimensional space of functions can often have many different norms, and this suggests that the theory will have great versatility. However, before the possibilities can be fully exploited, certain difficulties must be overcome, for while in finite dimensions the analytical properties of the space (for example, that a bounded sequence always has a convergent subsequence) are automatic consequences of the properties of real numbers, in infinite dimensions this is not so. Some of the most important ways of solving equations involve iterative procedures, and this raises the question of which sequences are guaranteed to converge. In finite dimensions the basic criterion is that of Cauchy, which says roughly that if the terms of a sequence get closer together then the sequence converges. This condition, called completeness, is not always satisfied for normed infinite dimensional spaces, but spaces which do have this property are the most important in theory and practice. They are the Banach spaces—the principal subject of this chapter.

In order to achieve as much generality and conceptual simplicity as possible the approach adopted in functional analysis is an axiomatic one. The inspiration for the axioms chosen is those properties which make finite dimensional spaces tractable. Thus the vector space axioms are derived from the algebraic laws for combining Cartesian vectors, the definition of norm from the properties of Euclidean distance, and the concept of completeness from an important property of real numbers. On the other hand, the notion of "basis" which is useful in finite dimensions, is not so fruitful in more general applications and plays no role in our development (except in the special case of Hilbert space).

A final remark concerns the conceptual framework in which infinite dimensional systems may be considered. A necessarily rather imprecise, but still useful distinction may be drawn between those properties which are primarily geometric and those which are primarily analytical. The geometric properties are those which are shared with geometry in three dimensions—examples might be the properties of straight lines or spheres. The analytical properties are typically those in which the convergence of sequences, completeness and so on play a leading role. The boundary between geometry and analysis is certainly not clearcut, and there is indeed a subtle interplay between their respective roles. Nonetheless, by regarding each function as a point in a vector space, a geometric picture may be formed which is often of considerable value—so long as it is recognized that there are limitations to its validity.

The aim of this chapter then is the development of the most basic parts of Banach space theory. In Section 2 some elements of the theory of vector spaces are recalled; there is no claim to comprehensiveness here, and only those topics which have a direct bearing on the sequel are covered. Next the idea of distance or norm is introduced, and a number of definitions (mostly familiar in the context of the real line) are given. In Section 4 we approach the heart of Banach space theory—the notion of completeness—and a number of examples of specific Banach spaces are listed; this list must be supplemented at various points in the sequel, and the important $\mathscr{L}_p$ spaces are not tackled until the necessary integration theory has been summarized in the next chapter. Finally, Hilbert spaces are introduced in the last section. These are Banach spaces with an additional "inner product" structure (modelled on the scalar product for Cartesian vectors) and their geometry bears in certain important respects a stronger resemblance to the geometry of Euclidean space.

As general references on the theory of vector spaces and linear operators we recommend Friedman (1970), Taylor (1958), and Lusternik and Sobolev (1974); the two thousand or so pages of Dunford and Schwartz (1958, 1963) contain much of what is known in this area. Although perhaps less useful from a strictly practical point of view, the beautiful introductory account of Simmons (1963) must also be mentioned.

1.2 Vector Spaces

The vector space axioms are suggested by the algebraic properties of vector addition and multiplication by scalars for three-dimensional Cartesian vectors.

1.2.1 Definition. Let $\mathscr{V}$ be a non-empty set, and suppose that any pair of elements $f, g \in \mathscr{V}$ can be combined by an operation called addition to give an element $f + g$ in $\mathscr{V}$. Assume that for any $f, g, h \in \mathscr{V}$:

(i) $f + g = g + f$;
(ii) $f + (g + h) = (f + g) + h$;
(iii) there is a unique element 0 (called zero) in $\mathscr{V}$ such that $f + 0 = f$ for all $f \in \mathscr{V}$;
(iv) for each $f \in \mathscr{V}$ there is a unique element $(-f)$ in $\mathscr{V}$ such that $f + (-f) = 0$.

In the following the scalars will either be the real numbers $\mathbb{R}$ or the complex numbers $\mathbb{C}$. Suppose any $f \in \mathscr{V}$ and any scalar α can be combined to give an element αf in $\mathscr{V}$, and assume that for any scalars α, β;

(v) $\alpha(f + g) = \alpha f + \alpha g$;
(vi) $(\alpha + \beta)f = \alpha f + \beta f$;
(vii) $(\alpha\beta)f = \alpha(\beta f)$;
(viii) $1 . f = f$.

Then $\mathscr{V}$ is called a **vector space** (or a complex vector space) if the scalar field is $\mathbb{C}$, or a **real vector space** if it is $\mathbb{R}$. The members $f, g, h, \ldots$ of $\mathscr{V}$ are known as **points**, **elements**, or **vectors** depending on which seems most appropriate in the context.

The extra generality in taking the scalar field to be the complex numbers is useful in dealing with spaces of functions, and seldom causes any difficulty. We emphasize that a vector space will *always* be complex unless it is specifically referred to as real.

Cartesian vectors have two further well known laws of combination—those of scalar and vector multiplication. The first of these has a useful generalization which will be considered in Section 1.5, but vector multiplication will not be discussed here.

Concepts such as "linear independence", "basis", and "dimension" play a major part in the theory of finite-dimensional vector spaces, but do not always have nice generalizations to infinite dimensions. The next definition summarizes most of what it is necessary to know about these notions for present purposes.

1.2.2 Definition. Let $\mathscr{V}$ be a vector space. A *finite* set $S = \{f_j\}_{j=1}^n$ of vectors in $\mathscr{V}$ is called **linearly dependent** iff there are scalars $\alpha_1, \ldots, \alpha_n$ not all of which are zero such that $\Sigma\alpha_j f_j = 0$, otherwise S is said to be **linearly independent**. An arbitrary set S of vectors in $\mathscr{V}$ is **linearly independent** iff every finite non-empty subset of S is linearly independent; otherwise it is **linearly dependent**.

If there is a positive integer n such that $\mathscr{V}$ contains n but not $n + 1$ linearly independent vectors, $\mathscr{V}$ is said to be **finite dimensional** with **dimension** n. $\mathscr{V}$ is **infinite dimensional** iff it is not finite dimensional. The finite set S of vectors in $\mathscr{V}$ is called a **basis** of $\mathscr{V}$ iff S is linearly independent and each element of $\mathscr{V}$ may be written as $\sum_1^n \alpha_j f_j$ for some $\alpha_1, \ldots, \alpha_n \in \mathbb{C}$ and $f_1, \ldots, f_n \in S$ (of course n is the dimension of $\mathscr{V}$).

Note that a basis has only been defined for a finite-dimensional space. Corresponding concepts are much less useful in infinite dimensions (see the remark after Example 1.4.20) except in the special case of Hilbert space (Section 1.5) where a valuable notion of basis can be defined.

Some examples of vector spaces are as follows.

1.2.3 Example. Let $\mathscr{V}$ be the set of ordered n-tuples (which may equivalently be regarded as finite sequences) of scalars $(f_1, \ldots, f_n)$, and let $f = (f_1, \ldots, f_n)$, $g = (g_1, \ldots, g_n)$ be arbitrary elements of $\mathscr{V}$. Then if the laws of combination of "componentwise" addition and multiplication by scalars

$$f + g = (f_1 + g_1, \ldots, f_n + g_n),$$

$$\alpha f = (\alpha f_1, \ldots, \alpha f_n) \qquad (\alpha \text{ a scalar}),$$

are adopted, a real vector space $\mathbb{R}^n$ or a (complex) vector space $\mathbb{C}^n$ are obtained depending on whether the scalars are $\mathbb{R}$ or $\mathbb{C}$. The f_i are called the components of f. $\mathbb{R}^3$ is the familiar space of three-dimensional Cartesian vectors.

1.2.4 Example. An infinite-dimensional space is obtained by generalizing $\mathbb{R}^n$ or $\mathbb{C}^n$ a little and taking infinite sequences $f = (f_n)$ as the elements of the space, but using the analogous componentwise laws of combination. This space will be denoted by the symbol ℓ, and is known as a *sequence space.*

1.2.5 Example. From the point of view of applications by far the most important vector spaces are those whose elements are functions. To illustrate the natural laws of combination, consider the set $\mathscr{V}$ of complex valued functions defined on an interval $[a, b]$. For $f, g \in \mathscr{V}$ and $\alpha \in \mathbb{C}$, define new functions $f + g$ and αf by requiring the following relations to hold for all $x \in [a, b]$:

$$(f + g)(x) = f(x) + g(x);$$

$$(\alpha f)(x) = \alpha f(x).$$

Of course $(f + g)(x)$, $(af)(x)$ denote the values of the functions $(f + g)$, αf respectively at x; these laws of combination are described as pointwise addition and multiplication by scalars. It is easy to check that the vector space axioms are satisfied, and $\mathscr{V}$ is a (complex) vector space. The real valued functions similarly form a real vector space. Obviously $\mathscr{V}$ is infinite dimensional.

In practice vector spaces of arbitrary functions are intractable, and some limitation on the class of function allowed is always imposed. One example is the space of bounded functions. Another is the particularly useful vector space $\mathscr{C}([a, b])$ of bounded continuous functions defined on $[a, b]$, and further examples will be introduced shortly.

It must be emphasized that in the previous example the points of $\mathscr{V}$ are functions. The geometrical point of view outlined in the introduction may be seen clearly here: the aim is to attempt to relate the properties of sets of functions with familiar geometric concepts, each function being thought of as a point in physical space. Pursuing this analogy, we next generalize the concepts of lines and planes through the origin in $\mathbb{R}^3$.

1.2.6 Definition. A **linear subspace** $\mathscr{M}$ of a vector space $\mathscr{V}$ is a non-empty subset of $\mathscr{V}$ which is itself a vector space under the laws of combination that hold in $\mathscr{V}$. $\mathscr{M}$ can contain the zero element only, in which case we shall write conventionally $\mathscr{M} = 0$.

The reason for including the word "linear" in the definition is to emphasize the purely algebraic character of $\mathscr{M}$, and to avoid confusion with a "closed subspace" (Definition 1.4.12) on which an additional analytical condition is imposed. A linear subspace necessarily contains zero; the generalization of an arbitrary line or plane is an *affine manifold* (any set $\{f + g : g \in \mathscr{M}\}$ where $f \in \mathscr{V}$ is fixed and $\mathscr{M}$ is a linear subspace), but this concept is less often useful.

1.2.7 Definition. Let S be a non-empty subset of a vector space $\mathscr{V}$. The set of all finite linear combinations (that is with a finite number of terms) of elements of S is called the **linear span** of S and is written $[S]$.

$[S]$ is obviously a linear subspace of $\mathscr{V}$. It is therefore clear that every vector space contains an abundance of linear subspaces, and the decomposition of $\mathscr{V}$ into linear subspaces may thus be usefully envisaged.

1.2.8 Definition. Let $\mathscr{M}$ and $\mathscr{N}$ be linear subspaces of the vector space $\mathscr{V}$, and suppose that $\mathscr{M}, \mathscr{N} \neq 0$. Assume that each $f \in \mathscr{V}$ can be written in the form $f = g + h$ for some $g \in \mathscr{M}$, $h \in \mathscr{N}$. Then we write $\mathscr{V} = \mathscr{M} + \mathscr{N}$, and say that $\mathscr{V}$ is the **vector sum** of $\mathscr{M}$ and $\mathscr{N}$. If in addition the vectors g and h are uniquely determined for every $f \in \mathscr{V}$, then $\mathscr{V}$ is said to be the **direct sum** of $\mathscr{M}$ and $\mathscr{N}$, and we write $\mathscr{V} = \mathscr{M} \oplus \mathscr{N}$.

1.2.9 Lemma. *Assume that $\mathscr{V} = \mathscr{M} + \mathscr{N}$. Then $\mathscr{V} = \mathscr{M} \oplus \mathscr{N}$ if and only if $\mathscr{M} \cap \mathscr{N} = 0$.*

1.2.10 Example. Let $\mathscr{V}$ be $\mathscr{C}([-1, 1])$ the vector space of continuous complex valued functions defined on $[-1, 1]$. Take two distinct points $x_1, x_2 \in [-1, 1]$, and for $i = 1, 2$ set $\mathscr{M}_i = \{f : f \in \mathscr{V}, f(x_i) = 0\}$. Obviously each $\mathscr{M}_i$ is a linear subspace of $\mathscr{V}$. Further $\mathscr{V} = \mathscr{M}_1 + \mathscr{M}_2$, but $\mathscr{V} \neq \mathscr{M}_1 \oplus \mathscr{M}_2$ since $\mathscr{M}_1 \cap \mathscr{M}_2 \neq 0$.

The odd and even functions form linear subspaces $\mathscr{M}_o$, $\mathscr{M}_e$ of $\mathscr{V}$, and in this case $\mathscr{V} = \mathscr{M}_o \oplus \mathscr{M}_e$.

If S_1 and S_2 are two sets, the set of ordered pairs $[f, g]$ with $f \in S_1$, $g \in S_2$ is denoted by $S_1 \times S_2$. This notation provides a convenient method of describing certain subsets. For example the square $0 \leqslant x, y \leqslant 1$ becomes $[0, 1] \times [0, 1]$. If $\mathscr{V}$ and $\mathscr{W}$ are vector spaces, $\mathscr{V} \times \mathscr{W}$ may be made into a vector space as follows.

1.2.11 Definition. Let $\mathscr{V}$ and $\mathscr{W}$ be vector spaces. For $f_1, f_2 \in \mathscr{V}$ and $g_1, g_2 \in \mathscr{W}$ set

$$[f_1, g_1] + [f_2, g_2] = [f_1 + f_2, g_1 + g_2],$$

$$\alpha[f, g_1] = [\alpha f_1, \alpha g_1] \qquad (\alpha \in \mathbb{C}).$$

The vector space $\mathscr{V} \times \mathscr{W}$ whose elements are $[f, g]$ is known as the **direct product** *of* $\mathscr{V}$ and $\mathscr{W}$. (The simplest example is $\mathbb{R}^2 = \mathbb{R} \times \mathbb{R}$).

The concept of convexity in $\mathbb{R}^3$ has a useful generalization.

1.2.12 Definition. A subset S of a vector space $\mathscr{V}$ is said to be **convex** iff $af + (1 - a)g$ lies in S for each $f, g \in S$ and for each a with $0 \leqslant a \leqslant 1$. It is equivalent to require $(af + bg)/(a + b)$ to be in S for each $f, g \in S$ and for all non-negative a, b not both zero. The **convex hull** co S of S in $\mathscr{V}$ is the smallest convex subset of $\mathscr{V}$ containing S.

1.3 Normed Vector Spaces

A vector space is a purely algebraic object, and if the processes of analysis are to be meaningful in it, a measure of distance must be supplied; the distance is known as a norm in this context. In this section it will be shown that many of the familiar analytical concepts (of which convergence is the most important) in $\mathbb{R}^3$ with the usual distance $|\cdot|$ generalize sensibly to a normed vector space, and some of the most frequently encountered examples of such spaces will be discussed. Often a vector space can be equipped with a variety of different norms; reasons for making particular choices will be discussed in the next section. It should be noted that there is no analogue for angle or perpendicularity in a general normed vector space, and we shall see, in an example, that as a consequence the geometry may be somewhat unusual. For certain norms a sensible definition of perpendicularity is possible (see Section 1.5), and then the geometry corresponds much more closely to that of Euclidean space.

Intuitively one expects distance to be a non-negative real number, to be symmetric, and to satisfy the triangle inequality. These considerations motivate the following definition.

1.3.1 Definition. Let X be a set. Associate with each pair of elements $f, g \in X$ a non-negative number $d(f, g)$ such that for all $f, g, h \in X$:

(i) $d(f, g) = 0$ iff $f = g$;
(ii) $d(f, g) = d(g, f)$;
(iii) $d(f, g) \leqslant d(f, h) + d(h, g)$ (the triangle inequality).

d is called a **metric** on X, and X equipped with a metric is known as a **metric space**.

In this definition X need not have any algebraic structure. For most vector spaces of interest it is possible to define a rather stronger concept incorporating a measure of distance from the origin. The notation $\|\cdot\|$ emphasizes that it is a generalization of the usual distance in $\mathbb{R}^3$.

1.3.2 Definition. Let $\mathscr{V}$ be a vector space, and suppose that to each element $f \in \mathscr{V}$ a non-negative number $\|f\|$ is assigned in such a way that for all $f, g \in \mathscr{V}$:

(i) $\|f\| = 0$ iff $f = 0$;
(ii) $\|\alpha f\| = |\alpha| \|f\|$ for any scalar α;
(iii) $\|f + g\| \leqslant \|f\| + \|g\|$ (the triangle inequality).

The quantity $\|f\|$ is called the **norm** of f, and $\mathscr{V}$ is known as a **normed vector space**. The symbols $\mathscr{V}, \mathscr{W}$ will henceforth always denote normed vector spaces.

A metric d is obviously obtained on setting $d(f, g) = \|f - g\|$, so all normed vector spaces are metric spaces. Normed vector spaces will play a much more prominent role here than metric spaces, the metric spaces mentioned being simply subsets of a normed vector space which are not linear subspaces. Note that while a vector space may often be equipped with more than one norm, the associated normed vector spaces are regarded as different unless the norms are the same.

The analogue of a solid sphere in $\mathbb{R}^3$ is usually called a ball in functional analysis. The terms open and closed introduced below are suggested by open and closed intervals of the real line, but are later put in a more general context.

1.3.3 Definition. Assume that a vector $f \in \mathscr{V}$ and a number r with $0 < r < \infty$ are given. The sets of points $S(f, r) = \{g : \|f - g\| < r\}$ and $\bar{S}(f, r) = \{g : \|f - g\| \leqslant r\}$ are called respectively the **open** and **closed balls** with centre f and radius r. The open and closed balls with centre the origin and radius unity are called respectively the open and closed **unit balls**.

1.3.4 Definition. A subset S of $\mathscr{V}$ is said to be **bounded** iff it is contained in some ball (of finite radius). If S is bounded, its **diameter** is the diameter of the closed ball of smallest radius containing S. The **distance**, written dist(f, S), of a point f from S is the number $\inf_{g \in S} \|f - g\|$†.

† Inf and sup (abbreviations for infimum and supremum) denote the greatest lower bound and least upper bound respectively.

1.3.5 Example. Let $\mathscr{V} = \mathbb{R}^n$. Define the *Euclidean norm* of $f = (f_1, \ldots, f_n)$ to be

$$\|f\|_2 = \left\{\sum_1^n f_i^2\right\}^{\frac{1}{2}}.$$

If $n = 3$ the norm is of course just the usual distance in $\mathbb{R}^3$. With the standard pictorial representation, the closed unit ball $\bar{S}(0, 1)$ is simply a solid sphere centre the origin and radius 1. $\bar{S}(0, 1)$ is not itself a vector space, but with $d(f, g) = \|f - g\|_2$ it is a metric space.

1.3.6 Example. This example emphasizes that the geometry in a normed vector space can have unusual features. With $\mathscr{V} = \mathbb{R}^2$, for $f = (f_1, f_2)$ define $\|f\|_1 = |f_1| + |f_2|$. Certainly $\mathscr{V}$ is then a normed vector space. However, the unit ball is square as is shown in the figure. An unpleasant consequence of

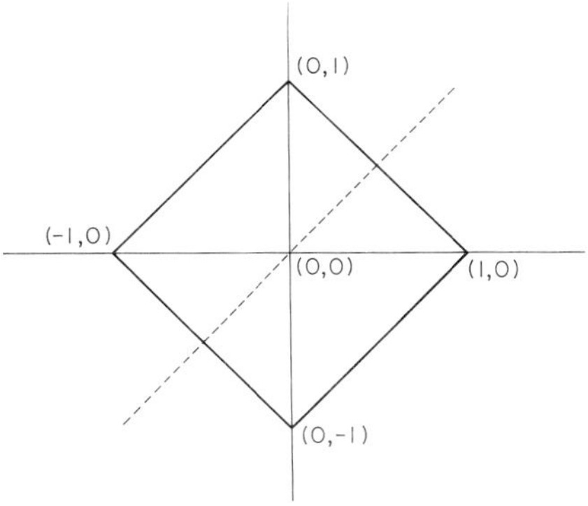

Fig. 1.1

this fact must be noted. In the usual two-dimensional space ($\mathbb{R}^2$ with the Euclidean norm), given a line l through the origin and a point P not on l, there is a unique point P' on l such that the distance PP' is a minimum. In vector space terminology this means that given a linear subspace $\mathscr{M}$ and a point f not in $\mathscr{M}$ there is a unique $g \in \mathscr{M}$ such that $\|f - g\|$ is a minimum. However, with the norm $\|\cdot\|_1$ the uniqueness is lost—as can be seen by taking $\mathscr{M}$ to be the line at the angle $\pi/4$ to the axis and $f = (1, 0)$. Then any $g \in \mathscr{M}$ has components (α, α) and $\|f - g\|_1 = |1 - \alpha| + |\alpha|$; the minimum of $\|f - g\|_1$ is unity, but this is attained for *any* α such that $0 \leqslant \alpha \leqslant 1$. In certain normed vector spaces, if $\mathscr{M}$ is infinite dimensional, the minimum distance may not be attained by any $g \in \mathscr{M}$.

Analysis in a normed vector space $\mathscr{V}$ is principally concerned with the properties of infinite sets of points. The most fundamental notion is that of the convergence of a sequence. With the help of this concept we shall be able to describe two types of subset of $\mathscr{V}$ on which interest will often centre.

1.3.7 Definition. Let (f_n) be a sequence in $\mathscr{V}$. The sequence is said to **converge** iff there is a vector $f \in \mathscr{V}$ such that $\lim \|f_n - f\| = 0$. f is called the **limit** of the sequence, and we write $f_n \to f$ or $\lim f_n = f$.

The limit is unique, for if $f_n \to f$ and $f_n \to g$, by the triangle inequality

$$\|f - g\| = \|f - f_n + f_n - g\| \leqslant \|f - f_n\| + \|f_n - g\|,$$

and the right-hand side tends to zero as $n \to \infty$, whence $f = g$.

1.3.8 Definition. Let S be a subset of $\mathscr{V}$, and define a new subset $\bar{S} \subset \mathscr{V}$ called the **closure** of S by requiring that $f \in \bar{S}$ iff there is a sequence of (not necessarily distinct) points of S converging to f. S is said to be **closed** iff $S = \bar{S}$.

If S_1, S_2 are two subsets of $\mathscr{V}$ and $S_1 \subset S_2$, then S_1 is said to be **closed in** S_2 iff S_1 is the intersection of a closed set with S_2. The closure of S_1 in S_2 is $\bar{S}_1 \cap S_2$.

1.3.9 Definition. A subset S of $\mathscr{V}$ is said to be **open** iff either of the following equivalent conditions holds:

(i) its complement $\mathscr{V} \backslash S$ is closed;
(ii) for each $f \in S$ there is an open ball centre f contained in S.

If $S_1 \subset S_2 \subset \mathscr{V}$, S_1 is said to be **open in** S_2 iff it is the intersection of an open set with S_2.

A **neighbourhood** of a point is any set which contains an open set which itself contains the point.

Since it is legitimate for the sequence in Definition 1.3.8 to consist of coincident points, a set containing just one point is closed. Note that a point f lies in the closure of S_1 in S_2 iff it lies in S_2 and is the limit of a sequence of points in S_1. A set need be neither open nor closed. It can in general also be both, although in a normed vector space only the whole space and the empty set have this property. A rather obvious remark is that an open ball is open and a closed ball is closed. Further properties are listed in Problems 1.3–1.7.

1.3.10 Definition. A point f is an **interior point** of $S \subset \mathscr{V}$ iff there is a neighbourhood of f contained in S. The **interior** S^0 of S is the set of interior points of S (and is open).

A point f is a **boundary point** of S iff every neighbourhood of f contains points of both S and its complement $\mathscr{V}\backslash S$. The **boundary** ∂S of S is the set of boundary points of S.

1.3.11 Example. We illustrate the concepts above in an easy case, more difficult examples will come later. Let a, b be the end points of a finite interval. Then $(a, b) = \{x : a < x < b\}$ is open, and $[a, b] = \{x : a \leqslant x \leqslant b\}$ is closed. $[a, b)$ is neither open nor closed. It is not open since a ball centre a will contain points not in $[a, b)$, and it is not closed since b is the limit of points in $[a, b)$ but $b \notin [a, b)$.

Suppose next that S is the set of rational points in $[a, b]$. Then again S is neither open nor closed, for if $c \in S$, a ball centre c will contain irrational points, while a sequence of rational points may have an irrational limit. Further S contains no open subset, hence its interior is empty. On the other hand its boundary is $[a, b]$ a set much larger than S itself. The intuitive picture of a boundary as a "thin set" is misleading here.

So far only finite dimensional normed vector spaces have been encountered. As the first infinite dimensional example, take the sequence space ℓ (Example 1.2.4) and define the (possibly infinite) quantities

$$\|f\|_p = \{\Sigma\, |f_n|^p\}^{1/p} \qquad (1 \leqslant p < \infty), \tag{1.3.1}$$

$$\|f\|_\infty = \sup |f_n|, \tag{1.3.2}$$

where $f \in \ell$ is the sequence (f_n). We shall show that $\|\cdot\|_p$ is a norm on the subset of ℓ consisting of those f such that $\|f\|_p$ is finite. Evidently the only difficulty is in verifying the triangle inequality. This is taken care of by Minkowski's inequality below; Hölder's inequality will be useful later.

Two numbers p, q with $1 \leqslant p, q \leqslant \infty$ are called *conjugate indices* iff $p^{-1} + q^{-1} = 1$; if $p = 1$, $q = \infty$.

1.3.12 Theorem. *Assume that $1 \leqslant p \leqslant \infty$, and let q be the conjugate index. Then for any $f, g \in \ell$ (infinite values being allowed):*

(i) $\|fg\|_1 \leqslant \|f\|_p \|g\|_q$ *(Hölder's inequality);*

(ii) $\|f + g\|_p \leqslant \|f\|_p + \|g\|_p$ *(Minkowski's inequality).*

Proof. (i) We may assume that $0 < \|f\|_p, \|g\|_q < \infty$, for otherwise the inequality is obvious. If $p = 1$,

$$\|fg\|_1 = \Sigma\, |f_n g_n| \leqslant (\Sigma\, |f_n|) \,.\, \sup|g_n| = \|f\|_1 \|g\|_\infty,$$

and a similar argument holds if $q = 1$. For $p, q > 1$ start by noting the elementary inequality $ab \leqslant p^{-1}a^p + q^{-1}b^q$ $(a, b \geqslant 0)$, from which it follows with $a = |f_n|/\|f\|_p$, $b = |g_n|/\|g\|_p$ that

$$\frac{|f_n g_n|}{\|f\|_p \|g\|_q} \leqslant p^{-1}\frac{|f_n|^p}{\|f\|_p^p} + q^{-1}\frac{|g_n|^q}{\|g\|_q^q}.$$

Summation over n gives

$$\frac{\|fg\|_1}{\|f\|_p \|g\|_q} \leqslant p^{-1}\frac{\Sigma |f_n|^p}{\|f\|_p^p} + q^{-1}\frac{\Sigma |g_n|^q}{\|g\|_q^q} = p^{-1} + q^{-1} = 1.$$

(ii) Observing that $(p-1)q = p$, and using Hölder's inequality, we have

$$\begin{aligned}\Sigma |f_n + g_n|^p &= \Sigma |f_n|\,|f_n + g_n|^{p-1} + \Sigma |g_n|\,|f_n + g_n|^{p-1} \\ &\leqslant \|f\|_p\{\Sigma |f_n + g_n|^{(p-1)q}\}^{1/q} + \|g\|_p\{\Sigma |f_n + g_n|^{(p-1)q}\}^{1/q},\end{aligned}$$

which yields the result if $\|f + g\|_p \neq 0, \infty$. In the first case the conclusion is obvious. In the second we deduce from the inequality

$$|f_n + g_n|^p \leqslant (2\max(|f_n|, |g_n|))^p \leqslant 2^p(|f_n|^p + |g_n|^p)$$

on summation that either $\|f\|_p$ or $\|g\|_p$ is infinite, whence the result. □

1.3.13 Definition. Take some $p \geqslant 1$. The normed vector space consisting of all $f = (f_1, f_2, \ldots)$ in ℓ such that $\|f\|_p < \infty$ is denoted by ℓ_p.

1.3.14 Example. The extra difficulties associated with convergence in an infinite dimensional space may be illustrated as follows.

Evidently $\mathbb{R}^n$ and $\mathbb{C}^n$ are both normed vector spaces when equipped with $\|\cdot\|_p$. It is not hard to show that no matter what $p \geqslant 1$ is chosen, a sequence $(f^{(k)})$ where $f^{(k)} = (f_1^{(k)}, \ldots, f_n^{(k)})$ is convergent iff $(f_j^{(k)})$ is convergent for each j—that is iff each sequence of components is convergent.

In ℓ_p this condition is not sufficient, for consider the sequence $(f^{(k)})$ in ℓ_∞ where $f^{(k)}$ is the vector whose first k components are zero and whose other components are unity. Certainly $\lim_{k\to\infty} f_j^{(k)} = 0$ for every j, and it follows that if $(f^{(k)})$ has a limit this must be zero, but $\|f^{(k)}\|_\infty = 1$ for each k, so the sequence is not convergent. Further, it is a simple exercise to construct a sequence convergent in ℓ_∞ but not in ℓ_1, and this shows that convergence in ℓ_p depends on the index p.

Normed vector spaces of continuous and differentiable functions are of great importance in applications, so the possibilities will be investigated in some detail.

1.3.15 Definition. Let Ω be a subset of $\mathbb{R}^n$, and suppose that f is a complex valued function defined on Ω. f is said to be **continuous at the point** $x_0 \in \Omega$ iff one of the following equivalent conditions holds:

(i) For each $\varepsilon > 0$ there exists $\delta > 0$ such that $|f(x) - f(x_0)| < \varepsilon$ whenever $x \in \Omega$ and $|x - x_0| < \delta$;
(ii) For each sequence (x_n) in Ω with limit x_0, $\lim f(x_n) = f(x_0)$.

f is said to be **continuous** iff it is continuous at every point of Ω.

1.3.16 Definition. f is said to be **uniformly continuous** on Ω iff for each $\varepsilon > 0$ there exists a $\delta > 0$ such that $|f(x) - f(x_0)| < \varepsilon$ whenever $x, x_0 \in \Omega$ and $|x - x_0| < \delta$.

For Ω the finite interval $[a, b]$, continuity at a or b is thus to be interpreted as continuity from the right or left respectively. Of course if f is continuous on $[a, b]$ it is bounded, but f need not be bounded if it is continuous only on (a, b). For functions of several variables, continuity in the sense of the definition is sometimes referred to as "joint continuity" to distinguish it from "separate continuity" which means that the function is continuous in each variable in turn when the other variables are fixed. For example, if f is a function of two variables, separate continuity requires only that $f(x, \cdot)$† and $f(\cdot, y)$ should be continuous for fixed x and y respectively. Uniform continuity is in general stronger than continuity since one δ must serve for every $x_0 \in \Omega$, but if $\Omega \subset \mathbb{R}^n$ is closed and bounded these concepts are in fact equivalent (Theorem 5.3.2).

1.3.17 Definition. Let Ω be a subset of $\mathbb{R}^n$. The vector space of bounded continuous complex valued functions defined on Ω is denoted by $\mathscr{C}(\Omega)$.

The space $\mathscr{C}(\Omega)$ may be normed in a number of ways. Set first

$$\|f\| = \sup_{x \in \Omega} |f(x)| . \tag{1.3.3}$$

Then $|f(x) + g(x)| \leqslant |f(x)| + |g(x)|$ for each x, and it follows that $\|f + g\| \leqslant \|f\| + \|g\|$. Evidently therefore $\|\cdot\|$ is a norm on $\mathscr{C}(\Omega)$.

1.3.18 Definition. The norm defined by (1.3.3) will be called the **sup norm**. Another term that is sometimes used is "uniform norm", but this will not be employed here to avoid confusion with the other uses of uniform in functional analysis.

† The notation $f(x, \cdot)$ will frequently be used. It means that x is fixed and the function is regarded as a function of its second argument only.

$\mathscr{C}(\Omega)$ with the sup norm is thus a normed vector space. This space is of the utmost importance and will be used frequently in the sequel. Another possible norm is that defined by

$$\|f\|_1 = \int_\Omega |f(x)|\,\mathrm{d}x. \tag{1.3.4}$$

For reasons which will become clear in the next section, this is not a suitable norm on $\mathscr{C}(\Omega)$, and the sup norm is the first choice in most problems involving continuous functions. A number of generalizations of $\mathscr{C}(\Omega)$ are suggested by a consideration of differential equations.

1.3.19 Example. One method (see Theorem 4.3.11) of solving the initial value problem for the equation

$$\frac{\mathrm{d}f(x)}{\mathrm{d}x} = \psi[x, f(x)] \tag{1.3.5}$$

on the interval $[a, b]$ is based on rewriting (1.3.5) as an integral equation. If a solution of this in $\mathscr{C}([a, b])$ can be found, then under mild conditions on ψ it is easy to show that the solution is differentiable and satisfies (1.3.5). Analogously for an elliptic partial differential equation on $\Omega \subset \mathbb{R}^n$, the use of a Green's function leads to an integral equation which may be tackled in $\mathscr{C}(\Omega)$.

1.3.20 Example. The initial value problem for the simultaneous equations

$$\frac{\mathrm{d}f_j(x)}{\mathrm{d}x} = \psi_j[x, f_1(x), \ldots, f_n(x)] \qquad (j = 1, \ldots, m) \tag{1.3.6}$$

may be treated similarly often without any appreciable increase in difficulty by slightly generalizing $\mathscr{C}([a, b])$.

1.3.21 Definition. Let f denote the $\mathbb{C}^m$ valued function $(f_1, \ldots, f_m)$ where for each j, f_j is a bounded continuous complex valued function defined on a subset Ω of $\mathbb{R}^n$. The set of such functions with the laws of combination

$$(f + g)(x) = (f_1(x) + g_1(x), \ldots, f_m(x) + g_m(x)),$$
$$(\alpha f)(x) = (\alpha f_1(x), \ldots, \alpha f_m(x)) \qquad (\alpha \in \mathbb{C}),$$

is a vector space, which will be denoted by $\mathscr{C}(\Omega, \mathbb{C}^m)$; $\mathscr{C}(\Omega, \mathbb{R}^m)$ of course denotes the corresponding real space of $\mathbb{R}^m$ valued functions. The sup norm is defined by

$$\|f\| = \max_{1 \leqslant j \leqslant m} \sup_{x \in \Omega} |f_j(x)|.$$

With $\Omega \subset \mathbb{R}$, the set of simultaneous equations (1.3.6) may then be written compactly as

$$\frac{\mathrm{d}f(x)}{\mathrm{d}x} = \psi[x, f(x)],$$

and the corresponding integral equation tackled in $\mathscr{C}(\Omega, \mathbb{C}^m)$—see Theorem 4.3.11.

1.3.22 Example. An alternative (and in some ways more attractive) method of dealing with a differential equation is based on a direct attack on the equation itself and avoids the intermediate stage of constructing an integral equation. Spaces of differentiable functions may therefore be expected to play a role. Because of the limiting processes involved in defining derivatives, difficulties are encountered if Ω is an arbitrary set; this explains the restriction on Ω in the next definition.

1.3.23 Definition. Let Ω be an open subset of $\mathbb{R}^n$, and let k be a positive integer. The vector space (with the usual laws of combination) consisting of all $\mathbb{C}^m$ valued functions defined on Ω such that all partial derivatives up to and including those of order k of all components are bounded and continuous is denoted by $\mathscr{C}^k(\Omega, \mathbb{C}^m)$. The vector space $\mathscr{C}^\infty(\Omega, \mathbb{C}^m)$ consists of those functions lying in $\mathscr{C}^k(\Omega, \mathbb{C}^m)$ for every $k \geqslant 0$, that is $\mathscr{C}^\infty(\Omega, \mathbb{C}^m) = \bigcap_{k=1}^{\infty} \mathscr{C}^k(\Omega, \mathbb{C}^m)$.

$\mathscr{C}^k(\overline{\Omega}, \mathbb{C}^m)$ consists of those continuous functions defined on $\overline{\Omega}$ which on Ω have bounded and *uniformly* continuous partial derivatives up to and including those of order k. (For $n > 1$ this avoids difficulties with the definition of derivatives on $\partial\Omega$ which may not be a smooth set). Also $\mathscr{C}^\infty(\overline{\Omega}, \mathbb{C}^m) = \bigcap_{k=1}^{\infty} \mathscr{C}^k(\overline{\Omega}, \mathbb{C}^m)$.

Finally, it is sometimes convenient to exclude the boundary from consideration. The spaces $\mathscr{C}_0^k(\Omega', \mathbb{C}^m)$ where $\Omega \subset \Omega' \subset \overline{\Omega}$ consist of those functions in $\mathscr{C}^k(\Omega, \mathbb{C}^m)$ each of which have bounded support† contained in the interior of Ω' (the support may vary from function to function).

The symbols $\mathscr{C}^k(\Omega), \ldots$ and $\mathscr{C}^k(\Omega, \mathbb{R}^m), \ldots$ denote of course the corresponding subsets of $\mathscr{C}(\Omega)$ and $\mathscr{C}(\Omega, \mathbb{R}^m)$ respectively.

For the finite interval $[a, b]$, a function is therefore in $\mathscr{C}^1([a, b])$ iff it has a continuous derivative on (a, b) and has left and right derivatives at b, a respectively which are the limits of the derivatives in the interior.

† The *support* of a function is the closure of the set on which it is non-zero.

The sup norm is of course a norm on $\mathscr{C}^k(\overline{\Omega})$. Another possibility may be illustrated by taking $\Omega = (a, b)$ and setting

$$\|f\|_{\mathscr{C}^k} = \sum_{j=0}^{k} \sup_{x \in [a, b]} |f^{(j)}(x)|, \tag{1.3.7}$$

where $f^{(j)}$ denotes the jth derivative of f; this norm turns out to be a much more suitable basis for analysis in $\mathscr{C}^k(\overline{\Omega})$. Corresponding norms may be defined when Ω is a subset of $\mathbb{R}^n$ for $n > 1$ by summing over the partial derivatives.

This section closes with illustrations in $\mathscr{C}([0, 1])$ of some of the main concepts defined above.

1.3.24 Example. Suppose first that $\mathscr{C}([0, 1])$ is equipped with the sup norm. If $g(x) = 2 - x^2$, the closed ball $\bar{S}(g, \frac{1}{2})$ centre g and radius $\frac{1}{2}$ consists of those continuous functions f whose graphs lie in the shaded region of Fig. 1.2. For

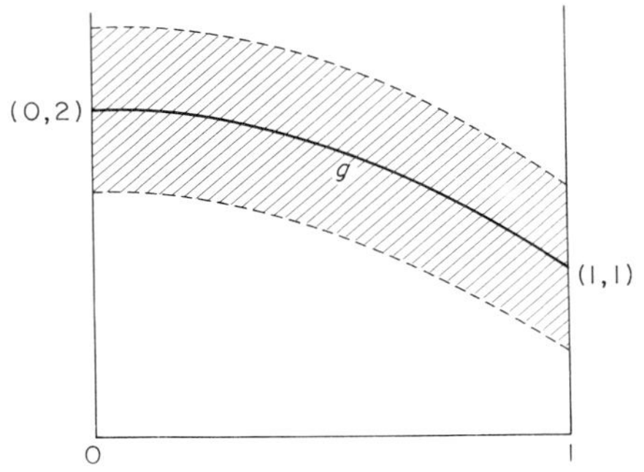

Fig. 1.2 The ball $\pi(g, \frac{1}{2})$ in $\mathscr{C}([0, 1])$ with the sup norm, consisting of all continuous functions whose graphs lie in the shaded region.

the analogous open ball $S(g, \frac{1}{2})$ the graphs must not meet the outer lines; its boundary $\partial S(g, \frac{1}{2})$ is the subset of $\bar{S}(g, \frac{1}{2})$ of functions whose graphs meet these lines.

If instead the norm $\|\cdot\|_1$ defined by (1.3.4) is adopted, no easy graphical description of the corresponding balls is possible, for these contain functions close to g for most x but which have long thin spikes.

A comparison of the concept of "closeness" for two functions f, g in the sup norm and in $\|\cdot\|_1$ shows that the sup norm bounds the difference $|f(x) - g(x)|$ for every x, whereas $\|\cdot\|_1$ only restricts the average value of this difference. Obviously convergence in sup norm is a very strong condition—

Suppose next that $\mathscr{V}$ is $\mathscr{C}([-1, 1])$ with the norm $\|f\|_1 = \int_{-1}^{1} |f(x)| \, dx$, and consider the sequence (f_n) with f_n as in the figure. An easy calculation confirms that (1.4.1) is certainly satisfied, whence (f_n) is Cauchy. However, (f_n) does not have a continuous limit. Indeed in a crude sense (f_n) converges to the function which is 1 for $-1 \leqslant x \leqslant 0$ and is zero for $0 < x \leqslant 1$, and this function is not continuous. Thus (f_n) is not convergent in $\mathscr{C}([-1, 1])$ equipped with the norm $\| \cdot \|_1$.

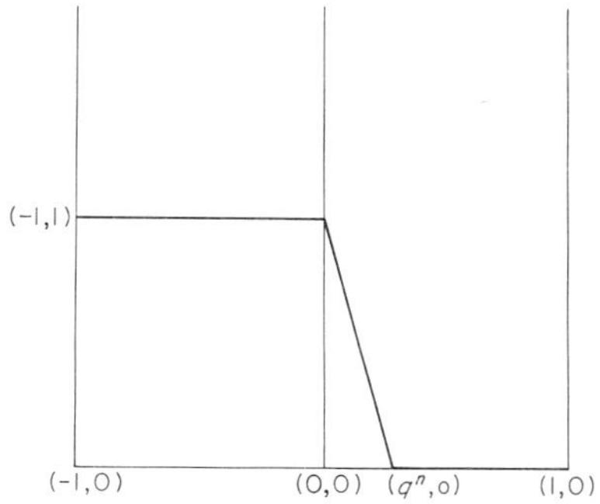

Fig. 1.3 The bold line represents f_n.

Two conclusions may be drawn from this example. First: *a Cauchy sequence in a normed vector space need not be convergent.* Secondly, if all Cauchy sequences are convergent, as expected convergence does indeed follow from (1.4.1), while if such a strong condition as (1.4.1) does not ensure convergence, then analysis in such a space is likely to present serious difficulties. Attention will therefore usually be restricted from here on to spaces in which all Cauchy sequences are convergent.

1.4.4 Definition. A set S in a normed vector space $\mathscr{V}$ is said to be **complete** iff each Cauchy sequence in S converges to a point of S. $\mathscr{V}$ itself is known as a complete normed vector space or a **Banach space** iff it is complete. Throughout the symbols $\mathscr{B}$ and $\mathscr{C}$ will *always* denote Banach spaces.

The simplest Banach spaces are $\mathbb{R}^n$ and $\mathbb{C}^n$—with any norm. Before giving further examples, we attempt to elucidate further the role of completeness.

An elementary result states that an absolutely convergent series of real numbers is convergent. The next lemma shows that this is essentially a consequence of the completeness of $\mathbb{R}$.

1.4.5 Definition. Let (f_n) be a sequence in $\mathscr{V}$. The series Σf_n is called **absolutely convergent** iff $\Sigma \|f_n\| < \infty$, and **convergent** iff $\sum_1^N f_n$ converges as $N \to \infty$ to an element f of $\mathscr{V}$, f being known as the **sum** of the series.

1.4.6 Lemma. *A normed vector space is complete if and only if every absolutely convergent series in it is convergent. In a Banach space the terms in the series Σf_n may be rearranged without affecting its sum if the series is absolutely convergent.*

Proof. The only difficulty comes in proving the if part of the first statement; we need to show that if every absolutely convergent series is convergent, then every Cauchy sequence is convergent. The first step is to show that every Cauchy sequence (f_n) has, under the stated conditions, a convergent subsequence. If $a_n = \sup_{m>n} \|f_n - f_m\|$, from the definition of a Cauchy sequence, $\lim a_n = 0$. There is therefore a subsequence, (a_{n_j}) say†, such that $a_{n_j} \leqslant j^{-2}$ for all j. Thus

$$\|f_{n_j} - f_{n_{j+1}}\| \leqslant j^{-2}$$

and it follows that the series Σg_j with $g_j = f_{n_j} - f_{n_{j+1}}$ is absolutely convergent, and so by assumption convergent. Since

$$f_{n_i} = \sum_{j=1}^{i} (f_{n_j} - f_{n_{j+1}}) = \sum_{j=1}^{i} g_j,$$

the subsequence (f_{n_i}) is convergent.

The second and final step is to prove that the original sequence (f_n) is itself convergent. Now if $\lim_i f_{n_i} = f$, then

$$\|f_n - f\| \leqslant \|f_n - f_{n_i}\| + \|f_{n_i} - f\|.$$

Since (f_n) is Cauchy, the first term on the right is arbitrarily small for large enough n and i, and the other term tends to zero as $i \to \infty$. Hence $\lim f_n = f$, and (f_n) is convergent. To prove that rearrangement is legitimate, use the argument employed for a series whose terms are complex numbers. □

† As frequently happens the proof requires the use of a subsequence of a sequence (f_n), and a comment on the notation to be used must be made. Sometimes the exact details of the subsequence do not matter (for example if all that is required is the existence of *some* subsequence) then it is common practice to use (f_n) for the subsequence itself, the remark "the subsequence still being denoted by (f_n)" being sufficient to alert the reader to the switch. In most other cases the subsequence will be denoted by (f_{n_j}), the terms being $f_{n_1}, f_{n_2}, \ldots$. Occasionally it will be necessary to take a chain of subsequences each of which is a subsequence of the previous subsequence, and in this case the kth subsequence will be denoted by $(f_{n,k})$, the terms being $f_{1,k}, f_{2,k}, \ldots$.

A Cauchy sequence (f_n) in a set S may be regarded as "potentially convergent" in that its terms get closer together as $n \to \infty$, a property possessed by convergent sequences. Whether or not the sequence fulfills its potentiality depends roughly on whether S is "large enough" or complete in the present terminology. Consider for example the subset $S = (0, 1]$ of $\mathbb{R}$. The sequence (n^{-1}) is evidently Cauchy, but its limit in $\mathbb{R}$ is 0 which does not lie in S. S is therefore not complete. On the other hand $\bar{S} = [0, 1]$ is complete. In view of the fact that the closure of S is complete, one may speculate that there is a relation between closed and complete sets, and this is confirmed by the next result.

1.4.7 Lemma. *A subset S of a Banach space $\mathscr{B}$ is complete if and only if S is closed (in $\mathscr{B}$).*

Proof. If S is closed, a Cauchy sequence (f_n) in S is convergent (since $\mathscr{B}$ is complete) to some point $f \in \mathscr{B}$, and from the definition of a closed set, $f \in S$. On the other hand suppose that S is complete. For any $f \in \bar{S}$ there is a sequence, (f_n) say, in S with $\lim f_n = f$. However, (f_n) is Cauchy, and as S is complete, $f = \lim f_n$ lies in S. Thus $\bar{S} \subset S$, which proves that S is closed. □

It should be noted that this result is true in general only if the set in which the closure is taken is itself complete. This may be seen by taking $S_2 = (0, 1]$ and choosing S_1 to be the set composed of the points n^{-1} for $n = 1, 2, \ldots$. Then S_1 is closed in S_2, but S_1 is not complete. However, the fact that closed = complete in a Banach space is extremely useful.

The setting for analysis throughout this account will be a Banach space. Our primary task must therefore be to show that an adequate supply of such spaces can be found. Attention is first focused on the spaces of continuous and differentiable functions discussed in the last section.

1.4.8 Theorem. *Let Ω be a subset of $\mathbb{R}^n$. The space $\mathscr{C}(\Omega)$ of bounded continuous complex valued functions with the sup norm is a Banach space.*

Proof. To show that $\mathscr{C}(\Omega)$ is complete, we shall regard it as a subset of the space $\mathscr{V}$ of bounded functions on Ω. With the sup norm $\mathscr{V}$ is obviously a normed vector space, so in view of Lemma 1.4.7 it is enough to prove two facts, first that $\mathscr{C}(\Omega)$ is closed (in $\mathscr{V}$), and secondly that $\mathscr{V}$ is complete.

To prove that $\mathscr{C}(\Omega)$ is closed, we verify that $\overline{\mathscr{C}(\Omega)} \subset \mathscr{C}(\Omega)$. Suppose that $f \in \overline{\mathscr{C}(\Omega)}$. Then from the definition of the closure, given $\varepsilon > 0$ there is a $g \in \mathscr{C}(\Omega)$ such that $\|f - g\| < \varepsilon$. f is therefore bounded, and also $|f(x_0) - g(x_0)| < \varepsilon$ for all $x_0 \in \Omega$. Now for each $x_0 \in \Omega$, since g is continuous, there is a $\delta > 0$ such that $|g(x) - g(x_0)| < \delta$ for $|x - x_0| < \delta$. Thus

$$\begin{aligned}|f(x) - f(x_0)| &= |[f(x) - g(x)] + [g(x) - g(x_0)] + [g(x_0) - f(x_0)]| \\ &\leqslant |f(x) - g(x)| + |g(x) - g(x_0)| + |g(x_0) - f(x_0)| \\ &< 3\varepsilon,\end{aligned}$$

and since ε is arbitrary, it follows that f is continuous. As f is also bounded, it is in $\mathscr{C}(\Omega)$, which proves that $\mathscr{C}(\Omega)$ is closed.

To show that $\mathscr{V}$ is complete, let (f_n) be a Cauchy sequence in $\mathscr{V}$. For any fixed $x \in \Omega$, $|f_n(x) - f_m(x)| \leqslant \|f_n - f_m\|$, and as (f_n) is Cauchy, $\lim_{m,n} \|f_n - f_m\| = 0$. Therefore for each fixed x, $(f_n(x))$ is a Cauchy sequence of real numbers, and as $\mathbb{R}$ is complete, this sequence has a finite limit, $f(x)$ say. f is thus the pointwise limit of (f_n)—that is $\lim f_n(x) = f(x)$ for each x— and is evidently a strong candidate for being the limit in norm of (f_n). That this is indeed the case follows from the fact that since (f_n) is Cauchy, for each $\varepsilon > 0$ there is an n_0 such that $\|f_n - f_m\| < \varepsilon$ whenever $m, n \geqslant n_0$, whence

$$\begin{aligned}\|f - f_n\| &= \sup_{x\in\Omega} |f(x) - f_n(x)| \\ &\leqslant \sup_{x\in\Omega}\sup_{m\geqslant n_0} |f_m(x) - f_n(x)| \\ &= \sup_{m\geqslant n_0}\sup_{x\in\Omega} |f_m(x) - f_n(x)| \\ &= \sup_{m\geqslant n_0} \|f_m - f_n\| \\ &\leqslant \varepsilon.\end{aligned}$$

Finally as the Cauchy sequence (f_n) is necessarily bounded (Problem 1.13) and $f_n \to f$, f is a bounded function and so lies in $\mathscr{V}$. □

1.4.9 Corollary. *Let Ω be a subset of $\mathbb{R}^n$. The space $\mathscr{C}(\Omega, \mathbb{C}^m)$ of bounded continuous functions mapping Ω into $\mathbb{C}^m$ when equipped with the sup norm (Definition 1.3.2.1) is a Banach space.*

The next question that arises is under what norms the spaces $\mathscr{C}^k(\Omega)$ of differentiable functions (Definition 1.3.23) are complete. Certainly the sup norm will not do (the reader may readily construct a sequence in $\mathscr{C}^1([0, 1])$ which converges in sup norm to a function whose derivative is not continuous), and the most obvious course of action is to adopt a stronger norm such as (1.3.7) restricting also the derivatives. The following specimen result is enough for present purposes; its proof is easily deduced from the previous theorem.

1.4.10 Lemma. *Let Ω be an open subset of $\mathbb{R}^n$, and consider functions $f = (f_1, \ldots, f_m)$ mapping Ω into $\mathbb{C}^m$. With the norm*

$$\|f\|_{\mathscr{C}^1} = \sum_{i=1}^{m} \sup_{x \in \Omega} |f_i(x)| + \sum_{i=1}^{m} \sum_{j=1}^{n} \sup_{x \in \Omega} \left| \frac{\partial f_i(x)}{\partial x_j} \right|,$$

$\mathscr{C}^1(\Omega, \mathbb{C}^m)$ is a Banach space.

Use of the sup norm imposes an extremely strong restriction on the convergence of sequences of functions, and in applications it is sometimes natural to use some notion of "average" convergence such as that measured by the norm $\|f\|_1 = \int_\Omega |f(x)|\, dx$. We have already noted that $\mathscr{C}(\Omega)$ is not complete in this norm, but it turns out that a rather larger class of "integrable functions" do form a Banach space $\mathscr{L}_1$ in $\|\cdot\|_1$. The $\mathscr{L}_p$ spaces of which $\mathscr{L}_1$ is an example are a second class of Banach spaces important in applications. Because of technical difficulties connected with the inadequacy of the Riemann integral in this context, the discussion of these spaces is deferred until the next chapter. We close our listing of Banach spaces temporarily with the remark that the ℓ_p spaces (Definition 1.3.13) are Banach spaces. The proof is an easier version of the argument to be used in Theorem 2.5.5 and is omitted.

1.4.11 Theorem. *Suppose that $1 \leqslant p \leqslant \infty$. The sequence space ℓ_p with the norm $\|\cdot\|_p$ is a Banach space.*

For the rest of this section some concepts useful in Banach space theory are discussed. In view of the importance of completeness, linear subspaces which are complete (or what is the same, closed, see Lemma 1.4.7) may be expected to play a significant part. Such a subspace is of course itself a Banach space on its own.

1.4.12 Definition. A linear subspace $\mathscr{M}$ of $\mathscr{B}$ which is closed (in $\mathscr{B}$) is called a **closed subspace**.

1.4.13 Definition. Let S be a subset of $\mathscr{B}$. The closed subspace $\overline{[S]}$ is called the **closed linear span** of S.

1.4.14 Example. Take $\mathscr{C}([a, b])$ with the sup norm. Fix some $x_0 \in [a, b]$ and let $\mathscr{M}$ be the linear subspace consisting of those functions in $\mathscr{C}([a, b])$ which are zero at x_0. Suppose (f_n) is a convergent sequence in $\mathscr{M}$ with limit f. Then $|f_n(x_0) - f(x_0)| \leqslant \|f_n - f\| \to 0$ as $n \to \infty$, and since $f_n(x_0) = 0$ for all n, therefore $f(x_0) = 0$. Thus $f \in \mathscr{M}$, and it follows that $\mathscr{M}$ is a closed subspace.

It may be noted that $\mathcal{M}$ contains no open subset of $\mathscr{C}([a, b])$, for such a subset would necessarily contain a ball of non-zero radius, and there would be some f in this ball with $f(x_0) \neq 0$. The interior of $\mathcal{M}$ is therefore empty and every $f \in \mathcal{M}$ is a boundary point. Of course if $\mathcal{M}$ is regarded as a Banach space in its own right, it is equipped with a plentiful supply of open sets, that is of sets open in $\mathcal{M}$. In fact every proper subspace of a Banach space has empty interior.

The idea of approximating a set, and in particular the whole space $\mathscr{B}$, by another smaller set may be made precise as follows.

1.4.15 Definition. Suppose that $S_1 \subset S_2 \subset \mathscr{B}$. S_1 is said to be **dense** in S_2 iff the closure of S_1 in S_2 is S_2 itself.

Dense sets are important in both theory and applications. The technique of showing that a certain property holds for a set S dense in $\mathscr{B}$ rather than for $\mathscr{B}$ itself, and proving independently (often by a continuity argument) that the property may be taken over to the whole of $\mathscr{B}$, will be used frequently in the theory. In applications the idea of a dense set lies at the heart of much of numerical analysis and approximation theory. For example, in finding a continuous solution of say an integral equation, it is enough to search for approximate solutions among the set S of continuous piecewise linear functions, for S is dense in $\mathscr{C}(\Omega)$, so the existence of an adequate approximation in S to the solution is guaranteed. Sets of polynomials, piecewise quadratic functions, and splines are further examples in common use. The following results give two of the most useful sets dense in $\mathscr{C}(\Omega)$ in the sup norm. The proof of the second is easy; for the first see Simmons (1963).

1.4.16 Weierstrass' Theorem. *Suppose that Ω is a closed bounded set in $\mathbb{R}^n$. The set of polynomials in n variables is dense in $\mathscr{C}(\Omega)$ in the sup norm.*

1.4.17 Lemma. *Let $[a, b]$ be a finite interval. The set of continuous piecewise linear functions is dense in $\mathscr{C}([a, b])$ in the sup norm. Indeed it is enough to restrict the set further to those functions piecewise linear on equal subdivisions of $[a, b]$.*

We have observed that $\mathscr{C}(\Omega)$ is complete in the sup norm but not in $\|\cdot\|_1$, and we next ask in what manner a norm may be modified without loss of completeness.

1.4.18 Definition. The two norms $\|\cdot\|_a$ and $\|\cdot\|_b$ on a vector space $\mathscr{V}$ are **equivalent** iff there exist strictly positive real numbers c_1, c_2 such that $c_1\|f\|_a \leqslant \|f\|_b \leqslant c_2\|f\|_a$ for all $f \in \mathscr{V}$.

1.4.19 Lemma. *If a sequence in a normed vector space is convergent, then it is convergent in any equivalent norm. If $\mathscr{B}$ is a Banach space, it is a Banach space in any equivalent norm.*

The proof is easy and is left as an exercise. In a finite dimensional space all norms are equivalent, and every finite dimensional normed vector space is a Banach space. Note, however, that the *rate* of convergence of a sequence may be considerably affected by the choice of norm—see Problem 1.11.

1.4.20 Example. Take $a > 0$ and for $f \in \mathscr{C}([0, 1])$ define

$$\|f\|_* = \sup_{x \in [0,1]} |e^{-ax} f(x)|.$$

Then

$$e^{-a} \sup_{x \in [0,1]} |f(x)| \leqslant \sup_{x \in [0,1]} |e^{-ax} f(x)| \leqslant \sup_{x \in [0,1]} |f(x)|,$$

whence we deduce that $\|\cdot\|_*$ and the sup norm are equivalent on $\mathscr{C}([0, 1])$. The use of an equivalent norm can sometimes significantly extend the applicability of an abstract argument—as indicated in Problem 4.15.

Our final remarks concern bases in Banach spaces. Definition 1.2.2 was for finite dimensional spaces only, and obviously needs modification if it is to be applied in a general context. However, producing such a modification presents problems. The most natural possibility is perhaps to say that the sequence (f_n) is a *basis* for $\mathscr{B}$ iff for each $f \in \mathscr{B}$ there is a unique sequence (α_n) of scalars such that $f = \Sigma \alpha_n f_n$ (note that this is *not* the same as saying that the closed linear span of $\{f_n\}$ is $\mathscr{B}$, which is a weaker assertion—see Problem 1.15). Briefly the position is that many Banach spaces have bases in this sense, but that these are not useful. An example from the theory of Fourier series highlights the difficulty: the sequence $(e^{inx})_{n=-\infty}^{n=\infty}$ is *not* a basis for $\mathscr{C}([0, 2\pi])$ with the sup norm. The idea of a basis will not enter into this account further except in Hilbert spaces, where useful bases are often readily available. There is, however, a concept loosely of the same type which is useful in Banach space theory.

1.4.21 Definitions. A Banach space is said to be **separable** iff it contains a countable dense set, that is iff there is a set $S = \{f_n\}_{n=1}^{n=\infty}$ in $\mathscr{B}$ such that for each $\varepsilon > 0$ and $f \in \mathscr{B}$ there is an $f_n \in S$ with $\|f - f_n\| < \varepsilon$.

1.4.22 Lemma. (i) *If Ω is a closed bounded subset of $\mathbb{R}^n$, $\mathscr{C}(\Omega)$ with the sup norm is separable.*

(ii) *ℓ_p is separable for $1 \leqslant p < \infty$.*

(iii) *ℓ_∞ is not separable.*

Proof. We prove (i) and leave the others as exercises. The polynomials with rational coefficients are a countable set and are dense in the set of all polynomials. By Weierstrass' theorem the polynomials are dense in $\mathscr{C}(\Omega)$. □

Most results in a separable space are valid in general, but their proofs may be more difficult, and we shall therefore occasionally assume separability when stating results. The penalty from the point of view of applications is minimal as most useful Banach spaces are separable.

1.5 Hilbert Space

Hilbert spaces are the simplest type of infinite dimensional normed spaces to play a significant role in functional analysis. Their relative simplicity is due to an additional structure—called an inner product—which is imposed on the space; a Banach space with an inner product is a Hilbert space. The inner product is itself a generalization of the scalar product of elementary Cartesian vector analysis. The scalar product is usually defined in terms of the components of the vector, but in accordance with the standard tactics in functional analysis, the algebraic properties of the scalar product are taken as axioms in the abstract context.

The presence of this additional algebraic structure much enriches the geometrical properties of the space. Most significantly, it is possible to define a notion of perpendicularity for two vectors, and the geometry corresponds in several fundamental respects with Euclidean geometry. The effect of the inner product on the analytical (as opposed to the geometric) properties is more subtle. Basically, as in Banach spaces, the main problems in Hilbert spaces are connected with their infinite dimensionality. However, in some respects there can be considerable simplification. This will be seen later in this section when bases of Hilbert space are considered. From the point of view of applications perhaps the most important fact is that a sensible definition of a self adjoint operator may be given in Hilbert space and a powerful body of theory based on this concept developed.

1.5.1 Definition. Let $\mathscr{V}$ be a vector space. An **inner product** is a complex valued function $(\cdot, \cdot)$ on $\mathscr{V} \times \mathscr{V}$ such that for all $f, g, h \in \mathscr{V}$ and $\alpha \in \mathbb{C}$ the following hold:

(i) $(f,f) \geqslant 0$, and $(f,f) = 0$ iff $f = 0$;
(ii) $(f, g + h) = (f, g) + (f, h)$;
(iii) $(f, g) = \overline{(g, f)}$, where the bar denotes the complex conjugate;
(iv) $(\alpha f, g) = \alpha(f, g)$.

A space $\mathscr{V}$ equipped with an inner product will be known as a **pre-Hilbert space** (the term *inner product space* is also used in the literature). If $\mathscr{V}$ is a real vector space, and the inner product is real valued, a **real pre-Hilbert space** is obtained.

1.5.2 Example. For real Cartesian vectors in $\mathbb{R}^3$ the inner product (usually known in this context as a scalar product and written $a \,.\, b$) is defined by

$$(a, b) = a_1b_1 + a_2b_2 + a_3b_3$$

for $a = (a_1, a_2, a_3)$ and $b = (b_1, b_2, b_3)$. It is very easy to check that (i)–(iv) above are satisfied, and three dimensional Euclidean space is thus pre-Hilbert. For the complex vector space $\mathbb{C}^n$, the appropriate inner product is, with $f = (f_1, \ldots, f_n)$, $g = (g_1, \ldots, g_n)$,

$$(f, g) = \sum_{j=1}^{n} f_j\bar{g}_j.$$

1.5.3 Example. An inner product may easily be constructed for $\mathscr{C}(\Omega)$ by setting

$$(f, g) = \int_{\Omega} f(x)\, \overline{g(x)}\, \mathrm{d}x.$$

The quantity $(f, f)^{\frac{1}{2}}$ is denoted by $\|f\|$. As the notation suggests $\|\cdot\|$ is a norm, but to prove this, the following inequality which is fundamental in Hilbert space theory is needed.

1.5.4 Schwartz's Inequality. *For any elements f, g of a pre-Hilbert space,*

$$|(f, g)| \leqslant \|f\|\,\|g\|. \tag{1.5.1}$$

Proof. The result is obvious if $g = 0$. It is therefore enough to show that $|(f, g/\|g\|)| \leqslant \|f\|$ for $g \neq 0$, which is the same as proving that $|(f, h)| \leqslant \|f\|$ for all unit vectors h. This follows from the relations (see Fig. 1.4)

$$\begin{aligned} 0 &\leqslant \|f - (f, h)h\|^2 \\ &= (f - (f, h)h, f - (f, h)h) \\ &= (f, f) - (f, h)(h, f) - \overline{(f, h)}(f, h) + (f, h)\overline{(f, h)} \\ &= \|f\|^2 - |(f, h)|^2. \end{aligned}$$

□

1.5.5 Theorem. *A pre-Hilbert space $\mathscr{H}$ is a normed vector space with the norm $\|f\| = (f, f)^{\frac{1}{2}}$.*

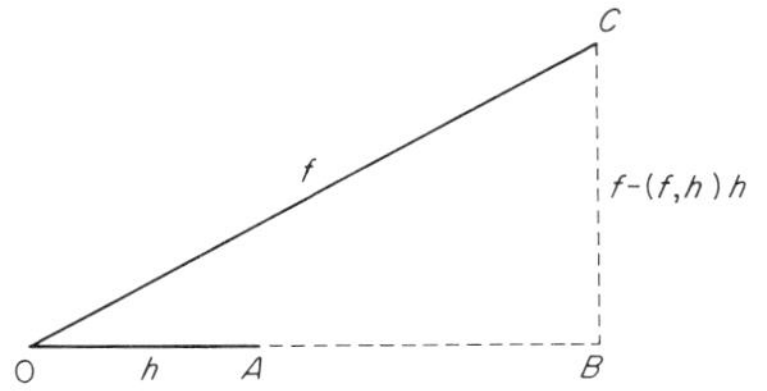

Fig. 1.4 In $\mathbb{R}^2$, OA, OC represent h, f respectively, while OB is $(f, h)h$. The relation proved in 1.5.4, $\|f - (f, h)h\|^2 = \|f\|^2 - |(f, h)|^2$, is simply Pythagoras' theorem.

Proof. The only axiom for a norm that is not obvious is the triangle inequality. From Schwartz's inequality,

$$(f, g) + (g, f) = 2\,\mathrm{Re}(f, g) \leqslant 2\|f\|\,\|g\|.$$

The triangle inequality is then deduced as follows:

$$\begin{aligned}\|f + g\|^2 &= (f + g, f + g)\\ &= \|f\|^2 + (f, g) + (g, f) + \|g\|^2\\ &\leqslant \|f\|^2 + 2\|f\|\,\|g\| + \|g\|^2\\ &= (\|f\| + \|g\|)^2.\end{aligned}$$ □

This result shows that the norm in a pre-Hilbert space is the natural extension of length in $\mathbb{R}^3$, and the generalization of some further familiar properties of $\mathbb{R}^3$ is given in Problem 1.17. A further analogy is with the angle between two vectors and with their perpendicularity (known as orthogonality in the present context). The angle θ between two vectors f, g in a *real* pre-Hilbert space may be defined in a manner consistent with the properties of the inner product by means of the relation $(f, g) = \|f\|\,\|g\| \cos\theta$ (note that Schwartz's inequality then says that $|\cos\theta| \leqslant 1$). This definition is not satisfactory in a complex space. Actually the magnitude of the angle does not enter significantly into Hilbert space theory except in the particular case when $(f, g) = 0$ corresponding to orthogonal vectors, but the last notion is fundamental and the following formal definition is based on it.

1.5.6 Definition. Let $\mathscr{H}$ be a pre-Hilbert space. Two vectors $f, g \in \mathscr{H}$ are said to be **orthogonal** iff $(f, g) = 0$; we write $f \perp g$. Given a set $S \subset \mathscr{H}$, the set of vectors in $\mathscr{H}$ orthogonal to every vector in S is known as the **orthogonal complement** of S, and is written $S^\perp$.

In $\mathbb{R}^3$ if $\mathscr{M}$ is a plane through the origin, $\mathscr{M}^\perp$ is the line through the origin perpendicular to $\mathscr{M}$. Two typical properties of $\mathbb{R}^3$, both of which extend to any finite dimensional pre-Hilbert space, are as follows. First there is a unique vector in $\mathcal{M}$ at minimum distance from a given vector in contrast to Banach spaces where neither the existence nor uniqueness of such a vector is assured. Secondly any vector f may be written uniquely in the form $f = g + h$ where $g \in \mathcal{M}$, $h \in \mathcal{M}^{\perp}$; in other words $\mathbb{R}^3$ may be decomposed into the direct sum of orthogonal subspaces. Although these properties may be regarded as geometric, difficulties essentially analytic in nature prevent their generalization to an infinite dimensional pre-Hilbert space. As with normed spaces, the assumption of completeness is the key to further progress.

1.5.7 Definition. A pre-Hilbert space which is complete with respect to the norm $\|f\| = (f,f)^{\frac{1}{2}}$ is called a **Hilbert space**. The symbol $\mathcal{H}$ will henceforth *always* denote a Hilbert space.

1.5.8 Example. $\mathbb{R}^n$ and $\mathbb{C}^n$ with the usual inner product (Example 1.5.2) are of course Hilbert spaces. The simplest example of an infinite dimensional Hilbert space is ℓ_2 (Definition 1.3.13) with the inner product

$$(f, g) = \Sigma f_n \bar{g}_n$$

where $f = (f_1, f_2, \ldots), g = (g_1, g_2, \ldots)$. Then $\|f\| = \{\Sigma |f_n|^2\}^{\frac{1}{2}}$, which is the original definition of the norm, and consequently ℓ_2 is complete by Theorem 1.4.11, and so is a Hilbert space when equipped with the inner product above.

1.5.9 Example. We have already noted that $\mathscr{C}(\Omega)$ is pre-Hilbert with the inner product $(f, g) = \int_{\Omega} f(x)\,\overline{g(x)}\,\mathrm{d}x$. However, $\mathscr{C}(\Omega)$ is not complete in the associated norm—as may be proved by using the sequence in Example 1.4.2. Unfortunately this situation is typical. It is often easy to make a vector space into a Banach space with one norm or a pre-Hilbert space with another, but it is only occasionally that the norms coincide.

Of course in applications the main Hilbert spaces of interest are those composed of sets of functions. However, because of technical problems with integration theory, the discussion of such spaces is deferred until the next chapter. For the moment we merely note that in the most useful Hilbert space $\mathscr{L}_2$, the inner product is based on that given above, but in order to achieve completeness other functions than those in $\mathscr{C}(\Omega)$ must be included.

We continue by generalizing the geometric properties described above under the assumption that the space is complete.

1.5.10 Theorem. *Let $\mathcal{M}$ be a closed subspace of the Hilbert space $\mathcal{H}$, and take any $f \in \mathcal{H}$. Then there is a unique element in $\mathcal{M}$ closest to f.*

Proof. Set $d = \text{dist}(f, \mathscr{M})$, and choose a sequence (g_n) in $\mathscr{M}$ such that $\lim\|f - g_n\| = d$. Then by the parallelogram law (Problem 1.17),

$$\|g_n - g_m\|^2 = \|(g_n - f) - (g_m - f)\|^2$$
$$= 2\|g_n - f\|^2 + 2\|g_m - f\|^2 - 4\|\tfrac{1}{2}(g_n + g_m) - f\|^2.$$

Since $(g_n + g_m)/2 \in \mathscr{M}$, the magnitude of the last term is not less than $4d^2$. Hence

$$\lim_{m,n\to\infty} \|g_n - g_m\| \leqslant 2d^2 + 2d^2 - 4d^2 = 0.$$

The sequence (g_n) is thus Cauchy, and as $\mathscr{M}$ is closed, convergent to an element, g say, of $\mathscr{M}$. Evidently $\|f - g\| = d$, which shows that the minimum distance is attained.

If there is another element $g' \in \mathscr{M}$ with $\|f - g'\| = d$, it is a simple deduction from the parallelogram law that

$$\|f - \tfrac{1}{2}(g + g')\|^2 = d^2 - \|g - g'\|^2.$$

Since $(g + g')/2 \in \mathscr{M}$, the left-hand side of this equation is not less than d^2. Therefore $g = g'$, and uniqueness follows. □

1.5.11 The Projection Theorem. *Let $\mathscr{M}$ be a closed subspace of the Hilbert space $\mathscr{H}$. Then $\mathscr{M}^\perp$ is also a closed subspace, and $\mathscr{H} = \mathscr{M} \oplus \mathscr{M}^\perp$. Further, in the decomposition $f = g + h$ where $g \in \mathscr{M}, h \in \mathscr{M}^\perp$, g is the element in $\mathscr{M}$ closest to f.*

Proof. From Problem 1.19, $\mathscr{M}^\perp$ is closed and $\mathscr{M} \cap \mathscr{M}^\perp = 0$. Therefore by Lemma 1.2.9, to show that $\mathscr{H} = \mathscr{M} \oplus \mathscr{M}^\perp$ it is enough to prove that $\mathscr{H} = \mathscr{M} + \mathscr{M}^\perp$. If $f \in \mathscr{M}$ the result is obvious, so take any $f \notin \mathscr{M}$, and let g be the point in $\mathscr{M}$ closest to f, the existence of g being guaranteed by the previous theorem. We shall show that $f - g \in \mathscr{M}^\perp$, whence $f = g + (f - g)$ with $g \in \mathscr{M}, f - g \in \mathscr{M}^\perp$.

For any $h \in \mathscr{M}$ and $\alpha > 0$, $g + \alpha h \in \mathscr{M}$, and therefore

$$\|f - g\|^2 \leqslant \|f - g - \alpha h\|^2 = \|f - g\|^2 - 2\,\text{Re}\,\alpha(h, f - g) + \alpha^2\|h\|^2.$$

It follows that $2\text{Re}\,\alpha(h, f - g) \leqslant \alpha^2\|h\|^2$, and dividing by α and letting $\alpha \to 0$, we conclude that $\text{Re}(h, f - g) \leqslant 0$. A repetition of the argument with $\alpha < 0$ proves that $\text{Re}(h, f - g) \geqslant 0$. Therefore $\text{Re}(h, f - g) = 0$, and similarly $\text{Im}(h, f - g) = 0$. Thus $(h, f - g) = 0$ for all $h \in \mathscr{M}$, that is $f - g \in \mathscr{M}^\perp$ as asserted. □

This decomposition of $\mathscr{H}$ into orthogonal subspaces is fundamental in the theory of self-adjoint operators and elsewhere.

It was stated in the previous section that the study of bases in Banach spaces is not profitable from the point of view of applications. In contrast there is a very satisfactory theory of bases in Hilbert spaces, and this we shall now outline. The discussion will be restricted throughout to a separable Hilbert space $\mathscr{H}$ to avoid certain technical difficulties. Since most useful Hilbert spaces are separable, little is lost by this restriction; for the general case, Friedman (1970, Section 6.4) may be consulted. The approach to be followed is motivated by the observation that in $\mathbb{R}^3$ a set $\mathscr{K}$ of mutually orthogonal unit vectors is a basis if and only if $\mathscr{K}^\perp = 0$.

1.5.12 Definition A set $\mathscr{K}$ of vectors in $\mathscr{H}$ is said to be **complete**† iff $\mathscr{K}^\perp = 0$, in other words iff $(f, \phi) = 0$ for all $\phi \in \mathscr{K}$ implies that $f = 0$. A countable set $\mathscr{K} = \{\phi_n\}_{n=1}^{n=\infty}$ is called **orthonormal** iff $(\phi_n, \phi_m) = \delta_{nm}$ for all $m, n \geqslant 1$. The numbers (f, ϕ_n) are known as the **Fourier coefficients** of f (with respect to $\mathscr{K}$), and the **Fourier series** of f is the formal series $\Sigma(f, \phi_n)\phi_n$.

1.5.13 Lemma. *For $\mathscr{K} = \{\phi_n\}$ an orthonormal set in $\mathscr{H}$, the following hold.*

(i) *Bessel's Inequality: for any $f \in \mathscr{H}$,*

$$\Sigma|(f, \phi_n)|^2 \leqslant \|f\|^2. \tag{1.5.2}$$

(ii) (*Best fit*). *For any sequence (α_n) of scalars, any positive integer m and any $f \in \mathscr{H}$,*

$$\left\| f - \sum_1^m \alpha_n \phi_n \right\| \geqslant \left\| f - \sum_1^m (f, \phi_n)\phi_n \right\|. \tag{1.5.3}$$

Proof. By orthonormality,

$$\left\| f - \sum_1^m (f, \phi_n)\phi_n \right\|^2 = \|f\|^2 - \sum_1^m |(f, \phi_n)|^2, \tag{1.5.4}$$

and (1.5.2) follows on letting $m \to \infty$. Again on using orthonormality,

$$\left\| f - \sum_1^m \alpha_n \phi_n \right\|^2 = \|f\|^2 - \sum_1^m |(f, \phi_n)|^2 + \sum_1^m |(f, \phi_n) - \alpha_n|^2$$

$$\geqslant \|f\|^2 - \sum_1^m |(f, \phi_n)|^2,$$

and (1.5.3) follows from (1.5.4). □

† No direct connection with the previous use of complete (Definition 1.4.4). The reader should be warned that the terminology in this area of Hilbert space theory is not uniform in the literature.

1.5.14 Lemma. *Let* $\mathscr{K} = \{\phi_n\}$ *be an orthonormal set, and suppose that* (α_n) *is a sequence of scalars. Then the series* $\Sigma\alpha_n\phi_n$ *is convergent iff* $\Sigma|\alpha_n|^2$ *is convergent. If the series is convergent, its sum is independent of the order of its terms and*

$$\|\Sigma\alpha_n\phi_n\| = \{\Sigma|\alpha_n|^2\}^{\frac{1}{2}}. \tag{1.5.5}$$

Proof. As $\{\phi_n\}$ is orthonormal,

$$\left\|\sum_{n=i}^{j} \alpha_n\phi_n\right\| = \sum_{n=i}^{j} |\alpha_n|^2, \tag{1.5.6}$$

and it follows that if $\Sigma|\alpha_n|^2$ is convergent the sequence $\left(\sum_{n=1}^{m} \alpha_n\phi_n\right)_{m=1}^{m=\infty}$ is Cauchy, and since $\mathscr{H}$ is complete, convergent. The necessity is obtained by reversing the argument. Setting $i = 1$ and letting $j \to \infty$ in (1.5.6) we obtain (1.5.5).

Let $f = \Sigma\alpha_{m_u}\phi_{m_u}$ be a rearrangement of the series $g = \Sigma\alpha_n\phi_n$. It is easy to show that if $\Sigma|\alpha_n|^2$ is convergent,

$$\|f\|^2 = \|g\|^2 = \Sigma|\alpha_n|^2, \tag{1.5.7}$$

$$\|f - g\|^2 = \|f\|^2 - 2\mathrm{Re}(f, g) + \|g\|^2. \tag{1.5.8}$$

But

$$(f, g) = \lim_{j\to\infty} \left(\sum_{n=1}^{j} \alpha_{m_n}\phi_{m_n}, \sum_{n=1}^{j} \alpha_n\phi_n\right) = \Sigma|\alpha_n|^2.$$

Hence from (1.5.7) and (1.5.8), $\|f - g\| = 0$. □

1.5.15 Lemma. *Let* $\mathscr{K} = \{\phi_n\}$ *be an orthonormal set. Then the series* $\Sigma(f, \phi_n)\phi_n$ *is convergent independently of the order of its terms to the sum* Pf, *say, and* $Pf = 0 (f \in [\mathscr{K}]^\perp)$, $Pf = f (f \in \overline{[\mathscr{K}]})$.

Proof. The first assertion is an immediate consequence of Bessel's inequality (1.5.2) and Lemma 1.5.14. Next note that all the Fourier coefficients are zero if $f \in \mathscr{K}^\perp$, and therefore if $f \in \overline{[\mathscr{K}]}^\perp = \mathscr{K}^\perp$, then $Pf = 0$. On the other hand if $f \in \overline{[\mathscr{K}]}$, by the definition of closure, for each $\varepsilon > 0$ there exist $\alpha_1, \ldots, \alpha_m \in \mathbb{C}$, $\phi_1, \ldots, \phi_m \in \mathscr{K}$ such that

$$\left\|f - \sum_{n=1}^{m} \alpha_n\phi_n\right\| < \varepsilon.$$

Hence by Lemma 1.5.13(ii),

$$\left\|f - \sum_{n=1}^{m} (f, \phi_n)\phi_n\right\| < \varepsilon.$$

The same result shows that the term on the left-hand side of this inequality does not increase as m increases, so on letting $m \to \infty$ we obtain $\|f - Pf\| < \varepsilon$. As ε is arbitrary this proves that $f = Pf$. □

Having completed these technical preparations, we may now deduce rather easily the main results on the basis.

1.5.16 Definition. An orthonormal set $\mathscr{K} = \{\phi_n\}$ is called an **orthonormal basis** of $\mathscr{H}$ iff for every $f \in \mathscr{H}$,

$$f = \Sigma(f, \phi_n)\phi_n.$$

This expansion is of course the Fourier series of f (Definition 1.5.12); note that by Lemma 1.5.14, a rearrangement of the order of its terms is allowed.

1.5.17 Theorem. *Suppose $\mathscr{H}$ is a separable Hilbert space, and assume that $\mathscr{K} = \{\phi_n\}$ is an orthonormal set in $\mathscr{H}$. Then the following conditions are equivalent:*

(i) *$\mathscr{K}$ is complete, that is $\mathscr{K}^{\perp} = 0$;*
(ii) $\overline{[\mathscr{K}]} = \mathscr{H}$;
(iii) *$\mathscr{K}$ is an orthonormal basis;*
(iv) *for any $f \in \mathscr{H}$*

$$\|f\|^2 = \Sigma|(f, \phi_n)|^2 \qquad (\textit{Parseval's formula}). \tag{1.5.9}$$

Proof. (i) ⇒ (ii). If $\overline{[\mathscr{K}]} \neq \mathscr{H}$, there is a point $f \in \mathscr{H}$ not in $\overline{[\mathscr{K}]}$. Therefore by the Projection Theorem 1.5.11, $f = g + h$ for some $g \in \overline{[\mathscr{K}]}$ and some non-zero $h \in \overline{[\mathscr{K}]}^{\perp}$. This implies that $\mathscr{K}^{\perp} = \overline{[\mathscr{K}]}^{\perp} \neq 0$, which contradicts (i). (ii) ⇒ (iii). This follows immediately from Lemma 1.5.15. (iii) ⇒ (iv). Simply let $m \to \infty$ in (1.5.4). (iv) ⇒ (i). If $f \in \mathscr{K}^{\perp}$, each term in the sum in (1.5.9) is zero, whence $f = 0$. □

1.5.18 Theorem. *A separable Hilbert space has an orthonormal basis.*

Proof. Let $\{f_n\}$ be a countable dense set, and apply the Gram–Schmidt procedure (Problem 1.21), rejecting at each stage the next f_n if it together with the previous orthonormal elements are not linearly independent. The new set, $\mathscr{K}$ say, is orthonormal, and since $\{f_n\}$ is dense in $\mathscr{H}$, then $\overline{[\mathscr{K}]} = \mathscr{H}$. Therefore, by the previous theorem, $\mathscr{K}$ is an orthonormal basis. □

1.5.19 Example. Consider the Hilbert space ℓ_2 (Example 1.5.8). Let e_n be the element whose mth component is δ_{mn}—essentially e_n is a unit vector along the "nth axis". Clearly $\{e_n\}$ is an orthonormal set. Now if $f = (f_1, f_2, \ldots)$ is any vector in ℓ_2, then $(f, e_n) = f_n$. Therefore $f = 0$ if $(f, e_n) = 0$ for $n \geqslant 1$. This proves that $\{e_n\}$ is complete, and it follows from Theorem 1.5.17 that it is a basis.

The last two theorems are an extremely satisfactory conclusion to the study of bases in Hilbert space. In particular the fact that if $\mathscr{K}$ is complete it is a basis which is in pleasing correspondence with the situation in finite dimensions; no easy analogue of this condition exists in an arbitrary Banach space. From a practical point of view the bases of $\mathscr{L}_2$ (Example 1.5.9) are of special importance. Most of these are best constructed using the self-adjointness of certain differential operators discussed later together with the Hilbert–Schmidt Theorem 7.5.1. In certain cases a direct proof based on Theorem 1.5.17 is possible; this technique is reviewed in detail by Higgins (1977).

Finally, in addition to the general references cited in the introduction to this chapter, we draw the reader's attention to an unusual and fascinating book—Halmos (1967)—in which a judicious choice of example and counter-example is used to develop Hilbert space theory.

Problems

1.1 Deduce from the definition of a norm that

$$\|f - g\| \geqslant \big|\|f\| - \|g\|\big|.$$

1.2 Show that an open or closed ball in a normed vector space is convex.

1.3 Deduce from Definition 1.3.8 that a point f lies in the closure of S_1 in S_2 iff $f \in S_2$ and f is the limit of a sequence in S_1.

1.4 Prove the equivalence of the two conditions in Definition 1.3.9 of open set.

1.5 Let $\mathscr{V}$ be a normed vector space. Show that an open ball is an open set and a closed ball is a closed set. Prove also that $\mathscr{V}$ itself (and consequently the empty set $\varnothing$) is both open and closed.

1.6 In a normed vector space show that the union of any class of open sets is open, and the intersection of a *finite* number of open sets is open. Show by constructing a counterexample in $\mathbb{R}$ that the intersection of an infinite number of open sets need not be open. Prove also that the intersection of any class of closed sets is closed, and the union of a finite number of closed sets is closed.

1.7 A subset of a normed vector space is open if and only if it is the union of open spheres.

1.8 A curious subset of $\mathbb{R}$ is the *Cantor set*. Take $[0, 1]$ and extract the open middle third $(\frac{1}{3}, \frac{2}{3})$. Repeat the process of extracting the open middle third of the remaining subintervals an infinite number of times. What is left is the Cantor set. Show that this is a closed subset of $\mathbb{R}$ and has empty interior.

1.9 Equip $\mathscr{C}([0, 1])$ with the norm $\|\cdot\|_1$ where $\|f\|_1 = \int_0^1 |f(x)|\,dx$. Set $f_n(x) = x^{\frac{1}{2}}(1 - x)^n$ for $n \geqslant 1$. Show that in the sense of Definition 1.4.5 Σf_n is absolutely convergent but not convergent. Deduce from Lemma 1.4.6 that with this norm $\mathscr{C}([0, 1])$ is not complete.

1.10 Consider the family $\{f_\alpha\}$ of functions in $\mathscr{C}([0, 1])$ where

$$f_\alpha(x) = 2\alpha x/(1 + \alpha^2 x^2) \qquad (0 \leqslant x \leqslant 1, 0 < \alpha < 1).$$

Show that $\|f_\alpha\| = 1$, $\|f_\alpha\|_1 = |\alpha|^{-1} \log(1 + \alpha^2)$, where $\|\cdot\|$ is the sup norm and $\|\cdot\|_1$ is as above. Deduce that these norms are not equivalent.

1.11 In $\mathbb{R}^n$ for $n > 1$ consider the sequence (f_k) where

$$f_k = (1, 2k^{-1}, 3k^{-1}, \ldots, nk^{-1}),$$

and set $f = (1, 0, 0, \ldots, 0)$. Show that

$$\|f_k - f\|_\infty = nk^{-1}, \qquad \|f_k - f\|_1 = k^{-1}[\tfrac{1}{2}n(n + 1) - 1],$$

where $\|\cdot\|_p$ denotes the norm defined by (1.3.1) and (1.3.2). Note that although $\lim f_k = f$ for any $p \geqslant 1$ (since all norms on $\mathbb{R}^n$ are equivalent), the rate of convergence is considerably affected by the choice of norm.

1.12 Let $\mathscr{B}$ be a Banach space, and suppose that $S_1 \subset S_2 \subset S_3 \subset \mathscr{B}$. Prove that S_1 is dense in S_2 iff for each $f \in S_2$ and $\varepsilon > 0$ there is a $g \in S_1$ such that $\|f - g\| < \varepsilon$. Deduce that if S_1 is dense in S_2 and S_2 is dense in S_3, then S_1 is dense in S_3.

1.13 Show that a Cauchy sequence in a normed vector space is bounded.

1.14 Equip $\mathscr{C}([0, 1])$ with the sup norm.

(i) Construct a sequence in the closed unit ball which converges pointwise (that is for each x) to a discontinuous limit.
(ii) Show that the bounded sequence (f_n), where $f_n(x) = nx/(1 + n^2x^2)$, converges pointwise but not in norm to the (continuous) function zero.
(iii) Show that convergence in norm implies pointwise convergence.

1.15 Equip $\mathscr{C}([0, 1])$ with the sup norm, and take the subset $S = \{x^n\}_{n=0}^{n=\infty}$. By Weierstrass' Theorem the closed linear span of S is $\mathscr{C}([0, 1])$. Is it true that every $f \in \mathscr{C}([0, 1])$ is expressible as a sum (convergent in sup norm) of the form $\Sigma \alpha_n x^n$? (Hint: consider $\exp(-1/x^2)$).

1.16 Take the sequence (f_n) in $\mathscr{C}^1([-1, 1])$ where $f_n(x) = (x^2 + n^{-2})^{\frac{1}{2}}$. Let $\|\cdot\|$ denote the sup norm and set

$$\|f\|_{\mathscr{C}^1} = \sup_{x \in [-1, 1]} |f(x)| + \sup_{x \in [-1, 1]} |f'(x)|.$$

Show that (f_n) converges in $\|\cdot\|$ to $|x|$, but is not convergent in $\|\cdot\|_{\mathscr{C}^1}$.

1.17 For f, g, h elements of a pre-Hilbert space and $\alpha, \beta \in \mathbb{C}$ prove:

(i) $(f, \alpha g) = \bar{\alpha}(f, g)$;
(ii) $(\alpha f + \beta g, h) = \alpha(f, h) + \beta(g, h)$;
(iii) $2\|f\|^2 + 2\|g\|^2 = \|f - g\|^2 + \|f + g\|^2$ (the parallelogram law);
(iv) $4(f, g) = \|f + g\|^2 - \|f - g\|^2 + i\|f + ig\|^2 - i\|f - ig\|^2$.

1.18 Show using (iv) of 1.17 that if $\mathscr{V}$ is a normed vector space with a norm $\|\cdot\|$ satisfying (iii) of 1.17, then $\mathscr{V}$ may be made into a pre-Hilbert space with an inner product $(\cdot, \cdot)$ such that $(f, f)^{\frac{1}{2}} = \|f\|$ for all $f \in \mathscr{V}$. Thus the parallelogram law characterizes the norm in pre-Hilbert space. Check that ℓ_∞ cannot be made into a pre-Hilbert space.

1.19 Let S be a subset of the Hilbert space $\mathscr{H}$. Show that $S^\perp$ is a closed subspace of $\mathscr{H}$, and prove that $S^\perp = \overline{[S]}^\perp$. Show also that $f \in S \cap S^\perp$ implies that $f = 0$.

1.20 A Banach space is said to be *uniformly convex* iff for each $\varepsilon > 0$ there is a $\delta > 0$ such that $\|f\| = \|g\| = 1$ and $\|\frac{1}{2}(f + g)\| > 1 - \delta$ imply that $\|f - g\| < \varepsilon$. Show that Hilbert space is uniformly convex, and give two examples of Banach spaces which are not.

Prove that uniform convexity ensures the uniqueness of a point in a linear subspace at minimum distance from a given point of $\mathscr{B}$. (The existence of such a point is not asserted).

1.21 Let $\{\psi_n\}_{n=1}^{n=\infty}$ be a linearly independent set of vectors in a pre-Hilbert space. The Gram–Schmidt orthogonalization process is a method for constructing inductively from this set a set of orthonormal vectors $\{\phi_n\}$ by setting $h_1 = \psi_1$, $\phi_1 = h_1/\|h_1\|$ and

$$h_n = \psi_n - \sum_{k=1}^{n-1} (\psi_n, \phi_k)\phi_k, \qquad \phi_n = h_n/\|h_n\|.$$

Verify that $\{\phi_n\}$ is an orthonormal set, and show that the linear spans of $\{\psi_n\}$ and $\{\phi_n\}$ are the same.

With the inner product $\int_{-1}^{1} f(x)\overline{g(x)}\,dx$ on $\mathscr{C}([-1, 1])$ carry out the Gram–Schmidt procedure for the first few terms of $1, x, x^2, \ldots$ and show that Legendre polynomials are produced.

1.22 Suppose that $\{\phi_n\}$ and $\{\psi_n\}$ are orthonormal sets in a Hilbert space, and assume that $\{\phi_n\}$ is complete. Prove that $\{\psi_n\}$ is complete if

$$\Sigma\|\phi_n - \psi_n\|^2 < 1.$$

Chapter 2

Lebesgue Integration and the $\mathscr{L}_p$ Spaces

2.1 Introduction

In the last chapter we tried to convince the reader that a complete normed vector space is a suitable setting for analysis, but we gave only a few examples of complete spaces of functions—$\mathscr{C}(\Omega)$ with the sup norm and related spaces—and no example of a Hilbert space of functions. To confine the analysis always to $\mathscr{C}(\Omega)$, useful as this space is, will severely limit the range of problems which can be tackled, and in particular will rule out the application of the powerful results of the abstract theory of operators on a Hilbert space. Our object here is to enlarge our catalogue of spaces by introducing the Banach spaces $\mathscr{L}_p$ and the Hilbert space $\mathscr{L}_2$. These together with $\mathscr{C}(\Omega)$ will provide a sufficiently wide range of spaces for most applications.

The most obvious approach to the construction of further Banach spaces of functions is to attempt to equip $\mathscr{C}(\Omega)$ with a different norm. For example in $\mathscr{C}([0, 1])$ it is reasonable to consider taking

$$\|f\|_1 = \int_0^1 |f(x)|\,\mathrm{d}x,$$

for $\|\cdot\|_1$ is indeed a norm, and $\|f\|_1$ represents a physically important concept—the average value of $|f|$. Unfortunately $\mathscr{C}([0, 1])$ is not complete in this norm (Example 1.4.2). One way round the difficulty is simply to take the "completion" of $\mathscr{C}([0, 1])$ in $\|\cdot\|_1$. This procedure certainly yields a Banach space, but it does not provide an intuitive guide to the properties of the functions which the space will contain, and indeed it is not even obvious whether the elements of the abstract completion will be functions at all. Another possibility is to enlarge the set of functions to include all those which are Riemann integrable. However, this strategy fails to produce a complete space—there are just not enough Riemann integrable functions—or to put it another way, a Cauchy sequence in $\|\cdot\|_1$ of continuous functions need not have a Riemann integrable limit. The last observation suggests the tactics to be adopted here, that an integral capable of dealing with a wider class of functions should be constructed.

The theory to be presented originated with the work of Lebesgue around the beginning of the century, and is now a standard part of classical real variable theory. Indeed except in the most elementary circumstances, the Lebesgue integral is markedly superior to the Riemann integral. For example, problems which involve integration together with a limiting process are often awkward with the Riemann integral, but are almost trivial when the Lebesgue integral is used. Thus while our primary motive for tackling the theory is that it is essential in defining the $\mathscr{L}_p$ spaces, the integral will be found useful in many other contexts. Unfortunately, from the point of view of the non-specialist the analysis leading to the integral may present something of a difficulty, for it is quite lengthy and requires techniques which are often unfamiliar, and which furthermore will not be useful in most other areas of interest. On the other hand the *results* of the theory are elegant and easy to apply. We have taken the view that an outline of the basic elements of the theory must be presented in order to give the reader some intuitive feel for the subject, but that proofs should usually be omitted since the arguments will not be needed subsequently. The main results are gathered together as theorems for reference, these can be used without following the underlying analysis which may thus be omitted by those with a pragmatic point of view.

In order to motivate the approach to be followed, let us recall one method for constructing the Riemann integral of a real valued function f over the finite interval $[a, b]$. The method is based on approximating the function by step functions.

2.1.1 Definition. Let S be a subset of a set X. The function χ_S defined by setting $\chi_S(x) = 1$ $(x \in S)$, $\chi_S(x) = 0$ $(x \notin S)$ is known as the **characteristic function** of S. A **step function** is a linear combination of a finite number of characteristic functions of intervals.

For $X = [a, b]$, consider the subintervals $S_1 = [x_0, x_1], S_2 = (x_1, x_2], \ldots, S_n = (x_{n-1}, x_n]$, where $x_0 = a$, $x_n = b$, and define the integral of the step function $g = \sum_1^n c_i \chi_{S_i}$ to be the "area under the graph":

$$\int_a^b g(x)\,\mathrm{d}x = \sum_1^n c_i(x_i - x_{i-1}).$$

This integral is called a *lower sum* (respectively an *upper sum*) for f iff $g(x) \leqslant f(x)$ (respectively $g(x) \geqslant f(x)$) for $x \in [a, b]$. If the supremum of all lower sums has certain properties (it must be equal to the infimum of upper sums), then it is called the *integral* of f. This places a restriction on f, but for such Riemann integrable functions the integral has the usual nice properties such as additivity: $\int (f + g) = \int f + \int g$.

The next step is suggested by adopting a slightly different notation in which the length of S_i is denoted by $\mu(S_i)$. The lower sum is then

$$\sum_1^n c_i \mu(S_i).$$

The basic idea in defining a more general integral is to extend the concept of the length of an interval (and of the area of a rectangle in $\mathbb{R}^2$, etc.) to a notion of size for much more general sets, the size of a set being known as its measure. Measure theory thus lies at the heart of integration theory and must be our first preoccupation.

The plan of this chapter is as follows. In the next section an appropriately broad collection of sets is defined and the idea of the measure of these sets is introduced. Then measurable functions (roughly those for which an integral may be defined) are considered. The integral is introduced in Section 4 and its basic properties listed. Section 5 is concerned with the $\mathscr{L}_p$ spaces. Finally certain essential results which are naturally stated in $\mathscr{L}_p$ are given in Section 6.

Both de Barra (1974) and Bartle (1966) are strongly recommended and may be referred to for proofs. Burkill (1951) presents a somewhat different point of view and is also useful.

2.2 The Measure of a Set

Measure theory is based on the idea of generalizing the length of an interval in $\mathbb{R}$ (the area of a rectangle in $\mathbb{R}^2$, etc.) to the "measure" of a subset; the more subsets which are "measurable", the more functions can be integrated. It is over-ambitious to attempt to define a well-behaved measure on all subsets, but we shall see that there is a very wide class of subsets on which there is a measure with acceptable properties. Although motivated by the properties of length, a fairly abstract approach to measure theory is profitable. For one thing this seems to clarify the outline of the argument by ridding it of inessential detail, while for another certain other useful measures are obtained without extra effort; one of these leads to the Lebesgue–Stieltjes integral, an extension of the Riemann–Stieltjes integral. However, our main goal is Lebesgue measure and the Lebesgue integral in $\mathbb{R}^n$, which is the direct descendant of the Riemann integral.

Throughout, X will denote a set (which may be thought of as a subset of $\mathbb{R}^n$), and $\mathscr{S}$ a class of subsets of X. We first ask what class of subsets a measure can reasonably be defined for. It will be felt intuitively that if two subsets have a measure, then so should their union, their intersection and their difference, and if a sequence consists of disjoint subsets each of which has a measure, then their union should have a measure, which should be equal to the sum of the measures of the subsets. This suggests the following definition.

2.2.1 Definition. A class $\mathscr{S}$ of subsets of X is said to be a **σ-algebra** iff

(i) $\varnothing, X \in \mathscr{S}$;
(ii) $S \in \mathscr{S} \Rightarrow X \backslash S \in \mathscr{S}$;
(iii) $S_1, S_2, \ldots \in \mathscr{S} \Rightarrow \bigcup_{n=1}^{\infty} S_n \in \mathscr{S}$.

A set in a σ-algebra is called a **measurable set**.

Briefly, it is required that $\varnothing, X \in \mathscr{S}$ and that $\mathscr{S}$ should be closed with respect to complements and countable unions; other ways of combining sets in $\mathscr{S}$ are indicated in Problem 2.1. It is obvious that for any set X a σ-algebra does exist, since the collection of all subsets is a σ-algebra—although not a very useful one. A more profitable approach is to take any class $\mathscr{S}$ of subsets of X and note that the intersection of all σ-algebras containing $\mathscr{S}$ is a σ-algebra.

2.2.2 Definition. Let $\mathscr{S}$ be any class of subsets of X. The smallest σ-algebra containing $\mathscr{S}$ is called the **σ-algebra generated by** $\mathscr{S}$, and will be denoted by $\mathscr{S}_\sigma$.

2.2.3 Definition. The σ-algebra generated by the class of open subsets of $\mathbb{R}^n$ is called the **σ-algebra of Borel sets**, each set being a **Borel set**. If X is a subset of $\mathbb{R}^n$, S is a **Borel set** in X iff $S = X \cap S_1$ where S_1 is a Borel set in $\mathbb{R}^n$.

As this σ-algebra is of particular importance, it would be pleasant if an easy description of the Borel sets could be given. Our first observation is that the σ-algebra is extremely large, containing as it does arbitrary countable unions and intersections of open sets. On the other hand, an explicit characterization of the Borel sets, even by a countable infinity of operators starting with open sets, is not possible. At first this may seem disturbing, but it may be noted by way of analogy that while it is possible to describe the open sets in $\mathbb{R}$ in simple terms (each such set is the union of a disjoint sequence of open intervals), this cannot be done in $\mathbb{R}^n$ for $n \geqslant 2$. Nonetheless the open sets are extremely useful. In fact the crucial point is that many sets are readily recognized as Borel (differences, unions and intersections of sequences of Borel sets being Borel).

2.2.4 Example. In $\mathbb{R}$ both the rationals and irrationals are Borel sets. For single points are measurable (take the complement of $(-\infty, a) \cup (a, \infty)$, and as the set of rationals is a countable union of single points, by the definition of a σ-algebra this set is measurable. The complement of this set is the set of irrationals which is therefore also a Borel set.

2.2.5 Lemma. *The σ-algebra of Borel sets is generated by each of the following*:
- (i) *the closed sets*;
- (ii) *the bounded open sets*;
- (iii) *the bounded closed sets*;
- (iv) *the open balls*;
- (v) *the closed balls.*

Proof. (i) The complement of an open set is closed;
- (ii) Any open S is the union of the sequence of open sets $S(0, n) \cap S$ for integer n.
- (iii) Similar.
- (iv) It is not quite enough to observe that every open set is a union of open balls, a *countable* union must be obtained, and this is done by taking the balls to have centres with rational coordinates and rational radii.
- (v) Similar. □

We come now to the definition of measure itself. Useful measures are always defined on σ-algebras, and recalling the model of length, it is natural to expect that a measure should be a non-negative extended real-valued function†, and that it should be additive in some sense.

2.2.6 Definition. A **measure** μ is a mapping from a σ-algebra $\mathscr{S}$ to $\overline{\mathbb{R}}^+$ with the properties:

† Obviously a set such as $(-\infty, \infty)$ should have measure infinity. It is therefore convenient to adopt the convention that the set $\overline{\mathbb{R}}$ of *extended real numbers* is $\mathbb{R} \cup \{-\infty, \infty\}$, and $\mathbb{R}^+ = \{x: 0 \leqslant x < \infty\}$, $\overline{\mathbb{R}}^+ = \mathbb{R}^+ \cup \{\infty\}$. The usual algebraic operations are allowed in $\overline{\mathbb{R}}$ except that the combinations $(\pm\infty) + (\mp\infty)$ and quotients such as ∞/∞ are not permitted. In contrast, we allow $\infty + \infty = \infty$, and the products $0 \,.\, \infty$, $-\infty$ which are given the value 0.

(i) the measure of the empty set is zero, that is $\mu(\varnothing) = 0$;
(ii) μ is *countably additive*, that is, if (S_n) is any sequence of disjoint sets in $\mathscr{S}$, then

$$\mu\left(\bigcup_1^\infty S_n\right) = \sum_1^\infty \mu(S_n).$$

The triple $(X, \mathscr{S}, \mu)$ is known as a **measure space**. (Some basic properties of measures are listed in Problem 2.2.)

2.2.7 Example. (Dirac measure, the δ-function). Let $\mathscr{S}$ be a σ-algebra of subsets of X, and let a be a point of X. Define

$$\delta_a(S) = \begin{cases} 1 & (a \in S), \\ 0 & (a \notin S). \end{cases}$$

δ_a is obviously a measure on $\mathscr{S}$. If $X = \mathbb{R}^n$, the δ-function is obtained.

This example reassures us that measures do exist, but the Dirac measure certainly does not yield the usual length in $\mathbb{R}$. It is to this case that we now turn.

It may be shown that no such measure can be defined on the σ-algebra of *all* subsets of $\mathbb{R}$. To find a measure analogous to length on the Borel sets $\mathscr{S}$ of $\mathbb{R}$, it is natural to start with intervals, say with end points a and b, and to take their measure to be $b - a$. We may then hope that on combining these basic sets by taking countable unions and so on, we will end up with a measure on $\mathscr{S}$. Unfortunately an uncountable infinity of such operations will be needed (recall that it is not even possible to construct all Borel sets by a countable process starting with open sets), and for this reason a rather different course will be pursued.

2.2.8 Definition. A class $\mathscr{S}$ of subsets of X is called an **algebra** iff

(i) $\varnothing, X \in \mathscr{S}$;
(ii) $S \in \mathscr{S} \Rightarrow X \setminus S \in \mathscr{S}$;
(iii) the union of a finite collection of sets in $\mathscr{S}$ is in $\mathscr{S}$.

Note that an algebra and a σ-algebra differ only in property (iii), an algebra being closed under *finite* unions. The following extends Definition 2.2.6.

2.2.9 Definition. A **measure** μ on an algebra $\mathscr{S}$ is a mapping from $\mathscr{S}$ to $\overline{\mathbb{R}}^+$ such that:

(i) $\mu(\varnothing) = 0$;
(ii) if $S_1, S_2, \ldots$ are disjoint sets in $\mathscr{S}$ and $\bigcup_1^\infty S_n \in \mathscr{S}$, then

$$\mu\left(\bigcup_1^\infty S_n\right) = \sum_1^\infty \mu(S_n).$$

2.2.10 Example. The procedure suggested above may now be carried through to define a measure on the algebra consisting of finite unions of intervals which may be open, half open, or closed (the empty interval and single points being included). Let the measure of a finite interval with end points a, b be $b - a$, and let the measure of an infinite interval be ∞. Define μ on any $S \in \mathscr{S}$ in the obvious way, for example by setting

$$\mu\left(\bigcup_1^j S_n\right) = \sum_1^j \mu(S_n)$$

for disjoint sets $S_1, \ldots, S_j$. We assert that μ is a measure on $\mathscr{S}$.

To prove this it is only necessary to verify (ii) of Definition 2.2.9, for which it is enough to show that if $S = \bigcup_1^\infty S_n$ where S and each S_n are intervals, then $\mu(S) = \sum_1^\infty \mu(S_n)$. We do this for a typical case in which S is finite with end points a, b, and each S_n has end points a_n, b_n. Clearly for any j,

$$\sum_1^j \mu(S_n) = \sum_1^j (b_n - a_n) \leqslant b - a = \mu(S),$$

whence $\sum_1^\infty \mu(S_n) \leqslant \mu(S)$. To prove the opposite inequality, take any $\varepsilon > 0$ and let $T \subset S$ be a closed interval with $\mu(S) - \mu(T) < \varepsilon$. Choose $\varepsilon_n > 0$ such that $\sum_1^\infty \varepsilon_n < \varepsilon$, and set $T_n = (a_n - \varepsilon_n, b_n + \varepsilon_n)$ for each n, giving $\mu(T_n) = \mu(S_n) + 2\varepsilon_n$. Obviously $T \subset S = \bigcup_1^\infty S_n = \bigcup_1^\infty T_n$. Since T is closed, by the Heine–Borel Theorem, $T \subset T_{n_1} \cup \ldots \cup T_{n_k}$ for some finite k. Therefore, as ε is arbitrary, the result follows from the inequalities:

$$\mu(S) - \varepsilon \leqslant \mu(T) \leqslant \sum_1^k \mu(T_{n_i}) = \sum_1^k \mu(S_{n_i}) + 2\sum_1^k \varepsilon_{n_i} \leqslant \sum_1^\infty \mu(S_n) + 2\varepsilon.$$

2.2.11 Example. A very similar procedure in $\mathbb{R}^2$ leads to a measure based on the area of rectangles, and higher dimensional analogues are readily constructed.

The above procedure yields a measure on $\mathscr{S}$. However, $\mathscr{S}$ is not a σ-algebra, and the next step is to define a measure $\bar{\mu}$ on a σ-algebra containing $\mathscr{S}$ such that μ and $\bar{\mu}$ coincide on $\mathscr{S}$. It is possible to define such a measure on

$\mathscr{S}_\sigma$ (called *Borel measure*), but this measure is inconvenient since subsets of sets of measure zero need not be measurable. If a slightly larger σ-algebra is used the problem disappears.

2.2.12 Theorem. *Let μ be a measure on an algebra $\mathscr{S}$. Then there is a measure $\bar{\mu}$ on a σ-algebra $\overline{\mathscr{S}}$ containing $\mathscr{S}$ such that:*

(i) *μ and $\bar{\mu}$ coincide on $\mathscr{S}$;*
(ii) *if $S_2 \in \overline{\mathscr{S}}$ and $\bar{\mu}(S_2) = 0$, then $S_1 \in \overline{\mathscr{S}}$ and $\bar{\mu}(S_1) = 0$ for any $S_1 \subset S_2$.*

Moreover, if there is a sequence (S_n) with $S_n \in \mathscr{S}$ and $\mu(S_n) < \infty$ such that $X = \bigcup_1^\infty S_n$, then $\bar{\mu}$ is uniquely determined on the smallest σ-algebra containing $\mathscr{S}$ and satisfying (ii).

The proof of this theorem is omitted (see Bartle, 1966; Chap. 9) but it may be helpful to an intuitive understanding if an outline of the method for constructing $\bar{\mu}$ is given. For any $Y \subset X$ define

$$\mu^*(Y) = \inf\left\{\sum_1^n \mu(S_i) : Y \subset S_1 \cup \ldots \cup S_n, S_1, \ldots, S_n \in \mathscr{S}\right\}.$$

This construction gives a definition of μ^* by bracketing Y from outside. μ^* has the property $\mu^*(Y_1 \cup Y_2) \leqslant \mu^*(Y_1) + \mu^*(Y_2)$ for all $Y_1, Y_2 \in X$, but equality does not hold in general even when Y_1 and Y_2 are disjoint. Thus μ^* is not a measure (although it is called an outer measure!). The strategy is to pick out a class of sets on which μ^* is additive—and so a measure. This is by no means easy, but the class turns out to consist of all sets Y which satisfy $\mu^*(S \cap Y) + \mu^*(S \backslash Y) = \mu^*(S)$ for every $S \in \mathscr{S}$. These sets form the σ-algebra of the theorem.

We have now reached our goal. In $\mathbb{R}^n$ we simply take the measure described in Examples 2.2.10 and 2.2.11 and extend this as in the previous theorem.

2.2.13 Definition. The σ-algebra produced from the length of intervals, area of rectangles, ... in $\mathbb{R}$, $\mathbb{R}^2$, ... by the above theorem is called the **σ-algebra of Lebesgue sets**, and the measure $\bar{\mu}$ is **Lebesgue measure**.

For practical purposes the distinction between the Lebesgue and Borel sets is rather slight, for every Lebesgue measurable set is the union of a Borel set (with the same measure) and a set of Lebesgue measure zero. The Borel sets are of course Lebesgue measurable. In fact it is very difficult to produce a set which is not Lebesgue measurable (it requires a sophisticated principle from set theory), and you would be very unlucky to encounter such an object.

2.2.14 Example. Consider the Lebesgue measure of the set S of rationals in an interval X of $\mathbb{R}$. The measure of a single point is zero, and S is a countable union of single points. Hence by Definition 2.2.6(ii), $\mu(S) = 0$. By taking complements, we deduce that the measure of the irrationals is the length of X.

2.2.15 Example. Another useful measure may be obtained from the theorem. Let γ be any non-decreasing function on $\mathbb{R}$ which is continuous on the right. For $a, b \in \mathbb{R}$ define

$$\gamma((a, b]) = \gamma(b) - \gamma(a),$$

$$\gamma((a, b)) = \lim_{x \to b-} \gamma(x) - \gamma(a),$$

with similar definitions for the other types of interval. Then γ is a measure on finite unions of intervals and so extends to what is called a *Lebesgue–Stieltjes* measure.

An important convention in measure theory must be remarked on. Sets of measure zero are reasonably regarded as small, indeed almost small enough to ignore, and it is therefore natural to adopt the following definition.

2.2.16 Definition. A property is said to hold **almost everywhere** iff the measure of the set on which it does not hold is zero. The abbreviation a.e. will be used here; another possibility is p.p. (presque partout).

2.2.17 Example. Let μ be Lebesgue measure on $\mathbb{R}$. Assume that $\mathbb{R} = S \cup S'$ where S and S' are disjoint, and suppose that $f(x) = 1$ when $x \in S$, $f(x) = 0$ when $x \in S'$. Then $f = 1$ a.e. if and only if $\mu(S') = 0$. For this to hold S' may of course be any finite set of points, but it may also be a countable set such as the rationals—and even some uncountable sets have measure zero (for example the Cantor set in Problem 1.8).

We close with a general remark emphasizing a point already hinted at. In integration theory the definition of many objects is non-constructive (as for Lebesgue measure). At first sight this may seem a serious disadvantage. However, we shall see that it is always possible to deduce as much about the object as is needed for practical purposes, and the non-constructive nature of the definition is not a hindrance.

2.3 Measurable Functions

One of the advantages of the measure theoretic approach to integration is that integrable functions are often easy to recognize. The key is the prior description of the slightly wider class of measurable functions.

As with measures the extra convenience of dealing with *extended* real valued functions is considerable. For example, with $X = [0, 1]$ the function $x^{-\frac{1}{2}}$ may be handled without giving special treatment to the point $x = 0$. A small price must be paid for this convenience. Recalling the previous remarks on $\overline{\mathbb{R}}$, we must be careful not to allow the expression $f + g$ if at some point a sum such as $\infty + (-\infty)$ should occur. Again $0 \,.\, \infty$ and $0 \,.\, -\infty$ are allowed and given the value zero. Complex valued functions are not allowed to take infinite values because of difficulties with the interpretation of such expressions as $\infty + i\infty$. With this proviso, a complex valued function may be uniquely decomposed in the form $f = g + ih$ with g, h real valued, and results may therefore be stated for the real case without loss of generality.

2.3.1 Definition. Let $\mathscr{S}$ be a σ-algebra. A function $f : X \to \overline{\mathbb{R}}$ is said to be **measurable** iff for every real α the set

$$f^{-1}((\alpha, \infty]) = \{x : x \in X, f(x) > \alpha\}$$

is in $\mathscr{S}$. A complex valued function $g + ih$ is measurable iff g and h are both measurable. It is equivalent in the definition to substitute any of $[\alpha, \infty]$, $[-\infty, \alpha)$ or $[-\infty, \alpha]$ for $(\alpha, \infty]$.

2.3.2 Example. Consider the constant function $f(x) = a$. Then $f^{-1}((\alpha, \infty]) = \varnothing$ if $\alpha \geqslant a$, and $f^{-1}((\alpha, \infty]) = X$ if $\alpha < a$. Since by definition any σ-algebra contains $\varnothing$ and X, f is measurable.

2.3.3 Example. Let S be any set in $\mathscr{S}$, and let χ be its characteristic function. Then $\chi^{-1}((\alpha, \infty]) = \varnothing$, S or X for $\alpha \geqslant 1$, $0 \leqslant \alpha < 1$ or $\alpha < 0$ respectively. Therefore χ is measurable. Note that it follows that for the Lebesgue σ-algebra on $\mathbb{R}$, the rather unpleasant function $f(x) = 1$ (x irrational), $f(x) = 0$ (x rational) is measurable.

2.3.4 Example. Take $\mathscr{S}$ to be the Lebesgue σ-algebra on $\mathbb{R}^n$, and let $f : \mathbb{R}^n \to \mathbb{R}$ be continuous. Since f takes only finite values, $f^{-1}((\alpha, \infty]) = f^{-1}((\alpha, \infty))$, and from a result (Lemma 3.2.9) which will be proved in the next chapter, $f^{-1}((\alpha, \infty))$ is open. Therefore, as $\mathscr{S}$ contains all open sets, we conclude that *a continuous function is measurable.*

2.3.5 Example. Assume that $X = [0, \infty)$ and let $\mathscr{S}$ be the Lebesgue σ-algebra. For the extended real valued function $f(x) = x^{-a} (a > 0)$, $f^{-1}((\alpha, \infty])$ is X or $[0, \alpha^{-1/a})$ for $\alpha \leqslant 0$ or $\alpha > 0$ respectively. Since both of these sets are in $\mathscr{S}$, x^{-a} is measurable.

What other functions are measurable? The observation that certain combinations and limits of sequences of measurable functions are measurable enormously increases the list. For example, let (f_n) be a sequence of measurable functions and define the function $\sup f_n$ by setting $(\sup f_n)(x) = \sup f_n(x)$ for each x; note that this function always exists as extended real valued functions are allowed. Then

$$\{x : \sup_n f_n(x) > a\} = \{x : f_n(x) > a \text{ for at least one } n\}$$
$$= \bigcup_{n=1}^{\infty} \{x : f_n(x) > a\},$$

and since $\mathscr{S}$ is closed with respect to countable unions, $\sup f_n$ is measurable. The next two theorems summarize the position.

2.3.6 Theorem. *Let $\mathscr{S}$ be a σ-algebra of subsets of X, and suppose that $f, g : X \to \overline{\mathbb{R}}$ are measurable. Then the following are measurable:*

(i) af $(a \in \mathbb{R})$;
(ii) $|f|$;
(iii) $f + g$ (*with the proviso that* $\pm\infty + (\mp\infty)$ *should not occur*);
(iv) fg;
(v) $\max\{f, g\}, \min\{f, g\}$ (*where* $\max\{f, g\} = \max\{f(x), g(x)\}$ *for* $x \in X$);
(vi) $f^+ = \max\{f, 0\}$, $f^- = \max\{-f, 0\}$.

2.3.7 Theorem. *With $\mathscr{S}$ as above, suppose that (f_n) is a sequence of measurable functions from X to $\overline{\mathbb{R}}$. Then the following are measurable:*

(i) $\sup f_n, \inf f_n$;
(ii) $\limsup f_n \ (= \lim_{n\to\infty} \sup_{r \geq n} f_r)$, $\liminf f_n$;
(iii) *the pointwise limit* $\lim f_n$ *when it exists.*

Consider finally the class of "simple functions" on which the definition of integration will be based. A simple function is a generalization of the step function (Definition 2.1.1) used in Riemann integration.

2.3.8 Definition. Let $\mathscr{S}$ be a σ-algebra of subsets of X. A function $f : X \to \mathbb{R}$ is said to be **simple** iff there are a finite number of sets $S_1, \ldots, S_n \in \mathscr{S}$ and real numbers $c_1, \ldots, c_n$ with

$$f = \sum_1^n c_i \chi_{S_i},$$

where χ_{S_i} is the characteristic function of S_i.

Obviously simple functions are measurable (recall Example 2.3.3). Sums, products and moduli of simple functions are simple; limits of sequences of simple functions need not be simple, although they are certainly measurable, by Theorem 2.3.7. The importance of simple functions arises from the fact that every measurable function can be approximated by a simple function.

2.3.9 Lemma. *Let $\mathscr{S}$ be a σ-algebra of subsets of X, and suppose that $f: X \to \overline{\mathbb{R}}$ is measurable. Then:*

(i) *if $f \geqslant 0$, there is a monotone increasing sequence of non-negative simple functions which tends pointwise to f;*
(ii) *if f is bounded, there is a sequence of simple functions tending to f in the sup norm.*

We note in conclusion that the class of measurable functions is extremely vast, and certainly includes any function likely to arise in applications. For example, the pointwise limits of sequences of continuous functions are measurable (by Example 2.3.4 and Theorem 2.3.7). Indeed the problem of finding a non-measurable function is as difficult as that of finding a non-measurable set.

2.4 Integration

The difficult part of the measure theoretic approach to integration is the construction of measure, and now that this has been carried out the definition of the integral is straightforward. This proceeds in a series of steps. First the natural definition of the integral of a simple function is given. Next, recalling that an arbitrary non-negative measurable function $f: X \to \overline{\mathbb{R}}$ is the limit of simple functions, we take the integral of f to be the supremum of the integrals of all simple functions not greater than f—for the Riemann integral the procedure is the same except that only step functions are allowed. Finally, the definition for real-valued but not necessarily positive f is obtained by splitting f into its positive and negative parts, and for complex valued f by splitting into real and imaginary parts. The integral on X with respect to μ will be denoted as appropriate by $\int_X f\,\mathrm{d}\mu$ or $\int f\,\mathrm{d}\mu$ or $\int f(x)\,\mathrm{d}\mu(x)$.

2.4.1 Definition. Let $(X, \mathscr{S}, \mu)$ be a measure space and suppose that f is measurable.

(i) For f a simple function, say $f = \sum_1^n c_i \chi_{S_i}$, set

$$\int f\,\mathrm{d}\mu = \sum_1^n c_i \mu(S_i),$$

the convention $0 \,.\, \infty = 0$ being used.

(ii) For f extended real valued and non-negative, set

$$\int f\,\mathrm{d}\mu = \sup_g\left\{\int g\,\mathrm{d}\mu : g \text{ is simple and } 0 \leqslant g(x) \leqslant f(x) \text{ for } x \in X\right\}.$$

(iii) Suppose that f is extended real valued. Take $f^{\pm}$ to be the positive and negative parts of f respectively, and set

$$\int f\,\mathrm{d}\mu = \int f^+\,\mathrm{d}\mu - \int f^-\,\mathrm{d}\mu,$$

provided at least one of the integrals on the right is finite (to avoid the expression $\infty - \infty$).

(iv) Suppose f is complex valued. If the integrals of $\operatorname{Re} f$ and $\operatorname{Im} f$ can be defined as in (iii) and are finite, set

$$\int f\,\mathrm{d}\mu = \int \operatorname{Re} f\,\mathrm{d}\mu + i\int \operatorname{Im} f\,\mathrm{d}\mu.$$

(v) For measurable S, let χ_S be the characteristic function of S. If the integral of $f\chi_S$ can be defined as above, set

$$\int_S f\,\mathrm{d}\mu = \int_X f\chi_S\,\mathrm{d}\mu.$$

If the integral can be defined in this manner we say that the *integral exists*. Note that the use of the convention $0\,.\,\infty = 0$ in the definition implies that the integral over S exists and is zero if $f(x) = 0$ for $x \in S$ (even if $\mu(S) = \infty$), and if $\mu(S) = 0$ (even if $f(x) = \infty$ for $x \in S$). Also, if f is measurable, the integral exists, although it may be infinite, so long as either f is non-negative or one of the integrals of its positive and negative parts is finite. In fact we shall usually deal with a somewhat more restricted class of function for which all the integrals in the above definition are finite (which will only be the case if f is finite almost everywhere). For these functions the integral has all the expected elementary properties.

2.4.2. Definition. Suppose f is an extended real valued or a complex valued function. Then f is said to be **integrable** (on X with respect to μ) iff it is measurable and $\int |f|\,\mathrm{d}\mu < \infty$. When X is a subset of $\mathbb{R}^n$, f is said to be **locally integrable** iff it is integrable on each $S \subset X$ closed in $\mathbb{R}^n$ and bounded.

2.4.3 Lemma. (i) *If f is integrable, then* $\left|\int f\,\mathrm{d}\mu\right| \leqslant \int |f|\,\mathrm{d}\mu$.

(ii) *Let f, g be extended real valued measurable functions, and suppose that $f \leqslant g$. Then* $\int f\,\mathrm{d}\mu \leqslant \int g\,\mathrm{d}\mu$.

(iii) *Let f, g be measurable. For scalar a, if either $f, g \geqslant 0$ and $a \geqslant 0$, or f, g are integrable, then*

$$\int (f+g)\,\mathrm{d}\mu = \int f\,\mathrm{d}\mu + \int g\,\mathrm{d}\mu, \quad \int af\,\mathrm{d}\mu = a\int f\,\mathrm{d}\mu.$$

It is important to recognize that for most purposes sets of measure zero may be ignored in integration theory. For example, suppose f, g are integrable and equal almost everywhere. With $S = \{x : f(x) \neq g(x)\}$ and $S' = X \backslash S$,

$$\int_X (f-g)\,\mathrm{d}\mu = \int_{S'} (f-g)\,\mathrm{d}\mu + \int_S (f-g)\,\mathrm{d}\mu = 0,$$

the first integral on the right being zero since $f = g$ on S', the second since $\mu(S) = 0$. Therefore two functions equal almost everywhere have the same integral, or to put it another way, altering a function on a set of measure zero does not change its integral. One consequence of this is that the usual notion of boundedness is unnecessarily strong for integration theory, boundedness almost everywhere being more appropriate.

2.4.4. Definition. Define the **essential supremum** of f to be

$$\text{ess sup} f = \inf\{k : f(x) \leqslant k \text{ a.e.}\}.$$

By far the most useful integrals in applications are obtained by choosing μ to be Lebesgue measure.

2.4.5 Definition. Take $X = \mathbb{R}^n$. If μ is Lebesgue measure, the integral is called the **Lebesgue integral**, and is usually written $\int f(x)\,\mathrm{d}x$ or $\int_S f(x)\,\mathrm{d}x$. When μ is Lebesgue–Stieltjes measure, the integral is known as the **Lebesgue–Stieltjes integral.**

To clarify the process of Lebesgue integration, consider a non-negative continuous function defined on $[0, 1]$. For a step function the Riemann and Lebesgue integrals coincide. Now the Riemann integral is defined in the same manner as the Lebesgue integral except that the supremum in Definition 2.4.1 (ii) is taken over step functions. Since these are certainly simple functions, the Riemann integral is not greater than the Lebesgue integral. The opposite inequality is obtained by applying the same argument to $a - f$, where $a > \sup f$, and we conclude that the values of the integrals are the same. An extension of this argument leads to the following reassuring result.

2.4.6. Theorem. *Suppose X is a finite interval in $\mathbb{R}$ and f is bounded. If f is Riemann integrable, then it is Lebesgue integrable and the two integrals have the same value.*

Roughly then if a function is properly Riemann integrable (or in many circumstances even improperly Riemann integrable, see Problem 2.6), then it is Lebesgue integrable and the integrals have the same value. On the other hand, the class of Lebesgue integrable functions is much larger than that of Riemann integrable functions. The comparison between the two integrals is clarified by noting that on a finite interval, for a bounded function to be Riemann integrable it is necessary and sufficient that the function be continuous almost everywhere, whereas a function is Lebesgue integrable if it is merely Lebesgue measurable and bounded almost everywhere. An example of a function which is Lebesgue but not Riemann integrable follows.

2.4.7. Example. Define $f(x) = 0$ (x rational), $f(x) = 1$ (x irrational). f is certainly not Riemann integrable. However, the rationals have measure zero (Example 2.2.4), whence $f = 1$ a.e., and since a set of measure zero does not contribute to the integral then $\int_0^1 f(x)\,dx = 1$. The same conclusion may be reached by noting that f is a simple function.

Our aim in the rest of this section is to show that the integral defined above, and in particular the Lebesgue integral, is an extremely tractable object. The superiority of the Lebesgue over the Riemann integral is most marked when a limiting process is involved. For the Riemann integral the exchange of the order of taking limits and integrating is a tricky problem whose solution often rests on uniform convergence arguments. As the next three results show, the difficulty disappears when the Lebesgue integral is used. These results are of central importance in integration theory.

2.4.8 The Monotone Convergence Theorem. *Let $(X, \mathscr{S}, \mu)$ be a measure space, and assume that (f_n) is a monotone increasing sequence of non-negative, extended real-valued measurable functions. Then each of the following integrals is defined (although not necessarily finite), and*

$$\lim \int f_n \, d\mu = \int \lim f_n \, d\mu.$$

The theorem remains true if the sequence is merely monotone increasing almost everywhere. To see this, let S be the set on which the sequence is monotone increasing, let χ_S be the characteristic function of S, and apply the theorem to $(f_n\chi_S)$. It is generally true that theorems in integration theory hold if "almost everywhere" is substituted for "everywhere", and we shall not remark on this in each case.

2.4.9 Example. As an application of the theorem consider the evaluation of the Lebesgue integral $\int_0^1 x^{-\frac{1}{2}}\,dx$. The function $x^{-\frac{1}{2}}$ is known to be measurable

(Example 2.3.5). Further, $\int_\varepsilon^1 x^{-\frac{1}{2}}\,dx$ is easily calculated when $\varepsilon > 0$, for by Theorem 2.4.6 it is equal to the Riemann integral. To complete the computation the Monotone Convergence Theorem is employed. Set

$$f_n(x) = \begin{cases} x^{-\frac{1}{2}} & (n^{-1} \leqslant x \leqslant 1), \\ 0 & (0 \leqslant x < n^{-1}). \end{cases}$$

Then (f_n) is a monotone increasing sequence of non-negative measurable functions whose limit is $x^{-\frac{1}{2}}$ a.e. (in fact except on the set $\{0\}$ which has measure zero), and

$$\int_0^1 x^{-\frac{1}{2}}\,dx = \int_0^1 \lim f_n(x)\,dx = \lim \int_0^1 f_n(x)\,dx = \lim 2(1 - n^{-\frac{1}{2}}) = 2.$$

For a more general result relating the improper Riemann integral and the Lebesgue integral, see Problem 2.6.

2.4.10 Fatou's Lemma. *Let $(X, \mathscr{S}, \mu)$ be a measure space, and assume that (f_n) is a sequence of extended real valued non-negative measurable functions. Then*

$$\int \liminf f_n\,d\mu \leqslant \liminf \int f_n\,d\mu.$$

2.4.11 The Lebesgue Dominated Convergence Theorem. *Assume that $(X, \mathscr{S}, \mu)$ is a measure space. Let (f_n) be a sequence of extended real-valued measurable functions, and suppose that $\lim f_n = f$. If there is an integrable function g such that $|f_n| \leqslant g$ $(n \geqslant 1)$, then f is integrable and*

$$\lim \int f_n\,d\mu = \int f\,d\mu.$$

2.4.12 Example. Consider the integrals

$$I_n = \int_0^\infty e^{-nx} x^{-\frac{1}{2}}\,dx.$$

We want to show that $\lim I_n = 0$. With $f_n(x) = e^{-nx} x^{-\frac{1}{2}}$, certainly $\lim_n f_n(x) = 0$ if $x \neq 0$, but the convergence is not uniform (since $\lim_{x\to 0} f_n(x) = \infty$), and using the Riemann integral it would be necessary to invent a delicate argument based on a subdivision of the range of integration to establish the result. In fact the proof is a trivial exercise if the Dominated Convergence Theorem is used. We simply note that $|f_n(x)| \leqslant e^{-x}x^{-\frac{1}{2}}$ for $x > 0$ and all n, and ignoring the single point $x = 0$ as it has measure zero, deduce directly from the theorem that $\lim I_n = 0$.

Consider next the properties of functions defined by integrals. The easiest such function is an indefinite integral. A well known theorem in calculus states that a real-valued function f on $\mathbb{R}$ has a continuous derivative if and only if f is an indefinite integral of some continuous function, g say, and then $f' = g$. Suppose now that g is merely locally Lebesgue integrable, what can be said about its indefinite integral f?

2.4.13 Definition. Let S be a finite interval. Then the real valued function f is said to be **absolutely continuous** on S iff for each $\varepsilon > 0$ there is a $\delta > 0$ such that $\sum_1^n |f(b_j) - f(a_j)| < \varepsilon$ for any finite set $\{[a_j, b_j]\}$ of disjoint intervals with total length less than δ. f is absolutely continuous on $\mathbb{R}$ iff it is absolutely continuous on every finite subinterval.

2.4.14 Theorem. *A real-valued function f defined on an interval in $\mathbb{R}$ is an indefinite integral of a locally Lebesgue integrable function, g say, iff f is absolutely continuous, in which case f is differentiable almost everywhere and* $f' = g$ (a.e.).

In the next two theorems X, Y are intervals and f is an extended real-valued function defined on the rectangle $X \times Y$.

2.4.15 Theorem. *Suppose that $f(x, \cdot)$ is measurable for each $x \in X$, and that $f(\cdot, y)$ is continuous for each $y \in Y$. Assume also that there is an integrable $g : X \to \overline{\mathbb{R}}$ such that $|f(x, y)| \leqslant g(y)$ for all $x \in X$, $y \in Y$. Then the following function is continuous:*

$$F(x) = \int_Y f(x, y)\, \mathrm{d}\mu(y) \tag{2.4.1}$$

Proof. By the Dominated Convergence Theorem, $F(x_n) \to F(x)$ as $x_n \to x$. □

2.4.16 Theorem. *Assume that $f(\cdot, y)$ is differentiable for each $y \in Y$, and that $f(x, \cdot)$ is integrable for each $x \in X$. Suppose that there is an integrable function $g : Y \to \overline{\mathbb{R}}$ such that $|\partial f/\partial x(x, y)| \leqslant g(y)$ for all $x \in X$, $y \in Y$. Then F defined by (2.4.1) is differentiable, and*

$$\frac{\mathrm{d}F}{\mathrm{d}x}(x) = \int_Y \frac{\partial f}{\partial x}(x, y)\, \mathrm{d}\mu(y).$$

Proof. If $\lim x_n = x$ and $x_n \neq x$ for any n, the functions g_n where

$$g_n(y) = [f(x_n, y) - f(x, y)]/(x_n - x),$$

are measurable. Since $\lim_n g_n(y) = \partial f/\partial x(x, y)$ for each y, by Theorem 2.3.7 $\partial f/\partial x(x, \cdot)$ is measurable for each x. Moreover, by the Mean Value Theorem $g_n(y) = \partial f/\partial x(z_n, y)$ for some z_n between x_n and x, and hence $|g_n| \leqslant g$ for all n. The result follows from the Dominated Convergence Theorem 2.4.11. □

In exchanging orders of integration matters are again much simplified. For the Lebesgue integral, which is denoted by $\iint f(x, y)\,dxdy$, over a measurable set in $\mathbb{R}^2$ the conclusion is as follows.

2.4.17 Theorem. *Assume that $f(\cdot, \cdot)$ is measurable, and suppose that one of the following conditions holds:*

(i) (*Tonelli*) $f \geqslant 0$;
(ii) (***Fubini***) *One of the integrals* $\iint |f(x, y)|\,dx\,dy$, $\int dx \int |f(x, y)|\,dy$, $\int dy \int |f(x, y)|\,dx$ *is finite.*

Then the functions $f(\cdot, y)$, $f(x, \cdot)$, $\int f(\cdot, y)\,dy$, $\int f(x, \cdot)\,dx$ *are measurable and*

$$\int dx \int f(x, y)\,dy = \int dy \int f(x, y)\,dx = \iint f(x, y)\,dx\,dy.$$

2.5 The $\mathscr{L}_p$ Spaces

We come now to the primary purpose of this chapter, which is the construction of a set of Banach spaces called the $\mathscr{L}_p$ spaces. The fact that the vital property of completeness can be achieved is a direct consequence of the civilized behaviour of the integral when limits are taken.

Throughout, a measure μ defined on a σ-algebra of subsets of a set X in $\mathbb{R}^n$ will be used. In most applications μ will simply be the Lebesgue measure. The functions will be extended real valued or complex valued.

2.5.1 Definition. Let f be measurable, and suppose that $p \geqslant 1$. Set

$$\|f\|_p = \left\{\int |f|^p\,d\mu\right\}^{1/p} \qquad (1 \leqslant p < \infty),$$

$$\|f\|_\infty = \text{ess sup}\,|f|.$$

The space $\mathscr{L}_p$ (or $\mathscr{L}_p(X)$ if X needs emphasis) is defined to be the set of measurable functions f such that $\|f\|_p < \infty$. Frequently X will be an interval of the real line with end points, a, b and μ will be Lebesgue measure, in which case the space will be denoted by $\mathscr{L}_p(a, b)$. Two functions f, g in $\mathscr{L}_p$ are said to be equal iff $f = g$ a.e.

Three points should be noted. First, since $\|f - g\|_p = 0$ iff $f = g$ a.e. (see Problem 2.13) for $\|\cdot\|_p$ to be a norm it is necessary to identify functions equal almost everywhere. Secondly, a non-negative function f is integrable only if f is finite almost everywhere. Thus all functions in $\mathscr{L}_p$ are finite almost everywhere; they can therefore have their values changed on a set of measure zero so as to be finite everywhere. Lastly, when $p = \infty$, if μ is the Lebesgue measure and f is continuous, then $\|f\|_\infty = \sup|f(x)|$. Thus the continuous bounded functions with the sup norm form a closed subspace of $\mathscr{L}_\infty$. If a function f in $\mathscr{L}_\infty$ is not continuous, the measure of the set for which $|f(x)| > \|f\|_\infty$ is zero. The following space will also occasionally be needed.

2.5.2 Definition. $\mathscr{L}_p^{\text{loc}}$ is the set of functions which lie in $\mathscr{L}_p(S)$ for every $S \subset X$ closed in $\mathbb{R}^n$ and bounded.

$\mathscr{L}_p$ may be thought of as the continuous analogue of the sequence space ℓ_p with integration replacing summation, and the basic Hölder and Minkowski inequalities are established simply by changing sums into integrals in the proof of Theorem 1.3.12.

2.5.3 Theorem. *Assume that $p \geqslant 1$ and let q be the conjugate index. With the notation of Definition 2.5.1 the following hold for any measurable f, g (infinite values being allowed):*

(i) $\|fg\|_1 \leqslant \|f\|_p \|g\|_q$ *(Hölder's inequality)*;

(ii) $\|f + g\|_p \leqslant \|f\|_p + \|g\|_p$ *(Minkowski's inequality)*.

2.5.4 Corollary (Young's inequality). *Suppose that $k : X \times X \to \mathbb{C}$ is measurable, and assume that*

$$\sup_{x \in X} \int |k(x, y)| \, \mathrm{d}\mu(y), \ \sup_{y \in X} \int |k(x, y)| \, \mathrm{d}\mu(x) \leqslant m < \infty.$$

If $f \in \mathscr{L}_p$ for some $p \geqslant 1$, then the function F, where

$$F(x) = \int k(x, y) \, f(y) \, \mathrm{d}\mu(y),$$

is in $\mathscr{L}_p$ and $\|F\|_p \leqslant m\|f\|_p$.

Proof. If $1 < p < \infty$ and q is the conjugate index,

$$\int |k(x, y)|\,|f(y)|\,\mathrm{d}\mu(y) = \int |k(x, y)|^{1/q}\,|k(x, y)|^{1/p}\,|f(y)|\,\mathrm{d}\mu(y)$$
$$\leqslant \left\{\int |k(x, y)|\,\mathrm{d}\mu(y)\right\}^{1/q} \left\{\int |k(x, y)|\,|f(y)|^p\,\mathrm{d}\mu(y)\right\}^{1/p}$$

by Hölder's inequality. Raising both sides to the pth power, integrating, and using Fubini's Theorem 2.4.17(ii) to change the order of integration, we obtain

$$\int\left\{\int |k(x, y)|\,|f(y)|\,\mathrm{d}\mu(y)\right\}^p \mathrm{d}\mu(x) \leqslant m^{p/q}\,\|f\|_p^p \sup_{y\in X} \int |k(x, y)|\,\mathrm{d}\mu(x)$$
$$\leqslant m^{p/q+1}\,\|f\|_p^p,$$

from which the result follows immediately. The easy cases $p = 1, \infty$ are left as exercises. □

Minkowski's inequality is obviously the triangle inequality in $\mathscr{L}_p$. It is therefore simple to verify that $\mathscr{L}_p$ is a normed vector space, and the subtlety comes in showing that it is complete, a result which on its own is known as the Riesz–Fischer Theorem. Hölder's inequality will be useful later; for the moment note that if $p = 2$ it is Schwartz's inequality (1.5.1) in the Hilbert space $\mathscr{L}_2$ of the next theorem.

2.5.5 Theorem. *If $p \geqslant 1$, under the norm $\|\cdot\|_p$, $\mathscr{L}_p$ is a Banach space. $\mathscr{L}_p$ is separable if $1 \leqslant p < \infty$. $\mathscr{L}_2$ is a Hilbert space when equipped with the inner product*

$$(f, g) = \int f\bar{g}\,\mathrm{d}\mu.$$

Proof. Completeness will be proved for $1 \leqslant p < \infty$, the case $p = \infty$ being left as an exercise. If (f_n) is a Cauchy sequence, there is a subsequence (f_{n_k}) such that

$$\|f_{n_{k+1}} - f_{n_k}\|_p < 2^{-k}.$$

Set

$$g_r = \sum_{k=1}^{r} |f_{n_{k+1}} - f_{n_k}|, \qquad g = \sum_{k=1}^{\infty} |f_{n_{k+1}} - f_{n_k}|,$$

and note that (g_r) is a monotone increasing sequence which tends pointwise to g. From Minkowski's inequality, $\|g_r\|_p < 1$, whence $\int g^p\,\mathrm{d}\mu = \lim \int g_r^p\,\mathrm{d}\mu$

$\leqslant 1$ by the Monotone Convergence Theorem 2.4.8. Therefore $g(x) < \infty$ a.e. (if $g = \infty$ on a set of non-zero measure its integral must be ∞), and so for almost all x the series

$$f_{n_1}(x) + \sum_{k=1}^{\infty} (f_{n_{k+1}}(x) - f_{n_k}(x))$$

is absolutely convergent, that is, the sequence $(f_{n_k}(x))$ converges almost everywhere to $f(x)$ say—$f(x)$ is defined only almost everywhere, but in view of the definition of $\mathscr{L}_p$ this does not matter.

To complete the proof we show that $f \in \mathscr{L}_p$ and $f_n \to f$ in $\mathscr{L}_p$ (it may not be true that $f_n \to f$ a.e.). Choose $\varepsilon > 0$. Since (f_n) is Cauchy, there is an n_0 such that for $m, n > n_0$,

$$\|f_n - f_m\|_p^p = \int |f_n - f_m|^p \, d\mu < \varepsilon.$$

Now taking $m = n_k$ and applying Fatou's Lemma 2.4.10, we conclude that for $n > n_0$,

$$\|f_n - f\|_p^p = \int \lim_k |f_n - f_{n_k}|^p \, d\mu \leqslant \liminf_k \int |f_n - f_{n_k}|^p \, d\mu \leqslant \varepsilon.$$

Hence $f_n - f \in \mathscr{L}_p$, and so $f \in \mathscr{L}_p$ and $\lim \|f_n - f\|_p = 0$. □

It will frequently be useful to know that certain sets of functions are dense in $\mathscr{L}_p$.

2.5.6 Theorem. *Let Ω be an open subset of $\mathbb{R}^n$, and let μ be Lebesgue measure. Then if $1 \leqslant p < \infty$, the following are dense in $\mathscr{L}_p(\Omega)$:*

(i) *the set of integrable simple functions;*

(ii) *the set $\mathscr{C}_0^\infty(\Omega)$ of $\mathscr{C}^\infty(\Omega)$ functions with bounded support contained in the interior of Ω.*

The observation that some of the elements of $\mathscr{L}_p$ are quite wild functions (for instance the function in Example 2.4.7) may prompt us to enquire whether the $\mathscr{L}_p$ spaces are suitable for applications where a continuous solution is often required. Sometimes indeed it is necessary to be content with a solution in $\mathscr{L}_p$, but more often, having found such a solution we are able to prove independently that it is continuous; several examples of this type of argument will later be encountered. The point is that the strategy of using $\mathscr{L}_p$ and then proving continuity may have considerable technical advantages over working with continuous functions throughout.

2.6 Applications

A number of miscellaneous results which are most naturally stated in $\mathscr{L}_p$ are now listed. Throughout the functions are defined on $\mathbb{R}^n$ and are complex valued.

Often results concerning the expansion of a function in terms of standard functions are most neatly formulated in $\mathscr{L}_2$. For example a number of well-known series of orthonormal functions converge in norm to an arbitrary function in $\mathscr{L}_2(a, b)$, whereas in $\mathscr{C}([a, b])$ the corresponding result is only true if additional conditions are imposed. We shall return at length to this topic in Chapter 10, where very general expansions are obtained from the Spectral Theorem for self-adjoint operators, but as an example, the $\mathscr{L}_2$ version of the well-known Fourier Integral Theorem is given here.

2.6.1 Theorem. *Suppose that $f \in \mathscr{L}_2(\mathbb{R}^n)$. Let Ω be a bounded cube centre the origin, and set*

$$\hat{f}_\Omega(\xi) = (2\pi)^{-n/2} \int_\Omega e^{i\xi \cdot x} f(x) \, dx, \tag{2.6.1}$$

where $\xi \cdot x$ denotes the scalar product in $\mathbb{R}^n$. Then as $\Omega \to \mathbb{R}^n$, $\hat{f}_\Omega$ converges in $\mathscr{L}_2(\mathbb{R}^n)$ to a function $\hat{f}$ called the Fourier transform of f; also $\|\hat{f}\|_2 = \|f\|_2$ (Parseval's formula), and $(\hat{f}, \hat{g}) = (f, g)$ (Plancherel's formula).

With the limit being taken in the same sense, the inversion formula is

$$f(x) = \lim_{\Omega \to \mathbb{R}^n} (2\pi)^{-n/2} \int_\Omega e^{-i\xi \cdot x} \hat{f}(\xi) \, d\xi.$$

Of course if $f \in \mathscr{L}_1(\mathbb{R}^n)$ instead of $\mathscr{L}_2(\mathbb{R}^n)$, no limiting process is necessary in defining $\hat{f}$, the integral in (2.6.1) simply being taken over the whole of $\mathbb{R}^n$. However, further conditions on f are needed if the inversion formula above is to be valid.

The convolution product of two functions is intimately related to the Fourier transform. The connection is described in Theorem 2.6.5 below after some basic properties of convolution have been listed.

2.6.2 Definition. The **convolution product** $f * g$ of f and g is defined formally by the equation

$$f * g(x) = \int_{\mathbb{R}^n} f(x - y) g(y) \, dy.$$

2.6.3 Theorem. *For $p \geq 1$, if $f \in \mathscr{L}_1(\mathbb{R}^n)$ and $g \in \mathscr{L}_p(\mathbb{R}^n)$, then $f * g \in \mathscr{L}_p(\mathbb{R}^n)$ and $\|f * g\|_p \leq \|f\|_1 \|g\|_p$.*

Proof. Take $k(x, y) = f(x - y), f(y) = g(y)$ in Young's inequality 2.5.4. □

2.6.4 Theorem. *Let k be a non-negative integer or infinity, and suppose that* $f \in \mathscr{C}_0^k(\mathbb{R}^n)$ *(see Definition 1.3.23). If* $g \in \mathscr{L}_p^{\mathrm{loc}}(\mathbb{R}^n)$ *for some* $p \geqslant 1$, *then* $f * g \in \mathscr{C}^k(\mathbb{R}^n)$ *and k differentiations under the integral sign are permissible.*

Proof. A straightforward application of Theorem 2.4.16. □

2.6.5 Theorem. *Suppose that* $f \in \mathscr{L}_1(\mathbb{R}^n)$, $g \in \mathscr{L}_2(\mathbb{R}^n)$. *Then* $\widehat{f * g} = (2\pi)^{n/2}\hat{f}.\hat{g}$ a.e. *If also* $h \in \mathscr{L}_2(\mathbb{R}^n)$, *then* $(f * g, h) = (\hat{f}.\hat{g}, \hat{h})$.

It is often convenient to approximate a function in $\mathscr{L}_p$ by a smooth function. This is constructed with the help of the convolution product.

2.6.6 Definition. Let $\bar{S}(0, \varepsilon)$ be the closed ball in $\mathbb{R}^n$ centre 0 and radius ε. Let j_ε be any non-negative continuous function with support in $\bar{S}(0, \varepsilon)$ and with $\int j_\varepsilon(x)\,\mathrm{d}x = 1$. Then the family $\{j_\varepsilon\}_{\varepsilon > 0}$ is called an **approximate identity**. j_ε is known as a **mollifier** iff in addition it is in $\mathscr{C}^\infty(\mathbb{R}^n)$.

2.6.7 Example. It is not quite obvious that mollifiers exist. To see that they do, define

$$k_\varepsilon(x) = \begin{cases} \exp[-1/(\varepsilon^2 - |x|^2)] & (|x| < \varepsilon), \\ 0 & (|x| \geqslant \varepsilon). \end{cases}$$

k_ε is constant on each sphere $|x| = r$, and it is easily verified that $\partial^n k_\varepsilon/\partial r^n$ tends exponentially to zero as $r \to \varepsilon -$. Therefore $k_\varepsilon \in \mathscr{C}^\infty(\mathbb{R}^n)$, and it is enough to take $j_\varepsilon = k_\varepsilon/\|k_\varepsilon\|_1$.

2.6.8 Theorem. *Assume that* $g \in \mathscr{L}_p(\mathbb{R}^n)$ *for some p such that* $1 \leqslant p < \infty$. *If* $\{j_\varepsilon\}$ *is an approximate identity,* $j_\varepsilon * g \to g$ *in* $\mathscr{L}_p(\mathbb{R}^n)$ *as* $\varepsilon \to 0$. *If further* j_ε *is a mollifier, then* $j_\varepsilon * g \in \mathscr{C}^\infty(\mathbb{R}^n)$.

Problems

2.1 Let $\mathscr{S}$ be a σ-algebra. Prove the following.

(i) If $S_1, S_2 \in \mathscr{S}$ and $S_2 \subset S_1$, then $S_1 \backslash S_2 \in \mathscr{S}$.
(ii) Countable intersections of sets in $\mathscr{S}$ lie in $\mathscr{S}$.

2.2 Let $(X, \mathscr{S}, \mu)$ be a measure space. Prove the following.

(i) If $S_1, S_2 \in \mathscr{S}$, $S_2 \subset S_1$ and $\mu(S_2) < \infty$, then $\mu(S_1 \backslash S_2) = \mu(S_1) - \mu(S_2)$.
(ii) If $S_i \nearrow S$ (that is $S_i \subset S_{i+1}$ for all i and $S = \cup S_i$) and the S_i and S are in $\mathscr{S}$, then $\mu(S_i) \nearrow \mu(S)$. [Hint: Consider $T_i = S_i - S_{i-1}$].

(iii) If $S_i \searrow S$ (that is $S_i \supset S_{i+1}$ and $S = \cap S_i$), $S_i, S \in \mathscr{S}$ and $\mu(S_1) < \infty$, then $\mu(S_i) \searrow \mu(S)$. By taking $S_i = [i, \infty) \subset \mathbb{R}$, show that the condition $\mu(S_1) < \infty$ is not superfluous.

2.3 Show that a set in $\mathbb{R}^n$ is Borel measurable iff its intersection with each closed and bounded set is Borel measurable.

2.4 Show that any monotone function on $\mathbb{R}$ is Borel measurable.

2.5 For $f: X \to \overline{\mathbb{R}}$, set $S_{\pm} = \{x : f(x) = \pm\infty\}$ respectively, and define $f_1(x) = f(x)$ $(x \notin S_+ \cup S_-)$, $f_1(x) = 0$ $(x \in S_+ \cup S_-)$. Then f is measurable iff $S_{\pm} \in \mathscr{S}$, and f_1 is a measurable function.

2.6 Take $X = [0, b)$ where b may be infinity. Suppose f is non-negative and Riemann integrable on $[0, a]$ for every $a < b$, and set

$$I_a = \int_0^a f(x)\, dx.$$

If $I = \lim_{a \to b-} I_a$ exists, prove that f is Lebesgue integrable on X with integral I.

2.7 Let (f_n) be a monotone increasing sequence of extended real-valued functions, and suppose that $\int f_1\, d\mu > -\infty$. Deduce from the Monotone Convergence Theorem 2.4.8 that $\lim \int f_n\, d\mu = \int \lim f_n\, d\mu$. Show by constructing a counterexample that the condition $\int f_1\, d\mu > -\infty$ cannot be dropped.

2.8 Deduce Fatou's Lemma 2.4.10 from the Monotone Convergence Theorem. Give a corresponding result with lim sup's and hence prove the Dominated Convergence Theorem.

2.9 (i) On $[0, 1]$ consider the functions (a) $nx/(1 + n^2x^2)$, and (b) $n^{\frac{3}{2}}x/(1 + n^2x^2)$. Show that in both cases $\lim \int_0^1 f_n(x)\, dx = 0$.

(ii) Prove the following relation (needed in dealing with the gamma function):

$$\lim \int_0^n (1 - x/n)^n\, x^{\alpha - 1}\, dx = \int_0^\infty e^{-x}\, x^{\alpha - 1}\, dx \qquad (\alpha > 0).$$

2.10 Let (f_n) be a sequence of real-valued, Lebesgue integrable functions. Show that if $\Sigma \int |f_n(x)|\, dx < \infty$, then $\Sigma f_n(x)$ converges for almost all x to a Lebesgue integrable function, and a change of order of summation and integration is justified, that is,

$$\int [\Sigma f_n(x)]\, dx = \Sigma \int f_n(x)\, dx.$$

2.11 Define $\gamma(x) = 0$ $(x < 0)$, $\gamma(x) = 1$ $(x \geqslant 0)$, and let γ be the associated Lebesgue–Stieltjes measure (Example 2.2.15). If $f : \mathbb{R} \to \mathbb{R}$ is continuous, show that $\int_{\mathbb{R}} f\, d\gamma = f(0)$.

2.12 Starting with the relation

$$\int_0^\infty e^{-tx}\,dx = t^{-1} \qquad (t > 0),$$

prove by differentiating that $\int_0^\infty x^n e^{-x}\,dx = n!$.

2.13 If $f:[0, 1] \to \mathbb{R}$ is Lebesgue measurable and $\int_0^1 |f(x)|\,dx = 0$, show that $f = 0$ a.e. (Hint: Take $S_n = \{x : |f(x)| \geqslant n^{-1}\}$, $S = \{x : |f(x)| \neq 0$, where $S = \cup S_n$, and integrate).

2.14 For the integrals of $(x^2 - y^2)/(x^2 + y^2)$ and $(1 - xy)^{-\alpha}$ on the square $0 \leqslant x, y \leqslant 1$, decide whether a change of order of integration is legitimate.

2.15 Suppose that f is measurable. Prove that $f \in \mathscr{L}_p$ iff $|f| \in \mathscr{L}_p$. Show also that if $g \in \mathscr{L}_p$ and $|f| \leqslant |g|$ a.e., then $f \in \mathscr{L}_p$.

2.16 Give a proof of Theorem 2.5.6(ii) based on Theorem 2.6.8.

Chapter 3

Foundations of Linear Operator Theory

3.1 Introduction

With an adequate list of specific Banach spaces now available, we may proceed with the development of an abstract theory of operators on these spaces designed to assist with the solution of particular equations arising in applications. A great variety of such equations may be written in the form $Af = g$ where A is a mapping of one Banach space into another, and if A is in some sense well behaved, much may be deduced about the solution by exploiting the structure of the spaces and in particular their completeness.

The questions that operator theory is best suited to answer are those of a qualitative nature, and those concerning general approximate procedures for solving the equation. Some of the most important of these are the following:

(i) Does the equation have a solution, and if so is it unique?

(ii) Is the equation stable in the sense that a small change in the "input" g results in a small change in the "output" f?

(iii) If the equation is linear, can the methods of linear operator theory in finite dimensions be taken over? In particular can a linear inverse A^{-1} be sensibly defined, and is A^{-1} well behaved?

(iv) If there is an operator A_0 which approximates A in some sense, is the solution f_0 of $A_0 f_0 = g$ a good approximation to the solution of $Af = g$?

(v) If the equation is a differential or integral equation, is a numerical solution obtained by some specific method close to the exact solution, and what is the error?

(vi) Are there valid iterative methods which will enable an initial guess of the solution to be systematically improved?

(vii) The familiar diagonalization of a Hermitian matrix provides a simple and illuminating characterization of a useful class of operator in finite dimensions. Is there an analogue of this for, say, linear differential equations?

(viii) The differential equation $f'' + \lambda f = 0$ with boundary conditions $f(0) = f(\pi) = 0$ has the set of solutions $\{\sin nx\}$ corresponding to $\lambda = n^2$ for $n = 1, 2, \ldots$. A wide class of function can be represented as a linear combination of these solutions, and this Fourier series representation is often useful. Are there generalizations to other differential equations, and to other types of equation?

The wish to answer questions of this nature will strongly influence the choice of theory to be described. In this chapter we shall lay down the foundations of linear operator theory on which the later treatment of both linear and nonlinear equations will depend.

The contents are arranged as follows. Section 2 consists of the definitions of basic operator terminology. The discussion of linear operators starts in Section 3. We enquire whether certain key results in finite dimensions generalize, and conclude that analytical aspects cannot be ignored if the generalization is to be successful. Therefore further restrictions of the spaces and operators must be assumed. The easiest and perhaps most useful restriction is that the operator should be a continuous mapping of one Banach space into another. Continuous linear operators are introduced in Section 4, and some of their basic properties are derived in Section 5. In Section 6 we are at last in a position to start to look at the problem of actually constructing solutions, and in particular of finding the inverse. The analysis is based on the method of successive substitution, or what is the same thing, the Neumann series, and applications are made to some standard problems. The investigation of the properties of the inverse leads naturally to a discussion of elementary spectral theory in Section 7. In the final section the weaker concept of "closed operator" is introduced in order to deal with differential operators, these operators not being continuous on the Banach spaces so far considered.

3.2 The Basic Terminology of Operator Theory

Let $\mathscr{V}$ and $\mathscr{W}$ be vector spaces. Let A be a mapping defined on some subset $D(A)$ of $\mathscr{V}$, and assume that A assigns to each element f of $D(A)$ a unique element Af in $\mathscr{W}$ (in the initial stages $D(A)$ will usually be the whole of $\mathscr{V}$).

3.2.1 Definition. The set $D(A)$ (sometimes denoted just by D if there is only one mapping under discussion) is called the **domain** of A. For $f \in D(A)$, the element Af is known as the **image** of f. Likewise the image $A(S)$ of a set $S \subset D(A)$ is the set of the images of all the elements of S. In particular the image of $D(A)$ is called the **range** of A and will be written as $R(A)$. The **pre-image** of a set $S_1 \subset \mathscr{W}$ is the set $A^{-1}(S_1) = \{f : f \in D(A), Af \in S_1\}$.

3.2.2 Definition. A is called an **operator** or a **function** from $\mathscr{V}$ into $\mathscr{W}$. The notation $A : S \to \mathscr{W}$ will mean that A is an operator with domain S and range in $\mathscr{W}$, and we say that A maps S *into* $\mathscr{W}$.

The following points arising from the definitions should be noted. First, an operator is always single valued in that it assigns exactly one element of its range to each element in its domain. Second, the statement that A is an operator from $\mathscr{V}$ into $\mathscr{W}$ allows the possibility that $D(A)$ is a proper subset of $\mathscr{V}$; in contrast $A : \mathscr{V} \to \mathscr{W}$ means that $D(L) = \mathscr{V}$. Lastly, although there is no strict distinction between "operator" and "function", it is customary to reserve "function" for the case when $\mathscr{V}$ and $\mathscr{W}$ are finite dimensional and to use "operator" otherwise. In view of its importance one particular type of operator is given a name of its own.

3.2.3 Definition. Let $\mathscr{V}$ be a complex (respectively real) vector space, and suppose that $\mathscr{W} = \mathbb{C}$ (respectively $\mathbb{R}$). Then an operator from $\mathscr{V}$ into $\mathscr{W}$ is known as a **functional**.

3.2.4 Example. The notation is illustrated in the space $\mathscr{C}([0, 1])$ by considering the familiar process of differentiation. Obviously not all continuous functions have derivatives, and we may therefore confine the operation of differentiation initially to very smooth functions, say to $\mathscr{C}^\infty([0, 1])$. For $f \in \mathscr{C}^\infty([0, 1])$ set $Af(x) = f'(x)$. Evidently the right-hand side of this is continuous for each f, and if $Af(x)$ denotes the value of the function Af at x, the above relation assumed to hold for all $f \in \mathscr{C}^\infty([0, 1])$ defines an operator from $\mathscr{C}([0, 1])$ into itself with domain $\mathscr{C}^\infty([0, 1])$. Thus we would write $A : \mathscr{C}^\infty([0, 1]) \to \mathscr{C}([0, 1])$.

Of course differentiation makes sense in the larger space $\mathscr{C}^1([0, 1])$, and it is therefore possible to define another operator, $\tilde{A}$ say, by setting $\tilde{A}f(x) = f'(x)$ for $f \in \mathscr{C}^1([0, 1])$. Although $\tilde{A}f = Af$ for $f \in D(A)$, since $D(A) \neq D(\tilde{A})$ we do not say that A and $\tilde{A}$ are equal, reserving that terminology for the case when $D(A) = D(\tilde{A})$. Rather $\tilde{A}$ is described as an extension of A.

3.2.5 Definition. Let A and $\tilde{A}$ be operators from $\mathscr{V}$ into $\mathscr{W}$. A and $\tilde{A}$ are said to be **equal** iff $D(A) = D(\tilde{A})$ and $Af = \tilde{A}f$ for all f in $D(A)$. $\tilde{A}$ is said to be an **extension** of A (written $A \subset \tilde{A}$), and A to be a **restriction** of $\tilde{A}$, iff $D(\tilde{A}) \supset D(A)$ and $Af = \tilde{A}f$ for all $f \in D(A)$. The extension is described as **proper** iff $D(\tilde{A}) \neq D(A)$.

A broad classification which is useful when considering the operator equation $Af = g$ is as follows.

3.2.6 Definition. Suppose A is an operator from $\mathscr{V}$ into $\mathscr{W}$. A is said to be **injective** iff for each $g \in R(A)$ there is exactly one $f \in D(A)$ such that $Af = g$. A is called **surjective** iff $R(A) = \mathscr{W}$, when we say that A maps $D(A)$ *onto* $\mathscr{W}$. A is known as **bijective** iff it is both injective and surjective. The terms "one-to-one", "onto", "one-to-one and onto" respectively are common alternatives in the literature.

3.2.7 Examples. Consider the following functions (operators) $\phi : \mathbb{R} \to \mathbb{R}$.

(i) $\phi(x) = \sin x$. Then $R(\phi) = [-1, 1]$ which is a proper subset of $\mathbb{R}$. Thus ϕ is not surjective. Since $0 = \phi(0) = \phi(\pi) = \ldots$, neither is ϕ injective.

(ii) $\phi(x) = x(x^2 - 1)$. Since $R(\phi) = \mathbb{R}$, ϕ is surjective, but $-1, 0, 1$ are all mapped to zero, hence certainly ϕ is not injective.

(iii) $\phi(x) = \tanh x$. Then $R(\phi) = (-1, 1)$ and ϕ is not surjective. It is injective since the equation $\tanh x = a$ $(a \in (-1, 1))$ has only one real solution.

(iv) $\phi(x) = x^3$. ϕ is obviously bijective.

In both linear and nonlinear operator theory, the class of continuous operators is the most useful. The following is a direct extension of the Definition 1.3.15 of continuity for a complex valued function.

3.2.8 Definition. Let $\mathscr{V}$ and $\mathscr{W}$ be normed vector spaces, and suppose that A is an operator from $\mathscr{V}$ into $\mathscr{W}$. A is said to be **continuous at the point** $f_0 \in D(A)$ iff one of the following pair of equivalent conditions holds:

(i) For each $\varepsilon > 0$ there is a $\delta > 0$ such that $\|Af - Af_0\| < \varepsilon$ if $f \in D(A)$ and $\|f - f_0\| < \delta$;

(ii) For every sequence (f_n) in $D(A)$ with limit f_0, $\lim Af_n = Af_0$.

A is said to be **continuous** iff it is continuous at every point of $D(A)$.

3.2.9 Lemma. *The operator A in the definition is continuous if and only if the pre-image of every open set in $\mathscr{W}$ is open in $D(A)$.*

Proof. We suppose first that A is continuous and show that if $S \subset \mathscr{W}$ is open then $A^{-1}(S)$ is open in $D(A)$. If $A^{-1}(S)$ is empty it is open so assume

$A^{-1}(S)$ is not empty, and let f_0 be any point in it. Then $Af_0 \in S$, and since S is open there is an open ball $S(Af_0, \varepsilon) \subset S$. By the definition of continuity, there is a $\delta > 0$ and a set $U = \{f : f \in D(A), \|f - f_0\| < \delta\}$ open in $D(A)$ such that $A(U) \subset S(Af_0, \varepsilon)$. Hence $A(U) \subset S$ and $U \subset A^{-1}(S)$. Since f_0 is an arbitrary point in $A^{-1}(S)$, this proves that $A^{-1}(S)$ is open in $D(A)$.

Assume on the other hand that S open $\Rightarrow A^{-1}(S)$ open in $D(A)$. Then for each $f_0 \in D(A)$ and $\varepsilon > 0$, the pre-image $A^{-1}(S(Af_0, \varepsilon))$ of the open ball $S(Af_0, \varepsilon)$ is open in $D(A)$. Hence for some $\delta > 0$, the set $U = \{f : f \in D(A), \|f - f_0\| < \delta\}$ is contained in the pre-image, whence $A(U) \subset S(Af_0, \varepsilon)$. By Definition 3.2.8(*i*) this establishes continuity at f_0. □

3.3 Some Algebraic Properties of Linear Operators

A linear operator is the vector space analogue of a function in one dimension represented by a straight line through the origin, that is a function $\phi : \mathbb{R} \to \mathbb{R}$ where $\phi(x) = \lambda x$ for some $\lambda \in \mathbb{R}$. In finite dimensions the theory of linear equations is highly developed and research interest is largely confined to the nonlinear case, but in an arbitrary vector space the situation is rather different, for although linear equations are evidently much more tractable than their nonlinear counterparts, the extra problems raised by infinite dimensionality can be of considerable difficulty. For the rest of this chapter linear operators only will be considered. Their algebraic properties are first discussed. In particular a preliminary examination of the operator equation $Lf = g$ is made, and some of the standard finite dimensional theory is reviewed with the aim of establishing what lines of enquiry may be fruitful in infinite dimensions.

3.3.1 Definition. Let $\mathscr{V}$ and $\mathscr{W}$ be vector spaces, and let $D(L)$ be a linear subspace of $\mathscr{V}$. An operator L from $\mathscr{V}$ into $\mathscr{W}$ with domain $D(L)$ is said to be **linear** iff

$$L(\alpha f + \beta g) = \alpha L f + \beta L g$$

for all $\alpha, \beta \in \mathbb{C}$ (or $\mathbb{R}$ if $\mathscr{V}$ and $\mathscr{W}$ are real spaces) and all $f, g \in D(L)$. (The restriction that $D(L)$ be a linear subspace is obviously necessary if the definition is to make sense; note that $R(L)$ is also a linear subspace).

Some of the wide variety of equations which may be written in the form $Lf = g$ with L a linear operator are described in the following examples.

3.3.2 Example. For the simultaneous algebraic equations

$$\sum_{j=1}^{n} \alpha_{ij} f_j = g_i \qquad (i = 1, \ldots, m), \tag{3.3.1}$$

define

$$(Lf)_i = \sum_{j=1}^{n} \alpha_{ij} f_j \qquad (i = 1, \ldots, m).$$

Then $L\colon \mathbb{C}^n \to \mathbb{C}^m$ is a linear operator and (3.3.1) becomes $Lf = g$. Conversely every linear operator $\mathbb{C}^n \to \mathbb{C}^m$ may be expressed in the above form by choosing bases for $\mathbb{C}^n$ and $\mathbb{C}^m$. The equations (3.3.1) may also be put in matrix form, but note that there is a distinction between the matrix, which depends on the bases chosen, and L, which does not.

3.3.3 Example. An obvious generalization is to an infinite system of equations. Proceeding formally for the moment, consider the relation

$$(Lf)_i = \sum_{j=1}^{\infty} \alpha_{ij} f_j \qquad (i = 1, 2, \ldots), \tag{3.3.2}$$

where α_{ij} is a given set of complex numbers which may be thought of as an infinite matrix. Tentatively L may be regarded as an operator from the sequence space ℓ into itself, but in contrast with the previous example, the sum on the right-hand side only makes sense for a certain subset of ℓ which depends on the precise behaviour of α_{ij} as $j \to \infty$. However, with this subset as the domain of L, (3.3.2) defines a linear operator from ℓ into itself.

3.3.4 Example. We come next to operators on spaces of functions. One of the simplest examples corresponds to multiplication by a given function τ, that is L is defined by requiring that the equation $Lf(x) = \tau(x) f(x)$ should hold for all f in some space $\mathscr{V}$. If τ is continuous, two obvious choices for $\mathscr{V}$ are $\mathscr{C}([0, 1])$ and $\mathscr{L}_p(0, 1)$, and L is then a linear operator mapping each of these spaces into itself.

3.3.5 Example. From the point of view of applications, differential and integral equations are of special importance, and specimens of these are next formulated as operator equations. Consider first the Fredholm integral equation

$$f(x) - \int_0^1 k(x, y) f(y)\, \mathrm{d}y = g(x) \qquad (0 \leqslant x \leqslant 1), \tag{3.3.3}$$

where k and g are given and f is the unknown function. For simplicity suppose that k and g are continuous complex valued functions, and define the integral operator K by requiring the relation

$$Kf(x) = \int_0^1 k(x, y) f(y)\, \mathrm{d}y$$

to hold for all $f \in \mathscr{C}([0, 1])$. Since Kf is a continuous function (by Theorem

2.4.15), K is a linear operator from $\mathscr{C}([0, 1])$ into itself. The original integral equation (3.3.3) may thus be written as $f - Kf = g$.

3.3.6 Example. Take lastly the differential equation

$$a_0 f''(x) + a_1 f'(x) + a_2 f(x) = g(x) \qquad (0 \leqslant x \leqslant 1),$$

where $a_0, a_1, a_2 \in \mathbb{C}$ and g is a given continuous function. In order for the left-hand side to make sense, f must be reasonably smooth, and it is natural (since g is continuous) to take the domain of the operator to be $\mathscr{C}^2([0, 1])$ and define L by requiring the relation

$$Lf(x) = a_0 f''(x) + a_1 f'(x) + a_2 f(x)$$

to hold on this domain. Of course Lf need not be differentiable, so L is a linear operator $\mathscr{C}^2([0, 1]) \to \mathscr{C}([0, 1])$. The differential equation then becomes $Lf = g$.

The first move in trying to solve the operator equation $Lf = g$ is modelled on the tactics in matrix theory: an attempt is made to define an inverse operator L^{-1}. Obviously the equation has a solution f for every $g \in R(L)$, and if the solution is unique, which will be the case iff L is injective, an operator L^{-1} with $D(L^{-1}) = R(L)$ may be defined by the relation $f = L^{-1}g$. It is easy to show that L^{-1} is a linear operator from $\mathscr{W}$ into $\mathscr{V}$. However, in general $R(L)$ is not the whole of $\mathscr{W}$, which reflects the fact that $Lf = g$ does not have a solution for all $g \in \mathscr{W}$. The situation is much more satisfactory if $R(L) = \mathscr{W}$, for then $Lf = g$ has a unique solution for all $g \in \mathscr{W}$. The possibilities may be summarized as follows:

(i) *L not injective.* No reasonable interpretation of L^{-1} as an operator in the sense of Definition 3.2.2 is possible. $Lf = g$ always has more than one solution if $g \in R(L)$.

(ii) *L injective but not surjective.* L^{-1} is a linear operator with domain $R(L)$. The equation $Lf = g$ has exactly one solution if $g \in R(L)$, but no solution otherwise.

(iii) *L bijective.* L^{-1} is a linear operator with domain $\mathscr{W}$, and $Lf = g$ has exactly one solution for each $g \in \mathscr{W}$.

To establish convenient criteria for deciding when this last nice possibility holds is one of the primary aims of the theory. Tentatively it may be hoped that a guide might be provided by the finite dimensional theory, and some of the main concepts which feature in this are next introduced.

3.3.7 Definition. Suppose that L is a linear operator from $\mathscr{V}$ into $\mathscr{W}$. Then we say that L has an inverse, or that the inverse of L exists, iff L is injective.

The inverse is the operator with domain $R(L)$ and range $D(L)$ defined by $f = L^{-1}g$ when $Lf = g$.

3.3.8 Definition. The **identity** operator, always denoted by I, is the operator from $\mathscr{V}$ onto itself such that $If = f$ for all $f \in \mathscr{V}$.

3.3.9 Definition. Suppose L is a linear operator from $\mathscr{V}$ into $\mathscr{W}$. The set $N(L) \subset D(L)$ of solutions of $Lf = 0$ is called the **null space** of L. ($N(L)$ is evidently a linear subspace, and $N(L) = 0$ if and only if L is injective).

Attention is now restricted temporarily to the most important case $\mathscr{V} = \mathscr{W}$. An easy result on the inverse is as follows.

3.3.10 Lemma. *If $L: \mathscr{V} \to \mathscr{V}$ is linear, the following hold:*

(i) *if L^{-1} exists and has domain $\mathscr{V}$, then $L^{-1}L = LL^{-1} = I$;*
(ii) *if there are linear operators $A, B: \mathscr{V} \to \mathscr{V}$ such that $AL = LB = I$, then L is bijective and $A = B = L^{-1}$.*

This lemma is of very limited use because of the difficulty of establishing the existence of A and B. In finite dimensions a standard method of investigating the equation $Lf = g$ is to study the null space. The proof of the following is based on showing that an injective operator preserves dimensions, see Halmos (1948, p. 41).

3.3.11 Theorem. *Suppose that $\mathscr{V}$ is finite dimensional, and let $L: \mathscr{V} \to \mathscr{V}$ be a linear operator. Then the following conditions are equivalent:*

(i) *L is bijective;*
(ii) *$N(L) = 0$, or what is the same, L is injective;*
(iii) *L is surjective.*

Thus if $N(L) = 0$, L^{-1} is defined on the whole of $\mathscr{V}$—the best possible situation—and $Lf = g$ has a unique solution for every $g \in \mathscr{V}$. The non-vanishing of the determinant of the matrix representation of L is of course sufficient to ensure that $N(L) = 0$. Unfortunately, as the following example shows, $N(L) = 0$ is *not* a sufficient condition in infinite dimensions.

3.3.12 Example. Let $L: \mathscr{C}([0, 1]) \to \mathscr{C}([0, 1])$ be the operator defined by

$$Lf(x) = \int_0^x f(t)\,dt \qquad (f \in \mathscr{C}([0, 1])).$$

Obviously $Lf(0) = 0$ for every $f \in \mathscr{C}([0, 1])$. Therefore $R(L)$ is a proper subset of $\mathscr{C}([0, 1])$ and L is not surjective. However, $N(L) = 0$, for differentiation shows that the only continuous solution of $Lf = 0$ is $f = 0$. Thus condition (ii) of Theorem 3.3.11 does not imply (iii) or (i).

The failure of Theorem 3.3.11 is a warning that familiar results about the inverse do not carry over in a straightforward manner from finite to infinite dimensions. In view of the importance of analytical considerations in infinite dimensions, it is natural to expect that further progress with linear operator theory cannot be based on purely algebraic arguments. Indeed a generalization of the finite dimensional results can only be achieved with the help of a complex and interesting theory in which analysis plays a leading part. The development of some of the most important parts of this theory will now be our main concern.

3.4 Continuity and Boundedness

The difficulties in linear operator theory in infinite dimensions arise essentially in two ways. The first is connected with the question of continuity; linear operators on finite dimensional normed vector spaces are necessarily continuous, but this is not the case when the space is infinite dimensional. The second problem comes from the complexity of the analytical properties of the space itself. The simplest operators for which substantial progress is possible are those whose domains are a Banach space and which in addition are continuous, and since such operators are also relatively common in applications, it is natural to commence with a detailed study of their properties.

Our first observations concern the simplification of the notion of continuity for linear as compared with nonlinear operators. A nonlinear function of even a single real variable may be continuous on certain parts of its domain but not on others. Also its gradient may be infinite although the function is continuous. The linearity excludes both these possibilities even in infinite dimensions. In the following analysis $\mathscr{B}$ and $\mathscr{C}$ as usual denote Banach spaces.

3.4.1 Lemma. *Suppose that L is a linear operator from $\mathscr{B}$ into $\mathscr{C}$. Then if L is continuous at some point $f \in D(L)$, it is continuous.*

Proof. If (f_n) is any sequence in $D(L)$ with limit f, then $Lf_n \to Lf$. Now if (g_n) is a sequence in $D(L)$ with limit g, also $g_n - g + f \to f$. Therefore $L(g_n - g + f) \to Lf$, from which it follows that $Lg_n - Lg + Lf \to Lf$, that is $Lg_n \to Lg$. □

3.4.2 Definition. Suppose that L is a linear operator from $\mathscr{B}$ into $\mathscr{C}$. L is said to be **bounded on** $D(L)$ iff there is a finite number m such that

$$\|Lf\| \leqslant m\|f\| \qquad (f \in D(L)). \tag{3.4.1}$$

(It is clear from the context that the norms here are those of $\mathscr{C}$ and $\mathscr{B}$ respectively, and it usually unnecessary to indicate this in the notation). L is said to be **bounded** iff also $D(L) = \mathscr{B}$. If L is not bounded on $D(L)$, it is said to be **unbounded**.

The infimum of all constants m such that (3.4.1) holds is denoted by $\|L\|$ and is called the operator norm† of L. An obviously equivalent definition of $\|L\|$ is

$$\|L\| = \sup_{\substack{f \in D(L) \\ f \neq 0}} \frac{\|Lf\|}{\|f\|} = \sup_{\substack{f \in D(L) \\ \|f\| = 1}} \|Lf\|.$$

Note that $\|Lf\| \leqslant \|L\|\,\|f\|$, with which the relation $|\phi(x)| = |\lambda x| = |\lambda|\,|x|$ for a linear operator $\phi : \mathbb{R} \to \mathbb{R}$ where $\phi(x) = \lambda x$ may be compared. $|\lambda|$ is a measure of the gradient of ϕ, and the norm of the operator L may therefore be thought of as its maximum gradient.

3.4.3 Theorem. *Suppose that L is a linear operator from $\mathscr{B}$ into $\mathscr{C}$. Then L is bounded on $D(L)$ if and only if it is continuous.*

Proof. If L is bounded, from (3.4.1) $f_n \to 0 \Rightarrow Lf_n \to 0$. Thus L is continuous at zero, and so by Lemma 3.4.1, L is continuous. On the other hand, if L is not bounded there is a sequence (g_n) such that $a_n = \|Lg_n\|/\|g_n\| \to \infty$. But if $f_n = g_n/(a_n\|g_n\|)$, then $\|f_n\| = a_n^{-1} \to 0$ and $\|Lf_n\| = 1$. Since $L0 = 0$, L is not continuous at zero, and therefore is not continuous. □

For the time being we shall be mainly concerned with continuous operators with domains the whole of a Banach space $\mathscr{B}$. However, it is sometimes convenient to define the operators first on a subset of $\mathscr{B}$, and the question then arises as to whether an operator which is continuous on $D(L)$ has an extension continuous on the whole of $\mathscr{B}$. The next result shows that this is indeed the case if the original domain is dense in $\mathscr{B}$. The argument used is itself of interest as it typifies a frequently employed technique which may be described as "extension by continuity".

3.4.4 Theorem. *Let $\mathscr{B}$ and $\mathscr{C}$ be Banach spaces, and suppose that L is a linear operator from $\mathscr{B}$ into $\mathscr{C}$ with domain dense in $\mathscr{B}$. Then if L is continuous on*

† It will be seen in the next section that the term "norm" is indeed used in a sense consistent with the original Definition 1.3.2.

$D(L)$, it has a unique continuous extension, $\tilde{L}$ say, to $\mathscr{B}$ itself, and this extension has the same norm as L.

Proof. Since $\overline{D(L)} = \mathscr{B}$, for each f in $\mathscr{B}$ there is a sequence (f_n) in $D(L)$ with $\lim f_n = f$. As (f_n) is convergent it is Cauchy. Hence given $\varepsilon > 0$, there is an n_0 such that for $m, n \geqslant n_0$, $\|f_n - f_m\| \leqslant \varepsilon/\|L\|$. Thus for $m, n \geqslant n_0$,

$$\|Lf_n - Lf_m\| = \|L(f_n - f_m)\| \leqslant \|L\|\|f_n - f_m\| \leqslant \varepsilon.$$

This shows that (Lf_n) is a Cauchy sequence in $\mathscr{C}$, and as $\mathscr{C}$ is complete there is an element $g \in \mathscr{C}$ with $\lim Lf_n = g$. It is easy to check that g is independent of the particular sequence used, and the extension $\tilde{L} : \mathscr{B} \to \mathscr{C}$ may therefore be defined by the relation $\tilde{L}f = g$. $\tilde{L}$ is obviously linear, and since

$$\|\tilde{L}f\| = \|g\| = \lim \|Lf_n\| \leqslant \lim \|L\|\|f_n\| = \|L\|\|f\|,$$

$\tilde{L}$ is in addition bounded. The last relation also shows that $\|\tilde{L}\| \leqslant \|L\|$, where $\|\tilde{L}\|$ denotes of course the norm of $\tilde{L}$ as an operator $\mathscr{B} \to \mathscr{C}$. On the other hand $\tilde{L}f = Lf$ for $f \in D(L)$, so certainly $\|\tilde{L}\| \geqslant \|L\|$. Therefore $\|\tilde{L}\| = \|L\|$.

If $\tilde{L}_1, \tilde{L}_2$ are two extensions, for any convergent sequence (f_n) in $D(L)$ with limit f,

$$\tilde{L}_1 f = \lim \tilde{L}_1 f_n = \lim Lf_n = \lim \tilde{L}_2 f_n = \tilde{L}_2 f,$$

and the uniqueness follows. □

The pleasant consequences that flow from the assumption of continuity will be discussed later. In the following examples, the specific linear operators considered in the previous section are re-examined with the aims firstly of establishing continuity, and secondly of obtaining estimates for the operator norm which is itself a quantity of particular importance in applications. It should be noted that in finite dimensions the choice of norm may be made purely on grounds of computational convenience since all linear operators are continuous, but that in infinite dimensions the definition of the operator itself, and certainly its continuity, depends critically on the choice of space. Thus when the abstract theory is to be applied in particular problems, this choice is crucial. Generally the need to produce a well behaved operator must be taken into account. Obviously also the functions in the space should satisfy any restriction such as continuity dictated by the problem. To achieve a satisfactory balance between these requirements is not always easy in practice.

3.4.5 Example. Both finite and infinite systems of algebraic equations (Examples 3.3.2 and 3.3.3) are most conveniently tackled in spaces with ℓ_p

norms. Consider first $L: \mathbb{C}^n \to \mathbb{C}^n$ defined by

$$(Lf)_i = \sum_{j=1}^{n} \alpha_{ij} f_j \qquad (i = 1, \ldots, n),$$

and suppose for a start that $\mathbb{C}^n$ is equipped with the norm $\|\cdot\|_\infty$ defined by (1.3.2). Then

$$\begin{aligned} \|Lf\|_\infty &= \sup_i \left| \sum_{j=1}^{n} \alpha_{ij} f_j \right| \\ &\leqslant \sup_i \sum_{j=1}^{n} |\alpha_{ij}||f_j| \\ &\leqslant \left(\sup_i \sum_{j=1}^{n} |\alpha_{ij}| \right) (\sup_j |f_j|) \\ &= m \|f\|_\infty \end{aligned}$$

say, where

$$m = \sup_i \sum_{j=1}^{n} |\alpha_{ij}|.$$

It follows from Definition 3.4.2. that $\|L\| \leqslant m$. Actually $\|L\| = m$. To prove this it is sufficient to show that there is an f such that $\|Lf\|_\infty \geqslant m\|f\|_\infty$. Now choose any integer k such that $m = \sum_{j=1}^{n} |\alpha_{kj}|$, the existence of such a k being obvious from the definition of m, and let f be the unit vector with jth component $\bar{\alpha}_{kj}/|\alpha_{kj}|$. Then

$$\|Lf\|_\infty = \sup_i \left| \sum_{j=1}^{n} \alpha_{ij} f_j \right| \geqslant \left| \sum_{j=1}^{n} \alpha_{kj} f_j \right| = \sum_{j=1}^{n} |\alpha_{kj}| = m\|f\|_\infty,$$

which proves the assertion.

For $1 < p < \infty$ the calculation goes as follows. With q the conjugate index, by Hölder's inequality 1.3.12,

$$\begin{aligned} \left| \sum_{j=1}^{n} \alpha_{ij} f_j \right| &\leqslant \left\{ \sum_{j=1}^{n} |\alpha_{ij}|^q \right\}^{1/q} \left\{ \sum_{j=1}^{n} |f_j|^p \right\}^{1/p} \\ &= \left\{ \sum_{j=1}^{n} |\alpha_{ij}|^q \right\}^{1/q} \|f\|_p. \end{aligned}$$

Hence

$$\begin{aligned}\|Lf\|_p^p &= \sum_{i=1}^{n} |(Lf)_i|^p \\ &= \sum_{i=1}^{n} \left| \sum_{j=1}^{n} \alpha_{ij} f_j \right|^p \\ &\leqslant \sum_{i=1}^{n} \left\{ \sum_{j=1}^{n} |\alpha_{ij}|^q \right\}^{p/q} \|f\|_p^p.\end{aligned}$$

Therefore

$$\|L\| \leqslant \left[\sum_{i=1}^{n} \left\{ \sum_{j=1}^{n} |\alpha_{ij}|^q \right\}^{p/q} \right]^{1/p}.$$

This provides an estimate for $\|L\|$. However, in contrast with the case $p = \infty$, it is not usually easy to find $\|L\|$ explicitly.

3.4.6 Example. For our first infinite dimensional example we return to the "infinite matrix" and take formally

$$(Lf)_i = \sum_{j=1}^{\infty} \alpha_{ij} f_j \qquad (i = 1, 2, \ldots). \tag{3.4.2}$$

Unless there is some restriction on the growth of α_{ij} as $j \to \infty$, it will be difficult to attach any sensible meaning to L as an operator on a Banach space. To illustrate the problem, let us enquire when L is an operator $\ell_\infty \to \ell_\infty$. For the right-hand side of (3.4.2) to be finite for all $f \in \ell_\infty$ it is evidently necessary that $\sum_j |\alpha_{ij}| < \infty$. However, this condition is by itself not enough to ensure that $Lf \in \ell_\infty$, and we might require that $m = \sup_i \sum_j |\alpha_{ij}| < \infty$, in which case certainly $L : \ell_\infty \to \ell_\infty$. In fact since $\|Lf\|_\infty \leqslant m\|f\|_\infty$, the condition $m < \infty$ enables us to assert further that L is bounded. For general p, sufficient conditions for boundedness are provided by the following result.

3.4.7 Theorem. *Let p and q be conjugate indices, and set*

$$\begin{aligned}|||\alpha|||_1 &= \sup_j \sum_i |\alpha_{ij}|, \\ |||\alpha|||_p &= \left[\sum_i \left\{ \sum_j |\alpha_{ij}|^q \right\}^{p/q} \right]^{1/p} \qquad (1 < p < \infty), \\ |||\alpha|||_\infty &= \sup_i \sum_j |\alpha_{ij}|.\end{aligned}$$

If for some p with $1 \leqslant p \leqslant \infty$ the quantity $|||\alpha|||_p$ is finite, then the linear *operator* $L: \ell_p \to \ell_p$ *defined by* (3.4.2) *is bounded and* $\|L\| \leqslant |||\alpha|||_p$. *In particular, for* $p = \infty$, $\|L\| = |||\alpha|||_\infty$.

Proof. Replace n by ∞ in Example 3.4.5. □

It may be added that also when $p = 1$, $\|L\| = |||\alpha|||_1$, see Example 6.5.4. Except when $p = 1, \infty$ it is not usually easy to find the exact value of the operator norm (from which necessary conditions for boundedness would follow), but alternative criteria for boundedness are given in Problem 3.4.

3.4.8 Example. Consider next the integral operator K defined formally by

$$Kf(x) = \int_a^b k(x, y) f(y) \, dy. \tag{3.4.3}$$

If the interval $[a, b]$ is finite only weak conditions on k are needed to ensure that K is defined and continuous on $\mathscr{C}([a, b])$ or the $\mathscr{L}_p$ spaces, but in the case of an infinite interval, as in the previous example, some restriction on the behaviour of $k(x, y)$ for large y must be imposed if the integral is to be finite. Two sample results follow.

3.4.9 Theorem. *Suppose* a *and* b *are finite and equip* $\mathscr{C}([a, b])$ *with the sup norm. Assume that* $k: [a, b] \times [a, b] \to \mathbb{C}$ *is continuous. Then* $K: \mathscr{C}([a, b]) \to \mathscr{C}([a, b])$ *is bounded and*

$$\|K\| \leqslant \sup_{x \in [a, b]} \int_a^b |k(x, y)| \, dy \leqslant (b - a) \sup_{x, y \in [a, b]} |k(x, y)|.$$

Proof. Kf is continuous if f is continuous (Theorem 2.4.15). The estimate for the norm is obtained from the relations:

$$|Kf(x)| \leqslant \left[\sup_{y \in [a, b]} |f(y)|\right] \int_a^b |k(x, y)| \, dy = \|f\| \int_a^b |k(x, y)| \, dy. \qquad \square$$

3.4.10 Theorem. *Suppose that* k *is measurable, and with* p, q *conjugate indices, set*

$$|||k|||_1 = \sup_{y \in [a, b]} \int_a^b |k(x, y)| \, dx,$$

$$|||k|||_p = \left[\int_a^b dx \left\{\int_a^b |k(x, y)|^q \, dy\right\}^{p/q}\right]^{1/p} \qquad (1 < p < \infty),$$

$$|||k|||_\infty = \sup_{x \in [a, b]} \int_a^b |k(x, y)| \, dy.$$

If for some p with $1 \leqslant p \leqslant \infty$ *the quantity* $|||k|||_p$ *is finite, then* $K : \mathscr{L}_p(a, b) \to \mathscr{L}_p(a, b)$ *is bounded, and* $\|K\| \leqslant |||k|||_p$.

Proof. Simply repeat the argument in Example 3.4.5 using Hölder's inequality for integrals (Theorem 2.5.3) instead of that for sums. □

As in the ℓ_p case, $\|K\| = |||k|||_p$ when $p = 1$ or $p = \infty$. Again except for these values of p the above conditions are not necessary for boundedness; an alternative criterion is given in Problem 3.5, and further results may be found in Zabreyko (1975).

In each of the last two examples, if the operator had range in the Banach space in question, it turned out to be bounded. It might be thought that this is true in general, but a counter-example to this hypothesis can (with some difficulty) be constructed; a linear operator $\mathscr{B} \to \mathscr{B}$ need not be bounded. The unboundedness of operators derived from differential equations presents a somewhat different problem, and one which from the point of view of applications is more significant.

3.4.11 Example. To illustrate the difficulty it is enough to examine the simple operator defined formally by $Lf(x) = f'(x)$, and enquire whether there is a reasonable choice of space on which $L : \mathscr{B} \to \mathscr{B}$ will be bounded. Let us fix our attention first on $\mathscr{C}([0, 1])$ with the sup norm, and take $D(L) = \mathscr{C}^{-1}([0, 1])$. Now for the sequence (f_n) where $f_n(x) = \sin n\pi x$, $Lf_n(x) = n\pi \cos n\pi x$ and $\|Lf_n\|/\|f_n\| = n\pi$. We conclude that L is not bounded on its domain, and a similar argument shows that neither is L bounded on the same domain in the $\mathscr{L}_p$ norms.

In view of the importance of differential equations in applications, this failure of boundedness raises very serious problems to whose solution great effort has been devoted. Two ways round the difficulty may be noted briefly. The first is to use the space $\mathscr{C}^k([0, 1])$ with the norm (1.3.7). Then indeed a kth order linear differential operator is continuous from $\mathscr{C}^k([0, 1])$ into $\mathscr{C}([0, 1])$. As it stands this approach is somewhat awkward, but a modification of the argument using certain Hilbert spaces will be the basis of the treatment of partial differential equations in Chapter 11. The second is to develop a theory of operators which satisfy a condition somewhat weaker than boundedness; these "closed" operators will be introduced in Section 3.8.

We close this section by defining certain classes of bounded operators which will be useful later. Suppose that $\mathscr{B}$ and $\mathscr{C}$ are Banach spaces and that

Denote this space by $\mathscr{L}(\mathscr{V}, \mathscr{W})$. $\mathscr{L}(\mathscr{V}, \mathscr{W})$ is a purely algebraic object, but it is easy to check that the operator norm (Definition 3.4.2) is indeed a norm on $\mathscr{L}(\mathscr{V}, \mathscr{W})$ if $\mathscr{V}$ and $\mathscr{W}$ are themselves normed spaces.

3.5.4 Definition. Let $\mathscr{B}$ and $\mathscr{C}$ be Banach spaces. The normed vector space of bounded linear operators from $\mathscr{B}$ into $\mathscr{C}$ with the operator norm is denoted by $\mathscr{L}(\mathscr{B}, \mathscr{C})$ or by $\mathscr{L}(\mathscr{B})$ when $\mathscr{B} = \mathscr{C}$.

The abbreviated terminology $L^{-1} \in \mathscr{L}(\mathscr{B}, \mathscr{C})$ will frequently be used to mean that L has a bounded inverse (Definition 3.5.2).

3.5.5 Theorem. *If $\mathscr{B}$ and $\mathscr{C}$ are Banach spaces, then $\mathscr{L}(\mathscr{B}, \mathscr{C})$ is a Banach space.*

Proof. The verification that $\mathscr{L}(\mathscr{B}, \mathscr{C})$ is a normed space is easy and is omitted. To prove completeness, we take a Cauchy sequence (L_n) of operators in $\mathscr{L}(\mathscr{B}, \mathscr{C})$ and prove that (L_n) has a limit which is a bounded linear operator and thus lies in $\mathscr{L}(\mathscr{B}, \mathscr{C})$.

The limit operator L is constructed as follows. For each $f \in \mathscr{B}$, since $\|L_n f - L_m f\| \leqslant \|L_n - L_m\| \|f\|$, the sequence $(L_n f)$ in $\mathscr{C}$ is Cauchy. As $\mathscr{C}$ is complete, $(L_n f)$ has a limit, g say, in $\mathscr{C}$. Set $Lf = g = \lim L_n f$. It is straightforward to check that L is a linear operator. To see that L is bounded, note that since $|\|L_n\| - \|L_m\|| \leqslant \|L_n - L_m\|$, the sequence $(\|L_n\|)$ of real numbers is Cauchy and thus has a finite limit, b say, and therefore

$$\|Lf\| = \lim \|L_n f\| \leqslant \lim \|L_n\| \|f\| = b\|f\|.$$

Finally we must prove that $\lim \|L_n - L\| = 0$. As (L_n) is Cauchy, for each $\varepsilon > 0$ there is an n_0 such that $\|L_m - L_n\| \leqslant \varepsilon$ for $m, n \geqslant n_0$. Thus for any $f \in \mathscr{B}$ and $m, n \geqslant n_0$, we have $\|L_n f - L_m f\| \leqslant \varepsilon \|f\|$, whence

$$\|L_n f - Lf\| = \lim_{m \to \infty} \|L_n f - L_m f\| \leqslant \varepsilon \|f\|.$$

Therefore $\|L_n - L\| \leqslant \varepsilon$ for $n \geqslant n_0$, and as ε is arbitrary it follows that $\lim \|L_n - L\| = 0$. □

The next result provides a means of recognizing bounded sets in $\mathscr{L}(\mathscr{B}, \mathscr{C})$, and is another of the central results in linear operator theory. For a proof see Friedman (1970, p. 139).

3.5.6 The Principle of Uniform Boundedness (The Banach–Steinhaus Theorem). *Assume that $\mathscr{B}$ and $\mathscr{C}$ are Banach spaces. For α in some set S, let $\{L_\alpha\}$ be a family of operators in $\mathscr{L}(\mathscr{B}, \mathscr{C})$, and suppose that given any $f \in \mathscr{B}$, the set $\{L_\alpha f\}$ of vectors is bounded. Then the family $\{L_\alpha\}$ is bounded in $\mathscr{L}(\mathscr{B}, \mathscr{C})$; that is there is an $m < \infty$ such that $\|L_\alpha\| \leqslant m$ for all $\alpha \in S$.*

The Banach space $\mathscr{L}(\mathscr{B}, \mathscr{C})$ is a convenient setting in which to tackle operator convergence.

3.5.7 Definition. Let (L_n) be a sequence of bounded linear operators mapping $\mathscr{B}$ into $\mathscr{C}$. We write $L_n \to L$ or $\lim L_n = L$ iff the sequence converges in the operator norm—that is in $\mathscr{L}(\mathscr{B}, \mathscr{C})$. The sequence is then said to **converge uniformly**.

It is easy to see that uniform convergence is equivalent to the following: there is a sequence (ε_n) of real numbers tending to zero such that $\|(L_n - L)f\| \leqslant \varepsilon_n \|f\|$ for all $f \in \mathscr{B}$. Note that (ε_n) is independent of f.

3.5.8 Example. Consider the integral operators K, K_n defined by

$$Kf(x) = \int_0^1 k(x, y)\, f(y)\, dy,$$

$$K_n f(x) = \int_0^1 k_n(x, y)\, f(y)\, dy.$$

Let Ω be the square $[0, 1] \times [0, 1]$, and assume that $k, k_n : \Omega \to \mathbb{C}$ are continuous. It is known from Theorem 3.4.9 that $K, K_n \in \mathscr{L}(\mathscr{C}([0, 1]))$, where $\mathscr{C}([0, 1])$ is equipped with the sup norm. If further $k_n \to k$ in $\mathscr{C}(\Omega)$ with the sup norm, by the same theorem

$$\|(K_n - K)f\| \leqslant \|k_n - k\| \, \|f\|, \tag{3.5.1}$$

and it follows that $K_n \to K$.

In the theory of integral equations, the technique of deducing the properties of an equation with kernel k from those of equations with approximating kernels k_n has been much used, see for example Tricomi (1957). One approach employs *degenerate kernels*—that is kernels of the form

$$k_n(x, y) = \sum_{j=1}^{n} \phi_j(x)\, \psi_j(y) \qquad (n < \infty).$$

The method will be successful if it is known that there is a sequence of degenerate kernels such that $K_n \to K$. A sample result is as follows.

3.5.9 Lemma. *Suppose that $k : [0, 1] \times [0, 1] \to \mathbb{C}$ is continuous. Then there is a sequence (k_n) of continuous degenerate kernels such that $K_n \to K$ in $\mathscr{L}(\mathscr{C}([0, 1]))$.*

Proof. By Weierstrass' Theorem 1.4.6 there is a sequence (p_n) of polynomials in two variables tending to k in $\mathscr{C}(\Omega)$, and each p_n is a degenerate kernel. The result follows from (3.5.1). □

Analogous results hold in the $\mathscr{L}_p$ spaces. Other types of approximating kernel such as double trigonometric series may also be used, and are sometimes more convenient in practice, see Atkinson (1974).

3.5.10 Example. Consider the initial value problem for a set of n simultaneous linear differential equations:

$$u_i'(t) = \sum_{j=1}^{n} a_{ij}u_j(t) \qquad (t > 0), \qquad u_i(0) = u_{0i} \qquad (i = 1, \ldots, n). \tag{3.5.2}$$

If u is regarded as a function $\mathbb{R} \to \mathbb{C}^n$, this system may be written as

$$u'(t) = Lu(t) \qquad (t > 0), \qquad u(0) = u_0, \tag{3.5.3}$$

where $u_0 \in \mathbb{C}^n$ and $L : \mathbb{C}^n \to \mathbb{C}^n$ is a linear operator. The system (3.5.2) is one example of an important class of problem which may be put in the form (3.5.3), and by substituting a Banach space for $\mathbb{C}^n$ and suitably interpreting L as a linear operator a great number of initial value problems may be expressed in this form; these problems range from a single equation to a partial differential equation. (3.5.3) is sometimes known as an abstract autonomous differential equation, the term autonomous indicating that L does not depend on t. For the moment we shall be interested in solving (3.5.3) when L is a bounded operator; an example where (3.5.3) represents a partial differential equation will be given in Section 9.7. The technique of solution, in which uniform operator convergence plays a central role, generalizes in a natural manner the method used in one dimension: it will be shown that the solution may be written as $e^{tL}u_0$, the operator e^{tL} being defined by a power series expansion.

Let $\mathscr{B}$ be a Banach space and take $\Omega = [0, \infty)$. The Banach space valued function $u : \Omega \to \mathscr{B}$ is said to be *differentiable* at $t \in \Omega$ iff there is a vector $u'(t)$—called the *derivative* of u at t—in $\mathscr{B}$ such that

$$\lim_{\substack{h \to 0 \\ t+h \in \Omega}} \left\| \frac{u(t+h) - u(t)}{h} - u'(t) \right\| = 0.$$

u is said to differentiable on Ω iff it is differentiable at each $t \in \Omega$.

Suppose that $L \in \mathscr{L}(\mathscr{B})$, and for $t \geqslant 0$ consider the series $\sum_{j=0}^{\infty} (tL)^j/j!$. Since

$$\sum_{j=0}^{\infty} \|(tL)^j\|/j! \leqslant \sum_{j=0}^{\infty} t^j\|L\|^j/j! < \infty,$$

the series is absolutely convergent in $\mathscr{L}(\mathscr{B})$, and as this space is complete, by Lemma 1.4.6 the sum of the series is an operator in $\mathscr{L}(\mathscr{B})$. We denote this operator by e^{tL}, that is we set

$$e^{tL} = \sum_{j=0}^{\infty} (tL)^j/j!.$$

By an argument closely analogous to that used in the elementary case when L is a complex number, it is easy to prove that e^{tL} is differentiable and that

(i) $\dfrac{d}{dt}(e^{tL}) = L\,e^{tL}$;

(ii) $e^{(t_1+t_2)L} = e^{t_1 L}\,e^{t_2 L}$.

It follows easily from (i) that $e^{tL}u_0$ is a solution of the initial value problem (3.5.3).

The suggestive form of the solution is valuable in practice. We shall see in Section 9.7 that under certain circumstances a useful interpretation in terms of eigenvalues may be given to the positivity of L, and hence a criterion for stability may be arrived at. The semigroup property (ii) is characteristic of autonomous equations; it implies essentially that the solution at $(t_1 + t_2)$ may be obtained also by first calculating the solution at t_1, and with this as the new initial condition proceeding through a further time t_2.

Uniform convergence is the most powerful type of operator convergence. A sequence (L_n) convergent in this sense is very tractable, and as we shall see, questions concerning the convergence of the sequence (L_n^{-1}) of inverses may often be answered satisfactorily. Uniform convergence is, however, sometimes too restrictive a requirement, and our next move is to introduce a less stringent but still useful criterion. A preparatory result is needed.

3.5.11 Lemma. *Let $\mathscr{B}$ and $\mathscr{C}$ be Banach spaces, and assume that (L_n) is a bounded sequence in $\mathscr{L}(\mathscr{B}, \mathscr{C})$. Suppose that $(L_n f)$ converges for all f in a dense subset S of $\mathscr{B}$. Then there is a unique $L \in \mathscr{L}(\mathscr{B}, \mathscr{C})$ such that* $\lim L_n f = Lf$ *for all $f \in \mathscr{B}$.*

Proof. Since the linear span $[S]$ of S consists of finite linear combinations of S, $(L_n f)$ converges for all $f \in [S]$. Define the (obviously linear) operator L on $[S]$ by setting $Lf = \lim L_n f$ for $f \in [S]$, and note that L is bounded on $[S]$ since

$$\|Lf\| \leqslant \lim \|L_n f\| \leqslant \sup \|L_n\| \, \|f\| \qquad (f \in [S]).$$

Therefore by Theorem 3.4.4, L has a continuous extension, still denoted by L, to $\mathscr{B}$. The proof will thus be complete if it can be shown that also $\lim L_n f = Lf$ for $f \in \mathscr{B} \backslash [S]$. Now since S is dense in $\mathscr{B}$, for each $f \in \mathscr{B}$ there is a sequence (f_j) in $[S]$ with $\lim f_j = f$, and

$$\|L_n f - Lf\| \leqslant \|(L_n - L)f_j\| + \|f - f_j\| \sup_n \|L_n - L\|.$$

Hence given any $\varepsilon > 0$, we may first choose j, and then n, sufficiently large to conclude that $\|L_n f - Lf\| < \varepsilon$. □

3.5.12 Corollary. *Assume that* (L_n) *is a sequence in* $\mathscr{L}(\mathscr{B}, \mathscr{C})$ *such that* $(L_n f)$ *is convergent for all* $f \in \mathscr{B}$. *Then there is a unique* $L \in \mathscr{L}(\mathscr{B}, \mathscr{C})$ such that $\lim L_n f = Lf$ *for all* $f \in \mathscr{B}$.

Proof. Since $(L_n f)$ is convergent it is bounded. Therefore by the Principle of Uniform Boundedness 3.5.6, (L_n) is a bounded sequence in $\mathscr{L}(\mathscr{B}, \mathscr{C})$. The result follows from the lemma. □

3.5.13 Definition. A sequence (L_n) of operators in $\mathscr{L}(\mathscr{B}, \mathscr{C})$ is said to **converge strongly** iff the sequence $(L_n f)$ converges for each $f \in \mathscr{B}$. The operator L (whose existence is assured by the corollary) such that $\lim L_n f = Lf$ $(f \in \mathscr{B})$ is called the **strong limit** of (L_n), and we write $L_n \underset{S}{\rightarrow} L$.

Thus if $L_n \underset{S}{\rightarrow} L$, for each f there is a sequence (ε_n) of real numbers tending to zero such that $\|(L_n - L)f\| \leqslant \varepsilon_n \|f\|$. The difference between this and uniform convergence is that the sequence (ε_n) may now depend on f. Thus evidently uniform convergence $L_n \rightarrow L$ implies strong convergence $L_n \underset{S}{\rightarrow} L$. The opposite implication is true in finite dimensions, but as the following example shows, not in infinite dimensions.

3.5.14 Example. Consider a sequence (L_n) of operators $\ell_2 \rightarrow \ell_2$ where

$$L_n(\alpha_1, \alpha_2, \ldots) = (0, \ldots, 0, \alpha_{n+1}, \alpha_{n+2}, \ldots).$$

Evidently for each f, $L_n f \rightarrow 0$, so $L_n \underset{S}{\rightarrow} 0$. However, $\|L_n\| \geqslant 1$ for $n \geqslant 1$, as may be seen by computing $\|L_n f_n\|$ where f_n is a vector all of whose components are zero except the $(n+1)$th. Therefore $L_n \nrightarrow 0$. Since the uniform and strong limits are the same when they both exist, it follows that (L_n) is not uniformly convergent.

3.5.15 Example. The distinction between strong and uniform convergence has serious practical implications when dealing with quadrature formulae. A standard approximate method for evaluating the integral $Q(f) = \int_0^1 f(x)$ dx is to subdivide $[0, 1]$ into $(n-1)$ subintervals, say $[x_j^{(n)}, x_{j+1}^{(n)}]$ for $j = 1, \ldots, (n-1)$, and to use the quadrature formulae

$$Q_n(f) = \sum_{j=1}^{n} w_j^{(n)} f(x_j^{(n)}), \tag{3.5.4}$$

where the $w_j^{(n)}$ are prescribed weights. The convergence of these approximations poses a question of obvious importance in applications.

To investigate this theoretically, we regard Q, Q_n as linear functionals (that is as linear operators with ranges in $\mathbb{C}$), and enquire in what manner Q_n converges to Q. A first observation is that certainly the convergence is not uniform, for with equal subintervals and f_n as in the figure, $Q_n(f_n) = 0$ for all n, but $Q(f_n) = 1$. A weaker but still useful result is the following.

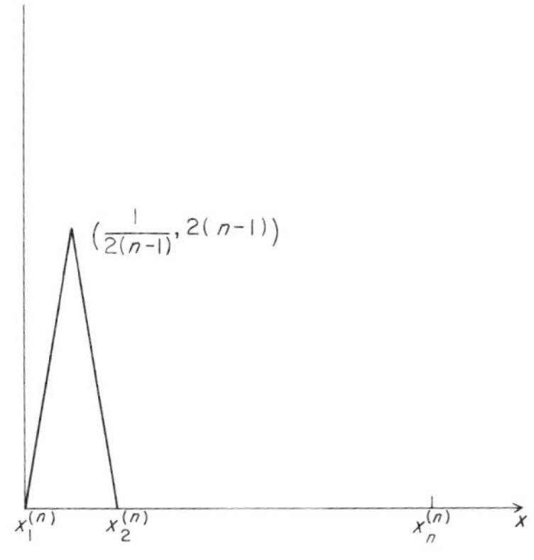

Fig. 3.1

3.5.16 Lemma. *Suppose that there is an $a < \infty$ such that*

$$\sum_{j=1}^{n} |w_j^{(n)}| \leqslant a \qquad (n \geqslant 1). \tag{3.5.5}$$

Equip $\mathscr{C}([0, 1])$ with the sup norm, and assume that there is a set S of functions dense in $\mathscr{C}([0, 1])$ such that $Q_n(f) \to Q(f)$ for each $f \in S$. Then $Q_n(f) \to Q(f)$ for each $f \in \mathscr{C}([0, 1])$.

Proof. The sequence $(\|Q_n\|)$ is bounded, for by (3.5.5),

$$|Q_n(f)| \leqslant \sum_{j=1}^{n} |w_j^{(n)}||f(x_j^{(n)})| \leqslant a\|f\|.$$

The result follows from Lemma 3.5.11 since $Q_n(f) \to Q(f)$ on a dense set. □

This lemma guarantees the convergence of many common quadrature formulae, and in particular of the trapezium rule, Simpson's rule, and Gaussian formulae. For the trapezium rule we divide $[0, 1]$ into equal subintervals of length h and set

$$Q_n(f) = \tfrac{1}{2}h[f(0) + 2f(h) + 2f(2h) + \dots + 2f((n-1)h) + f(1)].$$

Evidently (3.5.5) holds with $a = 1$. Take S to consist of all continuous functions which are linear on subintervals. This set is dense in $\mathscr{C}([0, 1])$ by Lemma 1.4.17, and it is easily checked that $Q_n(f) \to Q(f)$ for $f \in S$. Therefore the conditions of the previous lemma are satisfied; a similar argument obviously holds for Simpson's rule. The Gaussian formulae are exact for any given polynomial if n is large enough, and the polynomials are dense in $\mathscr{C}([0, 1])$ by Weierstrass' Theorem. Further, the weights are all positive, and it follows on calculating $Q_n(1)$ that (3.5.5) holds. The conditions of the lemma are again satisfied.

In practical terms the remarks above mean that the integral of a given continuous function f may be approximated to arbitrary accuracy, by the trapezium rule say, for choice of n large enough, but that nothing can be said about the value of n needed unless more is known about f. For this reason many standard error formulae assume additional smoothness for f and give the error in terms of the derivatives of f. However, if f happens to be the unknown function in an integral equation, this raises difficulties; more will be said on this topic in Section 7.6 when the numerical solution of integral equations is considered.

3.6 First Results on the Solution of the Equation $Lf = g$

With the preparations of the previous section complete, we are now in a position to obtain some useful results concerning the solution of the equation $Lf = g$. Questions (i) to (vi) posed in the introduction to this chapter motivate the general direction of the investigation. Thus we concentrate on finding conditions which ensure the existence of the inverse operator L^{-1} and its continuity, and on deriving approximate methods of solution, for example by considering a related operator L_0 which is in some sense a good approximation to L and whose inverse is assumed known.

All the results of this section are based on the successive substitution method for solving the equation

$$(I - M)f = g, \tag{3.6.1}$$

where $M \in \mathscr{L}(\mathscr{B})$; in spite of the simplicity of the idea, the results are useful in practice. A heuristic argument suggesting the procedure is as follows. Rewrite (3.6.1) as

$$f = g + Mf. \tag{3.6.2}$$

Then if M is "small", it is plausible that the term Mf may be neglected to first order; this yields the first approximation $f_0 = g$. This may be improved by substituting f_0 for f in the right-hand side of (3.6.2) giving the second approximation $f_1 = g + Mf_0$. Repetition of the argument leads to the sequence (f_n) of approximations:

$$f_0 = g, \qquad f_n = g + Mf_{n-1} \qquad (n \geqslant 1), \tag{3.6.3}$$

and it is plausible that $\lim f_n = f$, the solution of (3.6.1). We also note that formally

$$f = \sum_{n=0}^{\infty} M^n g = (I - M)^{-1} g, \tag{3.6.4}$$

so that we may tentatively write $(I - M)^{-1} = \sum_{n=0}^{\infty} M^n$. The successive substitution method is of course a standard technique for solving systems of algebraic equations and Fredholm integral equations; in the last case the series (3.6.4) is known as the *Neumann series*. In the next theorem recall that the statement $(I - M)^{-1} \in \mathscr{L}(\mathscr{B})$ means that $(I - M)$ is bijective and has a bounded inverse.

3.6.1 Theorem. *Suppose that $M \in \mathscr{L}(\mathscr{B})$ where $\mathscr{B}$ is a Banach space, and assume that $\|M\| < 1$. Then $(I - M)^{-1} \in \mathscr{L}(\mathscr{B})$. Also*

$$(I - M)^{-1} = \sum_{n=0}^{\infty} M^n, \tag{3.6.5}$$

the convergence of the series being in the operator norm (that is in $\mathscr{L}(\mathscr{B})$), and $\|(I - M)^{-1}\| \leqslant (1 - \|M\|)^{-1}$.

Proof. Since $\|M^n\| \leqslant \|M\|^n$ and $\|M\| < 1$, the series in (3.6.5) is absolutely convergent (Definition 1.4.5). Now $\mathscr{L}(\mathscr{B})$ is complete by Theorem 3.5.5, so it follows from Lemma 1.4.6 that the series is convergent with sum in $\mathscr{L}(\mathscr{B})$. A simple calculation shows that

$$(I - M)\left(\sum_{n=0}^{\infty} M^n\right) = I = \left(\sum_{n=0}^{\infty} M^n\right)(I - M),$$

and (3.6.5) follows from Lemma 3.3.10. The final estimate of the theorem is deduced from the inequalities

$$\left\|\sum_{n=0}^{\infty} M^n\right\| \leqslant \sum_{n=0}^{\infty} \|M^n\| \leqslant \sum_{n=0}^{\infty} \|M\|^n = (1 - \|M\|)^{-1}. \qquad \square$$

3.6.2 Corollary (The Neumann Series). *Suppose that the assumptions of the theorem hold. Then the equation $(I - M)f = g$ has exactly one solution f in $\mathscr{B}$. With $f_0 = g, f_n = g + Mf_{n-1} (n \geqslant 1)$, then $f = \lim f_n$, or equivalently,*

$$f = g + \sum_{n=1}^{\infty} M^n g,$$

and $\|f\| \leqslant (1 - \|M\|)^{-1}\|g\|$.

We next enquire what can be said about L^{-1} if the inverse L_0^{-1} of an approximation L_0 to L is known. Set $A = L_0 - L$, and note that

$$L = L_0 - A = L_0(I - L_0^{-1}A).$$

Suppose now that $\|A\| < \|L_0^{-1}\|^{-1}$. Then $\|L_0^{-1}A\| < 1$, and by Theorem 3.6.1, $(I - L_0^{-1}A)^{-1} \in \mathscr{L}(\mathscr{B})$. Therefore $L^{-1} \in \mathscr{L}(\mathscr{B})$ and

$$L^{-1} = [L_0(I - L_0^{-1}A)]^{-1} = (I - L_0 A)^{-1}L_0^{-1} = \sum_{n=0}^{\infty} (L_0^{-1}A)^n L_0^{-1}.$$

On subtracting L_0^{-1} and taking norms we deduce that

$$\|L^{-1} - L_0^{-1}\| \leqslant \|L_0^{-1}\| \sum_{n=1}^{\infty} \|L_0^{-1}\|^n \|A\|^n = \|L_0^{-1}\|^2 \|A\|[1 - \|L_0^{-1}\|\,\|A\|]^{-1}.$$

This proves the following theorem.

3.6.3 Theorem. *Let $\mathscr{B}$ be a Banach space, and suppose that L, L_0, $L_0^{-1} \in \mathscr{L}(\mathscr{B})$. If $\Delta = \|L - L_0\|\,\|L_0^{-1}\| < 1$, then $L^{-1} \in \mathscr{L}(\mathscr{B})$ and*

$$L^{-1} = \left[\sum_{n=0}^{\infty} (L_0^{-1}(L_0 - L))^n\right] L_0^{-1},$$

$$\|L^{-1} - L_0^{-1}\| \leqslant (1 - \Delta)^{-1}\,\Delta\|L_0^{-1}\|.$$

This shows that if L_0 has a bounded inverse and if the norm $\|L_0 - L\|$ of the difference between L_0 and L is small enough, then L also has a bounded inverse, and in addition supplies a series expansion for L^{-1} and an estimate for the norm of the difference between the inverses. The theorem is a sample, although an elementary one, of the type of result with which perturbation theory is concerned. The standard reference on this subject is Kato (1966).

The following examples illustrate the application of the theorems of this section.

3.6.4 Example. This is intended primarily to point out the relation between the restriction $\|M\| < 1$ and a condition often used in the numerical solution of linear simultaneous algebraic equations. If the equations have the form

$$f_i - \sum_{j=1}^{n} \alpha_{ij} f_j = g_i \qquad (i = 1, \ldots, n), \tag{3.6.6}$$

they are equivalent to the operator equation $(I - M)f = g$, with $M : \mathbb{C}^n \to \mathbb{C}^n$ the operator represented by the matrix $[\alpha_{ij}]$. Now it is known from Example 3.4.5 that if $\mathbb{C}^n$ has the ℓ_∞ norm, then

$$\|M\| = \sup_i \sum_{j=1}^{n} |\alpha_{ij}|. \qquad (3.6.7)$$

Thus by Corollary 3.6.2, for $\|M\| < 1$ the system (3.6.6) has a unique solution which is given by the standard iteration formula (3.6.3).

In fact this is a familiar result in numerical analysis, for the right-hand side of (3.6.7) is simply the maximum row sum of the matrix, the condition that this should be less than unity being known as *strict diagonal dominance*. It is well known that strict diagonal dominance ensures that the system (3.6.6) has a unique solution, and that this solution may be found by successive substitution. Note that if this condition is not satisfied, it may still be possible to establish the result by using another norm such as $\|\cdot\|_p$ on $\mathbb{C}^n$.

3.6.5 Example. It is clear that the analysis in the previous example will extend to an infinite system of linear simultaneous equations if n is replaced by infinity throughout. Such systems occur in various contexts, for example sometimes in tackling a partial differential equation by separation of variables. Usually a numerical method based on solving a truncated subsystem will be employed, and it is therefore important to be able to assess the validity and accuracy of the truncation approximation.

In order to illustrate a possible line of argument, consider the system

$$f_i - \sum_{j=1}^{\infty} \alpha_{ij} f_j = g_i \qquad (i = 1, 2, \ldots) \qquad (3.6.8)$$

in ℓ_2 (a similar technique may of course be applied in ℓ_p). Assume that $g \in \ell_2$ and that

$$|||\alpha|||_2 = \left\{ \sum_{i=1}^{\infty} \sum_{j=1}^{\infty} |\alpha_{ij}|^2 \right\}^{\frac{1}{2}} < 1. \qquad (3.6.9)$$

Then in an obvious notation (3.6.8) may be written as $(I - M)f = g$. By Theorem 3.4.7, $\|M\| < 1$, and by Corollary 3.6.2 the system will have a (unique) ℓ_2 solution. The corresponding truncated system is

$$\tilde{f}_i^{(n)} - \sum_{j=1}^{n} \alpha_{ij} \tilde{f}_j^{(n)} = g_i \qquad (i = 1, \ldots, n). \qquad (3.6.10)$$

In fact there is a slight difficulty in a theoretical attack on the system as it stands, for this is naturally formulated in a finite dimensional space rather than in ℓ_2. The difficulty can be avoided by a simple trick: we merely note that the solution of

$$\left.\begin{aligned} \tilde{f}_i^{(n)} - \sum_{j=1}^{n} \alpha_{ij}\tilde{f}_j^{(n)} &= g_i \qquad (i = 1, \ldots, n) \\ \tilde{f}_i^{(n)} &= g_i \qquad (i = n+1, \ldots) \end{aligned}\right\} \tag{3.6.11}$$

is $(\tilde{f}_1^{(n)}, \ldots, \tilde{f}_n^{(n)}, g_{n+1}, \ldots)$ where $(\tilde{f}_1^{(n)}, \ldots, \tilde{f}_n^{(n)})$ is the solution of (3.6.10), and consider (3.6.11) in ℓ_2 instead of (3.6.10). The questions which we seek to answer are: does (3.6.11) have a unique solution? If so, how good an approximation is this to the solution of the original system (3.6.8). The tactics will be to use Theorem 3.6.3.

To cast (3.6.11) into operator form, define a sequence (M_n) of operators by setting

$$(M_n f)_i = \begin{cases} \sum_{j=1}^{n} \alpha_{ij} f_j & (1 \leqslant i \leqslant n), \\ 0 & (n < i), \end{cases}$$

when (3.6.11) becomes $(I - M_n)\tilde{f}^{(n)} = g$. To use the theorem an estimate for $\|M - M_n\|$ is needed. Obviously the infinite matrix, $[\beta_{ij}]$ say, corresponding to $M - M_n$ is just that corresponding to M except that $\beta_{ij} = 0$ if $(i, j) \notin s(n)$, where $s(n)$ is the set of all pairs (i, j) of integers with both $i, j > n$. Then by Theorem 3.4.7,

$$\|M - M_n\| \leqslant \left\{ \sum_{i,\, j \in s(n)} |\alpha_{ij}|^2 \right\}^{\frac{1}{2}} = \varepsilon_n$$

say, where evidently $\varepsilon_n \to 0$ as $n \to \infty$. Therefore $\lim M_n = M$, and we deduce the following result from Theorem 3.6.3 on taking $L_0 = I - M$, $L = I - M_n$.

3.6.6 Theorem. *Suppose* $g \in \ell_2$ *and* $|||\alpha|||_2 < 1$. *Then* $\varepsilon_n \to 0$, *and there is an* n_0 *such that* $\varepsilon_n < 1 - |||\alpha|||_2$ *for* $n \geqslant n_0$. *For* $n \geqslant n_0$ *the approximate system* (3.6.11)—*or equivalently* (3.6.10)—*has exactly one solution* $\tilde{f}^{(n)}$, *and if* f *is the solution of* (3.6.8),

$$\|f - \tilde{f}^{(n)}\|_2 \leqslant \varepsilon_n (1 - |||\alpha|||_2)^{-1} (1 - |||\alpha|||_2 - \varepsilon_n)^{-1} \|g\|_2.$$

A reference on infinite systems of algebraic equations is Riesz (1913).

3.6.7 Example. An application of Corollary 3.6.2 together with an estimate for the norm such as that provided by Theorems 3.4.9 or 3.4.10 yield the familiar Neumann series for the integral equation

$$f(x) - \int_a^b k(x, y)\, g(y)\, \mathrm{d}y = g(x). \tag{3.6.12}$$

For instance, if k is measurable, $|||k|||_p < 1$ and $g \in \mathscr{L}_p(a, b)$, then (3.6.12) has exactly one $\mathscr{L}_p$ solution which is the sum in $\mathscr{L}_p(a, b)$ of the Neumann series $\Sigma\, g_n$ where $g_0 = g$ and

$$g_n(x) = \int_a^b k(x, y)\, g_{n-1}(y)\, dy \qquad (n \geqslant 1).$$

This example provides an opportunity of illustrating in a simple context a type of argument which is useful in a wide range of problems, and which will be employed in particular in dealing with partial differential equations in Chapter 11. Suppose a continuous solution of (3.6.12) is required. If k and g are continuous but $|||k|||_\infty > 1$, the convergence of the Neumann series in $\mathscr{C}([a, b])$ is not guaranteed by Corollary 3.6.2. However, if $|||k|||_p < 1$ for some p is it still possible to rescue the result by first working in $\mathscr{L}_p(a, b)$ and then using the following independent argument to deduce the existence of a continuous solution. Let $\bar{f}$ be the $\mathscr{L}_p$ solution; then $\bar{f} = g + K\bar{f}$ a.e., where K is the integral operator. But by Theorem 2.4.15, $K : \mathscr{L}_p(a, b) \to \mathscr{C}([a, b])$, and it follows that $g + K\bar{f} = f$ say, is a continuous function. Also $f = \bar{f}$ a.e. (and so in $\mathscr{L}_p$), and therefore $Kf = K\bar{f}$, whence $f = g + K\bar{f} = g + Kf$. Thus f is the required continuous solution.

3.6.8 Example. Another approximate method for treating (3.6.12) is based on finding a kernel k_0 close to k for which the equation

$$f_0(x) - \int_a^b k_0(x, y)\, f_0(y)\, dy = g(x)$$

can be solved. For example if $g \in \mathscr{L}_p(a, b)$, $|||k_0|||_p < 1$ and

$$\Delta = |||k - k_0|||_p\, (1 - |||k_0|||_p)^{-1} < 1,$$

then Theorem 3.6.3 predicts that (3.6.12) has an $\mathscr{L}_p$ solution f and that

$$\|f - f_0\|_p \leqslant (1 - \Delta)^{-1}\, \Delta(1 - |||k_0|||_p)^{-1}\, \|g\|_p.$$

One possibility is to take k_0 to be a degenerate kernel—see Example 3.5.8. This line of argument has interesting implications in integral equation theory; the error bounds are useful in the numerical treatment of integral equations, see Baker (1977, Chap. 4) and Atkinson (1976, Section 2.1).

3.7 Introduction to Spectral Theory

The analysis of the last section shows that the equation $(I - L)f = g$ is extremely tractable if $\|L\|$ is small. Unfortunately this condition is much too restrictive for most problems involving linear equations, and we shall therefore start the development of a theory in which larger values of $\|L\|$ are allowed. The central idea, which is familiar in finite dimensions, is to focus attention on the properties of the family of equations

$$(\lambda I - L)f = g \tag{3.7.1}$$

for λ a complex parameter. The theory is significantly simpler in a complex space, and the specific assumption that $\mathscr{B}$ is a *complex* Banach space will be made throughout this section.

3.7.1 Definition. Suppose that L is a (possibly unbounded) linear operator from $\mathscr{B}$ into $\mathscr{B}$. The set $\rho(L)$ of complex numbers for which $(\lambda I - L)^{-1} \in \mathscr{L}(\mathscr{B})$ is called the **resolvent set** of L. Its complement $\sigma(L)$ in $\mathbb{C}$ is the **spectrum** of L. For $\lambda \in \rho(L)$, the operator $R(\lambda; L) = (\lambda I - L)^{-1}$ is known as the **resolvent** of L.

Obviously $(I - L)$ has a bounded inverse if and only if $1 \in \rho(L)$. Thus in solving the equation $(I - L)f = g$ a fundamental question is: "What is the spectrum of L?" If L is bounded, by Theorem 3.6.1 $\sigma(L)$ is contained in a disc centre the origin and radius $\|L\|$. However, $\sigma(L)$ is often much smaller than this disc, and one of our long term aims will be to show that a much more accurate description of $\sigma(L)$ may be given at least for certain useful classes of operator.

If $\mathscr{B}$ is finite dimensional and $L : \mathscr{B} \to \mathscr{B}$, and if the equation $(\lambda I - L)f = 0$ has only the solution $f = 0$, then by Theorem 3.3.11, $\lambda \in \rho(L)$. On the other hand, in any Banach space, if the equation has a non-zero solution, $(\lambda I - L)$ is not injective and $(\lambda I - L)^{-1}$ is not even defined. With the aid of the following definition this result can be stated neatly.

3.7.2 Definition. Suppose that L is a linear operator from $\mathscr{B}$ into $\mathscr{B}$. A complex number λ is said to be an **eigenvalue** of L iff the equation $\lambda f - Lf$ has a non-zero solution. The corresponding non-zero solutions are called **eigenfunctions**, and the linear subspace spanned by these will be called the **eigenspace** corresponding to λ. The set $\sigma_p(L)$ of eigenvalues is known as the **point spectrum** of L.

In finite dimensions the spectrum is simply the set of eigenvalues of L. The crucial difference in infinite dimensions is that the spectrum can, and usually does, contain points other than eigenvalues, see for instance Example 3.3.12. For a bounded operator this may be put slightly differently. If $(\lambda I - L)$ is both injective and surjective, then by Theorem 3.5.3, $\lambda \in \rho(L)$. The first condition implies the second in finite but not in infinite dimensions, and it is in verifying the second condition that the main difficulty often arises in practice. The next definition completes the description of the spectrum, and allows us to summarize the above remarks briefly.

3.7.3. Definition. Suppose that L is a linear operator from $\mathscr{B}$ into $\mathscr{B}$. The set consisting of those $\lambda \in \sigma(L)$ for which $(\lambda I - L)$ is injective and $R(\lambda I - L)$ is dense (respectively not dense) in $\mathscr{B}$ will be called the **continuous spectrum** (respectively the **residual spectrum**).

3.7.4 Lemma. *Let $\mathscr{B}$ be a complex Banach space, and let $L : \mathscr{B} \to \mathscr{B}$ be a bounded linear operator. Then $\lambda \in \rho(L)$ if and only if $\lambda I - L$ is bijective. $\sigma(L)$ is the union of the point, continuous and residual spectra, which are disjoint sets. If $\mathscr{B}$ is finite dimensional $\sigma(L) = \sigma_p(L)$.*

For the rest of this section only bounded operators will be considered. The argument to be used is motivated by a one-dimensional analogy. If L, f, g are complex numbers, (3.7.1) will have a solution unless $\lambda = L$, and this point is the spectrum of the operator $\mathbb{C} \to \mathbb{C}$ corresponding to multiplication by L. The resolvent then corresponds to multiplication by $(\lambda - L)^{-1}$, which is an analytic function of λ on the resolvent set—that is all $\lambda \neq L$. The great power of analytic function theory suggests that an attempt be made to generalize this result to an operator L on a Banach space; it may be recalled that a similar type of argument is used in the theory of Fredholm integral equations (see Problem 3.25). Of course the theory must apply to the *operator valued* function $(\lambda I - L)^{-1}$ of λ, but the necessary extension does not turn out to present serious difficulties. We proceed as follows.

3.7.5 Definition. Suppose that $L \in \mathscr{L}(\mathscr{B})$. The function $R(\cdot\,; L) : \mathbb{C} \to \mathscr{L}(\mathscr{B})$ is said to be **analytic** at λ_0 iff the limit

$$\lim_{\lambda \to \lambda_0} \frac{R(\lambda; L) - R(\lambda_0; L)}{\lambda - \lambda_0}$$

exists in the operator norm (which of course automatically implies that the limit is independent of the manner in which $\lambda \to \lambda_0$).

3.7.6 Theorem. *Let $\mathscr{B}$ be a complex Banach space and suppose that $L \in \mathscr{L}(\mathscr{B})$. Then $\rho(L)$ is an open set, and for $\lambda, \lambda_0 \in \rho(L)$,*

$$R(\lambda_0; L) - R(\lambda; L) = (\lambda - \lambda_0)R(\lambda_0; L)R(\lambda; L). \tag{3.7.2}$$

Also, $R(\lambda; L)$ is an analytic function of λ in $\rho(L)$.

Proof. Take any $\lambda_0 \in \rho(L)$ and note that by Theorem 3.6.3 (applied to $\lambda_0 I - L$ and $\lambda I - L$), $\lambda \in \rho(L)$ if $|\lambda - \lambda_0| < \|L\|^{-1}\|(\lambda_0 I - L)^{-1}\|^{-1}$. Thus

λ_0 is contained in an open ball lying in $\rho(L)$, whence we deduce that $\rho(L)$ is open. Equation (3.7.2) is just the identity

$$\begin{aligned}(\lambda_0 I - L)^{-1} - (\lambda I - L)^{-1} &= (\lambda_0 I - L)^{-1}\,[(\lambda I - L) \\ &\qquad - (\lambda_0 I - L)]\,(\lambda I - L)^{-1} \\ &= (\lambda - \lambda_0)\,R(\lambda_0; L)\,R(\lambda; L).\end{aligned}$$

To prove the analyticity, we first remark that by the last assertion in Theorem 3.6.3, $\lim_{\lambda \to \lambda_0} R(\lambda; L) = R(\lambda_0; L)$. Therefore, from (3.7.2),

$$\begin{aligned}\lim_{\lambda \to \lambda_0} \frac{R(\lambda; L) - R(\lambda_0; L)}{\lambda - \lambda_0} &= -\lim_{\lambda \to \lambda_0} R(\lambda; L) R(\lambda_0; L) \\ &= -[R(\lambda_0; L)]^2. \qquad \square\end{aligned}$$

The spectrum of L can now be interpreted as the set of singularities of the analytic function $R(\lambda; L)$. Most of the standard results from the theory of analytic functions generalize just as easily. Since the modifications in the proofs consist merely of replacing moduli by norms, the details are omitted here; they may be found in Taylor (1958, Chap. 5). We note for future reference that the usual results on the Taylor and Laurent series hold and that Liouville's Theorem is valid.

Further information may now be obtained about the spectrum by using analyticity arguments. From Theorem 3.6.1, the series

$$R(\lambda; L) = \lambda^{-1} \Sigma\, \lambda^{-n} L^n \tag{3.7.3}$$

is convergent if $|\lambda| > \|L\|$. We may now interpret this series simply as the Laurent series of the resolvent, and conclude that the series will be absolutely convergent on the exterior of the smallest disc centre the origin which contains $\sigma(L)$. The factor which determines the convergence is thus not $\|L\|$ but the radius of this disc, and since this is usually less than $\|L\|$, the criterion for convergence may be improved.

3.7.7 Definition. The **spectral radius** $r_\sigma(L)$ of $L \in \mathscr{L}(\mathscr{B})$ is the number

$$r_\sigma(L) = \sup_{\lambda \in \sigma(L)} |\lambda|.$$

3.7.8 Theorem. *Let $\mathscr{B}$ be a complex Banach space, and assume that $L \in \mathscr{L}(\mathscr{B})$. Then the series* (3.7.3) *converges in $\mathscr{L}(\mathscr{B})$ to $R(\lambda; L)$ when $|\lambda| > r_\sigma(L)$ and diverges otherwise; the Neumann series for* (3.7.1) *converges if $|\lambda| > r_\sigma(L)$.*

3.7.9 Theorem. *If the dimension of the complex Banach space $\mathscr{B}$ is not zero and $L \in \mathscr{L}(\mathscr{B})$, $\sigma(L)$ is not empty.*

Proof. By Theorem 3.6.1, $\|R(\lambda; L)\| \leqslant (|\lambda| - \|L\|)^{-1}$ for $|\lambda| > \|L\|$, whence $\|R(\lambda; L)\| \to 0$ as $|\lambda| \to \infty$. Now if $\sigma(L)$ were empty, $R(\lambda; L)$ would be analytic and bounded over the whole complex plane, and by Liouville's Theorem $R(\lambda; L)$ would be a constant which must be the zero operator in view of its behaviour at infinity. But this is impossible since $R(\lambda; L)$ is surjective and by assumption $\mathscr{B}$ has at least one non-zero element. □

In order to obtain an expression for $r_\sigma(L)$, a preparatory result is needed. First make the natural definition of the polynomial function $p(L)$ of L by setting $p(L) = \sum_{j=0}^{n} \alpha_j L^j$. The proof of the following is straightforward and is left as an exercise.

3.7.10 The Spectral Mapping Theorem for Polynomials. *For $\mathscr{B}$ a complex Banach space and $L \in \mathscr{L}(\mathscr{B})$, the spectrum of $p(L)$ is the set $\{\mu : \mu = p(\lambda), \lambda \in \sigma(L)\}$. In other words $\sigma(p(L)) = p(\sigma(L))$.*

3.7.11 Theorem. *For $\mathscr{B}$ a complex Banach space and $L \in \mathscr{L}(\mathscr{B})$,*

$$r_\sigma(L) = \lim \|L^n\|^{1/n}.$$

Proof. Using Theorem 3.7.8 together with the standard expression for the radius of convergence of a power series, we find that

$$r_\sigma(L) = \limsup \|L^n\|^{1/n}. \tag{3.7.4}$$

We must show that the "lim sup" is an ordinary limit. By the Spectral Mapping Theorem, $\sigma(L^n)$ is composed of the nth powers of the points of $\sigma(L)$. Thus $r_\sigma(L^n) = [r_\sigma(L)]^n$. But $r_\sigma(L^n) \leqslant \|L^n\|$, whence

$$r_\sigma(L) = [r_\sigma(L^n)]^{1/n} \leqslant \|L^n\|^{1/n}. \tag{3.7.5}$$

Therefore

$$r_\sigma(L) \leqslant \liminf \|L^n\|^{1/n},$$

which combined with (3.7.4) gives the result. □

3.7.12 Example. To illustrate take $Lf(x) = \int_0^x f(t)\,dt$ and regard L as an operator $\mathscr{C}([0, 1]) \to \mathscr{C}([0, 1])$ where $\mathscr{C}([0, 1])$ has the sup norm. Obviously $\|L\| \geqslant 1$. However,

$$L^n f(x) = \frac{1}{(n-1)!} \int_0^x (x - t)^n f(t)\,dt,$$

from which it follows that $\|L^n\| \leqslant 1/(n - 1)!$, whence by Theorem 3.7.11, $r_\sigma(L) = 0$. The spectrum of L is thus the single point 0, and by Theorem 3.7.8

the Neumann series for the solution of the equation $(\lambda I - L)f = g$ converges for every non-zero λ. It is useful that an analogous result holds for a general Volterra integral operator, see Problem 3.24.

To classify the point zero in terms of the subdivision of the spectrum, note first that certainly L is injective—as may readily be proved by differentiating—and hence 0 is not in the point spectrum. Since $R(L)$ is contained in the proper closed subspace $\mathcal{M}$ consisting of continuous functions f with $f(0) = 0$, it follows that 0 is in the residual spectrum (Definition 3.7.3).

Analyticity arguments may be pressed further to yield a theory which can deal with general functions of operators, see Taylor (1958, Section 5.6) or Dunford and Schwartz (1958, Section 7.3). The advantage of this approach is hinted at by the utility of the "D-operator" method in elementary differential equations theory. This line of attack will not be pursued here except in the special case of a self-adjoint operator, which is dealt with from a somewhat different point of view in Chapter 9.

In order to proceed further with spectral theory, additional theoretical tools are needed. One of the most useful of these is the concept of the "adjoint" operator which will be discussed in Chapter 6. In practice many operators have certain simplifying properties apart from boundedness, and the use of the adjoint will enable a much more precise description of the spectrum to be given in at least two important cases—when the operator is "compact", and when it is "self-adjoint".

3.8 Closed Operators and Differential Equations

Almost all the results derived previously are applicable only to continuous operators. However, as noted in Example 3.4.11, differential operators are not continuous on the standard Banach spaces, and the theory of "closed" operators which we now introduce has been developed to deal with just this problem. This theory indeed presents technical difficulties greater than any encountered so far, but in view of the importance of differential equations in applications it is appropriate to point out at an early stage that in certain fundamental respects the qualitative properties of closed and continuous operators are similar. Perhaps the most striking example of such a property is the stability result for solutions of the equation $Lf = g$ when L is closed and bijective, which ensures that as for a continuous operator (Theorem 3.5.3) the solution depends continuously on the right-hand side.

The theory will be used only in the later attack on differential equations in the Hilbert space $\mathscr{L}_2$, and although a general treatment in Banach space may be supplied along similar lines, it is a little simpler to confine the analysis to Hilbert space. Throughout then $\mathscr{H}$ will be a Hilbert space, and L a linear operator from $\mathscr{H}$ into itself, the domain always being a linear subspace of $\mathscr{H}$. Recall that L is said to be unbounded iff it is not bounded on $D(L)$; an unbounded operator is discontinuous at every point of its domain (by Lemma 3.4.1 and Theorem 3.4.3). We start by giving the laws of combination for unbounded operators.

3.8.1 Definition. Let L, M be (unbounded) linear operators from $\mathscr{H}$ into $\mathscr{H}$, and set

$$D(\alpha L + \beta M) = D(L) \cap D(M),$$

$$(\alpha L + \beta M)f = \alpha Lf + \beta Mf \qquad (f \in D(\alpha L + \beta M))$$

for $\alpha, \beta \in \mathbb{C}$. For the product put

$$D(ML) = \{f : f \in D(L), Lf \in D(M)\},$$

$$(ML)f = M(Lf) \qquad (f \in D(ML)).$$

Note that in general $D(ML) \neq D(LM)$, and hence $ML \neq LM$. Extra care is thus needed in treating the inverse. For example it often happens that although L is unbounded, it has a bounded inverse, and then $D(LL^{-1}) = \mathscr{H}$ but $D(L^{-1}L) = D(L)$. Therefore $L^{-1}L \neq LL^{-1}$, and further $LL^{-1} = I$ but $L^{-1}L \subset I$ only. In dealing with unbounded operators it is often more convenient to work with vectors rather than with the operators themselves, and to write for example $L^{-1}Lf = f$ $(f \in D(L))$ rather than $L^{-1}L \subset I$.

The following example is intended to illustrate certain crucial differences between the usage of the term "operator" in functional analysis and in classical differential equation theory.

3.8.2 Example. Suppose operators corresponding to differentiation are to be considered in $\mathscr{L}_2(0, 1)$. An obvious possibility is to take $D(L) = \mathscr{C}^1([0, 1])$ and set $Lf = f'$ for $f \in D(L)$. However, there are alternatives. For example if an initial value problem is to be solved we may choose

$$D(L_1) = \{f : f \in \mathscr{C}^1([0, 1]), f(0) = 0\},$$

and take $L_1 f = f'$ for $f \in D(L_1)$. Then L, L_1 are linear unbounded operators, but L_1 is a proper restriction of L and certainly $L_1 \neq L$. This difference is not merely technical, but arises through the boundary conditions and so is closely related to the physics of the problem with which the operators are connected. Thus important properties such as the spectra of the operators will be different (see Problem 3.32).

On the other hand, in the classical theory a differential operator, l say, is often only a formal expression specifying the coefficients, and in particular contains no reference to boundary conditions. To emphasize this distinction the following definition is introduced.

3.8.3 Definition. Let p_r $(r = 0, 1, \ldots, n)$ be given functions on $\mathbb{R}$, and set

$$l = \sum_{r=0}^{n} p_r(x)\left(\frac{\mathrm{d}}{\mathrm{d}x}\right)^r.$$

l is called a **formal ordinary differential operator** of order n. If f is sufficiently smooth it makes sense to apply l to f and write

$$lf(x) = \sum_{r=0}^{n} p_r(x) f^{(r)}(x).$$

It is also convenient in Hilbert space to refer loosely to any operator L obtained from l by setting $Lf = lf$ for f in some specified domain as a *differential operator*.

For the rest of this section the subject will be closed operators.

3.8.4 Definition. Let $\mathscr{H}$ be a Hilbert space, and let L be a linear operator from $\mathscr{H}$ into $\mathscr{H}$. Consider the following conditions on the sequence (f_n):

(i) $f_n \in D(L)$ for all n;
(ii) (f_n) is convergent with limit f say;
(iii) (Lf_n) is convergent.

L is said to be **closed** iff for each sequence satisfying (i), (ii) and (iii), $f \in D(L)$ and $Lf = \lim Lf_n$.

The significance of this concept may perhaps be most easily grasped by comparing it with continuity. If L is continuous on $D(L)$, then (iii) is a consequence of (i) and (ii), whereas if L is only closed, (iii) is required as an additional condition rather than following as a consequence. Note also that for a closed operator L, the limit f of (f_n) lies in $D(L)$. Thus a continuous operator is closed only if $D(L)$ is closed, while this condition is obviously sufficient.

3.8.5 Lemma. *Suppose L is continuous on $D(L)$. Then L is closed if and only if $D(L)$ is closed.*

3.8.6 Example. To motivate the tactics that are commonly adopted in tackling ordinary differential equations, suppose the equation $f' = g$ for given $g \in \mathscr{L}_2(0, 1)$ is to be solved on $[0, 1]$ subject to the initial condition $f(0) = 0$. Let $D(L)$ consist of all $f \in \mathscr{L}_2(0, 1)$ satisfying the following conditions.

(i) f is absolutely continuous (Definition 2.4.13). By Theorem 2.4.14, f will be differentiable almost everywhere and f' will be locally Lebesgue integrable.
(ii) $f' \in \mathscr{L}_2(0, 1)$. This restriction is natural as a solution of $f' = g$ is required for all $g \in \mathscr{L}_2(0, 1)$. Mathematically it ensures that $R(L) \subset \mathscr{L}_2(0, 1)$.
(iii) $f(0) = 0$. Thus every $f \in D(L)$ satisfies the given initial condition.

Finally define $Lf = f'$ for all $f \in D(L)$. The above conditions on the domain are designed to ensure that f' makes sense and lies in $\mathscr{L}_2(0, 1)$ while restricting the domain as little as possible, and thus allowing the maximum chance for $Lf = g$ to have a solution. Indeed $D(L)$ is dense in $\mathscr{L}_2(0, 1)$, a fact that will later be important. To show that L fits into the theory, we must prove that L is closed.

Take any sequence (f_n) in $D(L)$ such that $f_n \to f$ and $Lf_n = f'_n$ is convergent with limit h, say; it must be established that $f \in D(L)$ and $f' = h$. From condition (iii),

$$f_n(x) = \int_0^x f'_n(t)\,dt. \tag{3.8.1}$$

But

$$\left| \int_0^x f'_n(t)\,dt - \int_0^x h(t)\,dt \right| \leqslant \|f'_n - h\|_2 \qquad \text{(by Schwartz's inequality)},$$

which tends to zero as $n \to \infty$. Using (3.8.1) we deduce that $f_n(x) \to \int_0^x h(t)\,dt$ in sup norm, and as by assumption $f_n \to f$ in $\mathscr{L}_2(0, 1)$, it follows that

$$f(x) = \int_0^x h(t)\,dt. \tag{3.8.2}$$

Since $h \in \mathscr{L}_2(0, 1)$, also $h \in \mathscr{L}_1(0, 1)$, and so h is locally integrable. Hence f is absolutely continuous. Also $f' = h$, whence $f' \in \mathscr{L}_2(0, 1)$. Finally, (3.8.2) shows that $f(0) = 0$. Thus $f \in D(L)$ and $Lf = g$.

Definition 3.8.4 shows clearly the distinction between closed and continuous operators, but is inconvenient in practice because of the complicated description of the sequence (f_n). A very much more convenient basis for the analysis is the new definition that is next introduced. The underlying idea is that the *simultaneous* convergence of the sequences (f_n) and (Lf_n) is what counts for closed operators, and this is best discussed in a single Hilbert space. The first move is to make the vector space $\mathscr{H} \times \mathscr{H}$ (Definition 1.2.11) into a Hilbert space. It is readily checked that $\mathscr{H} \times \mathscr{H}$ is indeed a Hilbert space when equipped with the inner product and norm defined as follows:

$$([f_1, g_1], [f_2, g_2]) = (f_1, f_2) + (g_1, g_2); \tag{3.8.3}$$

$$\|[f_1, g_1]\| = (\|f_1\|^2 + \|g_1\|^2)^{\frac{1}{2}}; \tag{3.8.4}$$

here f_1, g_1, f_2, g_2 are elements of $\mathscr{H}$, and $[f_1, g_1]$ denotes an element of $\mathscr{H} \times \mathscr{H}$. Noting that the graph of a real valued function ϕ of the real variable x is the set of pairs $[x, \phi(x)]$, we introduce the following terminology.

3.8.7 Definition. The **graph** $G(L)$ of L is the linear subspace of $\mathscr{H} \times \mathscr{H}$ consisting of all elements of the form $[f, Lf]$ with $f \in D(L)$. The *inverse graph* $G'(L)$ is similarly the set of all $[Lf, f]$ with $f \in D(L)$.

3.8.8 Lemma. *Let L be a linear operator from the Hilbert space $\mathscr{H}$ into $\mathscr{H}$. Then L is closed if and only if $G(L)$ is closed (in $\mathscr{H} \times \mathscr{H}$).*

Proof. If $([f_n, Lf_n]) \to [f, g]$ in $\mathscr{H} \times \mathscr{H}$, it follows from (3.8.4) that $f_n \to f$ and $Lf_n \to g$ (in $\mathscr{H}$). If L is closed this implies that $f \in D(L)$ and $Lf = g$. Hence $[f, Lf] \in G(L)$, which proves that $G(L)$ is closed. On the other hand, if $G(L)$ is closed, for any sequence $([f_n, Lf_n])$ convergent in $\mathscr{H} \times \mathscr{H}$, $f_n \to f \in D(L)$ and $Lf_n \to Lf$. Thus L is closed. □

3.8.9 Definition. The linear operator L from $\mathscr{H}$ into $\mathscr{H}$ is said to be **closed** iff its graph is closed.

In view of the lemma, this definition and Definition 3.8.4 are equivalent. Observe that the fact that $G(L)$ is closed does not imply either that $D(L)$ or $R(L)$ is closed in $\mathscr{H}$.

With the help of the graph it is fairly straightforward to extend several of the properties of bounded operators to closed unbounded operators. Consider first the extension of an operator by continuity (Theorem 3.4.4). In order to develop an analogue of this the concept of continuity on the domain must be generalized.

3.8.10 Definition. L is said to be **closable** iff it has a closed extension.

It may seem at first sight that every operator is closable since its graph necessarily has a closure. However, this may not be the graph of an operator (see Problem 3.30). The following result is useful in deciding whether an operator is closable.

3.8.11 Lemma. *In order that a linear subspace $\mathscr{M}$ of $\mathscr{H} \times \mathscr{H}$ be the graph of a linear operator, it is necessary and sufficient that no element of the form $[0, g]$ with $g \neq 0$ belongs to $\mathscr{M}$.*

Proof. Suppose $\mathcal{M}$ is the graph of an operator, L say. If $[0, g] \in \mathcal{M}$, then for some $f \in D(L)$, $0 = f$ and $g = Lf$. This is impossible as L is linear.

On the other hand assume that $[0, g] \in \mathcal{M} \Rightarrow g = 0$. Then if $f \in S = \{f : \exists g \text{ s.t. } [f, g] \in \mathcal{M}\}$ there is exactly one g such that $[f, g] \in \mathcal{M}$. For if g_1, g_2 are two such elements, then $[f, g_1] - [f, g_2] = [0, g_1 - g_2] \in \mathcal{M}$ (as $\mathcal{M}$ is a linear subspace), and it follows from the initial assumption that $g_1 = g_2$. It is easily checked that S is a linear subspace of $\mathcal{H}$, and we may thus define a linear operator L with $D(L) = S$ and $G(L) = \mathcal{M}$ by setting $Lf = g$. □

The lemma shows that every linear subspace of a graph is a graph. It follows that if L is closable, $\overline{G(L)}$ is the graph of a closed extension of L (since $\overline{G(L)}$ is a closed subspace of the graph of any closed extension of L). Therefore L has a closed extension, $\bar{L}$ say, with $G(\bar{L}) = \overline{G(L)}$, and $\bar{L}$ is evidently the minimal closed extension of L, that is, every closed extension of L is also an extension of $\bar{L}$.

3.8.12 Definition. If L is closable, its minimal closed extension $\bar{L}$ is called its **closure**.

The explicit calculation of the closure of certain operators is important in differential equation theory, and systematic methods for this will be developed in Section 10.2. For continuous operators the idea is familiar, as the following shows.

3.8.13 Lemma. *Every linear operator L from $\mathcal{H}$ into $\mathcal{H}$ which is bounded on $D(L)$ is closable. Its closure is identical with its extension by continuity to the closure of its domain.*

We turn now to a discussion of the inverse of a closed operator. The analysis is based on the following fundamental theorem. Note that the converse of this result is easy (Lemma 3.8.5), but for the proof of the theorem itself the Open Mapping Theorem is needed.

3.8.14 The Closed Graph Theorem. *Let $\mathcal{H}$ be a Hilbert space, and suppose that $L : \mathcal{H} \to \mathcal{H}$ is a closed linear operator. Then L is bounded.*

Proof. Define linear operators $P_1, P_2 : G(L) \to \mathcal{H}$ by setting

$$P_1[f, Lf] = f, \quad P_2[f, Lf] = Lf \qquad (f \in \mathcal{H}).$$

Then

$$\|P_1[f, Lf]\| = \|f\| \leqslant \{\|f\|^2 + \|Lf\|^2\}^{\frac{1}{2}} = \|[f, Lf]\|.$$

Hence P_1 is bounded, and by a similar argument so is P_2. Now clearly P_1 is bijective. Thus by Theorem 3.5.3 (which is a consequence of the Open Mapping Theorem), P_1^{-1} is bounded, and so therefore is $P_2P_1^{-1}$. Since $P_2P_1^{-1}f = P_2[f, Lf] = Lf$, then $P_2P_1^{-1} = L$, whence L is bounded. □

3.8.15 Lemma. *If L is an injective linear operator from $\mathscr{H}$ into itself, then L^{-1} is closed if and only if L is closed.*

Proof. Simply note that $G(L) = G'(L^{-1})$ (Definition 3.8.7). □

3.8.16 Theorem. *Let $\mathscr{H}$ be a Hilbert space, and assume that L is a closed injective linear operator from $\mathscr{H}$ onto $\mathscr{H}$. Then $L^{-1} \in \mathscr{L}(\mathscr{H})$.*

Proof. By the previous lemma L^{-1} is closed. Since $D(L^{-1}) = \mathscr{H}$, the result follows on applying the Closed Graph Theorem to L^{-1}. □

This generalization of Theorem 3.5.3 is the crucial stability result for closed operators promised previously. If it is known that the equation $Lf = g$ has exactly one solution for every $g \in \mathscr{H}$ (so that L is bijective), then if L is closed, the theorem ensures that the solution depends continuously on g. Phrased in terms of the spectrum, this result says that if $(\lambda I - L)$ is bijective and L is closed, then λ is in the resolvent set. This is exactly the conclusion if L is continuous (Lemma 3.7.4).

3.8.17 Example. Consider the differential equation

$$(p(x)f'(x))' + q(x)f(x) = g(x) \qquad (0 \leqslant x \leqslant 1), \tag{3.8.5}$$

together with the boundary conditions $f(0) = f(1) = 0$. We shall see later that if p, q are smooth and $p(x) \neq 0$ for $x \in [0, 1]$, the equation may be written in the form $Lf = g$ where L is a closed operator from $\mathscr{L}_2(0, 1)$ into itself. Theorem 3.8.16 then yields the following result: if (3.8.5) has exactly one solution for every $g \in \mathscr{L}_2(0, 1)$, then the solution depends continuously on the right-hand side, g. When (3.8.5) arises from a specific problem, this is the result usually to be expected on physical grounds. In the present case if the corresponding homogeneous equation has only the zero solution, then existence and uniqueness can be established without difficulty. However, in general the proof of surjectivity will present more of a problem, and we defer further discussion until Chapter 10.

The next two lemmas provide useful criteria for the existence of a bounded inverse.

3.8.18 Lemma. *Let L be a linear operator from the Hilbert space $\mathscr{H}$ into $\mathscr{H}$. Then the following hold:*

(i) *Suppose that there is an $m > 0$ such that $\|Lf\| \geqslant m\|f\|$ for all $f \in D(L)$. Then L is closed if and only if $R(L)$ is closed.*

(ii) *Assume that L is closed. Then $L^{-1} \in \mathscr{L}(\mathscr{H})$ if and only if $R(L)$ is dense in $\mathscr{H}$ and there is an $m > 0$ such that $\|Lf\| \geqslant m\|f\|$ for all $f \in D(L)$.*

Proof. (i) The condition $\|Lf\| \geqslant m\|f\|$ shows first that L is injective and so has an inverse L^{-1} with $D(L^{-1}) = R(L)$, and second that $\|L^{-1}g\| \leqslant m^{-1}\|g\|$ for $g \in D(L^{-1})$. Therefore L^{-1} is bounded on $R(L)$, and it follows from Lemma 3.8.5 that L^{-1} is closed if and only if $R(L)$ is closed. Lemma 3.8.15 then yields the result.

(ii) If $L^{-1} \in \mathscr{L}(\mathscr{H})$, then $\|L^{-1}f\| \leqslant \|L^{-1}\|\,\|f\|$, and with $g = L^{-1}f$ this proves that $\|Lg\| \geqslant \|L^{-1}\|^{-1}\|g\|$ for $g \in D(L)$. To prove the converse, note that from (i), $R(L)$ is closed, and since it is dense in $\mathscr{H}$, then $R(L) = \mathscr{H}$. As L is injective, $L^{-1} \in \mathscr{L}(\mathscr{H})$ by Theorem 3.8.16. □

3.8.19 Lemma. *Suppose that L is a closed injective linear operator from the Hilbert space $\mathscr{H}$ into $\mathscr{H}$. Assume that there is a linear operator M from $\mathscr{H}$ into $\mathscr{H}$ with $R(M) \subset D(L)$ and with domain dense in $\mathscr{H}$, and suppose that $LMf = f$ for all $f \in D(M)$. If M is bounded on its domain, then $L^{-1} \in \mathscr{L}(\mathscr{H})$ and $L^{-1} = \overline{M}$.*

Proof. Since $LMf = f$ for all $f \in D(M)$, $R(L)$ is dense in $\mathscr{H}$. Also $Mf = L^{-1}f$ on $D(M)$, and it follows that L^{-1} is bounded on $D(M)$. However, L^{-1} is closed (Lemma 3.8.15), so by Lemma 3.8.5 applied to L^{-1}, $D(L^{-1}) = R(L)$ is closed. Therefore $R(L) = \mathscr{H}$, and the result follows from Theorem 3.8.16. □

The above results emphasize the similarities in the properties of closed and continuous operators. However, it is obviously impossible to recover those results for continuous operators which depend explicitly on the use of the operator norm. For example, the main Perturbation Theorem 3.6.3. will not be applicable for unbounded operators. In fact the perturbation of one closed unbounded operator by another is a topic of central importance in quantum mechanics, and effective alternative methods have been devised to deal with the problems which arise, see Kato (1966). Problem 3.33 gives a flavour of the analysis required; the technique is to consider a perturbation small relative to the operator.

Problems

3.1 If a linear operator is not injective its inverse is not defined. In $\mathbb{R}^2$ show by constructing an example that $Lf = g$ may still have a solution for some g, and establish a criterion on g for this to happen.

3.2 Define the right and left shift operators $S_R, S_L : \ell_2 \to \ell_2$ as follows:

$$S_R(\alpha_1, \alpha_2, \ldots) = (0, \alpha_1, \alpha_2, \ldots), \; S_L(\alpha_1, \alpha_2, \ldots) = (\alpha_2, \alpha_3, \ldots).$$

Show that S_R is injective but not surjective, and that S_L is surjective but not injective. This is another example of the failure of Theorem 3.3.11 in infinite dimensions.

3.3 If $L : \mathscr{B} \to \mathscr{C}$, $M : \mathscr{C} \to \mathscr{D}$ where $\mathscr{B}, \mathscr{C}, \mathscr{D}$ are normed vector spaces, show that $\|LM\| \leqslant \|L\| \, \|M\|$.

3.4 Suppose that

$$m = \max\left\{\sup_i \sum_j |\alpha_{ij}|, \sup_j \sum_i |\alpha_{ij}|\right\} < \infty.$$

Establish an analogue of Young's inequality 2.5.4, and deduce that $L : \ell_p \to \ell_p \; (p \geqslant 1)$, where $(Lf)_i = \sum_j \alpha_{ij} f_j \; (i \geqslant 1)$, is bounded with $\|L\| \leqslant m$.

3.5 Let Ω be a finite or infinite interval with end points a, b, and suppose that $k : \Omega \times \Omega \to \mathbb{C}$ is measurable. Prove using Young's inequality that if

$$m = \max\left\{\sup_{x \in \Omega} \int_\Omega |k(x, y)| \, \mathrm{d}y, \sup_{y \in \Omega} \int_\Omega |k(x, y)| \, \mathrm{d}x\right\} < \infty,$$

and

$$Kf(x) = \int_\Omega k(x, y) \, f(y) \, \mathrm{d}y,$$

then for any $p \geqslant 1$, $K : \mathscr{L}_p(\Omega) \to \mathscr{L}_p(\Omega)$ is bounded and $\|K\| \leqslant m$.

3.6 Consider the integral operator K defined by

$$Kf(x) = \int_0^1 \frac{k(x, y)}{|x - y|^\alpha} f(y) \, \mathrm{d}y,$$

where k is continuous and α is a positive number. If $\mathscr{C}([0, 1])$ has the sup norm, show that $K \in \mathscr{L}(\mathscr{C}([0, 1]))$ *if* $\alpha < 1$. Give a condition on α which will ensure that $K : \mathscr{L}_p(0, 1) \to \mathscr{L}_p(0, 1)$ is bounded.

3.7 Let $\mathscr{C}((-\infty, \infty))$ be equipped with the sup norm. Let k be continuous, and assume that

$$\sup_x \int_{-\infty}^{\infty} |k(x, y)| \, \mathrm{d}y < \infty.$$

In the notation of Problem 3.5, show that $K \in \mathscr{L}(\mathscr{C}((-\infty, \infty)))$.

3.8 Prove that every separable Hilbert space is isometrically isomorphic to ℓ_2.

3.9 Prove that $\mathscr{L}(\mathscr{B}, \mathscr{C})$ in Definition 3.5.4 is a normed vector space.

3.10 Let $\mathscr{B}$ and $\mathscr{C}$ be Banach spaces, and assume that $L \in \mathscr{L}(\mathscr{B}, \mathscr{C})$. Show that if $R(L)$ is closed, there is an m such that for each $g \in R(L)$ there exists an $f \in \mathscr{B}$ with $Lf = g$ and $\|f\| \leqslant m \|g\|$. (Use Theorem 3.5.1).

3.11 Let (L_n) be a sequence of operators in $\mathscr{L}(\mathscr{B}, \mathscr{C})$. Show that $L_n \to L$ iff there is a sequence (ε_n) of real numbers tending to zero such that $\|(L_n - L)f\| \leqslant \varepsilon_n \|f\|$ for all $f \in \mathscr{B}$.

3.12 Let (μ_n) be a sequence in $\mathscr{C}([0, 1])$ equipped with the sup norm, and suppose that $\mu_n \to \mu$. Define $M_n f(x) = \mu_n(x) f(x)$, $Mf(x) = \mu(x) f(x)$. Show that $M_n \to M$. Is this result necessarily true if $\mu_n \to \mu$ pointwise only?

3.13 Let $\mathscr{B}$ be a Banach space. Suppose that for each $n \geqslant 1$, $L_n, L_n^{-1} \in \mathscr{L}(\mathscr{B})$, and assume that $\lim L_n = L \in \mathscr{L}(\mathscr{B})$. Show by constructing a counterexample that L^{-1} need not exist. Prove that a necessary and sufficient condition for $L^{-1} \in \mathscr{L}(\mathscr{B})$ is that the sequence $(\|L_n^{-1}\|)$ be bounded, and if this condition is satisfied show that $L_n^{-1} \to L^{-1}$.

3.14 Suppose that $L \in \mathscr{L}(\mathscr{B}, \mathscr{C})$ with $\mathscr{B}, \mathscr{C}$ Banach spaces. Show that L^{-1} exists and is bounded on $R(L)$ iff there is an $m > 0$ such that $\|Lf\| \geqslant m \|f\|$ $(f \in \mathscr{B})$. If this inequality holds prove that $R(L)$ is closed, and decide whether necessarily $R(L) = \mathscr{C}$.

3.15 For $\mathscr{H}$ a Hilbert space and $L \in \mathscr{L}(\mathscr{H})$, assume that there is an $m > 0$ such that $|(Lf, f)| \geqslant m \|f\|^2$ $(f \in \mathscr{H})$. Show that $L^{-1} \in \mathscr{L}(\mathscr{H})$.

3.16 Suppose that $k \in \mathscr{L}_1(-\infty, \infty)$. Define $K, H_\alpha\colon \mathscr{L}_2(-\infty, \infty) \to \mathscr{L}_2(-\infty, \infty)$ as follows:

$$Kf(x) = \int_{-\infty}^{\infty} k(x - y) f(y)\, dy,$$

$$K_a f(x) = \begin{cases} \displaystyle\int_{-a}^{a} k(x - y) f(y)\, dy & (|x| \leqslant a), \\ 0 & (|x| > a). \end{cases}$$

Show that K_a does not converge uniformly to K as $a \to \infty$.

3.17 Suppose that the assumptions of the previous problem hold. If $\|k\|_1 < 1$, the equation

$$f_a(x) - \int_{-a}^{a} k(x - y) f_a(y)\, dy = g(x) \qquad (-a < x < a) \tag{$*$}$$

may be easily solved by use of the Fourier transform for $g \in \mathscr{L}_2(-\infty, \infty)$ if $a = \infty$. It appears plausible that f_∞ should be an approximation for large a to the solution of $(*)$, but the lack of uniform convergence seems to rule out any argument based on Theorem 3.6.3. However, prove that with $\|\cdot\|_{2,a}$ the norm of $\mathscr{L}_2(-a, a)$,

$$\|f_a - f_\infty\|_{2,a} \leqslant (1 - \|k\|_1)^{-1} \|k\|_1 \left\{ \int_{|x| > a} |f_\infty(x)|^2\, dx \right\}^{\frac{1}{2}}.$$

3.18 Let $\mathbb{C}^2$ have the norm $\|\cdot\|_\infty$. If $L\colon \mathbb{C}^2 \to \mathbb{C}^2$ has the matrix representation

$$\begin{bmatrix} 1 & a \\ 0 & 1 \end{bmatrix}$$

verify that $r_\sigma(L) = \lim \|L^n\|^{1/n}$.

3.19 Let $\mathbb{C}^3$ have the norm $\|\cdot\|_\infty$. Assume that $L: \mathbb{C}^3 \to \mathbb{C}^3$ has the matrix representation

$$\begin{bmatrix} 0 & 3 & 2 \\ 0 & 0 & 1 \\ -1 & 0 & 0 \end{bmatrix}.$$

Show that $r_\sigma(L) = 3^{\frac{1}{2}}$ whereas $\|L\| = 5$. Compare this result with the estimates from Theorem 3.7.11 : $\|L^2\|^{\frac{1}{2}} = 5^{\frac{1}{2}}$, $\|L^3\|^{\frac{1}{3}} = 13^{\frac{1}{3}}$, $\|L^4\|^{\frac{1}{4}} = 13^{\frac{1}{4}}$.

3.20 Let $S_L : \ell_2 \to \ell_2$ be the left shift (Problem 3.2). Prove that the point spectrum of L is the interior of the unit disc, that its continuous spectrum is the edge of this disc, and that its residual spectrum is empty.

3.21 Prove the Spectral Mapping Theorem 3.7.10 for polynomials.

3.22 Suppose that $L \in \mathscr{L}(\mathscr{B})$ with $\mathscr{B}$ a Banach space. Show that $\|R(\lambda; L)\| \geqslant 1/d(\lambda)$, where $d(\lambda)$ is the distance of λ from $\sigma(L)$.

3.23 A slight generalization of the successive substitution method is useful if a good first guess of the solution is available. If $M \in \mathscr{L}(\mathscr{B})$ with $\mathscr{B}$ a Banach space, and $r_\sigma(M) < 1$, show that the sequence (f_n) where $f_n = g + Mf_{n-1}$ $(n \geqslant 1)$ converges to the solution of $f - Mf = g$ for any initial guess f_0.

3.24 Equip $\mathscr{C}([0, a])$ with the sup norm. For continuous complex valued k, define the *Volterra integral operator* $K : \mathscr{C}([0, a]) \to \mathscr{C}([0, a])$ by setting

$$Kf(x) = \int_0^x k(x, y) f(y)\, dy.$$

Set $m(x) = \sup_{0 \leqslant y \leqslant x} |k(x, y)|$, and show that

$$|K^n f(x)| \leqslant [xm(x)]^n \|f\|/n! \qquad (0 \leqslant x \leqslant a).$$

Deduce that $r_\sigma(K) = 0$. Therefore $\sigma(K)$ is the single point zero, and by Theorem 3.7.11 the Neumann series for the Volterra integral equation $(\lambda I - K)f = g$ converges for all $\lambda \neq 0$.

3.25 Let $\mathscr{C}([-1, 1])$ have the sup norm. Consider the integral equation

$$\lambda f(x) - \int_{-1}^1 xy\, f(y)\, dy = g(x) \qquad (-1 \leqslant x \leqslant 1),$$

where $g \in \mathscr{C}([-1, 1])$, and denote the integral operator by K. Show that for $\lambda \neq 0, \frac{2}{3}$,

$$R(\lambda; K)g(x) = \lambda^{-1}\left[g(x) + \frac{3}{3\lambda - 2} \int_{-1}^1 xy\, g(y)\, dy \right].$$

$R(\cdot\,; K)$ is obviously analytic except at $\lambda = 0, \frac{2}{3}$, which points therefore constitute the spectrum of K. This is essentially the operator theoretic version of the classical Fredholm result, $3xy/(3\lambda - 2)$ being related to the "resolvent kernel".

3.26 Suppose $M \in \mathscr{L}(\mathscr{B})$ with $\mathscr{B}$ a Banach space, and assume that λ lies outside a convex set containing $\sigma(M)$. Set $\lambda I - M = \mu I - N$ where $\mu = \lambda - \alpha$, $N = M - \alpha I$.

Show that an $\alpha \in \mathbb{C}$ may be chosen in such a way that the modified Neumann series $\Sigma(\mu^{-1}N)^n$ is convergent. In terms of complex variable theory, the new series is the analytic continuation for $|\lambda| < r_\sigma(M)$ provided by the expansion in powers of $(M - \alpha I)/(\lambda - \alpha)$.

This is sometimes a useful technique (related to over-relaxation) for accelerating the rate of convergence of the Neumann series (see Problem 7.15 and Hutson *et al.* [1972] for applications). It also shows that a modified Neumann series can sometimes be used even if $|\lambda| < r_\sigma(M)$. For example if the spectrum is real (which will be the case if the operator is "self-adjoint"—see Chapter 6), every λ with $\operatorname{Im} \lambda \neq 0$ can be treated in this way.

In the remaining problems $\mathscr{H}$ is a Hilbert space and L, M are linear operators from $\mathscr{H}$ into itself.

3.27 Assume that L is closed. Prove that:

(i) $L + A$ is closed if $A \in \mathscr{L}(\mathscr{H})$.
(ii) The null space $N(L)$ of L is closed.

3.28 (i) $L \subset M$ iff $G(L) \subset G(M)$. (ii) A linear subspace of a graph is itself a graph.

3.29 Show that L is closable iff $g = 0$ for every sequence (f_n) such that $f_n \in D(L)$, $f_n \to 0$ and $Lf_n \to g$.

3.30 (Reed and Simon, 1972). The following is an example of an operator that is not closable. Let $\{\phi_n\}$ be an orthonormal basis of the separable Hilbert space $\mathscr{H}$. Let e be a vector in $\mathscr{H}$ which is not in the linear span, say $\mathscr{M}$, of the ϕ_n. Define $D(L)$ to be the direct sum of $\mathscr{M}$ and the one dimensional subspace $\{ae : a \in \mathbb{C}\}$, and for any positive integer N and any $a, a_1, \ldots, a_N \in \mathbb{C}$ set

$$L\left(ae + \sum_{n=1}^{N} a_n\phi_n\right) = ae.$$

Show that $(0, e) \in \overline{G(L)}$, and deduce that L is not closable.

3.31 If $L : \mathscr{H} \to \mathscr{H}$ and $(Lf, g) = (f, Lg)$ for all $f, g \in \mathscr{H}$, then L is bounded.

3.32 Differential operators associated with the same formal operator may have quite different properties. Let $\mathscr{H} = \mathscr{L}_2(0, 1)$ and take $l = i\,d/dx$. Let $\mathscr{A}$ be the set of absolutely continuous functions with first derivatives in $\mathscr{L}_2(0, 1)$. Set

$$D(L_1) = \{f : f \in \mathscr{A}\}, \qquad L_1 f = lf \ (f \in D(L_1)),$$
$$D(L_2) = \{f : f \in \mathscr{A}, f(0) = 0\}, \qquad L_2 f = lf \ (f \in D(L_2)).$$

Then $L_1 f = L_2 f$ for $f \in D(L_1) \cap D(L_2)$, which set is dense in $\mathscr{L}_2(0, 1)$. Show, however, that $\sigma(L_1)$ is the whole complex plane, while $\sigma(L_2)$ is empty (calculate $R(\lambda; L_2)$).

3.33*. *Perturbation by closed operators* (compare Theorem 3.6.3). The operator M will be said to be *L-bounded* iff $D(M) \supset D(L)$ and there are non-negative constants a, b such that

$$\|Mf\| \leqslant a\|f\| + b\|Lf\| \qquad (f \in D(L)).$$

Assume that L is closed, and that M is L-bounded with constant $b < 1$.

(i) Prove that $L + M$ is closed.
(ii) If L has a bounded inverse and $a\|L^{-1}\| + b < 1$, prove that $(L + M)^{-1} \in \mathscr{L}(\mathscr{B})$, and show that

$$\|(L + M)^{-1}\| \leqslant \|L^{-1}\|(1 - a\|L^{-1}\| - b)^{-1},$$

$$\|(L + M)^{-1} - L^{-1}\| \leqslant \|L^{-1}\|(a\|L^{-1}\| + b)(1 - a\|L^{-1}\| - b)^{-1}$$

Results of this type are basic in perturbation theory for differential operators, see Kato (1966, Section 4.1).

Chapter 4

Introduction to Nonlinear Operators

4.1 Introduction

Until relatively recently the theory of nonlinear equations consisted principally of a collection of isolated results on particular problems. However, in the last few years great advances have been made, and rather general results are now available for several wide classes of equation. In this development functional analysis has played an important role. Indeed it is perhaps in this context that the contribution of functional analytic methods in applications has been most valuable. In this and later chapters some of the parts of the theory of nonlinear operators which have been most useful will be outlined.

In the study of linear operators on a Banach space, the very comprehensive general principles that are known for the finite dimensional case are of great help in suggesting fruitful lines of enquiry. The difficulties are thus connected almost solely with the transition from finite to infinite dimensions, and so are essentially analytic in character. For nonlinear operators it is also natural to look first to the finite dimensional case for guidance. However, finite dimensional nonlinear problems are themselves often of considerable difficulty. Indeed a very active area of research is concerned with just such problems, and many of the main advances are recent; the standard reference here is Ortega and Rheinboldt (1970). The study of nonlinear operators in finite dimensions may be classed as an essentially geometric theory as it is concerned with the "shapes" of the functions. With this understanding we may summarize by saying that the theory of nonlinear operators on a Banach space has both a geometric and an analytic part, and the emphasis on the geometric part is relatively greater than in the linear theory.

The following tactics for proposing methods of solution of nonlinear operator equations are thus commonly adopted. On the basis of geometrical intuition, formulate a method in low dimensional spaces, being careful to ensure that this is meaningful in a Banach space; it may be added that the conceptual difficulties of dealing with functions even from $\mathbb{R}^2$ into itself are such that it is frequently useful to look first to the one-dimensional case for inspiration. Next, attempt to check whether the method is valid for an arbitrary finite dimensional space. Lastly, take the analytical step of bridging the gap between finite and infinite dimensions; an insight from the linear theory can sometimes be useful at this stage.

From the point of view of applications it is natural to look first for general constructive methods of solution, which for present purposes will be interpreted as including numerical methods based on iterative procedures. It is probably fair to say that constructive methods are known for almost all types of linear equation, but for nonlinear equations the position is far less satisfactory. For certain classes of equation there are indeed very efficient constructive methods available. However, these do not work for many equations arising in practice, and it is often necessary to rely on qualitative techniques which are concerned with questions of existence, uniqueness, stability and so on, but which nonetheless can provide illuminating insight into the behaviour of the system. A considerable part of the nonlinear theory consists of a study of such techniques.

The account of nonlinear operators starts in the next section with a listing of some typical problems for differential and integral equations, and a unified formulation for these is then derived in terms of the "fixed points" of an operator. In the remainder of this chapter the simplest results of the theory are studied. No analytical idea of greater depth than the completeness of a Banach space is involved, and the level of difficulty is comparable with that of the last chapter. Two related methods will be discussed, both generalizing well known algorithms for solving equations in one dimension. The first is based on the idea of finding a solution by successive substitution starting with a plausible initial guess—a procedure which has already been successfully used for linear operators in Chapter 3. This leads to the famous Contraction Mapping Principle (or Banach Fixed Point Theorem). The second method is the infinite dimensional version of the Newton algorithm. The formulation of this depends on the development of an appropriate definition of the "derivative" of an operator. This motivates the study in Section 4 of the Fréchet derivative.

Both of the methods above may be used to establish existence and uniqueness and both are constructive. When applicable they thus provide answers as complete as could be wished for. Indeed we shall reach conclusions very similar to those in Section 3.6 for the equation $(I - L)f = g$ where $\|L\| < 1$. In a sense then the theory of this chapter may be regarded as parallelling the previous linear theory. As with the latter the restrictions on the operator are very severe, but the discussion of more subtle methods must be postponed until certain further concepts in Banach space theory are introduced.

The position with regard to references is less satisfactory than that for linear operators, and there is no single comprehensive book on the nonlinear theory. Taken together the three volumes of Krasnoselskii (1964(a), (b), 1972) are probably the nearest to a complete account, and several of the considerable recent advances are described by Berger (1977). Also useful are Smart (1974) which provides a brief overview of fixed point theorems, Vilenkin (1972) and Krasnoselskii (1958) which contain an outline of many of the main developments without giving much detail, and Saaty (1967) which emphasizes the applications. For the topics in this chapter see Rall (1969) and Krasnoselskii (1972).

4.2 Preliminaries

In this section some introductory remarks are made concerning the formulation of nonlinear problems which is most convenient from the point of view of abstract operator theory. It will be useful to have some specific examples in mind to motivate the development, so we will start by describing a few typical problems involving differential and integral equations.

4.2.1 Example. An important system in applications is the initial value problem for a system of ordinary differential equations. With f taking values in $\mathbb{C}^n$ and $\psi : [0, \infty) \times \mathbb{C}^n \to \mathbb{C}^n$ this may be written as

$$\left.\begin{aligned} f'(t) &= \psi[t, f(t)] \qquad (t \geqslant 0), \\ f(0) &= \alpha, \end{aligned}\right\} \tag{4.2.1}$$

a solution with continuous first derivative being required. On integration (4.2.1) becomes

$$f(t) = \alpha + \int_0^t \psi[s, f(s)]\, \mathrm{d}s, \tag{4.2.2}$$

a nonlinear Volterra integral equation. If ψ is continuous and $f \in \mathscr{C}([0, t_0], \mathbb{C}^n)$ is a solution of (4.2.2), the right-hand side of (4.2.2) has continuous first derivative. Hence $f \in \mathscr{C}^1([0, t_0], \mathbb{C}^n)$, and (4.2.1) may be recovered by differentiation. Thus (4.2.1) and (4.2.2) are equivalent, but (4.2.2) is sometimes easier to tackle.

4.2.2 Example. Of equal importance is the boundary value problem for an ordinary or partial differential equation. With the help of a Green's function this can often be recast as an integral equation, which again may be easier to handle. A relatively simple example is the system

$$\left.\begin{aligned} -f''(x) + \omega f(x) &= \psi[x, f(x)] \qquad (0 \leqslant x \leqslant 1), \\ f(0) = f(1) &= 0, \end{aligned}\right\} \tag{4.2.3}$$

where $\psi : [0, 1] \times \mathbb{C} \to \mathbb{C}$ is continuous, a solution in $\mathscr{C}^2([0, 1])$ being required. If $\omega > -\pi^2$ this may be rewritten as the integral equation

$$f(x) = \int_0^1 k(x, y)\, \psi[y, f(y)]\, \mathrm{d}y. \tag{4.2.4}$$

Here k is the Green's function, which may of course be found explicitly. For example, in the simplest case $\omega = 0$,

$$k(x, y) = \begin{cases} y(1 - x) & (0 \leqslant y \leqslant x \leqslant 1), \\ x(1 - y) & (0 \leqslant x \leqslant y \leqslant 1). \end{cases} \tag{4.2.5}$$

In general the k corresponding to any $\omega > -\pi^2$ is continuous on $[0, 1] \times [0, 1]$ and is non-negative.

4.2.3 Example. The equation

$$f(x) = \int_\Omega k(x, y, f(y))\, \mathrm{d}y + g(x), \tag{4.2.6}$$

where k, g are given and Ω is a closed subset of $\mathbb{R}^n$, is known as a Urysohn integral equation. A special case of this which has been extensively studied is the Hammerstein integral equation

$$f(x) = \int_\Omega k(x, y)\, \psi[y, f(y)]\, \mathrm{d}y. \tag{4.2.7}$$

The importance of these equations is due not a little to the fact that they include the integral equations derived in the previous two examples as particular cases.

4.2.4 Example. One type of problem must be picked out for special mention. For μ a parameter consider the integral equation

$$f(x) = \mu \int_0^1 k(x, y)\, [f(y)]^2\, \mathrm{d}y.$$

It is obvious that zero is a solution for any μ, but in practice interest centres on those values of μ for which there is a non-zero solution. By analogy with the linear case this may be called an eigenfunction problem, μ (the reciprocal of an eigenvalue) being known as a characteristic value. Such equations are of considerable importance in applications, but present special mathematical difficulties because of the presence of the trivial zero solution. There are also crucial qualitative differences between the linear and nonlinear eigenfunction problems. Some methods which have been devised to deal with these problems are described in Chapter 14.

To illustrate the formulation of these examples as problems in operator theory, consider (4.2.6) and proceeding first formally define

$$Af(x) = \int_{\Omega} k(x, y, f(y))\,\mathrm{d}y + g(x) \qquad (x \in \Omega).$$

Then (4.2.6) becomes

$$f = Af. \tag{4.2.8}$$

4.2.5 Definition. A point $\bar{f}$ is said to be a **fixed point** of the operator A iff $\bar{f} = A\bar{f}$.

The statement that A has a fixed point is thus just another way of saying that $f = Af$ has a solution. However, the first formulation has the advantage of conveying a suggestive geometrical picture—that is every fixed point is invariant under the application of A; this point of view is emphasized in an interesting introductory article by Shinbrot (1969). In the current literature on nonlinear operator theory the fixed point formulation is usually adopted, and many of the results appear in the form of fixed point theorems.

A study of the following simple example brings to light certain features of nonlinear problems which will influence the treatment of nonlinear operator theory.

4.2.6 Example. The Hammerstein equation

$$f(x) = \int_0^1 \psi[f(y)]\,\mathrm{d}y + a, \tag{4.2.9}$$

(where a is a real number and ψ is continuous) is, in the above notation, $f = Af$ with A defined formally by

$$Af(x) = \int_0^1 \psi[f(y)]\,\mathrm{d}y + a.$$

For a precise definition of A, the first requirement is an appropriate Banach space. For a linear equation there are usually a number of obvious possibilities, but here the choice must be made with rather more care. For example, if $\psi(z) = z^2$, A cannot be defined on $\mathscr{L}_p(0, 1)$ for $1 \leqslant p < 2$ because $[f(\cdot)]^2$ will not be integrable for all f in this space, while if $\psi[z] = \exp(z)$, no value of p will suffice. Thus the choice of an $\mathscr{L}_p$ space is far from ideal, but if such a space must be used the rate of growth of the nonlinearity has to be taken into account. With $\mathscr{C}([0, 1])$ this difficulty disappears, and it is usually easier to work with this space where possible.

The second remark concerns the algebraic structure of A. In contrast with the linear case, it does not turn out to be advantageous to define A so that it vanishes at the origin and study the equation $f = Af + g$. It should therefore be noted that in general $A0 \neq 0$ in the nonlinear theory.

To illustrate the third point, suppose that $\psi[z] = |z|^2/2$. If a is small the method of successive substitution may be expected to work since the contribution from the integral will be less than in the linear case $\psi[z] = z/2$. However, for large a the iterates will increase rapidly because of the nonlinearity, and convergence cannot be expected—indeed the equation may not have a solution (compare Problem 4.2). Clearly the treatment of the equation must take into account the region of interest in the Banach space, and it is usually appropriate to restrict the search for a solution to a bounded subset (often some closed ball) of the space even if the operator can conveniently be defined on the whole space.

The following abstract framework is thus adopted for the study of nonlinear problems. With $\mathscr{B}$ and $\mathscr{C}$ Banach spaces, a specific subset D of $\mathscr{B}$ is chosen and an operator $A : D \to \mathscr{C}$ is considered; in the fixed point problem of course $\mathscr{B} = \mathscr{C}$. It is obvious that even in one dimension progress will be difficult unless A is continuous, and in the sequel this condition is always imposed on A. A notion of boundedness may be defined by requiring for example that

$$\sup_{f \in D} \|Af\| < \infty, \tag{4.2.10}$$

but in contrast with the linear case boundedness and continuity are not equivalent (even in one dimension). Of the two concepts, continuity is by far the more important. Conditions for the continuity of integral operators may be found in Krasnoselskii (1964a) or Zabreyko (1975), and for reference the following are quoted.

4.2.7 Lemma. *Suppose $r < \infty$ and let d_r be the disc $\{z : z \in \mathbb{C}, |z| \leqslant r\}$. Assume that $[a, b]$ is a finite interval. If $k : [a, b] \times [a, b] \times d_r \to \mathbb{C}$ is continuous, the Urysohn operator $A : D \to \mathscr{C}([a, b])$ defined by*

$$Af(x) = \int_a^b k(x, y, f(y))\, dy$$

is continuous, where D is the closed ball $\bar{S}(0, r)$ in $\mathscr{C}([a, b])$ with the sup norm.

4.2.8 Lemma. *Let $[a, b]$ be a finite interval, and let $k : [a, b] \times [a, b] \to \mathbb{C}$ and $\psi : [a, b] \times \mathbb{C} \to \mathbb{C}$ be continuous. Assume that there are real numbers $p \geqslant 1$ and α, β such that*

$$|\psi[x, z]| \leqslant \alpha + \beta |z|^p \qquad (x \in [a, b], z \in \mathbb{C}).$$

Then the Hammerstein operator A where

$$Af(x) = \int_a^b k(x, y)\, \psi[y, f(y)]\, dy$$

is a continuous operator mapping $\mathscr{L}_p(a, b)$ into itself.

The concept of the inverse, although less useful than in the linear case, will also occasionally be needed.

4.2.9 Definition. Let $\mathscr{B}$ and $\mathscr{C}$ be Banach spaces, and let D be a subset of $\mathscr{B}$. Assume that $A : D \to \mathscr{C}$ is injective. The **inverse** $A^{-1} : R(A) \to D$ of A is the operator which maps each $g \in R(A)$ to the preimage f of g (that is to the unique solution of $Af = g$).

A is said to be a **homeomorphism** of $D_0 \subset D$ onto $R_0 \subset R(A)$ iff $A : D_0 \to R_0$ is a bijection and $A : D_0 \to R_0$ and $A^{-1} : R_0 \to D_0$ are continuous.

It is clear then that each of the above examples (under reasonable conditions on the given functions) may be formulated as a search for the fixed points of a continuous operator $A : D \to \mathscr{B}$, where D is some subset of a Banach space $\mathscr{B}$, a convenient choice of which will often be $\mathscr{C}(\Omega)$. The simplest of the fixed point theorems is the subject of the next section.

4.3 The Contraction Mapping Principle

Perhaps the most obvious technique for the approximate computation of a root of a polynomial is the successive substitution method; it is almost certainly the oldest with a history stretching back over two thousand years. Application of this technique to an essentially infinite dimensional situation appears to have been made first by Liouville, who used it for initial value problems for ordinary differential equations. The earliest abstract result in this area is due to Banach in 1922, and is variously known as the Banach Fixed Point Theorem and the Contraction Mapping Principle. As the result is constructive, it has been found very useful in applications, and we shall start our account of the nonlinear theory with a fairly detailed examination of the procedure together with some of its extensions and applications.

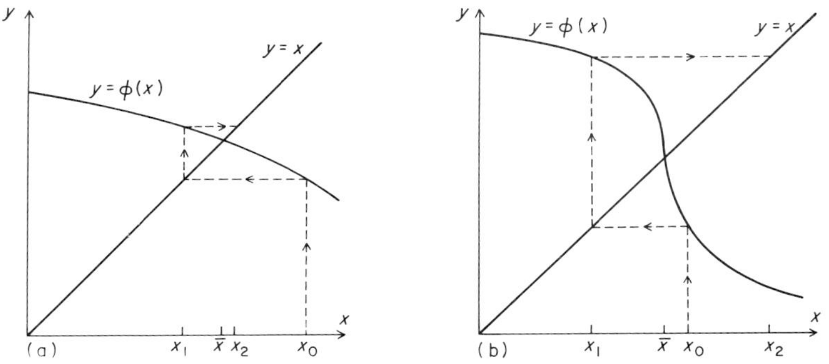

Fig. 4.1

Consider first a one dimensional example with $D = [0, 1]$ and $\phi : D \to \mathbb{R}$ continuous. An initial remark (obvious from inspection of a graph) is that if ϕ maps D into itself then certainly A has a fixed point in D, whereas if this condition is violated a fixed point may or may not exist. Whether or not this condition is enough to ensure the existence of a fixed point in higher dimensions is far from obvious. We shall return to this question in Chapter 8, but here we shall pursue a rather different line of enquiry, and assuming that $\phi(D) \subset D$ ask how a fixed point could be constructed by successive substitution. Define a sequence (x_n) with $x_0 \in D$ and $x_{n+1} = \phi(x_n)$ for $n \geqslant 0$. Then if the graph of ϕ is as in Fig. 4.1(a) the sequence converges to a fixed point for any initial guess, whereas in Fig. 4.1(b) the sequence does not converge (unless x_0 is itself the fixed point). In the first case we observe that $|\phi'(x)| < 1$ on D, and further graphical experiment suggests that this condition is enough to ensure convergence. Actually it is not hard to guess that it is unnecessary for ϕ to be differentiable, so long as its rate of growth is limited. A condition slightly weaker than differentiability is that there should be a number $q < 1$ such that

$$|\phi(x) - \phi(y)| \leqslant q|x - y| \qquad (x, y \in D); \tag{4.3.1}$$

for obvious reasons such a function is called a "contraction". Of course if ϕ is differentiable a legitimate choice of q is

$$q = \sup_{x \in D} |\phi'(x)|. \tag{4.3.2}$$

We are thus led to formulate the following "Contraction Mapping Principle", which it should be noted is meaningful in a general Banach space: *If D is closed and $\phi : D \to D$ is a contraction, then ϕ has a unique fixed point $\bar{x}$ in D, and $\bar{x} = \lim x_n$.*

In assessing the prospects for the validity of this principle in an arbitrary Banach space, an examination of an analogous linear problem suggests that the analytical difficulties will not be insurmountable. For take $D = \mathscr{B}$ and suppose $Af = Lf + h$ for $f \in \mathscr{B}$, where $h \in \mathscr{B}$ and $L \in \mathscr{L}(\mathscr{B})$. Then

$$\|Af - Ag\| = \|Lf - Lg\| \leqslant \|L\| \|f - g\| \qquad (f, g \in D),$$

and if $\|L\| < 1$, the analogue of (4.3.1) is satisfied. On the other hand it is known (Corollary 3.6.2) that if $\|L\| < 1$ the equation $f = Lf + h\ (= Af)$ has a unique solution which is given by successive substitution. This is just the conclusion above.

The Contraction Mapping Principle is thus supported by heuristic arguments. The rigorous analysis which follows establishes its validity conclusively. We start with some definitions the last of which generalizes (4.3.1). Throughout D will be a subset of the Banach space $\mathscr{B}$ and A maps D into $\mathscr{B}$.

4.3.1 Definition. A is said to satisfy a **Lipschitz condition** on D with **Lipschitz constant** q iff there is a $q < \infty$ such that

$$\|Af - Ag\| \leqslant q\|f - g\| \qquad (f, g \in D).$$

In one dimension a function which satisfies a Lipschitz condition is absolutely continuous, and hence differentiable almost everywhere. It is convenient to have the following terminology available when D is unbounded.

4.3.2 Definition. A will be said to satisfy a **local Lipschitz condition** iff for each bounded $S \subset D$, A satisfies a Lipschitz condition on S with Lipschitz constant q_S (which may depend on S).

4.3.3 Definition. A will be called a **contraction** iff it satisfies a Lipschitz condition with Lipschitz constant $q < 1$.

4.3.4 The Contraction Mapping Principle. *Suppose that A maps the closed subset D of the Banach space $\mathscr{B}$ into D and is a contraction. Then A has exactly one fixed point, $\bar{f}$ say, in D. Further, for any initial guess $f_0 \in D$, the successive approximations $f_{n+1} = Af_n$ $(n \geqslant 0)$ converge to $\bar{f}$, and the following estimate for the convergence rate holds:*

$$\|\bar{f} - f_n\| \leqslant q^n(1 - q)^{-1}\|Af_0 - f_0\| \tag{4.3.3}$$

Proof. Since A is a contraction,

$$\|f_n - f_{n+1}\| = \|Af_{n-1} - Af_n\| \leqslant q\|f_{n-1} - f_n\|,$$

and it follows from Lemma 1.4.3 that for $n > m$,

$$\|f_m - f_n\| \leqslant q^n(1 - q)^{-1}\|Af_0 - f_0\|.$$

This proves that (f_n) is Cauchy. Since D is closed and (f_n) lies in D, (f_n) converges to some $\bar{f}$ say in D. As A is continuous, $A\bar{f} = \lim Af_n = \lim f_{n+1} = \bar{f}$. That is $\bar{f}$ is a fixed point. To prove uniqueness, suppose that $\bar{g}$ is another fixed point. Then

$$\|\bar{f} - \bar{g}\| = \|A\bar{f} - A\bar{g}\| \leqslant q\|\bar{f} - \bar{g}\|,$$

and since $q < 1$, it follows that $\bar{f} = \bar{g}$. □

4.3.5 Corollary. *Let f_0 be a given point in $\mathscr{B}$. Suppose that A is a contraction with Lipschitz constant q on $\bar{S}(f_0, r)$ where*

$$r \geqslant (1 - q)^{-1}\|Af_0 - f_0\|.$$

Then A has a unique fixed point $\bar{f}$ in $\bar{S}(f_0, r)$, and $\bar{f}$ is the limit of the sequence (f_n) in the theorem. Further, the estimate (4.3.3) for the convergence rate holds.

Proof. For $f \in \bar{S}(f_0, r)$,

$$\|Af - f_0\| \leqslant \|Af - Af_0\| + \|Af_0 - f_0\| \leqslant q\|f - f_0\| + (1 - q)r.$$

Since $\|f - f_0\| \leqslant r$, this proves that $A(\bar{S}(f_0, r)) \subset \bar{S}(f_0, r)$, and the result follows from the Contraction Mapping Principle. □

This special case of the Contraction Mapping Principle is useful since the most convenient choice of D is usually a closed ball. The next result, which may be compared with Theorem 3.6.3 for the linear case, shows that fixed points are stable under continuous perturbations of the operator.

4.3.6 Theorem. *Let $\mathscr{B}$ and $\mathscr{C}$ be Banach spaces. Take closed $D \subset \mathscr{B}$, and an arbitrary $E \subset \mathscr{C}$, and assume that $A: D \times E \to D$ is continuous. Suppose that there is a $q < 1$ such that $A(\cdot, g)$ is a contraction with Lipschitz constant q for each $g \in E$. For given $g \in E$ let $\bar{f}(g)$ be the unique fixed point of $A(\cdot, g)$. Then $\bar{f}(\cdot)$ is continuous; that is $\lim_{g \to g_0} \bar{f}(g) = \bar{f}(g_0)$ for any $g_0 \in E$.*

Proof. The existence and uniqueness of $\bar{f}$ is of course a consequence of the Contraction Mapping Principle. To prove continuity, let g_0 be any point in E. Then

$$\begin{aligned}\|\bar{f}(g) - \bar{f}(g_0)\| &= \|A(\bar{f}(g), g) - A(\bar{f}(g_0), g_0)\| \\ &\leqslant \|A(\bar{f}(g), g) - A(\bar{f}(g_0), g)\| + \|A(\bar{f}(g_0), g) - A(\bar{f}(g_0), g_0)\| \\ &\leqslant q\|\bar{f}(g) - \bar{f}(g_0)\| + \|A(\bar{f}(g_0), g) - A(\bar{f}(g_0), g_0)\|.\end{aligned}$$

Hence

$$\|\bar{f}(g) - \bar{f}(g_0)\| \leqslant (1 - q)^{-1}\|A(\bar{f}(g_0), g) - A(\bar{f}(g_0), g_0)\|.$$

By the assumed continuity of A, the right-hand side tends to zero as $g \to g_0$, and therefore $\bar{f}(g) \to \bar{f}(g_0)$. □

In applications of the Contraction Mapping Principle, it is often natural to define the operator on the whole of $\mathscr{B}$ and then to look for an appropriate subset D of $\mathscr{B}$ on which to apply the theorem. The following elementary example emphasizes the importance of the choice of D.

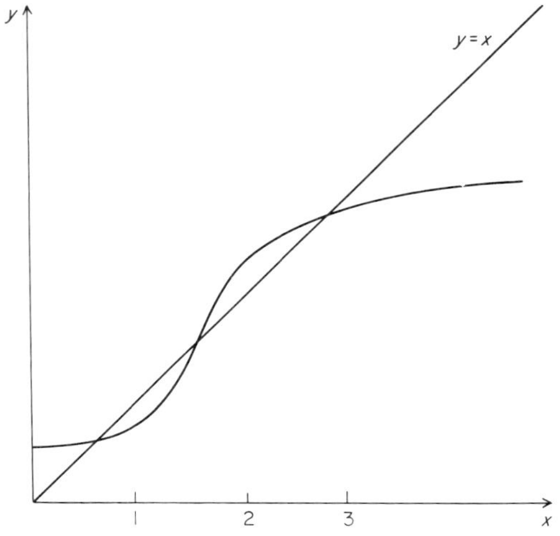

Fig. 4.2

4.3.7 Example. Consider the function $\phi : \mathbb{R} \to \mathbb{R}$ in Fig. 4.2. Obviously ϕ maps $D = [0, 1]$ into itself and is a contraction on D. We can conclude that A has a unique fixed point in D. However, there are other fixed points in $\mathbb{R}$—uniqueness is only guaranteed in D.

On the other hand, if D is chosen to be say $[0, 2]$, there are two fixed points in D, and certainly the conditions of the theorem cannot be satisfied, for otherwise uniqueness would be violated.

Lastly note that the successive substitution method may work even if the conditions of the Contraction Mapping Principle are not satisfied. For example with $D = [0, 3]$ convergence of the iterates to the smallest and largest fixed points in D are obtained for $x_0 = 0$, $x_0 = 3$ respectively.

When attempting to use the Contraction Mapping Principle the extremely restrictive nature of the contraction condition is often only too obvious. It is therefore important to note that there are a number of devices available for increasing the scope of the principle. Essentially there are three main possibilities. The first, illustrated in Problems 4.8–11, is based on rearranging the equation. Second, an equivalent norm may be used, see Problem 4.14. Lastly, it can happen that A^n is a contraction although A itself is not; this is a useful observation which motivates Theorem 4.3.10 below.

For a typical application of the Contraction Mapping Principle consider the initial value problem for a system of ordinary differential equations:

$$\left.\begin{aligned} f'(t) &= \psi[t, f(t)] \qquad (t \geqslant 0), \\ f(0) &= \alpha, \end{aligned}\right\} \tag{4.3.4}$$

a solution $f : \mathbb{R} \to \mathbb{C}^n$ with continuous first derivative being required. The system may be rewritten as $f = Af$ where

$$Af(t) = \alpha + \int_0^t \psi[s, f(s)]\, ds. \tag{4.3.5}$$

The choice of suitable conditions on ψ is influenced by the following examples.

4.3.8 Example. In one dimension the equation

$$f'(t) = 2[f(t)]^{\frac{1}{2}} \qquad (t \geqslant 0)$$

with initial condition $f(0) = 0$ has a solution $f(t) = 0$ $(0 \leqslant t \leqslant a)$, $f(t) = (t - a)^2$ $(t > a)$ for every $a \geqslant 0$. Thus on any interval $[0, t_0]$ there are an infinity of solutions.

4.3.9 Example. In one dimension consider the equation

$$f'(t) = 1 + [f(t)]^2 \qquad (t \geqslant 0),$$

again with $f(0) = 0$. On $[0, \frac{1}{2}\pi)$ this has the solution $f(t) = \tan t$, but $\tan t$ becomes infinite at $\frac{1}{2}\pi$, and there is no smooth solution on $[0, t_0]$ if $t_0 \geqslant \frac{1}{2}\pi$.

It is obvious that there will be difficulties in applying the Contraction Mapping Principle in examples of this type. In the first, because of the lack of uniqueness, the principle cannot be expected to work in $\mathscr{C}([0, t_0])$ for any t_0 whatsoever. In the second, global existence (that is for all $t_0 > 0$) fails, and at best it may be possible to establish a local result for some $t_0 < \frac{1}{2}\pi$. Thus it is clear that the continuity of ψ is not enough for either global existence or uniqueness. In the following a limitation on the rate of growth of ψ rules out both the above unpleasant possibilities.

Assume that $\psi : [0, \infty) \times \mathbb{C}^n \to \mathbb{C}^n$ is continuous, and impose in addition the Lipschitz condition that there exist $t_0 > 0$ and m such that

$$|\psi(t, z_1) - \psi(t, z_2)| \leqslant m|z_1 - z_2| \qquad (0 \leqslant t \leqslant t_0, z_1, z_2 \in \mathbb{C}^n).$$

For choice of space take $\mathscr{C}([0, t_0], \mathbb{C}^n)$ with the sup norm. Then by the Mean Value Theorem for integrals $\|Af - Ag\| \leqslant mt_0\|f - g\|$. If $mt_0 < 1$, the crucial contraction condition is satisfied, and existence readily follows. If $mt_0 \geqslant 1$, the argument may be modified to hold on $\mathscr{C}([0, t_1], \mathbb{C}^n)$ for some (smaller) t_1. However, this is an essentially local result. In fact a much stronger conclusion is valid.

A hint as to how we might proceed comes from the observation that if $\psi[t, z]$ is linear in z, A is a linear Volterra operator, and the iterates A^n are more tractable than A itself (Problem 3.24). The following extension of the Contraction Mapping Principle is designed to take advantage of similar behaviour in the nonlinear case.

4.3.10 Theorem (Caccioppoli). *Let f_0 be a given point in the Banach space $\mathscr{B}$, and set $D = S(f_0, r)$. Assume that $A : \mathscr{B} \to \mathscr{B}$, and define $q_0 = 1$,*

$$q_n = \sup_{\substack{f, g \in D \\ f \neq g}} \frac{\|A^n f - A^n g\|}{\|f - g\|} \qquad (n \geqslant 1).$$

Suppose the series $\sum_0^\infty q_n$ is convergent and let ρ be its sum. Then if $r \geqslant \rho\|Af_0 - f_0\|$, A has a unique fixed point $\bar{f}$ in D, and $\bar{f} = \lim f_n$ where $f_n = Af_{n-1}$ for $n \geqslant 1$. Further,

$$\|f_n - \bar{f}\| \leqslant \left(\rho - \sum_0^{n-1} q_k\right)\|Af_0 - f_0\|.$$

Proof. We have

$$\|f_{n+1} - f_n\| = \|A^n f_1 - A^n f_0\| \leqslant q_n\|f_1 - f_0\| = q_n\|Af_0 - f_0\|.$$

A straightforward calculation then shows that for $n > m$,

$$\|f_n - f_m\| \leqslant \|f_1 - f_0\| \sum_m^{n-1} q_k. \tag{4.3.6}$$

Since $\sum_0^\infty q_k$ is convergent, (f_n) is Cauchy, and hence convergent with limit $\bar{f}$, say. Setting $m = 0$ in (4.3.6) we obtain

$$\|f_n - f_0\| \leqslant \|f_1 - f_0\| \sum_0^{n-1} q_k \leqslant \|f_1 - f_0\|\rho \leqslant r.$$

Hence (f_n) lies in $\bar{S}(f_0, r)$, and since this set is closed, $\bar{f} \in \bar{S}(f_0, r)$. Since A is continuous $\bar{f}$ is a fixed point. The convergence estimate follows from (4.3.6) on letting $n \to \infty$. Finally, for uniqueness, if $\bar{f}, \bar{g}$ are fixed points of A,

$$\|\bar{f} - \bar{g}\| = \|A^n\bar{f} - A^n\bar{g}\| \leqslant q_n\|\bar{f} - \bar{g}\|,$$

and since $q_n \to 0$ as $n \to \infty, \bar{f} = \bar{g}$. □

4.3.11 Theorem (Picard). *Suppose* $\psi : [0, t_0] \times \mathbb{C}^n \to \mathbb{C}^n$ *is continuous, and assume that the following Lipschitz condition holds:*

$$|\psi(t, z_1) - \psi(t, z_2)| \leqslant m|z_1 - z_2| \qquad (0 \leqslant t \leqslant t_0, z_1, z_2 \in \mathbb{C}^n). \qquad (4.3.7)$$

Then the system (4.3.4) has exactly one solution $\bar{f}$ *in* $\mathbb{C}^1([0, t_0], \mathbb{C}^n)$. $\bar{f}$ *is the limit in sup norm of the sequence* (f_n) *where* $f_0 = \alpha$ *and*

$$f_n(t) = \alpha + \int_0^t \psi[s, f_{n-1}(a)]\,\mathrm{d}s \qquad (n \geqslant 1).$$

Further

$$\|\bar{f} - \alpha\| \leqslant \mathrm{e}^{mt_0} \sup_{0 \leqslant t \leqslant t_0} \left| \int_0^t \psi[s, \alpha]\,\mathrm{d}s \right|.$$

Proof. Equip $\mathscr{C}([0, t_0], \mathbb{C}^n)$ with the sup norm, and define the operator A from this space into itself by (4.3.5). Obviously the result will follow by differentiation if it can be shown that A has a unique fixed point in $\mathscr{C}([0, t_0], \mathbb{C}^n)$. We apply the previous theorem, an estimate for q_n being obtained from the inequality

$$|A^n f(t) - A^n g(t)| \leqslant \frac{(mt)^n}{n!}\|f - g\| \qquad (0 \leqslant t \leqslant t_0) \qquad (4.3.8)$$

established as follows. The result is obvious for $n = 1$. If it is valid for some $n \geqslant 1$, then

$$|A^{n+1}f(t) - A^{n+1}g(t)| \leqslant \int_0^t |\psi[s, A^n f(s)] - \psi[s, A^n g(s)]|\,\mathrm{d}s,$$

$$\leqslant m\int_0^t |A^n f(s) - A^n g(s)|\,\mathrm{d}s \qquad \text{(from (4.3.7))}$$

$$\leqslant m\|f - g\| \int_0^t \frac{(ms)^n}{n!}\,\mathrm{d}s \qquad \text{(by hypothesis)}$$

$$= \frac{(mt)^{n+1}}{(n+1)!}\|f - g\|.$$

(4.3.8) follows by induction. Hence $q_n \leqslant (mt_0)^n/n!$, and

$$\rho \leqslant \sum_0^\infty \frac{(mt_0)^n}{n!} = e^{mt_0}.$$

This proves the theorem, the estimate for the solution being obtained on choosing $r = \rho\|Af_0 - f_0\|$. □

It is clear that for the global existence and uniqueness just proved to be valid a very strong condition must be imposed on ψ. There is naturally much interest in investigating the equation under weaker restrictions on ψ. No attempt will be made here to review the extensive literature on this topic (see Flett (1979)), but two standard types of result may be noted. In the first (Problem 4.15) a local Lipschitz condition holds as in Example 4.3.9, and local existence may be proved by use of the Contraction Mapping Principle. In the second only the continuity of ψ is assumed as in Example 4.3.8. The Contraction Mapping Principle cannot then be used, but it will be shown later (Example 8.2.6) that local existence follows from another fixed point theorem.

In summary, the Contraction Mapping Principle is the easiest of the significant results in nonlinear operator theory, and may be regarded as an analogue of the Neumann series result (Corollary 3.6.2) for bounded linear operators. Its advantages are that it is simple to apply, it guarantees uniqueness, and it is constructive with estimates available for the maximum error at each iteration. However, a price must be paid for these nice properties, for in practice the convergence is often slow, and also the conditions on the operator are rather restrictive—even if all available devices for extending the applicability of the principle are used. In Section 8.3 alternative conditions assuring the convergence of the successive substitution method will be derived.

There is a substantial body of literature on the Contraction Mapping Principle. From the point of view of applications, Rall (1969) and Krasnoselskii (1972) are recommended. The principle is applied to problems arising in the theory of nonlinear oscillations by Holtzman (1970), and this reference provides an interesting study of the scope and limitations of the method.

4.4 The Fréchet Derivative

The intuitive basis for the method of successive substitution is a simple geometrical construction for a fixed point of a function ϕ in one dimension. An advantage of this method is that no restriction on ϕ of differentiability is required. However, if ϕ is differentiable a variety of more subtle results may be established with the help of the poweful machinery of differential calculus.

It is thus natural to enquire whether an analogous approach to nonlinear operators in a Banach space is possible. Evidently the basic requirement is a viable concept of derivative. There are in fact a number of possible definitions, but the simplest and most useful is that of "Fréchet derivative". As many results in elementary calculus generalize naturally in terms of this concept, the Fréchet derivative plays an important role in operator theory. In this section the derivative is defined, and some specimen results in the associated infinite dimensional calculus proved. The Fréchet derivative finds an immediate application in the discussion of Newton's method in Section 4.5.

The definition of the Fréchet derivative is motivated by the observation that if $\phi : \mathbb{R} \to \mathbb{R}$ is smooth,

$$\phi(y + h) - \phi(y) - ah = o(|h|) \tag{4.4.1}$$

for small $|h|$, where the real number a is the derivative of ϕ at y. Geometrically this simply represents the fact that the tangent at y is an approximation to ϕ near y. If a linear operator L is defined by setting $Lx = ax$ for $x \in \mathbb{R}$, the relation (4.4.1) has an immediate formal generalization to higher dimensions:

$$\phi(y + h) - \phi(y) - Lh = o(\|h\|).$$

The Fréchet derivative is interpreted as the linear operator L. In one dimension the derivative may be thought of either as the real number a, or as an operator corresponding to multiplication by a. With the latter interpretation in mind it is natural to denote L by $\phi'(y)$ in higher dimensions also. For an operator A on a Banach space the definition to be adopted is then as follows.

4.4.1 Definition. Suppose that $\mathscr{B}$ and $\mathscr{C}$ are Banach spaces. Let D be an open subset of $\mathscr{B}$, and let A be an operator mapping D into $\mathscr{C}$. A is said to be **Fréchet differentiable** at the point $g \in D$ if there exists an operator $L \in \mathscr{L}(\mathscr{B}, \mathscr{C})$ such that

$$\lim_{\|h\| \to 0} \|A(g + h) - Ag - Lh\|/\|h\| = 0,$$

the limit being required to exist as $h \to 0$ in any manner†. The operator L, often denoted by $A'(g)$, is called the **Fréchet derivative** of A at g.

Higher Fréchet derivatives may be defined in a similar manner, see Rall (1969, p. 108), Berger (1977, p. 72), but will not be needed here.

4.4.2 Example. The Fréchet derivative of any bounded linear operator $L : \mathscr{B} \to \mathscr{C}$ at any point is L itself. This follows from the identity $\|L(g + h) - Lg - Lh\| = 0$.

† That is, given $\varepsilon > 0$ there exists a $\delta > 0$ such that for $\|h\| < \delta$,

$$\|A(g + h) - Ag - Lh\| \leqslant \varepsilon\|h\|.$$

4.4.3 Example. Let D be an open subset of $\mathbb{R}^n$, and assume that $\phi: D \to \mathbb{R}^n$ is in $\mathscr{C}^1(D, \mathbb{R}^n)$. Represent points in $\mathbb{R}^n$ by column vectors, and for $y = (y_1, \ldots, y_n)^T \in D$ set $\phi(y) = (\phi_1(y), \ldots, \phi_n(y))^T$. If h is small, by Taylor's theorem,

$$\phi(y + h) = \phi(y) + \phi'(y)h + r,$$

where h, r are column vectors, $\phi'(y)$ is the Jacobian matrix (with components $\partial\phi_i/\partial y_j(y)$), and $\|r\| = o(\|h\|)$. The Fréchet derivative is thus simply the Jacobian matrix.

4.4.4 Example. Suppose that the operator $\phi : \mathbb{R}^2 \to \mathbb{R}$ is defined by $\phi(y) = y_1^2 + y_1 y_2 + y_2^2$ where $y = (y_1, y_2)^T$. Then with $h = (h_1, h_2)^T$,

$$\phi(y + h) = \phi(y) + (2y_1 + y_2)h_1 + (y_1 + 2y_2)h_2 + r,$$

where $\|r\| = o(\|h\|)$. The Fréchet derivative of ϕ at $(y_1, y_2)^T$ is thus represented by the 1×2 matrix

$$[2y_1 + y_2, y_1 + 2y_2].$$

This emphasizes that the Fréchet derivative is an operator with domain and range respectively in the same spaces as those of A.

4.4.5 Example. For an infinite dimensional example consider the Urysohn operator A where

$$Af(x) = \int_0^1 k(x, y, f(y))\,dy.$$

Denote the partial derivative of k with respect to its last argument by $\partial k/\partial u$, and suppose that $k, \partial k/\partial u : [0, 1] \times [0, 1] \times \mathbb{C} \to \mathbb{C}$ are continuous. Let $\mathscr{C}([0, 1])$ be equipped with the sup norm. Then the Fréchet derivative of A at $g \in \mathscr{C}([0, 1])$ is the bounded linear operator $L: \mathscr{C}([0, 1]) \to \mathscr{C}([0, 1])$ defined by

$$Lh(x) = \int_0^1 \frac{\partial k}{\partial u}(x, y, g(y))\, h(y)\,dy \qquad (h \in \mathscr{C}([0, 1])).$$

To prove this note that

$$\left| k(x, y, g(y) + h(y)) - k(x, y, g(y)) - \frac{\partial k}{\partial u}(x, y, g(y))\, h(y) \right|$$

$$= \left| \int_0^1 \left\{ \frac{\partial k}{\partial u}(x, y, g(y) + th(y)) - \frac{\partial k}{\partial u}(x, y, g(y)) \right\} h(y)\,dt \right|$$

$$\leqslant \|h\| \int_0^1 \left| \frac{\partial k}{\partial u}(x, y, g(y) + th(y)) - \frac{\partial k}{\partial u}(x, y, g(y)) \right| dt.$$

The integral tends to zero uniformly in x and y as $\|h\| \to 0$ since $\partial k/\partial u$ is uniformly continuous on compact subsets of $[0, 1] \times [0, 1] \times \mathbb{C}$. Therefore,

$$\|A(g + h) - Ag - Lh\|$$

$$= \sup_{x\in[0,1]} \left| \int_0^1 \left\{ k(x, y, g(y) + h(y)) - k(x, y, g(y)) - \frac{\partial k}{\partial u}(x, y, g(y))h(y) \right\} \mathrm{d}y \right|$$

$$= o(\|h\|)$$

as $h \to 0$, and the result follows from Definition 4.4.1.

4.4.6 Example. The differential operator $A : \mathscr{C}^1([0, 1]) \to \mathscr{C}([0, 1])$ where

$$Af(x) = f'(x) + [f(x)]^2$$

evidently has Fréchet derivative $A'(g) : \mathscr{C}^1([0, 1]) \to \mathscr{C}([0, 1])$ where

$$(A'(g)h)(x) = h'(x) + 2g(x)h(x).$$

The motivation for the introduction of the Fréchet derivative is that it may be used as the basis of an infinite dimensional calculus. Two typical results are now described. The first, the proof of which must be postponed (see Problem 6.12), is a generalization of the Mean Value Theorem in which the ordinary derivative is replaced by the Fréchet derivative $A'(g)$. Of course here $A'(g)$ is an operator, and the operator norm—that is the norm of $\mathscr{L}(\mathscr{B}, \mathscr{C})$—must be used; similarly the continuity of $A'(\cdot)$ will refer to continuity in the operator norm.

4.4.7 Lemma. *Suppose that $\mathscr{B}$ and $\mathscr{C}$ are Banach spaces. Let D be a convex subset of $\mathscr{B}$ and assume that $A : D \to \mathscr{C}$ is Fréchet differentiable at every point of D. Then*

$$\|Af - Ag\| \leqslant \|f - g\| \sup_{h\in D} \|A'(h)\|.$$

In other words (cf. (4.3.2)) A satisfies a Lipschitz condition with constant

$$q = \sup_{h\in D} \|A'(h)\|.$$

This bound provides a useful method for estimating the Lipschitz constant in applications of the Contraction Mapping Principle. Evidently the smaller the maximum of $\|A'(h)\|$ on D, the more rapidly will the iterates converge to the fixed point $\bar{f}$. The following says a little more: the asymptotic rate of convergence (that is near $\bar{f}$) is determined by the Fréchet derivative at $\bar{f}$.

4.4.8 Lemma. *Let D be an open subset of the Banach space $\mathscr{B}$, and assume that $A : D \to \mathscr{B}$ has a fixed point $\bar{f}$ in D. Suppose that A is Fréchet differentiable at $\bar{f}$ with $\|A'(\bar{f})\| < 1$. Then given any ε with $0 < \varepsilon < 1 - \|A'(\bar{f})\|$, there is an open ball $S(\bar{f}, \delta)$ such that if $f_0 \in S(\bar{f}, \delta)$, the iterates $f_n = Af_{n-1} (n \geqslant 1)$ also lie in $S(\bar{f}, \delta)$, $\lim f_n = \bar{f}$, and*

$$\|f_n - \bar{f}\| \leqslant (\|A'(\bar{f})\| + \varepsilon)^n \|f_0 - \bar{f}\|.$$

Proof. Choose any ε as above. Then from the definition of the Fréchet derivative, there is a $\delta > 0$ such that for any $f \in S(\bar{f}, \delta)$,

$$\|Af - A\bar{f} - A'(\bar{f})(f - \bar{f})\| \leqslant \varepsilon\|f - \bar{f}\|.$$

Therefore

$$\begin{aligned}\|Af - \bar{f}\| &= \|Af - A\bar{f}\| \\ &\leqslant \|Af - A\bar{f} - A'(\bar{f})(f - \bar{f})\| + \|A'(\bar{f})(f - \bar{f})\| \\ &\leqslant (\|A'(\bar{f})\| + \varepsilon)\|f - \bar{f}\| \\ &\leqslant \delta.\end{aligned}$$

That is $Af \in S(\bar{f}, \delta)$. It follows by induction that if $f_0 \in S(\bar{f}, \delta)$ so does f_n for $n \geqslant 1$. The above inequality with f_n replacing f shows that

$$\|f_{n+1} - \bar{f}\| \leqslant (\|A'(\bar{f})\| + \varepsilon)\|f_n - \bar{f}\|.$$

Repeated application of this relation gives the final result, from which the convergence follows. □

Next consider operator equations of the form $A(f, g) = 0$ when for some $g = g_0$ a solution f_0 is known. If A is well-behaved, it is plausible that for g near to g_0 there will be a solution f near to f_0. In other words the solution f is a continuous function of the parameter g. In finite dimensions there are well known implicit function theorems dealing with this situation. An infinite dimensional version in which A maps a subset of $\mathscr{B} \times \mathscr{C}$ into $\mathscr{D}$ is now proved. For given $g \in \mathscr{C}$ denote the Fréchet derivative at f of A with respect to its first argument by $A_1(f, g)$; for given f, g, $A_1(f, g)$ is a linear operator mapping $\mathscr{B}$ into $\mathscr{D}$.

4.4.9 Implicit Function Theorem. *Let $\mathscr{B}, \mathscr{C}, \mathscr{D}$ be Banach spaces, and let f_0, g_0 be given points in $\mathscr{B}, \mathscr{C}$ respectively. For given $a, b > 0$ let $D \subset \mathscr{B} \times \mathscr{C}$ be the set*

$$\{[f, g]: \|f - f_0\| \leqslant a, \|g - g_0\| \leqslant b\}.$$

Suppose that $A : D \to \mathscr{D}$ satisfies the following conditions:

(i) *A is continuous;*

(ii) *$A_1(\cdot, \cdot)$ exists and is continuous in D (in the operator norm);*

(iii) $A_1(f_0, g_0)$ *has an inverse in* $\mathscr{L}(\mathscr{D}, \mathscr{B})$;
(iv) $A(f_0, g_0) = 0$.

Then there are neighbourhoods U *of* g_0 *and* V *of* f_0 *such that the equation* $A(f, g) = 0$ *has exactly one solution* $\bar{f} \in V$ *for every* $g \in U$. $\bar{f}$ *depends continuously on* g.

Proof. If in D we define

$$B(f, g) = f - [A_1(f_0, g_0)]^{-1} A(f, g),$$

it is clear that the solutions of $A(f, g) = 0$ and $f = B(f, g)$ are identical. The theorem will be proved by applying the Contraction Mapping Principle to B. Note first that since

$$B_1(f, g) = I - [A_1(f_0, g_0)]^{-1} A_1(f, g),$$

$B_1(\cdot\,, \cdot)$ is continuous in the operator norm (from (ii)). Now $B_1(f_0, g_0) = 0$, so for some $\delta > 0$ there is a $q < 1$ such that

$$\|B_1(f, g)\| \leqslant q \tag{4.4.2}$$

for $\|f - f_0\| \leqslant \delta$, $\|g - g_0\| \leqslant \delta$. It follows from Lemma 4.4.7 that $B(\cdot\,, g)$ is a contraction. It is a consequence of (i) that $B(f_0, \cdot)$ is continuous. Therefore since $B(f_0, g_0) = f_0$, there is an ε with $0 < \varepsilon \leqslant \delta$ such that

$$\|B(f_0, g) - f_0\| \leqslant (1 - q)\delta \tag{4.4.3}$$

for $\|g - g_0\| \leqslant \varepsilon$. The existence of a unique fixed point in $\bar{S}(f_0, \delta)$ follows from Corollary 4.3.5, and the continuity from Theorem 4.3.6. □

Under slightly stronger conditions more information may be obtained about the size of the neighbourhoods.

4.4.10 Corollary. *Suppose the conditions of the theorem hold. Choose* δ *such that for* $f \in \bar{S}(f_0, \delta)$,

$$\|I - [A_1(f_0, g_0)]^{-1} A_1(f, g_0)\| < \tfrac{1}{4}. \tag{4.4.4}$$

Assume that for $f \in \bar{S}(f_0, \delta)$, A_1 *satisfies the Lipschitz condition*

$$\|A_1(f, g) - A_1(f, g_0)\| \leqslant \rho \|g - g_0\|. \tag{4.4.5}$$

Set $\|[A_1(f_0, g_0)]^{-1}\| = m_0$, *and take* t_0 *such that*

$$t_0 m_0 \rho < \tfrac{1}{4}, \tag{4.4.6}$$

$$m_0 \|A(f_0, g)\| < \tfrac{1}{4}\delta \qquad (g \in \bar{S}(g_0, t_0)). \tag{4.4.7}$$

Then for any $g \in \bar{S}(g_0, t_0)$, *the equation* $A(f, g) = 0$ *has exactly one solution* $\bar{f}$ *in* $\bar{S}(f_0, \delta)$. $\bar{f}$ *depends continuously on* g.

Proof. The existence of such δ, t_0 follows from (ii) and (i) respectively. Therefore it is only necessary to verify (4.4.2) and (4.4.3). For $f \in \bar{S}(f_0, \delta)$,

$$\begin{aligned}\|B_1(f,g)\| &= \|I - [A_1(f_0,g_0)]^{-1}[A_1(f,g) - A_1(f,g_0) + A_1(f,g_0)]\| \\ &\leqslant \|I - [A_1(f_0,g_0)]^{-1}A_1(f,g_0)\| \\ &\quad + \|[A_1(f_0,g_0)]^{-1}[A_1(f,g_0) - A_1(f,g)]\| \\ &\leqslant \tfrac{1}{4} + m_0\rho t_0 \qquad \text{(by (4.4.4) and (4.4.5))} \\ &< \tfrac{1}{2} \qquad \text{(by (4.4.6)).}\end{aligned}$$

This checks (4.4.2). Also

$$\|B(f_0,g) - f_0\| = \|[A_1(f_0,g_0)]^{-1}A(f_0,g)\| < \tfrac{1}{4}\delta$$

by (4.4.7), and (4.4.3) follows. □

4.5 Newton's Method for Nonlinear Operators

A well known and very effective technique for the solution of an equation in one dimension is Newton's method. Although this has a history only relatively less venerable than the method of successive substitution, it is but recently that a higher dimensional analogue of the method has come into common usage, and its first systematic treatment seems to be due to Kantorovich as late as 1948. It is possible that this slow development may not be unconnected with conceptual difficulties in formulating an appropriate generalization. At any rate it is certainly true that the notation of operator theory suggests naturally the form which the Newton sequence should take in general, with the concept of the Fréchet derivative illuminating the geometrical analogy between the one-dimensional and infinite dimensional versions. With the help of this theory, a sound basis for the technique has been developed, and the method is now a well established tool in the solution of nonlinear differential and integral equations. In the discussion which follows, attention is concentrated on two questions of practical relevance: the speed of convergence, and the sensitivity to the initial guess.

Consider first Newton's method in one dimension. As usually stated this is a technique for solving the equation $\psi(x) = 0$ by successive linearization of ψ. A simple geometrical argument leads to the Newton sequence (x_n) where

$$x_{n+1} = x_n - [\psi'(x_n)]^{-1}\psi(x_n), \tag{4.5.1}$$

the derivative being assumed non-zero in the relevant range. In the development of the theory it is convenient to look at this sequence from a slightly different point of view. First set

$$\phi(x) = x - [\psi'(x)]^{-1}\psi(x),$$

and note that the solutions of $\psi(x) = 0$ and $x = \phi(x)$ are the same. Further, if the successive substitution method is applied to ϕ, the sequence obtained is identical with the original Newton sequence. In other words it is equivalent to study the zeros of ψ by Newton's method and the fixed points of ϕ by the method of successive substitution. The advantage of the second formulation is that use can be made of previous results.

The generalization of Newton's method to higher dimensions is suggested by the form of (4.5.1). $\psi'(x_n)$ is simply interpreted as the Fréchet derivative of ψ at x_n and $[\psi'(x_n)]^{-1}$ as its inverse. The formulation in terms of the fixed points of ϕ then also goes through exactly as before. Let then D be an open subset of the Banach space $\mathscr{B}$, and assume that $B : D \to \mathscr{B}$ is Fréchet differentiable. With $f_0 \in D$, the sequence (f_n) such that

$$f_{n+1} = f_n - [B'(f_n)]^{-1}Bf_n \qquad (n \geq 1), \tag{4.5.2}$$

is called a *Newton sequence*. The analogue of ϕ defined above is the operator A where

$$Af = f - [B'(f)]^{-1}Bf. \tag{4.5.3}$$

Obviously the Newton sequence for B and the sequence obtained by successive substitution applied to A are the same.

In one dimension it is easy to see by differentiation that if $\bar{x}$ is a zero of ψ, then $\phi'(\bar{x}) = 0$, and it follows from Lemma 4.4.8 that the rate of convergence is very rapid. This argument generalizes readily, and Theorem 4.5.2 below confirms that the great advantage of speed of convergence for Newton's method is retained in infinite dimensions, the asymptotic rate being better than that given by any power law. Further discussion of the convergence rate may be found in Flett (1979).

4.5.1 Lemma. *Let D be an open subset of the Banach space $\mathscr{B}$. For the operator $B : D \to \mathscr{B}$ assume that there is an $\bar{f} \in D$ such that $B\bar{f} = 0$. Suppose that B is Fréchet differentiable in D and that $B'(\cdot)$ is continuous (in the operator norm). Suppose also that $[B'(\bar{f})]^{-1} \in \mathscr{L}(\mathscr{B})$. Then in some neighbourhood of $\bar{f}$, $[B'(f)]^{-1} \in \mathscr{L}(\mathscr{B})$, and $[B'(\cdot)]^{-1}$ is continuous at $\bar{f}$. Further, A defined by (4.5.3) is Fréchet differentiable at $\bar{f}$ and $A'(\bar{f}) = 0$.*

Proof. The existence and continuity of the inverses are consequences of Theorem 3.6.3. Thus in some neighbourhood of $\bar{f}$,

$$\begin{aligned} Af - A\bar{f} = {} & [B'(f)]^{-1}\{B'(\bar{f})(f - \bar{f}) - Bf + B\bar{f}\} \\ & + [B'(f)]^{-1}\{B'(f) - B'(\bar{f})\}(f - \bar{f}) \end{aligned}$$

identically, whence

$$\frac{\|Af - A\bar{f}\|}{\|f - \bar{f}\|} \leqslant \|[B'(f)]^{-1}\| \left\{ \frac{\|Bf - B\bar{f} - B'(\bar{f})(f - \bar{f})\|}{\|f - \bar{f}\|} + \|B'(f) - B'(\bar{f})\| \right\}.$$

It follows that $\|Af - A\bar{f}\|/\|f - \bar{f}\| \to 0$ as $f \to \bar{f}$, which proves that $A'(\bar{f}) = 0$. □

4.5.2 Theorem. *Let B be as in the previous lemma. Then given any* ε *with* $0 < \varepsilon < 1$, *there is a ball* $S(\bar{f}, \delta)$ *such that*

(i) *If* $f_0 \in S(\bar{f}, \delta)$, *the Newton sequence defined by* (4.5.2) *lies in* $S(\bar{f}, \delta)$, *and* $\lim f_n = \bar{f}$.
(ii) $\|f_n - \bar{f}\| \leqslant \varepsilon^n \|f_0 - \bar{f}\|$.

Proof. $A'(\bar{f}) = 0$ by the previous lemma. The result follows from Lemma 4.4.8. □

This theorem shows that, in contrast with the successive substitution method, no upper bound is needed on the rate of growth of the operator for Newton's method to converge—so long as a sufficiently good initial guess f_0 can be found. Unfortunately the method is very sensitive to the choice of f_0. Since the theorem does not provide a criterion for assessing this choice, it is of somewhat limited use. The following result due to Kantorovich, which supplies such a criterion and in addition proves existence, is of much greater practical significance.

4.5.3 Theorem. *Let* $\mathscr{B}$ *be a Banach space. Given* $r > 0$ *and* $f_0 \in \mathscr{B}$, *suppose that* $B : \bar{S}(f_0, r) \to \mathscr{B}$ *is Fréchet differentiable, and assume that* $[B'(f_0)]^{-1} \in \mathscr{L}(\mathscr{B})$. *Suppose also that B satisfies the following Lipschitz condition in* $\bar{S}(f_0, r)$:

$$\|B'(f) - B'(g)\| \leqslant p\|f - g\|.$$

Assume that

$$b_0 \geqslant \|B'(f_0)\|^{-1}, \eta_0 \geqslant \|f_1 - f_0\| = \|[B'(f_0)]^{-1} B f_0\|,$$

and set $h_0 = b_0 p \eta_0$. *If*

$$h_0 \leqslant \tfrac{1}{2}, \qquad [1 - (1 - 2h_0)^{\frac{1}{2}}] h_0^{-1} \eta_0 \leqslant r,$$

the Newton sequence with initial guess f_0 *converges to a solution* $\bar{f}$ *of* $Bf = 0$ *in* $\bar{S}(f_0, r)$.

In order to motivate the statement of the theorem, consider $B = \psi : \mathbb{R} \to \mathbb{R}$ and assume for simplicity that ψ has continuous second derivative and $\psi(x_0) > 0$. Then the Lipschitz condition is satisfied with $p = \sup|\psi''(x)|$ in

$\bar{S}(x_0, r)$. Define the quadratics

$$Q_{\pm}(x) = \pm\tfrac{1}{2}p(x - x_0)^2 + \psi'(x_0)(x - x_0) + \psi(x_0).$$

Then $Q_-(x) \leqslant \psi(x) \leqslant Q_+(x)$. Thus $\psi(x) = 0$ will have a solution if Q_+ (and so necessarily Q_-) has real roots, and a necessary and sufficient condition for this is

$$[\psi'(x_0)]^2 \geqslant 2p\psi(x_0),$$

which is just the condition $h_0 \leqslant \frac{1}{2}$ of the theorem. Further, the smallest root, x_1 say, satisfies

$$x_1 - x_0 = [1 - (1 - 2h_0)^{\frac{1}{2}}]h_0^{-1}\eta_0 = r_0,$$

say. If therefore $r \geqslant r_0$, x_1 lies in $\bar{S}(x_0, r)$. This yields the last condition of the theorem. Finally it is clear that the sequence (x_n) then lies in $\bar{S}(x_0, r)$.

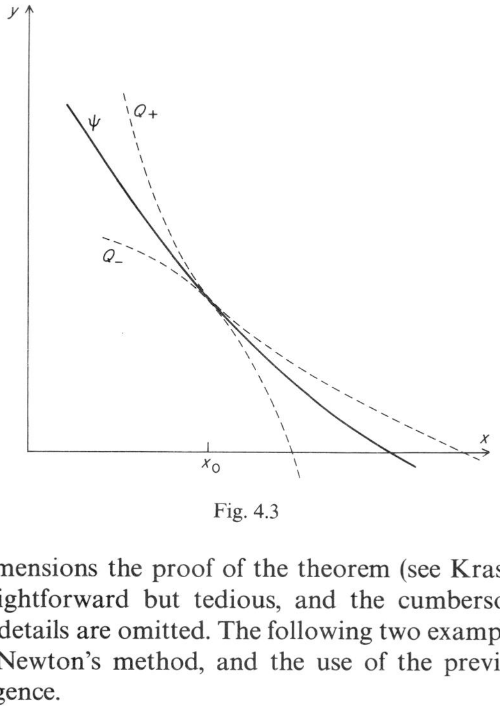

Fig. 4.3

In infinite dimensions the proof of the theorem (see Krasnoselskii (1972, p. 142)) is straightforward but tedious, and the cumbersome and rather unenlightening details are omitted. The following two examples illustrate the application of Newton's method, and the use of the previous theorem in proving convergence.

4.5.4 Example. Newton's method with initial guess $x_1 = x_2 = 0$ will be applied to the following pair of simultaneous equations:

$$x_1^2 + x_2^2 + 30x_2 - 31 = 0,$$

$$x_2^2 - 2x_2 - 16x_1 + 1 = 0.$$

If points in $\mathbb{R}^2$ are represented by column vectors, the equations are $\psi(x) = 0$, where $x = (x_1, x_2)^T$ and

$$\psi(x) = \begin{bmatrix} x_1^2 + x_2^2 + 30x_2 - 31 \\ x_2^2 - 2x_2 - 16x_1 + 1 \end{bmatrix}.$$

The l_∞ norm $\|\cdot\|$, where $\|x\| = \max(|x_1|, |x_2|)$ will be used.

In this simple case explicit calculation of the various quantities involved is possible. Thus

$$\psi'(x) = 2\begin{bmatrix} x_1 & x_2 + 15 \\ -8 & x_2 - 1 \end{bmatrix},$$

$$[\psi'(x)]^{-1} = \frac{1}{2[x_1(x_2 - 1) + 8(x_2 + 15)]}\begin{bmatrix} x_2 - 1 & -x_2 - 15 \\ 8 & x_1 \end{bmatrix}$$

and the Newton sequence $(x^{(n)})$ may readily be written down; we find for example that $x^{(1)} = (-2, 31)^T/30$. Let us apply Theorem 4.5.3. From the expression for the operator norm given in Example 3.4.5,

$$\|[\psi'(x^{(0)})]^{-1}\| \leqslant \frac{1}{15 \times 16}\max[(1 + 15), 8] = \frac{1}{15} = b_0$$

say. Also

$$\psi'(x) - \psi'(y) = 2\begin{bmatrix} x_1 - y_1 & x_2 - y_2 \\ 0 & x_2 - y_2 \end{bmatrix},$$

whence $\|\psi'(x) - \psi'(y)\| \leqslant 4\|x - y\|$. Therefore we may take $p = 4$. Finally, $\eta_0 = \|x^{(1)} - x^{(0)}\| = 31/30$. Thus $h_0 = 62/225 \leqslant \frac{1}{2}$ as required. Also

$$[1 - (1 - 2h_0)^{\frac{1}{2}}]h_0^{-1}\eta_0 = \tfrac{1}{4}[15 - 101^{\frac{1}{2}}] < 1{\cdot}25.$$

It follows from the theorem that the equations have a solution in the ball centre 0 and radius 1·25 to which the Newton sequence converges. This solution is obviously $x_1 = 0, x_2 = 1$.

A similar computation with $(0, -1)^T$ as initial guess shows that the condition $h_0 \leqslant \frac{1}{2}$ of the theorem is not satisfied. However, it may be checked independently (because of the simplicity of the example) that the sequence still converges. This emphasizes that the conditions of the theorem are certainly not necessary for the convergence of Newton's method.

4.5.5 Example. Newton's method is perhaps the most rapidly convergent technique for computing the solution of boundary value problems. As an example consider the system

$$f''(x) - [f(x)]^2 = g(x),$$

$$f(0) = f(1) = 0,$$

where g is a given continuous function, a solution in $\mathscr{C}^2([0, 1])$ being required. If $\mathscr{C}([0, 1])$ with the sup norm is used, this system is equivalent to the equation $Bf = 0$ where

$$Bf(x) = f(x) + \int_0^1 k(x, y)\{[f(y)]^2 + g(y)\}\, dy$$

and k is the Green's function (4.2.5). Thus Newton's method may be used on the operator B.

First consider the application of Theorem 4.5.3 to prove convergence, and suppose for simplicity that the initial guess $f_0 = 0$ is chosen. From Example 4.4.5,

$$B'(f)h(x) = h(x) + 2\int_0^1 k(x, y)\, f(y)\, h(y)\, dy.$$

Since $B'(f_0) = I$, there is no difficulty with inverting the Fréchet derivative, and b_0 may be chosen to be unity. For the linear integral operator K, where

$$Kh(x) = \int_0^1 k(x, y)\, h(y)\, dy,$$

it is easy to check using the relation $\|K\| = |||k|||_\infty$ in Example 3.4.8 that $\|K\| = \frac{1}{8}$. Hence choose $\eta_0 = \|g\|/8$. Also

$$\|B'(f_1)h - B'(f_2)h\| \leqslant 2\|h\|\,\|K\|\,\|f_1 - f_2\| \leqslant \|h\|\,\|f_1 - f_2\|/4,$$

and we may take $p = \frac{1}{4}$. Therefore $h_0 = b_0 p \eta_0 = \|g\|/32$. Thus if $\|g\| \leqslant 16$, by the theorem the equation has a solution to which the Newton sequence converges. Further, this solution lies in a ball centre 0 and radius $4[1 - (1 - \|g\|/16)^{\frac{1}{2}}]$.

In order to compute the solution by applying Newton's method directly to B, it would be necessary at each stage to solve a linear integral equation in order to obtain $[B'(f_n)]^{-1}Bf_n$, a tedious procedure in practice. However, differentiation shows that it is equivalent to solve the sequence of linear boundary value problems

$$f''_{n+1}(x) - 2f_n(x)f_{n+1}(x) = g(x) - [f_n(x)]^2,$$
$$f_{n+1}(0) = f_{n+1}(1) = 0.$$

This alternative is usually preferred in practice since efficient numerical techniques are available for such systems. In this form Newton's method is often known as quasilinearization—see Bellman and Kalaba (1965).

In a general assessment of Newton's method as a computational technique there are two difficulties which must be considered. The most inconvenient is the sensitivity to the choice of an initial guess. This is apparent even in finite dimensions when solving large systems of nonlinear algebraic equations, for the sequence often diverges unless f_0 is rather close to a solution, and other methods may have to be used to provide an adequate choice of f_0. Therefore results such as Theorem 4.5.3 giving global convergence criteria are of considerable practical interest. Unfortunately, as in Example 4.5.4, this theorem can be unduly pessimistic in its predictions, so it is interesting to note that as with the method of successive substitution (see Section 8.3) less exacting criteria can sometimes be obtained by the use of monotonicity arguments. In the infinite dimensional case this approach is discussed by Vandergraft (1967), and Mooney and Roach (1976).

The second difficulty arises in the evaluation of the Newton sequence itself. At each iteration the Fréchet derivative must be computed and a linear equation involving this must be solved. In practice these calculations sometimes turn out to be excessively time consuming, and a number of modified Newton methods have been devised to circumvent this difficulty. In one of the simplest of these the sequence defined by

$$f_{n+1} = f_n - [B'(f_0)]^{-1} B f_n$$

is used, and only one calculation of the Fréchet derivative is needed. Unfortunately some of the advantage of speed of convergence is lost. For a discussion of this group of methods Dennis (1971) may be consulted.

Nonetheless, although there are difficulties associated with the use of Newton's method, its speed of convergence is a factor which weighs heavily in its favour, and the method is much used as a computational technique. In both finite and infinite dimensions there remain interesting problems connected with the amelioration of the difficulties without undue sacrifice of efficiency.

Problems

4.1 Take $\omega > -\pi^2$, and calculate the Green's function for $(-\mathrm{d}^2/\mathrm{d}x^2 + \omega)$ on $[0, 1]$ with homogeneous Dirichlet conditions $f(0) = f(1) = 0$.

4.2 Discuss the role played by the real parameter a in the solution of the Hammerstein integral equation

$$f(x) - \int_0^1 [f(y)]^2 \, \mathrm{d}y = a \qquad (0 \leqslant x \leqslant 1).$$

4.3 Let D be the open unit ball in the Banach space $\mathscr{B}$, and assume that $A : D \to \mathscr{B}$ is continuous. If $f \in D$ show that A is bounded in the sense of (4.2.10) on some neighbourhood of f.

4.4 The operator $A : l_2 \to l_2$ is defined by

$$A(f_1, f_2, f_3, \ldots) = (f_1, f_2^2, f_3^3, \ldots).$$

Show that A is continuous but is not bounded on any ball $\bar{S}(0, r)$ if $r > 1$.

4.5 Let $\mathscr{B}$ be a Banach space. Suppose that $g \in \mathscr{B}$ and $L \in \mathscr{L}(\mathscr{B})$ are given, and consider the operator A where $Af = Lf + g$ for $f \in \mathscr{B}$. Assume that A maps some closed ball into itself.

(i) Prove that $\|L\| \leqslant 1$.
(ii) Let $\mathscr{B} = c_0$—the sup normed space of sequences which tend to zero at infinity. Show that if $g = (1, 0, 0, \ldots)$ and L is the right shift (Problem 3.2), A maps $\bar{S}(0, 1)$ into itself but has no fixed point in this ball.
(iii) Prove that A has a fixed point in the ball if $\mathscr{B}$ is a Hilbert space.

4.6 (Kakutani). Let D be the closed unit ball in the real sequence space l_2. On D define the operator A by setting

$$A(f_1, f_2, \ldots) = ([1 - \|f\|^2]^{\frac{1}{2}}, f_1, f_2, \ldots).$$

Prove that A maps D into itself and is continuous but has no fixed point. Thus even in Hilbert space, a continuous mapping of the closed unit ball into itself need not have a fixed point if the space is infinite dimensional.

4.7 Show by constructing a counterexample that it is not legitimate to weaken the Lipschitz condition to $\|Af - Ag\| < \|f - g\|$ in the Contraction Mapping Principle. For a discussion see Smart (1974, Chap. 5).

4.8 To illustrate the effect of a rearrangement of the equation on the convergence of the method of successive substitution, consider the calculation of the positive root of $x^2 = 3$.

(i) First rewrite the equation as $x = 3x^{-1}$. Unless $x_0 = \sqrt{3}$ the sequence does not converge.
(ii) Next put the equation in the form $x = \frac{1}{3}(3 - x^2) + x$. With $x_0 = 1$ calculate the first few terms of the sequence. For various x_0 discuss the application of the Contraction Mapping Principle to the proof of convergence.

4.9 The positive solution of $e^x - x - 2 = 0$ is required. Rewrite the equation as $x = x - a[e^x - 2 - x]$. Discuss the effect of the choice of a on the Lipschitz constant, on the rate of convergence of the successive substitution method, and on the ball for which the function is a contraction.

4.10 Define $\phi(x) = x + (x - 1)^2$ for $x \in \mathbb{R}$. Show that ϕ is not a contraction on any interval including 1. Prove that the sequence (x_n) with $x_0 = \frac{1}{2}$, $x_{n+1} = \phi(x_n)$ $(n \geqslant 1)$ nonetheless converges to the fixed point $x = 1$ of ϕ.

4.11 The non-zero solution of the integral equation

$$f(x) = \int_0^1 [f(t)]^2 \, dt \qquad (0 \leqslant x \leqslant 1),$$

is to be calculated by the successive substitution method. Describe an appropriate rearrangement of the equation.

4.12 Let D be a subset of the Banach space $\mathscr{B}$ and consider $A : D \to D$. For given $n > 1$ assume that A^n has exactly one fixed point $\bar{f}$ in D. Show that $\bar{f}$ is the unique fixed point of A in D.

4.13 Consider the Hammerstein integral equation

$$f(x) - \int_0^1 k(x, y)\, \psi[f(y)]\, \mathrm{d}y = g(x) \qquad (0 \leqslant x \leqslant 1),$$

where k, g, h are real valued and continuous. Equip $\mathscr{C}([0, 1])$ with the sup norm. In an obvious notation the equation may be written as $f - K\Psi f = g$ where $K, \Psi : \mathscr{C}([0, 1]) \to \mathscr{C}([0, 1])$ and K is linear.

(i) Suppose ψ satisfies a Lipschitz condition with constant m on $\mathbb{R}$. Show that if $m\|K\| < 1$ existence and uniqueness follow from the Contraction Mapping Principle, and obtain a bound for the solution.
(ii) If $\psi(z) = z^2$ then (i) will not work. Discuss this case under appropriate restrictions on $\|g\|$.

4.14 For real a define

$$\|f\|_* = \sup_{0 \leqslant t \leqslant t_0} |e^{-at} f(t)|.$$

Show that $\|\cdot\|_*$ and the sup norm are equivalent norms on $\mathscr{C}([0, t_0])$. Using $\|\cdot\|_*$ deduce Theorem 4.3.11 directly from the Contraction Mapping Principle.

4.15 Theorem 4.3.11 does not apply to the equation

$$f'(t) = 1 + [f(t)]^2 \qquad (t \geqslant 0)$$

with initial condition $f(0) = 0$. Prove a local result for the interval $[0, t_0]$ choosing t_0 as large as possible.

4.16 Using the Contraction Mapping Principle, show that

$$f''(x) + \mu\psi[f(x)] = g(x),$$
$$f(0) = f(1) = 0,$$

where ψ satisfies a local Lipschitz condition and g is continuous, has a solution in $\mathscr{C}^2([0, 1])$ for sufficiently small $|\mu|$.

4.17 Let Ω be a triangle bounded by lines parallel to the axes and with base a finite subinterval of $x + y = 0$. Suppose that homogeneous Cauchy conditions are prescribed on $x + y = 0$ for the hyperbolic partial differential equation

$$\frac{\partial^2 u}{\partial x \partial y} = f\left(x, y, u, \frac{\partial u}{\partial x}, \frac{\partial u}{\partial y}\right),$$

where f is real valued and continuous and satisfies the Lipschitz condition

$$|f(x, y, u, p, q) - f(x, y, u_1, p_1, q_1)| \leqslant L(|u - u_1| + |p - p_1| + |q - q_1|)$$

for all real $x, y \in \Omega$ and u, u_1, p, p_1, q, q_1. First rewrite the equation as an integral equation, and then use the Contraction Mapping Principle to prove local existence and uniqueness—that is on a neighbourhood in Ω of the initial line.

4.18 The Poisson–Boltzmann Equation of electrolyte solution theory is (in slightly modified form)

$$f''(x) = \sinh[f(x)] \qquad (x \geqslant 0)$$

with $f(0) = \alpha$ and $f(x) \to 0$ as $x \to \infty$. After rearrangement write this as an integral equation by using a Green's function. Prove existence for α suitably restricted by using the Contraction Mapping Principle on $\mathscr{C}([0, \infty))$—with an appropriate norm.

4.19 Describe the application of Newton's method to the system

$$f''(x) - [f(x)]^3 = g(x) \qquad (0 \leqslant x \leqslant 1),$$
$$f(0) = f(1) = 0,$$

with initial guess zero, giving conditions on g (assumed continuous) based on Theorem 4.5.3 for convergence.

4.20 Suppose $k : [0, 1] \times [0, 1] \to \mathbb{C}$ is continuous, and define the operator K on $\mathscr{C}([0, 1])$, equipped with the sup norm, by setting

$$Kf(x) = \int_0^1 k(x, y) f(y) \, dy.$$

Deduce from the Implicit Function Theorem 4.4.9 that if $\lambda \in \rho(K)$, there is a $\delta > 0$ such that the integral equation

$$\lambda f(x) = \int_0^1 k(x, y) \{f(y) + [f(y)]^2\} \, dy + g(x)$$

has a solution in $\mathscr{C}([0, 1])$ for every $g \in \mathscr{C}([0, 1])$ such that $\|g\| \leqslant \delta$.

Chapter 5

Compact Sets in Banach Spaces

5.1 Introduction

The analysis of the previous two chapters shows that if the "gradient" of an operator A on a Banach space is not large, most of the principal results for operator equations involving A readily carry over from finite to infinite dimensions. Unfortunately the restriction on the gradient is not satisfied by many of the operators in applications. On the other hand such operators often have compensating properties. In order to exploit these, it will be necessary to study the structure of Banach spaces more deeply before continuing with operator theory. In this chapter we shall concentrate on a certain class of well-behaved "compact subsets", the definition of a compact subset being motivated by a useful property of real numbers.

The proofs of many of the standard results for real valued functions of a real variable depend in an essential way on the Heine–Borel Theorem. This theorem states the following: If S is a closed and bounded subset of $\mathbb{R}$, from any class of open sets whose union contains S, a finite subclass (that is a subclass containing a finite number of sets) may be selected whose union also contains S. Most inconveniently the theorem is false if an infinite dimensional Banach space is substituted for $\mathbb{R}$. Our strategy here is to take the *conclusion* of the theorem as the defining property of a compact set. Although compact sets are not as plentiful, or as easy to identify, as closed and bounded sets, it turns out that they are in reasonable supply in the most useful Banach spaces of functions. As a consequence there is a large class of operators whose range (although not domain) have certain compactness properties.

Since, by their construction, compact sets behave much like closed and bounded sets in finite dimensions, it will be possible to establish a theory for such operators similar in many respects to that in finite dimensions.

The above definition of compactness raises some immediate problems, for it does not give a conceptually simple picture of compact sets, nor does it provide a convenient test for identifying these sets. In order to clarify the situation the following tactics will be adopted: first a number of different descriptions of compact sets are given; then the role of compactness in extending some important but simple results from finite to infinite dimensions is examined; finally easy criteria are derived for recognizing compact sets in the main spaces of functions, these criteria being essential for the exploitation of compactness in operator theory.

5.2 Definitions

The first definition of compactness is motivated by the Heine–Borel Theorem.

5.2.1 Definition. A class of sets in a Banach space is said to **cover** a given set S iff each point of S lies in at least one of the sets. If the diameter of each set in a cover of S is not greater than ε, the class is called an **ε-cover** of S.

5.2.2 Definition. Let S be a subset of a Banach space. S is said to be **compact** iff every class of open sets which covers S has a finite subclass which also covers S. S is said to be **relatively compact** iff its closure $\bar{S}$ is compact†.

Thus the Heine–Borel Theorem asserts that a closed and bounded set in $\mathbb{R}$ is compact. It is easy to prove that a compact set is closed and bounded, therefore "closed and bounded = compact" in $\mathbb{R}$, and the characterization of compact sets is transparent. In a general Banach space it remains true that a necessary condition for a set to be compact is that it should be closed and bounded, but this condition is not sufficient if the space is infinite dimensional.

Let us now adopt a different point of view. Recall that closed bounded sets in $\mathbb{R}$ have the characteristic property that every sequence in such a set has a subsequence converging to a point of the set.

5.2.3 Definition. A subset S of a Banach space is said to be **sequentially compact** iff every sequence in S contains a convergent subsequence with limit in S. S is said to be **relatively sequentially compact** iff its closure $\bar{S}$ is sequentially compact.

† Unfortunately the use of the terms compact and relatively compact is not uniform in the literature, compact being sometimes used for what we have called relatively compact.

It is a very useful fact that in a Banach space (though not in certain more general spaces) compactness and sequential compactness are equivalent, see Friedman (1970, p. 108) for example. The remaining results in the following theorem summarizing the situation are easy and are left as exercises.

5.2.4 Theorem. *Let S be a subset of the Banach space $\mathscr{B}$. Then:*

(i) *S is compact (respectively relatively compact) if and only if it is sequentially compact (respectively relatively sequentially compact);*
(ii) *if S is compact, it is closed and bounded;*
(iii) *if S is closed and bounded and $\mathscr{B}$ is finite dimensional, then S is compact.*

5.2.5 Lemma. *The closed unit ball in a normed vector space $\mathscr{V}$ is compact if and only if $\mathscr{V}$ is finite dimensional.*

This lemma shows that compact sets are harder to find in an infinite dimensional space, and that in particular closed balls are not an adequate substitute for closed intervals in $\mathbb{R}$. Rather than prove the lemma (see Friedman (1970, p. 133)), we shall give an example which shows how the loss of compactness is associated with the infinite dimensionality.

5.2.6 Example. Consider the infinite dimensional sequence space ℓ_∞ (Definition 1.3.13) of vectors $f = (f_1, f_2, \ldots)$ with norm given by

$$\|f\|_\infty = \sup_i |f_i|.$$

Let $f^{(n)}$ be the vector all of whose components are zero except for the nth which is unity, and denote the ith component of $f^{(n)}$ by $f_i^{(n)}$. Evidently the set $\{f^{(n)}\}$ lies in the closed unit ball of ℓ_∞. However, if $n \neq m$,

$$\|f^{(n)} - f^{(m)}\|_\infty = \sup_i |f_i^{(n)} - f_i^{(m)}| = 1.$$

Thus the vectors $f^{(n)}$ all lie at unit distance from each other, and it is clear that no subsequence can be convergent. The $f^{(n)}$ may be regarded as unit vectors along the nth axis; the lack of compactness is a direct result of the fact that there are an infinite number of axes.

At the expense of yet another definition, further clarification can be obtained.

5.2.7 Definition. A subset S of a Banach space is called **totally bounded** iff, for each $\varepsilon > 0$, there is a finite set of open balls which form an ε-cover of S.

It is easy to see that every totally bounded set is bounded. However, total boundedness is a much stronger condition, as is shown by the following theorem.

5.2.8 Theorem. *A subset S of a Banach space $\mathscr{B}$ is relatively compact if and only if it is totally bounded.*

Proof. Suppose first that S is compact. For any $\varepsilon > 0$ take the class of open balls $S(f, \varepsilon)$ of radii ε and centres f for all $f \in S$. This class evidently covers S, and since S is compact a finite subclass which covers S may be selected. Thus S is totally bounded.

It is next shown that if S is totally bounded, it is relatively sequentially compact, and so relatively compact (Theorem 5.2.4(i)). The proof employs a useful device known as a diagonalization argument. Let (f_n) be any infinite sequence in S. As S is totally bounded, there is a finite 1-cover of S, say $\{S(g_1, \frac{1}{2}), \ldots, S(g_i, \frac{1}{2})\}$. At least one of these balls contains an infinite subsequence, $(f_{n,1})$ say, of (f_n). Take next a finite $\frac{1}{2}$-cover of S and as before extract an infinite subsequence, $(f_{n,2})$ say, of $(f_{n,1})$ contained in one of these balls. We can proceed in this way to find for each m an infinite subsequence, $(f_{n,m})$ say, of $(f_{n,m-1})$ such that $(f_{n,m})$ is contained in a ball of diameter m^{-1}. Now consider the *diagonal* sequence $(f_{n,n})$. Then $(f_{j,j})_{j=n}^{\infty}$ is a subsequence of $(f_{j,n})_{j=n}^{\infty}$, and so is contained in a ball of diameter n^{-1}. Therefore

$$\|f_{n,n} - f_{m,m}\| \leqslant 1/\min(n, m).$$

Thus $(f_{n,n})$ is Cauchy, and hence is convergent as $\mathscr{B}$ is complete. This shows that (f_n) has a convergent subsequence, and proves relative sequential compactness. □

An obvious deduction is that a closed set is totally bounded if and only if it is compact. Thus in infinite dimensions "closed and totally bounded = compact" replaces "closed and bounded = compact" in finite dimensions. One may form a rough picture of a relatively compact set S as "approximately finite dimensional" in the sense that for any $\varepsilon > 0$ there is a finite number of points $f_1, \ldots, f_n$ such that every point of S is within a distance ε of one f_j.

5.3 Some Consequences of Compactness

An approach which helps to clarify the concept of compactness is to examine how familiar results for real-valued functions on closed and bounded intervals in $\mathbb{R}$ generalize when the domain of the functions is taken to be infinite dimensional. This approach is pursued in detail in the illuminating expository article by Hewitt (1960). Three examples in which the compactness of the domain is crucial for the generalization are given here.

The first result shows that the well-known equivalence of continuity and uniform continuity on a closed and bounded interval extends to a compact set.

5.3.1 Definition. Let F be a functional with domain $S \subset \mathscr{B}$, that is F is a mapping from S to $\mathbb{C}$. F is said to be **uniformly continuous** on S iff, for each $\varepsilon > 0$ there is a $\delta > 0$ such that $|F(f) - F(g)| < \varepsilon$ for all $f, g \in S$ with $\|f - g\| \leqslant \delta$.

5.3.2 Theorem. *Let F be a continuous functional on a compact subset S of a Banach space $\mathscr{B}$. Then F is uniformly continuous on S.*

Proof. If the result is false, there are an $\varepsilon > 0$ and sequences (f_n) and (g_n) in S such that

$$|F(f_n) - F(g_n)| \geqslant \varepsilon \tag{5.3.1}$$

for all n, but $\|f_n - g_n\| < n^{-1}$. Since S is sequentially compact (Theorem 5.2.4(i)), (f_n) has a convergent subsequence, (f_{n_i}) say, with limit in S. For the same reason there is a convergent subsequence of (g_{n_i}) with limit in S. Thus renumbering the sequences we conclude that (5.3.1) holds for some convergent sequences (f_n) and (g_n), and as $\|f_n - g_n\| \to 0$ obviously the sequences have a common limit, f say, in S. But

$$|F(f_n) - F(g_n)| \leqslant |F(f_n) - F(f)| + |F(f) - F(g_n)|,$$

and as F is continuous, the right-hand side tends to zero as $n \to \infty$. This contradicts (5.3.1). □

An elementary but useful result states that every real valued continuous function on a closed bounded interval attains its maximum and minimum. Again there is an extension to compact subsets.

5.3.3 Theorem. *Let F be a real valued continuous functional on a compact subset S of the Banach space $\mathscr{B}$. Then F is bounded and attains both its supremum and infimum.*

Proof. Set $m = \sup_{f \in S} F(f)$; at this stage m may be either finite or infinite, but in either case there is a sequence (f_n) in S with $m = \lim F(f_n)$. As S is compact, there is a convergent subsequence, which will still be denoted by (f_n), with limit $f \in S$. But F is continuous, hence also $\lim F(f_n) = F(f)$, which is of course finite. Therefore m is finite, and $m = F(f)$. The proof for the infimum is similar. □

Both of the above theorems are false if "closed and bounded" is substituted for "compact". In both cases it is easy to construct counterexamples. We do this for the second, leaving the reader to tackle the first.

5.3.4 Example. Equip $\mathscr{C}([0, 1])$ with the sup norm, and let $\mathscr{M}$ be the closed subspace consisting of those functions in $\mathscr{C}([0, 1])$ vanishing at $x = 1$. Take S to be the closed unit ball in $\mathscr{M}$. The functional defined by

$$F(f) = \int_0^1 |f(x)|\, dx$$

is evidently continuous on S, and $F(f) \leqslant 1$ for $f \in S$. Choosing $f_n(x) = 1 - x^n$ for $n \geqslant 1$, we see that $\lim F(f_n) = 1$, whence $\sup_{f \in S} F(f) = 1$. However, it is clear that $F(f) = 1$ for f continuous and $\|f\| \leqslant 1$ if and only if either $f(x) = 1$ or $f(x) = -1$. Since neither of these functions lies in S, F does not attain its supremum.

This example illustrates a difficulty which commonly occurs in optimization theory. Evidently the problem of finding extrema is much more delicate in infinite than in finite dimensions.

The final example is concerned with the existence of a fixed point of an operator $A : \mathscr{B} \to \mathscr{B}$. A common heuristic argument asserts that if a sequence (f_n) of guesses can be found such that the "residuals" $(Af_n - f_n)$ tend to zero, then (f_n) will have a convergent subsequence whose limit is a fixed point of A. On the face of it this is a reasonable conclusion and is one which is certainly useful in practice. However, if the problem is infinite dimensional there is no reason why (f_n) should have a convergent subsequence, and the result is not true in general. The situation is saved if (f_n) lies in a compact set, an assumption which, as we shall see, is true for a large class of operators that are important in applications.

5.3.5 Theorem. *Let S be a compact subset of the Banach space $\mathscr{B}$, and suppose that $A : S \to \mathscr{B}$ is continuous. Assume that for some sequence (f_n) in S, $\lim \|Af_n - f_n\| = 0$. Then (f_n) has a convergent subsequence with limit $\bar{f} \in S$, and $\bar{f}$ is a fixed point of A.*

Proof. The existence of a convergent subsequence, again denoted by (f_n), is assured by the compactness of S. Then if $f_n \to \bar{f}$, since A is continuous,

$Af_n \to A\bar{f}$. The fact that $\bar{f}$ is a fixed point of A follows from the inequality

$$\|A\bar{f} - \bar{f}\| \leqslant \|A\bar{f} - Af_n\| + \|Af_n - f_n\| + \|f_n - \bar{f}\|$$

when the condition $\lim \|Af_n - f_n\| = 0$ is used. □

5.4 Some Important Compact Sets of Functions

If compact sets are to be useful in dealing with operators on specific function spaces, readily applicable criteria for recognizing these sets must be available. It is a pleasant fact that such criteria do exist both for $\mathscr{C}(\Omega)$ and for the $\mathscr{L}_p$ spaces on which much of the analysis in applications is centred. The situation is easiest in $\mathscr{C}(\Omega)$, whose relatively compact subsets can be completely described in terms of the rather simple concept of "equicontinuity". As a bonus, the fact that it is reasonably easy to form a picture of equicontinuous and so of relatively compact sets in a concrete case helps to elucidate the notion of compactness.

In the following discussion Ω denotes a subset of $\mathbb{R}^n$ and $\mathscr{C}(\Omega)$ is the Banach space of continuous complex valued functions on Ω equipped with the sup norm.

5.4.1 Definition. A set $S \subset \mathscr{C}(\Omega)$ is said to be **equicontinuous** (*uniformly equicontinuous* in some texts) iff, for each $\varepsilon > 0$ there exists a $\delta > 0$ such that $|f(x) - f(y)| < \varepsilon$ for all $x, y \in \Omega$ with $|x - y| < \delta$ and for all $f \in S$.

Notice that for a given ε, the same δ can be chosen for every f in S. Equicontinuity means roughly that the degree of continuity is independent both of the position in the set Ω and of the functions in the set S. To clarify the idea consider the following two easy examples.

5.4.2 Example. The set $\{f_n\}$ of functions defined by

$$f_n(x) = nx/(1 + n^2x^2) \qquad (0 \leqslant x \leqslant 1, n = 1, 2, \ldots)$$

evidently lies in the closed unit ball of $\mathscr{C}([0, 1])$. If this set were equicontinuous, it would be possible to choose $y = 0$, and for each $\varepsilon > 0$ find a $\delta > 0$ such that $|f_n(x)| < \varepsilon$ whenever $x < \delta$ and $n \geqslant 1$. However, f_n has a maximum at $x = n^{-1}$ and $f_n(n^{-1}) = \frac{1}{2}$. δ would therefore necessarily have to be less than n^{-1}, and so cannot be chosen independent of n. The set is not equicontinuous.

5.4.3 Example. In the previous example the derivative of f_n at $x = 0$ is n, which conflicts with the "uniform smoothness" required for equicontinuity. An obvious way to generate a set $\{f_n\}$ of functions with derivatives uniformly bounded with respect to n is to use the smoothing operator of integration. We now prove that this has the required effect.

Let S be the closed unit ball of $\mathscr{C}([0, 1])$. For each $f \in S$ define

$$g(x) = \int_0^x f(t)\,\mathrm{d}t,$$

and denote the set of all such g by S_1. Then for every $g \in S_1$,

$$|g(x) - g(y)| \leqslant \left| \int_y^x f(t)\,\mathrm{d}t \right| \leqslant |x - y| \|f\| \leqslant |x - y|.$$

The right-hand side is independent of g. Thus for each $\varepsilon > 0$, if we choose $\delta = \varepsilon$ then $|g(x) - g(y)| < \varepsilon$ for $|x - y| < \delta$. S_1 is therefore equicontinuous.

It is easy to extend this argument to the set of all g of the form

$$g(x) = \int_0^1 k(x, y) f(y)\,\mathrm{d}y,$$

where $f \in S$ and k is continuous. Hence the integral operator maps bounded sets in $\mathscr{C}([0, 1])$ into bounded equicontinuous sets. The treatment in Chapter 7 of integral equations by operator theoretic methods is based on this fact used in conjunction with the next theorem.

5.4.4 The Arzelà–Ascoli Theorem. *Let Ω be a bounded subset of $\mathbb{R}^n$, and let S be a subset of $\mathscr{C}(\Omega)$. Then S is relatively compact if and only if it is bounded and equicontinuous.*

Proof. The necessity is easy, and less useful, and the reader is recommended to supply the argument. To establish sufficiency we shall show that any sequence (f_n) in S has a convergent subsequence, and so prove that S is relatively sequentially compact.

Let (x_m) be a dense sequence in Ω—the n-vectors with rational components will do. The first step is to show by a diagonalization argument that there is a subsequence $(f_{n,n})$ of (f_n) which is pointwise convergent on the set $\{x_m\}$, that is $\lim_{n\to\infty} f_{n,n}(x_m)$ exists for each m. To do this we note that since S is bounded, the sequence of complex numbers $(f_n(x_1))$ is bounded, and therefore there is a subsequence, $(f_{n,1})$ say, such that $(f_{n,1}(x_1))$ is convergent. Repeating the argument, we find that there are subsequences $(f_{n,m})$ of $(f_{n,m-1})$ such that $(f_{n,m}(x_m))$ is convergent for each m. Consider now the diagonal sequence $(f_{n,n})$. For each m, $(f_{n,n}(x_m))$ is a subsequence of $(f_{n,m}(x_m))$ after the mth term, and so is convergent as required.

For simplicity in notation set $g_n = f_{n,n}$. The final step is to go from pointwise convergence on $\{x_m\}$ to uniform convergence on Ω by proving that

(g_n) is a Cauchy sequence in $\mathscr{C}(\Omega)$. We start by noting that, because of the equicontinuity of $\{g_n\}$, given $\varepsilon > 0$ there is a $\delta > 0$ such that $|g_n(x) - g_n(y)| < \frac{1}{3}\varepsilon$ for all $|x - y| < \delta$ and for all n. Now Ω can be covered by a finite number, k say, of closed balls of radius $\frac{1}{2}\delta$. Since $\{x_m\}$ is dense in Ω, we may select an x_m in each of these balls, and then every $x \in \Omega$ is within a distance δ of one of the x_m. Let us renumber and call the points selected in this way $x_1, \dots, x_k$. From the conclusion of the previous paragraph, $(g_n(x_j))$ converges for each j with $1 \leqslant j \leqslant k$, the convergence being obviously uniform with respect to the finite set $x_1, \dots, x_k$. In other words there is an n_0 such that for $n, m > n_0$,

$$|g_n(x_j) - g_m(x_j)| < \tfrac{1}{3}\varepsilon \qquad (1 \leqslant j \leqslant k). \tag{5.4.1}$$

To extend this to the whole of Ω, use is made of the fact that every $x \in \Omega$ is within a distance δ of one x_j, whence because of equicontinuity the value of $g_n(x)$ is nearly $g_n(x_j)$. Thus for $|x - x_j| < \delta$,

$$|g_n(x) - g_m(x)| \leqslant |g_n(x) - g_n(x_j)| + |g_n(x_j) - g_m(x_j)| + |g_m(x_j) - g_m(x)|.$$

As noted above the first and last terms on the right-hand side are less than $\frac{1}{3}\varepsilon$, as is also the middle term, by (5.4.1). Hence $|g_n(x) - g_m(x)| < \varepsilon$ for $n, m > n_0$ and $x \in \Omega$. This proves that (g_n) is Cauchy, and the result follows as $\mathscr{C}(\Omega)$ is complete. □

This theorem provides the required description of the relatively compact subsets of $\mathscr{C}(\Omega)$. Note that the restriction that Ω be bounded cannot in general be removed. It is occasionally useful to known that the result also holds if Ω is a relatively compact subset of a Banach space, see Problem 5.11.

In the $\mathscr{L}_p$ spaces a number of criteria for compactness may be given. The condition in the following theorem (Problem 5.10) can be regarded as "equicontinuity on average".

5.4.5 Theorem. *Assume that* $p \geqslant 1$, *and let* S *be a bounded subset of* $\mathscr{L}_p(0, 1)$. *Extend the domain of the functions in* S *to* $\mathbb{R}$ *by defining them to be zero outside* $[0, 1]$. *Then* S *is relatively compact if for each* $\varepsilon > 0$, *there is a* $\delta > 0$ *such that for all* $f \in S$ *and* $|h| < \delta$,

$$\int_0^1 |f(x + h) - f(x)|^p \, dx < \varepsilon.$$

Problems

5.1 Prove that a compact subset of a Banach space is closed, bounded and separable.

5.2 Let S be a subset of a Banach space. Prove that:

(i) if S is totally bounded, it is bounded;

(ii) S is totally bounded iff $\bar{S}$ is totally bounded.

5.3 Assume that the subset S of a Banach space is compact. Show that $S_1 \subset S$ is compact iff it is closed.

5.4 Suppose that the sequence (f_n) is contained in a relatively compact subset of a Banach space, and assume that every convergent subsequence of (f_n) has the same limit f. Prove that (f_n) is convergent.

5.5 Consider the sequence space $\ell_p(1 \leqslant p < \infty)$ of vectors $f = (f_1, f_2, \ldots)$. Prove that a bounded set $S \subset \ell_p$ is relatively compact if for each $\varepsilon > 0$ there is an n_0 such that

$$\sum_{n=n_0}^{\infty} |f_n|^p < \varepsilon \qquad (f \in S).$$

5.6 Prove necessity in the Arzelà–Ascoli Theorem.

5.7 Two sets $\{f_n\}, \{g_n\}$ are defined by setting $f_n(x) = \sin nx$, $g_n(x) = 1 - x^n$ for $n \geqslant 1$. Examine the relative compactness of these as subsets of $\mathscr{C}([0, 1])$ equipped with the sup norm. Compare with the position in $\mathscr{L}_p(0, 1)$ where $1 \leqslant p < \infty$.

5.8 Equip $\mathscr{C}([0, 1])$ with the sup norm. Let k be continuous on $[0, 1] \times [0, 1]$, and define $K : \mathscr{C}([0, 1]) \to \mathscr{C}([0, 1])$ by setting

$$Kf(x) = \int_0^1 k(x, y) f(y) \, \mathrm{d}y.$$

Let S be the closed unit ball in $\mathscr{C}([0, 1])$. Show that $K(S)$ is bounded and equicontinuous, and deduce that $K(S)$ is relatively compact.

5.9 For k a non-negative integer, define a norm on $\mathscr{C}^k([0, 1])$ by setting

$$\|f\|_{\mathscr{C}^k} = \sum_{j=0}^{k} \sup_{x \in [0, 1]} |f^{(j)}(x)|.$$

Prove that if $k > h$ the unit ball of $\mathscr{C}^k([0, 1])$ is a relatively compact subset of $\mathscr{C}^h([0, 1])$.

5.10 Prove Theorem 5.4.5. For fixed h, establish relative compactness in $\mathscr{C}([0, 1])$—and hence in $\mathscr{L}_p(0, 1)$—for the set $\{f_h : f \in S\}$ where

$$f_h(x) = (2h)^{-1} \int_{x-h}^{x+h} f(t) \, \mathrm{d}t,$$

by using the Arzelà–Ascoli Theorem. Show that $\|f - f_h\|_p$ is small for small h, and deduce the result from total boundedness arguments.

5.11 Let Ω be a subset of the Banach space $\mathscr{B}$ with norm $\|\cdot\|$. Show that the set $\mathscr{C}(\Omega)$ of complex valued bounded continuous functions on Ω is a Banach space in the sup norm $\|\cdot\|_{\mathscr{C}(\Omega)}$ where

$$\|\phi\|_{\mathscr{C}(\Omega)} = \sup_{x \in \Omega} |\phi(x)|.$$

A set S in $\mathscr{C}(\Omega)$ is said to be equicontinuous iff for each $\varepsilon > 0$ there is a $\delta > 0$ such that $|\phi(x) - \phi(y)| < \varepsilon$ for all $x, y \in \Omega$ with $\|x - y\| < \delta$ and for all $\phi \in S$. Show that the Arzelà–Ascoli Theorem holds in $\mathscr{C}(\Omega)$ if Ω is relatively compact.

Chapter 6

The Adjoint Operator

6.1 Introduction

In the theory of linear algebraic equations in finite dimensions and of linear differential and integral equations, a crucial role is played by the adjoint equation. The related concept of the "adjoint" of a linear operator is equally important in the abstract theory. To motivate the definition of the adjoint operator, consider the integral equation

$$f(x) - \int_0^1 k(x, y) f(y) \, dy = h(x),$$

with k real-valued and continuous, and the adjoint equation

$$g(x) - \int_0^1 k(y, x) g(y) \, dy = h(x),$$

in the real Hilbert space $\mathscr{H} = \mathscr{L}_2(0, 1)$. In an obvious notation these equations may be written as $(I - L)f = h$ and $(I - L^*)g = h$ respectively, with L and L^* linear operators, and since L^* arises from the adjoint equation it is natural that L^* should be regarded as the adjoint of L. To cast the definition of L^* into abstract form, note that

$$\int_0^1 g(x) \, dx \int_0^1 k(x, y) f(y) \, dy = \int_0^1 f(x) \, dx \int_0^1 k(y, x) g(y) \, dy,$$

from which it follows that L^* must satisfy the equation

$$(Lf, g) = (f, L^*g) \qquad (f, g \in \mathscr{H}). \tag{6.1.1}$$

On the other hand, it is easy to see that given L, L^* is uniquely determined by (6.1.1). This suggests that (6.1.1) should be used to define the adjoint, and indeed this is the definition appropriate for Hilbert space.

An attempt to transfer this definition as it stands to a Banach space meets with an immediate difficulty—the lack of an inner product. To overcome this problem it is necessary to go rather more deeply into Banach space theory, and to introduce a new space $\mathscr{B}^*$, called the dual, related to the original space $\mathscr{B}$. An "outer product" $\langle \cdot, \cdot \rangle$ sharing some of the properties of $(\cdot, \cdot)$ may then be defined on $\mathscr{B} \times \mathscr{B}^*$ ($(\cdot, \cdot)$ being of course defined on $\mathscr{H} \times \mathscr{H}$), and formally extending (6.1.1) we set

$$\langle Lf, g \rangle = \langle f, L^*g \rangle \qquad (f \in \mathscr{B}, g \in \mathscr{B}^*).$$

A linear operator L^* on $\mathscr{B}^*$ into itself is obtained by this means, L^* being called the adjoint of L. As L^* is defined on $\mathscr{B}^*$ rather than on $\mathscr{B}$, it is not quite as convenient an object to deal with as the adjoint in Hilbert space. Nonetheless the use of the adjoint opens the way for a considerable further advance in linear operator theory.

Sections 2–4 consist of a review of the principal properties of the dual, and are a preparation for the discussion of the adjoint. The adjoint of a continuous operator is next defined, and this is then used in proving certain results concerning the solution of the operator equation $Lf = g$ which will be crucial in the sequel. The simplification of these results for a self-adjoint operator on a Hilbert space is next discussed. Finally, the adjoint of an unbounded operator is considered, the motivation being applications to differential equations.

6.2 The Dual of a Banach Space

This section starts with the definition of the dual. In the particular case when $\mathscr{B}$ is a Hilbert space, the dual turns out to be $\mathscr{B}$ itself. The situation is not so simple in general, but we shall see that the duals of several of the most important concrete Banach spaces have a simple explicit form. The remainder of the section consists of a study of the dual $\mathscr{B}^*$ of an abstract Banach space $\mathscr{B}$. The most significant result is the Hahn–Banach Theorem. For an operator $L \in \mathscr{L}(\mathscr{B})$, the domain of the adjoint L^* of L is $\mathscr{B}^*$. Therefore, in order that L^* should yield useful information about L, it is vital that $\mathscr{B}^*$ should be "as large as" $\mathscr{B}$ itself. That this is so is a consequence of the Hahn–Banach Theorem.

To motivate the definition of the dual, consider a Hilbert space $\mathscr{H}$ (for convenience assumed temporarily real) with inner product $(\cdot, \cdot)$. Fix an element $h \in \mathscr{H}$, and consider the mapping f^* (this notation does not imply any relation between f and f^*) such that $f^*(f) = (f, h)$ for all $f \in \mathscr{H}$. Evidently f^* is a linear operator from $\mathscr{H}$ into $\mathbb{R}$, and f^* is continuous by Schwarz's inequality. That is f^* is an element of $\mathscr{L}(\mathscr{H}, \mathbb{R})$. Further analysis shows that the spaces $\mathscr{H}$ and $\mathscr{L}(\mathscr{H}, \mathbb{R})$ are actually isometrically isomorphic, and may thus be identified with each other. It is therefore natural to write $h = f^*$ and to regard f^* both as an element of $\mathscr{H}$ and of $\mathscr{L}(\mathscr{H}, \mathbb{R})$. In a Banach space the argument is reversed: the starting point is $\mathscr{L}(\mathscr{B}, \mathbb{R})$ (or $\mathscr{L}(\mathscr{B}, \mathbb{C})$ if $\mathscr{B}$ is complex) which is called the dual $\mathscr{B}^*$ of $\mathscr{B}$. In contrast with the Hilbert space case, $\mathscr{B}^*$ cannot be identified with $\mathscr{B}$, but $\mathscr{B}^*$ is itself a Banach space, and further there is an "outer product" on $\mathscr{B} \times \mathscr{B}^*$ with several of the properties of the inner product on $\mathscr{H} \times \mathscr{H}$.

Let $\mathscr{B}$ be a complex Banach space. If $\mathscr{B}$ is real simply replace $\mathbb{C}$ by $\mathbb{R}$ in the following.

6.2.1 Definition. The elements of the space $\mathscr{L}(\mathscr{B}, \mathbb{C})$ of continuous linear operators from $\mathscr{B}$ into $\mathbb{C}$ are called **continuous linear functionals**.

6.2.2 Definition. The space $\mathscr{L}(\mathscr{B}, \mathbb{C})$ of continuous linear functionals on $\mathscr{B}$ is called the **dual** of $\mathscr{B}$ and is denoted by $\mathscr{B}^*$. Throughout $f, g, \ldots$ and $f^*, g^*, \ldots$ will denote elements of $\mathscr{B}$ and $\mathscr{B}^*$ respectively (no relation between f and f^* is implied). For $f \in \mathscr{B}$ and $f^* \in \mathscr{B}^*$, $f^*(f)$ will denote the complex number assigned to f by the mapping f^*.

6.2.3 Theorem. *If $\mathscr{B}$ is a Banach space, then $\mathscr{B}^*$ is also a Banach space, and*

$$|f^*(f)| \leqslant \|f^*\| \, \|f\| \qquad (f \in \mathscr{B}, f^* \in \mathscr{B}^*).$$

The norms are those of $\mathscr{B}^*$ and $\mathscr{B}$ respectively. It is seldom necessary to adopt a separate notation for these norms since it is usually clear from the context which is intended. The theorem itself is a special case of Theorem 3.5.5 with $\mathscr{C} = \mathbb{C}$.

For most of the common Banach spaces the dual may be identified (under an isometric isomorphism) with a known space. The following examples are sufficient for our purposes; Dunford and Schwartz (1958, Chap. 4) may be consulted for further information.

6.2.4 Example. Suppose that $\mathscr{B}$ is $\mathbb{R}^3$ equipped with the ℓ_p norm $\|\cdot\|_p$. For given $h = (h_1, h_2, h_3)$ the relation

$$f^*(f) = f_1 h_1 + f_2 h_2 + f_3 h_3 \qquad (f = (f_1, f_2, f_3) \in \mathscr{B})$$

defines a linear functional f^* on $\mathscr{B}$, and by analogy with the inner product we write also $f^*(f) = \langle f, f^* \rangle$. If p and q are conjugate indices, by Hölder's inequality (Theorem 1.3.12), $|f^*(f)| \leqslant \|h\|_q \|f\|_p$, and it follows that f^* is continuous. Conversely it is easy to show that to every continuous linear functional f^* on $\mathbb{R}^3$ (with the norm $\|\cdot\|_p$) there corresponds an element h of $\mathbb{R}^3$ (with the norm $\|\cdot\|_q$). Since the correspondence is actually an isometric isomorphism, it is convenient to set $h = f^*$ and to allow f^* to denote both the linear functional and an element of $\mathbb{R}^3$. The dual of $\mathbb{R}^3$ with $\|\cdot\|_p$ is (isometrically isomorphic to) $\mathbb{R}^3$ itself with $\|\cdot\|_q$. In fact any norm on $\mathbb{R}^3$ is equivalent to $\|\cdot\|_q$. Further, regarded as sets $\mathbb{R}^3$ and its dual are equal. The concept of duality is therefore of limited interest in finite dimensions.

6.2.5 Example. The contrast in infinite dimensions is illustrated by considering the $\mathscr{L}_p$ spaces. Suppose Ω is an interval in $\mathbb{R}$. Take any p with $1 \leqslant p < \infty$, let q be the conjugate index, and for given $h \in \mathscr{L}_q(\Omega)$ set

$$f^*(f) = \int_\Omega f(x)\, h(x)\, \mathrm{d}x \qquad (f \in \mathscr{L}_p(\Omega)). \tag{6.2.1}$$

Then

$$|f^*(f)| = \left| \int_\Omega f(x)\, h(x)\, \mathrm{d}x \right| \leqslant \|f\|_p \|h\|_q$$

by Hölder's inequality (Theorem 2.5.3). It follows that $f^* : \mathscr{L}_p(\Omega) \to \mathbb{C}$ is a continuous linear operator, and therefore $f^* \in \mathscr{L}(\mathscr{L}_p(\Omega), \mathbb{C})$. It may be proved (see Friedman, 1970; Section 4.14) that $\|f^*\| = \|h\|_q$, and that further every continuous linear operator $\mathscr{L}_p(\Omega) \to \mathbb{C}$ may be written as in (6.2.1) for some $h \in \mathscr{L}_q(\Omega)$. The mapping $h \to f^*$ of $\mathscr{L}(\mathscr{L}_p(\Omega), \mathbb{C})$ onto $\mathscr{L}_q(\Omega)$ is thus again an isometric isomorphism, and we conclude that the dual of $\mathscr{L}_p(\Omega)$ is $\mathscr{L}_q(\Omega)$; it is again convenient to set $h = f^*$ and to regard f^* both as an element of $\mathscr{L}_p(\Omega)^*$ and of $\mathscr{L}_q(\Omega)$. The conclusion is superficially similar to that of the previous example, but there is a crucial difference, for $\mathscr{L}_p(\Omega)$ and $\mathscr{L}_q(\Omega)$ are essentially different spaces.

6.2.6 Theorem. *Suppose that $1 \leqslant p < \infty$, and let q be the conjugate index. Let Ω be a finite or infinite interval. Then $\mathscr{L}_p(\Omega)^* = \mathscr{L}_q(\Omega)$, and for any continuous linear functional f^* on $\mathscr{L}_p(\Omega)$ there is a function f^* in $\mathscr{L}_q(\Omega)$ such that*

$$f^*(f) = \int_\Omega f(x) f^*(x)\, \mathrm{d}x \qquad (f \in \mathscr{L}_p(\Omega)).$$

It should be noted that this result does not hold for $p = \infty$, that is $\mathscr{L}_\infty(\Omega)^* \neq \mathscr{L}_1(\Omega)$. In the Hilbert space case $p = 2$, $\mathscr{L}_2(\Omega)$ is its own dual.

6.2.7 Example. For the sequence space ℓ_p an analogous result is valid. That is $\ell_p^* = \ell_q$ if $1 \leqslant p < \infty$ and p, q are conjugate indices. Again the result is false for $p = \infty$. The representation of the linear functional f^* is in an obvious notation

$$f^*(f) = \Sigma f_j f_j^*.$$

We proceed to derive the main properties of the dual. The development rests on the Hahn–Banach Theorem. Recall that a continuous linear functional defined on a dense linear subspace $\mathscr{M}$ of $\mathscr{B}$ has an extension with the same norm to the whole of $\mathscr{B}$ (Theorem 3.4.4). The Hahn–Banach Theorem asserts that this result is true even if $\mathscr{M}$ is not dense (but it should be noted that in contrast the extension may not be unique). This is a much deeper result, and the argument used in the proof is more subtle than the easy continuity argument of Theorem 3.4.4.

6.2.8 The Hahn–Banach Theorem. *Let $\mathscr{M}$ be a linear subspace of the Banach space $\mathscr{B}$, and suppose f^* is a continuous linear functional defined on $\mathscr{M}$. Then f^* may be extended to a continuous linear functional on $\mathscr{B}$ whose norm is the same as that of f^*.*

The result will be proved under the assumption that $\mathscr{B}$ is separable (see Friedman (1970, p. 150) for the general case). This dispenses with the use of transfinite induction (Zorn's Lemma) without affecting the basic structure of the proof. To avoid tedious manipulation with complex numbers, $\mathscr{B}$ is assumed real; the complex case is recovered by taking $f^*(f) = \operatorname{Re} f^*(f) - i \operatorname{Re} f^*(if)$. Finally, for a functional f^* with domain D the norm is denoted by $\|\cdot\|$, where of course

$$\|f^*\| = \sup_{f \in D,\ \|f\| = 1} |f^*(f)|.$$

Proof. The first step is to show that if $\mathscr{M} \neq \mathscr{B}$, f^* may be extended without change of norm to a slightly larger (by one dimension) subspace $\mathscr{M}_1$ say. Take any f_0 not in $\mathscr{M}$ and let $[f_0]$ denote its linear span. Set $\mathscr{M}_1 = \mathscr{M} \oplus [f_0]$, and note that any element f_1 of $\mathscr{M}_1$ has the unique representation $f_1 = f + \alpha f_0$ with $f \in \mathscr{M}$ and $\alpha \in \mathbb{R}$ (Lemma 1.2.9). For real a the functional f_1^* defined on $\mathscr{M}_1$ by setting

$$f_1^*(f_1) = f^*(f) + a\alpha$$

is clearly linear and equal to f^* on $\mathscr{M}$. f_1^* is thus an extension of f^*. We will prove that there is an a for which f_1^* is bounded and for which the norms of f^* and f_1^* (regarded as operators on $\mathscr{M}$ and $\mathscr{M}_1$ respectively) are the same. To do this it is enough to show that there is an a such that $|f_1^*(f_1)| \leqslant \|f^*\| \|f_1\|$ $(f_1 \in \mathscr{M}_1)$—for since $f_1^* = f^*$ on $\mathscr{M}_1$ it is obvious that $\|f^*\| \leqslant \|f_1^*\|$—or equivalently that

$$|f^*(f) + \alpha a| \leqslant \|f^*\| \|f + \alpha f_0\| \qquad (f \in \mathscr{M}, \alpha \in \mathbb{R}).$$

On dividing through by α (when $\alpha = 0$ the inequality is obvious) and setting $g = \alpha^{-1} f \in \mathscr{M}$, the last relation becomes

$$|f^*(g) + a| \leqslant \|f^*\| \|g + f_0\|$$

which will certainly hold if for all $g_1, g_2 \in \mathscr{M}$,

$$-f^*(g_1) - \|f^*\| \|g_1 + f_0\| \leqslant a \leqslant -f^*(g_2) + \|f^*\| \|g_2 + f_0\|. \quad (6.2.2)$$

Now

$$\begin{aligned} f^*(g_2) - f^*(g_1) &= f^*(g_2 - g_1) \\ &\leqslant \|f^*\| \|g_2 - g_1\| \\ &\leqslant \|f^*\| \|g_2 + f_0\| + \|f^*\| \|g_1 + f_0\|, \end{aligned}$$

whence

$$-f^*(g_1) - \|f^*\| \|g_1 + f_0\| \leqslant -f^*(g_2) + \|f^*\| \|g_2 + f_0\|.$$

Thus the upper bound (over all $g_1 \in \mathscr{M}$) of the left-hand side of this inequality is not greater than the lower bound (over all $g_2 \in \mathscr{M}$) of the right-hand side. Therefore (6.2.2) will hold for any a between these bounds. We conclude that as asserted, there is an f_1^* which extends f^* from $\mathscr{M}$ to $\mathscr{M}_1$ without change of norm.

The proof is completed by repeated extension of f^*. Since $\mathscr{B}$ is separable, there is a sequence, (f_j) say, which is dense in $\mathscr{B}$. Set $\mathscr{M}_0 = \mathscr{M}$ and $\mathscr{M}_{j+1} = \mathscr{M}_j \oplus [f_j]$ for $j \geqslant 0$. By the above argument there are linear functionals $f^*, f_1^*, f_2^*, \ldots$ with the same norms on the linear subspaces $\mathscr{M} = \mathscr{M}_0 \subset \mathscr{M}_1 \subset \mathscr{M}_2 \subset \ldots$ respectively. Let $\mathscr{M}_\infty$ be the set theoretic union of the $\mathscr{M}_j$; $\mathscr{M}_\infty$ is clearly a linear subspace, and since (f_j) is dense in $\mathscr{B}$, $\bar{\mathscr{M}}_\infty = \mathscr{B}$. Define the linear functional f_∞^* on $\mathscr{M}_\infty$ by requiring that $f_\infty^* = f_j^*$ on $\mathscr{M}_j$. Then $\|f_\infty^*\| = \|f^*\|$. Finally, recall that by Theorem 3.4.4, f_∞^* may be extended continuously without change of norm to $\bar{\mathscr{M}}_\infty = \mathscr{B}$. f_∞^* is the required extension. □

Four useful corollaries of the Hahn–Banach Theorem follow. The first is set as an exercise (Problem 6.1), and the others are easy deductions from it.

6.2.9 Corollary. *Suppose that $\mathscr{M}$ is a linear subspace of $\mathscr{B}$, and let g be any element of $\mathscr{B}$ not in $\bar{\mathscr{M}}$. Then there is a continuous linear functional f^* on $\mathscr{B}$ such that*

(i) $f^*(f) = 0 \ (f \in \mathscr{M})$;
(ii) $f^*(g) = 1$;
(iii) $\|f^*\| = 1/\text{dist}(g, \mathscr{M})$.

6.2.10 Corollary. *For each non-zero f in $\mathscr{B}$, there is an $f^* \in \mathscr{B}^*$ with $\|f^*\| = 1$ and $f^*(f) = \|f\|$.*

6.2.11 Corollary. *If $f^*(f) = 0$ for all $f^* \in \mathscr{B}^*$, then $f = 0$.*

6.2.12 Corollary. *For any* $f \in \mathscr{B}$,

$$\|f\| = \sup_{\|f^*\|=1} |f^*(f)|.$$

We asserted that it would be possible to construct an outer product $\langle \cdot, \cdot \rangle$ sharing some of the properties of the inner product $(\cdot, \cdot)$ of Hilbert space. $\langle \cdot, \cdot \rangle$ is defined by setting $\langle f, f^* \rangle = f^*(f)$, and is a mapping from $\mathscr{B} \times \mathscr{B}^*$ into $\mathbb{C}$ with the following properties. For $f, g \in \mathscr{B}$, $f^*, g^* \in \mathscr{B}^*$ and $a, b \in \mathbb{C}$:

(i) $\langle af + bg, f^* \rangle = a\langle f, f^* \rangle + b\langle g, f^* \rangle$;
$\langle f, af^* + bg^* \rangle = a\langle f, f^* \rangle + b\langle f, g^* \rangle$;
(ii) $|\langle f, f^* \rangle| \leqslant \|f\| \|f^*\|$;
(iii) if $\langle f, f^* \rangle = 0$ for all $f^* \in \mathscr{B}^*$, then $f = 0$ (Corollary 6.2.11).

Note the formal similarity of the algebraic relations (i) with those for the inner product (the missing "bars" for a complex space will be commented on in Section 6.4), and of (ii) with Schwartz's inequality. In Hilbert space the analogue of (iii) just expresses the fact that $\mathscr{H}^\perp = 0$. Here (iii) may be interpreted as showing that the dual is "sufficiently large"; it should be remarked that this crucial property depends on the Hahn–Banach Theorem.

The dual of $\mathscr{B}^*$ itself is sometimes useful.

6.2.13 Definition. The dual $\mathscr{B}^{**}$ of $\mathscr{B}^*$ is called the **second dual** of $\mathscr{B}$.

At first sight $\mathscr{B}^{**}$ may seem a somewhat obscure object. However, it is closely related to $\mathscr{B}$ itself. To see this, take any $f \in \mathscr{B}$ and consider the element $\hat{f}$ (evidently unique) in $\mathscr{B}^{**}$ defined by the relation $\hat{f}(f^*) = f^*(f)$ $(f^* \in \mathscr{B}^*)$. Since each f specifies uniquely some $\hat{f}$, the relation $\hat{f} = Kf$ defines an operator $K : \mathscr{B} \to \mathscr{B}^{**}$ often called the *natural imbedding* of $\mathscr{B}$ in $\mathscr{B}^{**}$. The space $\hat{\mathscr{B}} = R(K)$ is obviously a linear subspace of $\mathscr{B}^{**}$, and using Corollary 6.2.12 the following is readily deduced.

6.2.14 Lemma. *K is an isometric isomorphism between $\mathscr{B}$ and $\hat{\mathscr{B}}$. $\hat{\mathscr{B}}$ is a closed subspace of $\mathscr{B}^{**}$.*

In view of this result it is conventional to identify $\mathscr{B}$ and $\hat{\mathscr{B}}$ and to speak of $\mathscr{B}$ as a closed subspace of $\mathscr{B}^{**}$. Actually it is often the case that $\hat{\mathscr{B}} = \mathscr{B}^{**}$ in this sense.

6.2.15 Definition. $\mathscr{B}$ is said to be **reflexive** iff $\mathscr{B} = \mathscr{B}^{**}$.

6.2.16 Example. Evidently all finite dimensional normed spaces are reflexive. More importantly, if $1 < p < \infty$ the $\mathscr{L}_p$ spaces are reflexive (by Theorem 6.2.6), and the same is true of ℓ_p.

The following is an example of the type of result in which $\mathscr{B}^{**}$ may be used to prove a property of $\mathscr{B}$ and $\mathscr{B}^*$ that cannot easily be established directly.

6.2.17 Lemma. *Let $\{f_\alpha\}$ be a set of elements in a Banach space $\mathscr{B}$. Suppose that for each $f^* \in \mathscr{B}^*$ there is an m such that $\sup_\alpha |f^*(f_\alpha)| \leqslant m$. Then $\sup_\alpha \|f_\alpha\| < \infty$.*

Proof. Instead of $\{f_\alpha\}$ consider $\{\hat{f}_\alpha\}$. Each f_α may be regarded as an operator on $\mathscr{B}^*$. Hence

$$\sup_\alpha |\hat{f}_\alpha(f^*)| = \sup_\alpha |f^*(f_\alpha)| \leqslant m.$$

Therefore by the Principle of Uniform Boundedness 3.5.6, $\sup_\alpha \|\hat{f}_\alpha\| < \infty$. Since $\hat{f}_\alpha = Kf_\alpha$ and K is an isometry, $\|\hat{f}_\alpha\| = \|f_\alpha\|$, and the result follows. □

A detailed study of reflexivity is unnecessary for the present purpose, and we shall do no more than quote the following (Friedman (1970, Chapt. 4)) which is occasionally useful.

6.2.18 Lemma. (i) *A closed subspace of a reflexive Banach space is reflexive.*
(ii) *A Banach space is reflexive iff its dual is reflexive.*
(iii) *If $\mathscr{B}$ is reflexive, then $\mathscr{B}^*$ is separable iff $\mathscr{B}$ is separable.*

The utility of the concept of "orthogonal complement" was apparent in Section 1.5. A formal analogue in Banach space suggests itself if the outer product is used.

6.2.19 Definition. Given sets $S \subset \mathscr{B}$ and $S^* \subset \mathscr{B}^*$, the sets

$$S^\perp = \{f^* \in \mathscr{B}^*: \quad \langle f, f^* \rangle = 0, f \in S\},$$
$$S^{*\perp} = \{f \in \mathscr{B}: \quad \langle f, f^* \rangle = 0, f^* \in S^*\},$$

are known as the **orthogonal complements** of S and S^* respectively. The term *annihilator* is also sometimes used.

The analogy with Hilbert space is not as close as could be wished, for $S^\perp$ is a subset of $\mathscr{B}^*$ rather than of $\mathscr{B}$. Again the extra geometrical complexity of Banach space is apparent. Basic properties, which are obviously essential if the orthogonal complement is to be useful, are $\mathscr{B}^\perp = 0$ and $\mathscr{B}^{*\perp} = 0$; the first is obvious, but the second depends on the Hahn–Banach Theorem (Corollary 6.2.12). The orthogonal complement must be treated with some caution. For example it is possible to have a *proper* closed subspace $\mathscr{M}^*$ of $\mathscr{B}^*$ which is such that $\mathscr{M}^{*\perp} = 0$. The easy proof of the following is left as an exercise.

6.2.20 Lemma. *Suppose that $\mathscr{B}$ is a Banach space; let S and S^* be subsets of $\mathscr{B}$ and $\mathscr{B}^*$ respectively. Then $S^\perp$ and $S^{*\perp}$ are closed subspaces of $\mathscr{B}^*$ and $\mathscr{B}$ respectively. The orthogonal complements of the closed linear spans of S and S^* are $S^\perp$ and $S^{*\perp}$ respectively.*

6.3 Weak Convergence

It is sometimes convenient to have available a less severe criterion for the convergence of a sequence. The utility of the definition of "weak convergence" is a consequence of the civilized nature of the dual, which is itself assured by the Hahn–Banach Theorem. Weak convergence has direct applications in certain contexts, for example in optimization theory, but here it will be employed principally as a technical device.

6.3.1 Definition Let $\mathscr{B}$ be a Banach space. A sequence (f_n) in $\mathscr{B}$ is said to be **weakly convergent** iff there is an element f (called the **weak limit**) in $\mathscr{B}$ such that

$$\lim f^*(f_n) = f^*(f)$$

for all $f^* \in \mathscr{B}^*$. We write then $f_n \rightharpoonup f$.

6.3.2 Lemma. (i) *A weakly convergent sequence cannot have two weak limits.*
(ii) *A convergent sequence is weakly convergent to the same limit.*

Proof. (i) If the sequence has two weak limits f and g, for all $f^* \in \mathscr{B}^*, f^*(f) = f^*(g)$—or equivalently $f^*(f - g) = 0$. Hence $f = g$ by Corollary 6.2.11 to the Hahn–Banach Theorem.

(ii) If $f_n \to f$, as f^* is a continuous operator on $\mathscr{B}$,

$$\lim|f^*(f_n) - f^*(f)| = \lim|f^*(f_n - f)| \leqslant \|f^*\| \lim\|f_n - f\| = 0. \qquad \square$$

6.3.3 Example. In finite dimensional spaces weak and strong convergence are equivalent. This is also true in some infinite dimensional spaces (ℓ_1 is an example), but is usually false. For example, consider in $\mathscr{L}_2(0, 1)$ the sequence (f_n) where $f_n(x) = 2^{\frac{1}{2}} \sin n\pi x$. $\{f_n\}$ is an orthonormal set, and it follows from Bessel's inequality (Lemma 1.5.13(i)) that $\lim(f_n, f) = 0$ for any $f \in \mathscr{L}_2(0, 1)$. Since the dual of $\mathscr{L}_2(0, 1)$ is $\mathscr{L}_2(0, 1)$ itself (Theorem 6.2.6), this proves that $f_n \rightharpoonup 0$. However, a routine calculation shows that $\|f_n - f_m\|_2 = 2^{\frac{1}{2}}$ for $n \neq m$. Therefore (f_n) is not strongly convergent.

A reasonable question at this stage is whether a general Banach space can be renormed in such a way that the original weak convergence becomes just convergence in the new norm. Unfortunately the answer is in the negative. This makes the study of some of the deeper properties of weak convergence a rather difficult task, which is best carried out in the context of a topological vector space, see Dunford and Schwartz (1958, Chap. 2). However, it is unnecessary to introduce this new concept in order to derive the simple results which will be needed here.

A useful alternative criterion for weak convergence is as follows.

6.3.4 Lemma. *A sequence (f_n) in a Banach space $\mathscr{B}$ is weakly convergent if both the following conditions hold:*

(i) *the sequence is bounded;*
(ii) *there is an $f \in \mathscr{B}$ such that* $\lim f^*(f_n) = f^*(f)$ *for all f^* in a set S^* dense in $\mathscr{B}^*$.*

Proof. From (i) there is an m such that $\|f_n\| \leqslant m$ for $n \geqslant 1$. By (ii), for any $f^* \in \mathscr{B}^*$, there is a $g^* \in S^*$ such that $\|f^* - g^*\| < \varepsilon$. Hence

$$\begin{aligned}\lim|f^*(f_n) - f^*(f)| &= \lim|(f^* - g^*)(f_n) + g^*(f_n) - f^*(f)| \\ &\leqslant \varepsilon m + |g^*(f) - f^*(f)| \\ &\leqslant \varepsilon(m + \|f\|).\end{aligned}$$

Since ε is arbitrary, the result follows. □

The next lemma gives some elementary properties of weak convergence. The first two properties are the same as for ordinary convergence, and the third is a partial replacement for the norm continuity of convergent sequences.

6.3.5 Lemma. *Suppose $f_n \rightharpoonup f$ in the Banach space $\mathscr{B}$. Then*

(i) *the sequence is bounded;*
(ii) *f belongs to the closure of the linear span of $\{f_n\}$;*
(iii) $\|f\| \leqslant \liminf\|f_n\|$.

Proof. (i) This follows immediately from Lemma 6.2.17. (ii) Assume the contrary. By Corollary 6.2.9 there is an $f^* \in \mathscr{B}^*$ such that $f^*(f_n) = 0$ for $n \geqslant 1$, and $f^*(f) = 1$. Since $f_n \rightharpoonup f$, $f^*(f) = \lim f^*(f_n) = 0$. This gives a contradiction. (iii) For any $f^* \in \mathscr{B}^*$,

$$|f^*(f)| = \lim|f^*(f_n)| \leqslant \liminf\|f^*\|\,\|f_n\| = \|f^*\| \liminf\|f_n\|.$$

Corollary 6.2.12 yields the result. □

Associated with weak convergence, there is a natural concept of weak sequential compactness†. The main result concerning this is the Banach–Alaoglu Theorem, a slightly weaker version of which is presented below (for the general case see Friedman (1970, Section 4.12)). Recall that the corresponding result for convergence in norm is false unless the space is finite dimensional.

6.3.6 Definition. A subset S of a Banach space is said to be **relatively weakly sequentially compact** iff every sequence in S contains a weakly convergent subsequence. S is said to be **weakly sequentially compact** iff in addition the weak limits of the subsequences lie in S.

6.3.7 Theorem (Banach–Alaoglu). *The closed unit ball in a separable reflexive Banach space is weakly sequentially compact.*

Proof. Take any sequence (f_n) in $\bar{S}(0, 1)$. It will first be proved by a diagonalization argument that there is a subsequence $(f_{n,n})$ such that $(f^*(f_{n,n}))$ is convergent for all f^* in a set S^* dense in $\mathscr{B}^*$.

By assumption, $\mathscr{B}$ is separable, whence so also is $\mathscr{B}^*$ (Lemma 6.2.18). Therefore there is a sequence, (f_m^*) say, which is dense in $\mathscr{B}^*$. Now $|f_1^*(f_n)| \leqslant \|f_1^*\| \|f_n\|$, and it follows that the sequence $(f_1^*(f_n))$ of complex numbers is bounded, and thus has a convergent subsequence, say $(f_1^*(f_{n,1}))$. By repetition of this argument there are subsequences $(f_{n,m})$ of $(f_{n,m-1})$ such that $(f_m^*(f_{n,m}))$ is convergent for each m. Consider the diagonal sequence $(f_{n,n})$. For each m, $(f_m^*(f_{n,n}))$ is a subsequence of $(f_m^*(f_{n,m}))$ after the mth term, and so is convergent.

To extend the convergence to all $f^* \in \mathscr{B}^*$, identify $f_{n,n}$ with the element $\hat{f}_{n,n}$ of $\mathscr{B}^{**}$ by setting $\hat{f}_{n,n}(f^*) = f^*(f_{n,n})$ for all $f^* \in \mathscr{B}^*$. By assumption, $\mathscr{B}^{**} = \mathscr{B}$, therefore $\mathscr{B}$ is the set of continuous linear functionals on $\mathscr{B}^*$. Thus by Lemma 3.5.11, $(\hat{f}_{n,n}(f^*))$—and so also $(f^*(f_{n,n}))$—is convergent for all $f^* \in \mathscr{B}^*$. In other words $(f_{n,n})$ is weakly convergent. Lemma 6.3.5(iii) shows that its weak limit lies in $\bar{S}(0, 1)$. □

6.4 Hilbert Space

The analysis in the last two sections will now be re-examined briefly in the special context of Hilbert space. The first step is to justify the assertion that the dual is the space itself—that is Hilbert space is self-dual. For each g in

† The allied notion of "weak compactness" will not be needed here. It is nonetheless fair to warn readers that weak compactness and weak sequential compactness are not equivalent (cf. Theorem 5.2.4(i)).

the complex Hilbert space $\mathscr{H}$, define an operator $g^*: \mathscr{H} \to \mathbb{C}$ by requiring the equation $g^*(f) = (f, g)$ to hold for all $f \in \mathscr{H}$. Obviously $\|g^*\| \leqslant \|g\|$ by Schwartz's inequality—the norms being those of $\mathscr{L}(\mathscr{H}, \mathbb{C})$ and $\mathscr{H}$ respectively—and it follows that g^* is a continuous linear functional on $\mathscr{H}$, and thus lies in $\mathscr{L}(\mathscr{H}, \mathbb{C})$ which is the dual $\mathscr{H}^*$ of $\mathscr{H}$. The important remaining question, that is whether every element of $\mathscr{H}^*$ is given in this way by an element of $\mathscr{H}$, is answered affirmatively by the following result.

6.4.1 The Riesz Representation Theorem. *Corresponding to every element g^* of the dual $\mathscr{H}^*$ of a Hilbert space $\mathscr{H}$, there is a unique element g of $\mathscr{H}$ such that $g^*(f) = (f, g)$ for all $f \in \mathscr{H}$. Also $\|g^*\| = \|g\|$.*

Proof. The existence of such a g is first established. Let N be the null space of g^*. If $N = \mathscr{H}$ the result is obvious (simply take $g = 0$), so assume that $N \neq \mathscr{H}$. Since N is closed, it is a consequence of the Projection Theorem 1.5.11 that there is a non-zero $g_0 \in N^{\perp}$. Thus $\alpha = g^*(g_0) \neq 0$ as $g_0 \notin N$. Now

$$g^*(f - \alpha^{-1} g^*(f) g_0) = g^*(f) - g^*(f) \,.\, \alpha^{-1} g^*(g_0) = 0 \qquad (f \in \mathscr{H}).$$

Hence $f - \alpha^{-1} g^*(f) g_0 \in N$, and $(f - \alpha^{-1} g^*(f) g_0, g_0) = 0$. On rearrangement this equation becomes $g^*(f)\|g_0\|^2 = (f, \bar{\alpha} g_0)$. The required relation $g^*(f) = (f, g)$ is obtained on choosing $g = \bar{\alpha} g_0/\|g_0\|^2$.

To prove uniqueness, suppose there is another element g' with $g^*(f) = (f, g')$ for $f \in \mathscr{H}$. Then $(f, g - g') = 0$, and it follows on choosing $f = g - g'$ that $\|g - g'\|^2 = 0$. That is $g = g'$.

Lastly, to show that $\|g^*\| = \|g\|$, note first that

$$\|g\|^2 = |(g, g)| = |g^*(g)| \leqslant \|g^*\| \, \|g\|,$$

whence $\|g\| \leqslant \|g^*\|$. On the other hand,

$$\|g^*\| = \sup_{\|f\|=1} |g^*(f)| = \sup_{\|f\|=1} |(f, g)| \leqslant \|g\|. \qquad \square$$

This proves that there is a bijection, J say, between $\mathscr{H}$ and $\mathscr{H}^*$ which preserves norms, and so is an isometry. The algebraic structure of J is described by the relation

$$J(\alpha f + \beta g) = \bar{\alpha} Jf + \bar{\beta} Jg \qquad (f, g \in \mathscr{H}, \alpha, \beta \in \mathbb{C}). \tag{6.4.1}$$

Since complex conjugates occur on the right-hand side of this equation, J is not strictly linear but *anti-linear*; it is usually called an *anti-isomorphism* (of course J is an isomorphism if $\mathscr{H}$ is a real space). The explanation for the presence of these complex conjugates is the definition adopted for the inner product, from which (6.4.1) follows. For technical reasons connected with the adjoint, this definition is more convenient, and the slight complication in using an anti-isomorphism rather than an isomorphism will not obtrude in the sequel. It is conventional then to identify $\mathscr{H}$ with $\mathscr{H}^*$, in the sense of correspondence under the isometric anti-isomorphism J.

In developing the theory of the dual of a Banach space, a fundamental part was played by the Hahn–Banach Theorem. It is interesting to note that this is a much less deep result in Hilbert space. Indeed a proof based on the Riesz Representation Theorem and the Projection Theorem 1.5.11 may readily be supplied.

We remark finally that the definition of weak convergence in Hilbert space may be formulated in terms of the inner product, and so is conceptually a little simpler. The analogues of Definition 6.3.1 and Theorem 6.3.7 are as follows.

6.4.2 Definition. A sequence (f_n) in the Hilbert space $\mathscr{H}$ is said to be **weakly convergent** iff there is an $f \in \mathscr{H}$ such that

$$\lim(f_n, g) = (f, g) \qquad (g \in \mathscr{H}).$$

6.4.3 Theorem. *The closed unit ball in Hilbert space is weakly sequentially compact.*

Proof. It is only necessary to remove the separability assumption from Theorem 6.3.7. Let (f_n) be a sequence in $\bar{S}(0, 1)$, and let $\mathscr{M}$ be the closed linear span of $\{f_n\}$. $\mathscr{M}$ is a separable Hilbert space, and therefore by Theorem 6.3.7 there is a subsequence, still denoted by (f_n), and an $f \in \mathscr{M}$ such that $\lim(f_n, g) = (f, g)$ for all $g \in \mathscr{M}$. To complete the proof simply note that since $\mathscr{H} = \mathscr{M} \oplus \mathscr{M}^\perp$, the last equation is valid for all $g \in \mathscr{H}$. □

6.5 The Adjoint of a Bounded Linear Operator

With the concept of the dual now available, we may proceed with the primary purpose of this chapter—a discussion of the adjoint of a linear operator on a Banach space. Bounded operators, for which the analysis is relatively simple, are first tackled.

Let $\mathscr{B}$ and $\mathscr{C}$ be Banach spaces, and suppose that $L \in \mathscr{L}(\mathscr{B}, \mathscr{C})$. For given $g^* \in \mathscr{C}^*$ consider the mapping $h^* : \mathscr{B} \to \mathbb{C}$ which assigns to each element $f \in \mathscr{B}$ the complex number $g^*(Lf)$. Since L is continuous and linear, it is obvious that h^* is also continuous and linear, and hence lies in $\mathscr{B}^*$. Thus to each $g^* \in \mathscr{C}^*$ there corresponds a unique element $h^* \in \mathscr{B}^*$ such that $g^*(Lf) = h^*(f)$ for all $f \in \mathscr{B}$. If we write $h^* = L^*g^*$ this equation becomes $g^*(Lf) = L^*g^*(f)$†. Note that L^* maps $\mathscr{C}^*$ into $\mathscr{B}^*$. In order to emphasize the

† The use of the asterisk in two different senses here should cause no confusion, for elements of the dual and adjoints are always denoted by lower case letters and capitals respectively.

analogy with Hilbert space, we remark that the above may alternatively be written as $\langle Lf, g^* \rangle = \langle f, L^* g^* \rangle$, where the outer products are those of $\mathscr{C} \times \mathscr{C}^*$ and $\mathscr{B} \times \mathscr{B}^*$ respectively.

6.5.1 Definition. Suppose $L : \mathscr{B} \to \mathscr{C}$ is a bounded linear operator. The relation

$$g^*(Lf) = L^*g^*(f)$$

required to hold for all $f \in \mathscr{B}$ and all $g^* \in \mathscr{C}^*$ defines an operator L^* from $\mathscr{C}^*$ into $\mathscr{B}^*$ called the **adjoint** of L.

6.5.2 Theorem. *Let $\mathscr{B}$ and $\mathscr{C}$ be Banach spaces, and assume that $L \in \mathscr{L}(\mathscr{B}, \mathscr{C})$. Then $L^* \in \mathscr{L}(\mathscr{C}^*, \mathscr{B}^*)$, and $\|L^*\| = \|L\|$.*

Proof. The linearity is obvious. To prove the remaining assertions note that for any $f \in \mathscr{B}$, $g^* \in \mathscr{C}^*$,

$$|L^*g^*(f)| = |g^*(Lf)| \leqslant \|g^*\| \|Lf\| \leqslant \|g^*\| \|L\| \|f\|.$$

Hence

$$\|L^*g^*\| = \sup_{\|f\|=1} |L^*g^*(f)| \leqslant \|L\| \|g^*\|.$$

This shows that L^* is bounded, and that $\|L^*\| \leqslant \|L\|$. On the other hand, by Corollary 6.2.12,

$$\|Lf\| = \sup_{\|g^*\|=1} |g^*(Lf)| = \sup_{\|g^*\|=1} |L^*g^*(f)| \leqslant \sup_{\|g^*\|=1} \|L^*g^*\| \|f\| = \|L^*\| \|f\|,$$

whence $\|L\| \leqslant \|L^*\|$. Therefore $\|L^*\| = \|L\|$. □

6.5.3 Example. Consider first the familiar finite dimensional case where the operator may be represented by a matrix. Define $L : \mathbb{C}^n \to \mathbb{C}^n$ by setting

$$(Lf)_i = \sum_{j=1}^{n} \alpha_{ij} f_j \qquad (i = 1, \ldots, n).$$

Let the space $\mathbb{C}^n$ equipped with the ℓ_p norm be denoted by $\ell_p^{(n)}$. Then the dual of $\ell_1^{(n)}$ is $\ell_\infty^{(n)}$. With $g^* = (g_1^*, \ldots, g_n^*)$ any element of $[\ell_1^{(n)}]^*$,

$$\begin{aligned} g^*(Lf) &= \sum_{i=1}^{n} (Lf)_i g_i^* \\ &= \sum_{i=1}^{n} g_i^* \sum_{j=1}^{n} \alpha_{ij} f_j \\ &= \sum_{j=1}^{n} f_j \sum_{i=1}^{n} \alpha_{ij} g_i^* \\ &= (L^*g^*)(f), \end{aligned}$$

if

$$(L^*g^*)_j = \sum_{i=1}^{n} \alpha_{ij} g_i^*.$$

Thus the adjoint operator $L^* : \ell_\infty^{(n)} \to \ell_\infty^{(n)}$ of $L : \ell_1^{(n)} \to \ell_1^{(n)}$ corresponding to the matrix $[\alpha_{ij}]$ is represented by the transpose matrix $[\alpha_{ij}]^T$. However, it should be noted that even in finite dimensions the adjoint depends on the norm of the space and thus has more force than the algebraic "transpose". For example, the relation $\|L\| = \|L^*\|$ will only hold if the dual is correctly selected.

6.5.4 Example. For an infinite dimensional analogue of the operator in the previous example, take formally

$$(Lf)_i = \sum_{j=1}^{\infty} \alpha_{ij} f_j \qquad (i = 1, 2, \ldots). \tag{6.5.1}$$

If $\sup_j \sum_{i=1}^{\infty} |\alpha_{ij}|$ is finite, L is a bounded operator from ℓ_1 into itself and

$$\|L\| \leqslant \sup_j \sum_{i=1}^{\infty} |\alpha_{ij}| = m,$$

say. A calculation similar to that in Example 6.5.3 (the change of order of summation is easily justified) shows that the adjoint $L^* : \ell_\infty \to \ell_\infty$ is defined by

$$(L^*g^*)_j = \sum_{i=1}^{\infty} \alpha_{ij} g_i^* \qquad (j = 1, 2, \ldots).$$

The adjoint is thus obtained formally as before by taking the transpose of the infinite matrix.

It may be remarked that with the help of the adjoint it is possible to deduce that actually $\|L\| = m$. This follows from the relation $\|L\| = \|L^*\|$ (Theorem 6.5.2) on using the known value for $\|L^*\|$ (Theorem 3.4.7).

6.5.5 Example. Finally consider the integral operator K where

$$Kf(x) = \int_0^1 k(x, y) f(y) \, dy, \tag{6.5.2}$$

and assume for simplicity that $k : [0, 1] \times [0, 1] \to \mathbb{C}$ is continuous. Then evidently $K : \mathscr{L}_1(0, 1) \to \mathscr{L}_1(0, 1)$ is bounded. For $f \in \mathscr{L}_1(0, 1)$ and $g^* \in \mathscr{L}_\infty(0, 1)$, by Fubini's Theorem 2.4.17.

$$\int_0^1 g^*(x) \, dx \int_0^1 k(x, y) f(y) \, dy = \int_0^1 f(y) \, dy \int_0^1 k(x, y) g^*(x) \, dx.$$

The adjoint is therefore the operator $K^*: \mathscr{L}_\infty(0, 1) \to \mathscr{L}_\infty(0, 1)$ where

$$K^*g^*(y) = \int_0^1 k(x, y)\, g^*(x)\, dx.$$

The class of bounded operators from a Hilbert space $\mathscr{H}$ into itself is particularly important. Rather than use Definition 6.5.1 again, which would yield an adjoint mapping $\mathscr{H}^*$ into $\mathscr{H}^*$, it is more convenient to exploit the identification of $\mathscr{H}^*$ with $\mathscr{H}$ and define the adjoint on $\mathscr{H}$ itself.

6.5.6 *Definition.* Let $\mathscr{H}$ be a Hilbert space, and suppose that $L \in \mathscr{L}(\mathscr{H})$. The relation

$$(Lf, g) = (f, L^*g)$$

required to hold for all $f, g \in \mathscr{H}$ defines the bounded linear operator $L^* \in \mathscr{L}(\mathscr{H})$ called the (Hilbert space) **adjoint** of L.

A further minor difference between the definitions should be noticed, and this is that in the present case $(\alpha L)^* = \bar{\alpha}L^* (\alpha \in \mathbb{C})$, whereas in Banach space $(\alpha L)^* = \alpha L^*$; the explanation is that for the inner product, $(f, \alpha g) = \bar{\alpha}(f, g)$, but $\langle f, \alpha g\rangle = \alpha\langle f, g\rangle$ for the outer product. As it is always clear from the context which adjoint is being used, L^* will simply be referred to as the adjoint of L in both cases.

Since the adjoint of an operator L in $\mathscr{L}(\mathscr{H})$ itself lies in $\mathscr{L}(\mathscr{H})$, L^* is a more convenient object to deal with than the adjoint in Banach space. In particular there is a natural definition of "self-adjointness", whereas it is difficult to find a meaningful formulation of such a concept in Banach space. As we shall see, self-adjoint operators have a number of simplifying properties. For these reasons problems which may be tackled in Hilbert space are often relatively tractable.

6.5.7 *Definition.* Suppose $L \in \mathscr{L}(\mathscr{H})$ where $\mathscr{H}$ is a Hilbert space. L is said to be **self-adjoint** iff $L = L^*$.

6.5.8 *Example.* If $\Sigma\Sigma|\alpha_{ij}|^2 < \infty$, the operator $L: \ell_2 \to \ell_2$ defined by (6.5.1) is bounded. Since the inner product in ℓ_2 is $(f, g) = \Sigma f_i \bar{g}_i$,

$$\begin{aligned}(Lf, g) &= \sum_{i=1}^{\infty} (Lf)_i \bar{g}_i,\\ &= \sum_{i=1}^{\infty}\sum_{j=1}^{\infty} \alpha_{ij} f_j \bar{g}_i,\\ &= \sum_{j=1}^{\infty} f_j \overline{\sum_{i=1}^{\infty} \bar{\alpha}_{ij} g_i},\\ &= (f, L^*g),\end{aligned}$$

where

$$(L^*g)_j = \sum_{i=1}^{\infty} \bar{\alpha}_{ij} g_i.$$

The adjoint $L^* : \ell_2 \to \ell_2$ is therefore represented by the conjugate transpose of the infinite matrix. L is self-adjoint iff $\alpha_{ij} = \bar{\alpha}_{ji}$ $(i, j = 1, 2, \ldots)$, that is iff the matrix is *Hermitian*.

6.5.9 Example. The integral operator $K : \mathscr{L}_2(0, 1) \to \mathscr{L}_2(0, 1)$ defined by (6.5.2) is bounded, and its adjoint is given by

$$K^*g(y) = \int_0^1 \overline{k(x, y)}\, g(x)\, \mathrm{d}x.$$

K is self-adjoint iff the kernel is *Hermitian*: $k(x, y) = \overline{k(y, x)}$, or in the real case *symmetric*: $k(x, y) = k(y, x)$.

The first application of the adjoint is in the derivation of criteria for the solubility of the operator equation $Lf = g$. Assuming for simplicity that $g \in \mathscr{B}$ and $L \in \mathscr{L}(\mathscr{B})$, we may summarize the conclusions that were reached in Chapter 3 as follows. If L is bijective the equation will have a unique solution $f = L^{-1}g$ for any g, and since L^{-1} is continuous (Theorem 3.5.3) analytical problems do not arise to complicate the issue. If L is surjective there is always a solution, but this need not be unique, while if L is injective there is a (unique) solution if and only if $g \in R(L)$. In finite dimensions, L is bijective if it is injective, and there are a number of other powerful results of this type available (see Problem 6.16). In infinite dimensions the situation is more complex, and it may happen for example that L is injective but not surjective or surjective but not injective (see Problem 3.2). Nonetheless the finite dimensional results may be partially recovered as in the following theorem—for other results of this nature see Problem 6.17.

6.5.10 Theorem. *Let $\mathscr{B}$ and $\mathscr{C}$ be Banach spaces, and suppose that $L \in \mathscr{L}(\mathscr{B}, \mathscr{C})$. Then*

(i) $\overline{R(L)} = N(L^*)^{\perp}$;

(ii) *if $R(L)$ is closed, so is $R(L^*)$, and $R(L^*) = N(L)^{\perp}$.*

Proof. The proof of (i) is relatively simple and is left as an exercise. (ii) Since $N(L)^{\perp}$ is closed (Lemma 6.2.20) it is enough to show that $R(L^*) = N(L)^{\perp}$. The inclusion $R(L^*) \subset N(L)^{\perp}$ is easy, for if $g^* \in R(L^*)$, there is an f^* such that $L^*f^* = g^*$, and

$$g^*(f) = L^*f^*(f) = f^*(Lf) = 0 \qquad (f \in N(L)),$$

whence $g^* \in N(L)^{\perp}$.

To prove the opposite inclusion, take any $f^* \in N(L)^{\perp}$, and note that for each $g \in R(L)$ the complex number $f^*(f)$, where f is any vector such that $Lf = g$, is uniquely defined (for if $Lf_1 = Lf_2 = g$, then $f_1 - f_2 \in N(L)$ and $f^*(f_1 - f_2) = 0$). The relation $h^*(Lf) = f^*(f)$ $(f \in \mathscr{B})$ thus defines a functional h^* on $R(L)$. h^* is obviously linear. To prove that it is continuous, recall (Problem 3.10) that there is an m such that for any $g \in R(L)$ there exists an f with $Lf = g$ and with $\|f\| \leqslant m\|g\|$, whence

$$|h^*(g)| = |f^*(f)| \leqslant \|f^*\|\,\|f\| \leqslant m\|f^*\|\,\|g\|.$$

By the Hahn–Banach Theorem 6.2.8, h^* has an extension, $\tilde{h}^*$ say, to the whole of $\mathscr{C}$, and

$$L^*\tilde{h}^*(f) = \tilde{h}^*(Lf) = h^*(Lf) = f^*(f) \qquad (f \in \mathscr{B}).$$

It follows that $L^*\tilde{h}^* = f^*$, that is $f^* \in R(L^*)$. □

This result is an existence theorem. It is less satisfactory than the corresponding finite dimensional relation $R(L) = N(L^*)^{\perp}$, because only information on the closure of $R(L)$ rather than on $R(L)$ itself is obtained. Thus further restrictions on the operator are usually necessary before the theorem can usefully be applied. An example of such a case is encountered in the next chapter; there the range is closed and a powerful existence principle is deduced from the theorem.

The final result notes the connection between the inverses of L and L^*. The proof is closely analogous to that which will be used in Theorem 6.7.7 and is therefore omitted.

6.5.11 Theorem. *Let $\mathscr{B}$ and $\mathscr{C}$ be Banach spaces. The operator $L \in \mathscr{L}(\mathscr{B}, \mathscr{C})$ has a bounded inverse if and only if L^* has a bounded inverse, in which case $(L^{-1})^* = L^{*-1}$.*

6.6 Bounded Self-adjoint Operators—Spectral Theory

Self-adjoint operators form one of the simplest and at the same time most useful classes of linear operator. In this section attention is focused on the equation $(\lambda I - L)f = g$ when L is bounded and self-adjoint.

Theorem 6.5.10 is an attempt to characterize the range of a bounded linear operator in terms of the null space of its adjoint, and is relevant to the problem of deciding whether L has an inverse and whether the inverse is bounded. If L is replaced by $(\lambda I - L)$, the theorem yields information about the spectrum $\sigma(L)$. When L is self-adjoint, results in this area are much simplified and strengthened. For general $L \in \mathscr{L}(\mathscr{B})$ about all that can be said is that $\sigma(L)$ is contained in the disc $|\lambda| \leqslant \|L\|$. However, if L is self-adjoint, $\sigma(L)$ is real (compare the corresponding result for a Hermitian matrix) and estimates for its extent in terms of quantities which are relatively easy to evaluate are available.

Throughout this section $\mathscr{H}$ will be a *complex* Hilbert space (some of the results are false if $\mathscr{H}$ is real), $L : \mathscr{H} \to \mathscr{H}$ a bounded self-adjoint operator, and $\lambda = \mu + i\nu$ a complex number. The analysis starts with a technical result.

6.6.1 Lemma. *For all* $f \in \mathscr{H}$, $\|(\lambda I - L)f\| \geqslant |\text{Im}\, \lambda|\, \|f\|$.

Proof. By direct calculation,

$$\|(L - \lambda I)f\|^2 = \|(L - \mu I)f\|^2 + \nu^2\|f\|^2 + i\nu((L - \mu I)f, f) - i\nu(f, (L - \mu I)f).$$

As L is self-adjoint so is $(L - \mu I)$, and the last two terms cancel. The result follows immediately, since the first term on the right-hand side is non-negative. □

This lemma shows that the eigenvalues of L are real. We want to prove the stronger assertion that the whole spectrum—which may of course contain points other than eigenvalues—is real.

6.6.2 Lemma. *The closed subspaces* $N(\lambda I - L)$ *and* $\overline{R(\lambda I - L)}$ *are orthogonal complements of each other, and* $\mathscr{H} = N(\lambda I - L) \oplus \overline{R(\lambda I - L)}$.

Proof. This follows from Theorem 6.5.10. For if λ is real, $(\lambda I - L) = (\lambda I - L)^*$, and $\overline{R(\lambda I - L)} = N(\lambda I - L)$. On the other hand if $\nu \neq 0$, by the previous lemma $N(\bar{\lambda} I - L) = 0$, and since $(\lambda I - L)^* = (\bar{\lambda} I - L)$,

$$\overline{R(\lambda I - L)} = N(\bar{\lambda} I - L)^\perp = 0^\perp = \mathscr{H}.$$ □

6.6.3 Theorem. *Let* $\mathscr{H}$ *be a complex Hilbert space, and suppose that* $L : \mathscr{H} \to \mathscr{H}$ *is a bounded self-adjoint operator. Then* $\sigma(L)$ *is real, and*

$$\|(\lambda I - L)^{-1}\| \leqslant |\text{Im}\, \lambda|^{-1} \qquad (\text{Im}\, \lambda \neq 0).$$

Proof. As observed above, $\overline{R(\lambda I - L)} = \mathscr{H}$ if $\nu \neq 0$. Further, $\|(\lambda I - L)f\| \geqslant |\nu|\,\|f\|$, by Lemma 6.6.1. It follows from Lemma 3.8.18 that $(\lambda I - L)^{-1} \in \mathscr{L}(\mathscr{H})$, that is $\lambda \in \rho(L)$. The final assertion is obtained simply by setting $g = (\lambda I - L)f$ in the previous inequality. □

If $\mathscr{H}$ is finite dimensional $\sigma(L)$ evidently consists of eigenvalues only. The following is a partial substitute for this.

6.6.4 Theorem. *Let $\mathscr{H}$ be a complex Hilbert space, and suppose that $L : \mathscr{H} \to \mathscr{H}$ is bounded and self-adjoint. Then*

(i) *$\lambda \in \rho(L)$ iff there is an $m > 0$ such that*

$$\|(\lambda I - L)f\| \geqslant m\|f\| \qquad (f \in \mathscr{H}); \tag{6.6.1}$$

(ii) *$\lambda \in \sigma(L)$ iff there is a sequence (f_n) with*

$$\|f_n\| = 1, \lim \|(\lambda I - L)f_n\| = 0.$$

Proof. If $\lambda \in \rho(L)$,

$$\|f\| = \|(\lambda I - L)^{-1}(\lambda I - L)f\| \leqslant \|(\lambda I - L)^{-1}\| \|(\lambda I - L)f\|,$$

and (6.6.1) follows on taking $m^{-1} = \|(\lambda I - L)^{-1}\|$. The sufficiency of (6.6.1) is established by the argument used in proving the last theorem with m replacing $|v|$. (ii) is an obvious consequence of (i). □

The second part of this theorem provides an insight into the difference in the nature of the spectra in the finite and infinite dimensional cases. If $\mathscr{H}$ is finite dimensional the set $\{f_n\}$ is relatively sequentially compact, and there is a subsequence, denoted also by (f_n), which converges to a limit, f say. It follows that $Lf_n \to Lf$, and that f is an eigenfunction and λ an eigenvalue. Thus the fact that every point of the spectrum is an eigenvalue is recovered. For $\mathscr{H}$ infinite dimensional this argument breaks down because (f_n) need not have a convergent subsequence. On the other hand λ may loosely be regarded as an approximate eigenvalue. For even if there is no non-zero f with $(\lambda I - L)f = 0$, there is always a sequence (f_n) such that this equation holds approximately in the sense that $\|(\lambda I - L)f_n\| \to 0$.

A useful estimate for the extent of the spectrum is next derived. Note that since L is self-adjoint, the number (Lf, f) is real for any $f \in \mathscr{H}$.

6.6.5 Definition. For L self-adjoint set

$$m_- = \inf_{\|f\|=1} (Lf, f), \qquad m_+ = \sup_{\|f\|=1} (Lf, f).$$

6.6.6 Theorem. *Let $\mathscr{H}$ be a complex Hilbert space, and assume that $L : \mathscr{H} \to \mathscr{H}$ is bounded and self-adjoint. Then $\sigma(L)$ lies in the interval $[m_-, m_+]$.* Both *m_- and m_+ are in $\sigma(L)$.*

Proof. If $\lambda > m_+$,

$$(\lambda - m_+)\|f\|^2 \leqslant |((\lambda I - L)f, f)| \leqslant \|(\lambda I - L)f\| \|f\|.$$

Therefore $\|(\lambda I - L)f\| \geqslant (\lambda - m_+)\|f\|$, and by Theorem 6.6.4(i), $\lambda \in \rho(L)$. Similarly $\lambda \in \rho(L)$ if $\lambda < m_-$.

To prove the second assertion let $\lambda = m_-$. By the definition of m_-, there is a sequence (f_n) with $\|f_n\| = 1$ and $\lim((\lambda I - L)f_n, f_n) = 0$. Now from Problem 6.23 with $f = f_n$, $L = (\lambda I - L)$ and $g = (\lambda I - L)f_n$,

$$\|(\lambda I - L)f_n\|^4 \leqslant |((\lambda I - L)f_n, f_n)| \, \|\lambda I - L\|^3 \, \|f_n\|^2. \tag{6.6.2}$$

Hence $\lim\|(\lambda I - L)f_n\| = 0$, and it follows from Theorem 6.6.4(ii) that $\lambda \in \sigma(L)$. A similar proof holds for m_+.

6.6.7 Theorem. *Let $\mathscr{H}$ be a complex Hilbert space, and assume that L is bounded and self-adjoint. Then* $\|L\| = r_\sigma(L) = \max(|m_-|, |m_+|)$.

Proof. The second equality is a consequence of the previous theorem. To prove that $\|L\| = r_\sigma(L)$, note that by Theorem 3.7.11.

$$r_\sigma(L) = \lim\|L^n\|^{1/n}.$$

Therefore if it can be shown that $\|L^k\| = \|L\|^k$ for $k = 2^n$ and $n = 1, 2, \ldots,$ the result will follow. Now

$$\|Lf\|^2 = (Lf, Lf) = (L^2 f, f) \leqslant \|L^2\| \, \|f\|^2,$$

so $\|L\|^2 \leqslant \|L^2\|$. On the other hand, $\|L^2 f\| \leqslant \|L\| \, \|Lf\| \leqslant \|L\|^2 \, \|f\|$, whence $\|L^2\| \leqslant \|L\|^2$. Therefore $\|L^2\| = \|L\|^2$. Since $L^2, L^4, \ldots$ are also self-adjoint, the same argument shows that $\|L^k\| = \|L\|^k$ for $k = 2^n$. □

For general bounded L, $r_\sigma(L)$ may be strictly less than $\|L\|$, and it may even happen, as with the Volterra operator in Problem 3.24, that the spectrum consists of the single point zero. If L is self-adjoint both these possibilities are ruled out by the previous theorem.

6.6.8 Example. Several of the points raised in this section are illustrated by considering the integral equation $(\lambda I - K)f = g$, where

$$Kf(x) = (2\pi)^{-\frac{1}{2}} \int_{-\infty}^{\infty} k(x - y) f(y) \, dy \qquad (x \in \mathbb{R}).$$

If k is an even real valued function in $\mathscr{L}_1(-\infty, \infty)$, K is a bounded self-adjoint operator on $\mathscr{L}_2(-\infty, \infty)$. K is conveniently studied by means of the Fourier transform. By Theorem 2.6.1,

$$(Kf, f) = (\hat{k}\hat{f}, \hat{f}), \tag{6.6.3}$$

whence $m_- \geqslant k_- = \inf \hat{k}(t)$, $m_+ \leqslant k_+ = \sup \hat{k}(t)$, the "inf" and "sup" being over all $t \in \mathbb{R}$. We conclude from Theorem 6.6.6 that $\sigma(K) \subset [k_-, k_+]$. In fact $\sigma(K) = [k_-, k_+]$. To prove this let λ be any point in $[k_-, k_+]$, and note that

$$((\lambda I - K)^2 f, f) = ((\lambda - \hat{k})^2 \hat{f}, \hat{f}). \tag{6.6.4}$$

Since $\hat{k}$ is continuous, there is a t_0 such that $\hat{k}(t_0) = \lambda$. Define

$$\hat{f}_n(t) = \begin{cases} (\frac{1}{2}n)^{\frac{1}{2}} & (t \in [t_0 - n^{-1}, t_0 + n^{-1}]), \\ 0 & (t \notin [t_0 - n^{-1}, t_0 + n^{-1}]). \end{cases}$$

A brief calculation shows that $\lim((\lambda - \hat{k})\hat{f}_n, \hat{f}_n) = 0$, and it follows from (6.6.4) that $\lim((\lambda I - K)^2 f_n, f_n) = 0$. Therefore by (6.6.2), $\lim \|(\lambda I - K)f_n\| = 0$. Since $\|f_n\| = \|\hat{f}_n\| = 1$, it is a consequence of Theorem 6.6.4 that $\lambda \in \sigma(K)$.

Whether λ is in the point or continuous spectrum (the residual spectrum is empty, see Problem 6.3.1) depends on the measure of the set $\Lambda = \{t : \hat{k}(t) = \lambda\}$. To see this note that $(\lambda - \hat{k}(t))\hat{f}(t) = 0$ a.e. if f is a solution of $(\lambda I - K)f = 0$. If Λ has measure zero, $\hat{f} = 0$ a.e., and λ is not in the point spectrum and must be in the continuous spectrum. On the other hand, if Λ has positive measure, an eigenfunction may be constructed by defining $\hat{f} = 1$ on a compact subset of Λ of non-zero measure and $\hat{f} = 0$ otherwise. Then λ is in the point spectrum.

6.7 The Adjoint of an Unbounded Linear Operator in Hilbert Space

The definition of the adjoint will now be extended to unbounded operators. The discussion will be restricted to operators from a Hilbert space into itself, an approach broad enough to include our main application which is to operators connected with ordinary differential equations. Some extra care is needed in the unbounded case since the domains of the operators are proper subsets of the Hilbert space. Indeed a satisfactory theory is only possible if the domains are dense, but as this condition is satisfied in most important cases, and in particular for most differential operators, the restriction is not serious; if the domain is dense, most of the basic properties of the adjoint carry over almost unchanged.

Throughout $\mathscr{H}$ will be a complex Hilbert space and L will be a linear operator from $\mathscr{H}$ into $\mathscr{H}$ with domain $D(L)$ dense in $\mathscr{H}$†. Suppose that M is a linear operator such that for all $f \in D(L)$ and $g \in D(M)$,

$$(Lf, g) = (f, Mg). \tag{6.7.1}$$

If L were bounded and if this relation were to hold for all f, g in $\mathscr{H}$, then M would be the uniquely defined adjoint of L. However, in the unbounded case (6.7.1) by itself does not define M uniquely, for (6.7.1) does not specify $D(M)$—and indeed is satisfied by any restriction of M. It is plausible although not obvious that of all the operators satisfying (6.7.1) there will be one with

† It is important to bear in mind that the specification of a domain is an essential part of the definition of an unbounded operator.

a domain which is maximal (in the sense of set inclusion). It is this operator, L^* say, which provides the required generalization of the adjoint. The choice of $D(L^*)$ is clarified as follows.

Let $D(L^*)$ be the set of $g \in \mathscr{H}$ such that there exists an h in $\mathscr{H}$ with $(f, h) = (Lf, g)$ for all $f \in D(L)$. Given g, the element h is uniquely determined, for if there is a k such that $(f, k) = (Lf, g)$, then $(f, h - k) = 0$, and as $D(L)$ is dense, $h = k$. (Note that unless $D(L)$ is dense this definition does not make sense). Now set $h = L^*g$. It is easy to check that L^* is linear, and clearly $(Lf, g) = (f, L^*g)$ for all $f \in D(L)$ and $g \in D(L^*)$. Thus (6.7.1) is satisfied with $M = L^*$, and further every M satisfying this equation is a restriction of L^*. Therefore, as asserted above, $D(L^*)$ is maximal. In most cases of interest $D(L^*)$ itself is dense in $\mathscr{H}$.

6.7.1 Definition. Suppose L is a linear operator from $\mathscr{H}$ into $\mathscr{H}$ with dense domain. Let $D(L^*)$ be the set of elements g such that there is an h with $(Lf, g) = (f, h)$ for all $f \in D(L)$. Let L^* be the operator with domain $D(L^*)$ and with $L^*g = h$ on $D(L^*)$, or equivalently assume that

$$(Lf, g) = (f, L^*g) \qquad (f \in D(L),\, g \in D(L^*)).$$

L^* is called the **adjoint** of L.

In classical differential equation theory the term "adjoint" is used for a certain formal operator (called the *formal adjoint* here to fit in with Definition 3.8.3), and thus implies merely a description of the coefficients. The following example emphasizes that the operator theoretic adjoint is a much deeper concept in which the boundary conditions are also involved in an essential manner.

6.7.2 Example. Consider the formal differential operator $l = i\mathrm{d}/\mathrm{d}x$ on the interval $[0, 1]$. Let $\mathscr{A}$ be the linear subspace of $\mathscr{H} = \mathscr{L}_2(0, 1)$ consisting of absolutely continuous functions with derivatives in $\mathscr{L}_2(0, 1)$. Take $D(L) = \mathscr{A}$ and define $Lf = if'$ on $D(L)$. An integration by parts shows that for $f \in D(L)$,

$$\begin{aligned}(Lf, g) &= \int_0^1 if'(x)\,\bar{g}(x)\,\mathrm{d}x \\ &= i[f(1)\,\bar{g}(1) - f(0)\,\bar{g}(0)] + \int_0^1 f(x)\overline{(ig)'(x)}\,\mathrm{d}x \\ &= i[f(1)\,\bar{g}(1) - f(0)\,\bar{g}(0)] + (f, Mg)\end{aligned}$$

say, if $Mg = ig'$ and $D(M)$ is any set of smooth functions. Equation (6.7.1) will be satisfied if the square bracketed term vanishes, that is if $g(0) = g(1) = 0$.

However, this condition alone is not sufficient to specify $D(M)$. Noting that $D(L^*)$ must be maximal, we may speculate that a reasonable candidate for $D(L^*)$ is the set $\mathscr{A}' = \{g : g \in \mathscr{A}, g(0) = g(1) = 0\}$. To prove that indeed $D(L^*) = \mathscr{A}'$ we proceed as follows.

Obviously $\mathscr{A}' \subset D(L^*)$, so the result will follow if the opposite inclusion can be verified. Take any $g \in D(L^*)$ and set $L^*g = h$. Then for all $f \in D(L)$,

$$\int_0^1 f(x)\,\bar{h}(x)\,dx = (f, h) = (f, L^*g) = (Lf, g) = \int_0^1 if'(x)\,\bar{g}(x)\,dx. \qquad (6.7.3)$$

Now put $k(x) = \int_0^x h(t)\,dt$, and note that $k' = h, k(0) = 0$, and $k \in \mathscr{A}$. On integration by parts and substitution for h, (6.7.3) becomes

$$\int_0^1 f'(x)\,[i\bar{g}(x) + \bar{k}(x)]\,dx = f(1)\,g(1). \qquad (6.7.4)$$

If $v \in \mathscr{H}$ and $f(x) = \int_x^1 v(t)\,dt$, then $f \in D(L), f' = v$ and $f(1) = 0$. Therefore from (6.7.4), $(v, -ig + k) = 0$ for all $v \in \mathscr{H}$. This shows that $k = ig$, and it follows that $g \in \mathscr{A}$ and $g(0) = 0$. To prove that $g(1) = 0$ note that (6.7.4) is simply $f(1)\,k(1) = 0$ $(f \in D(L))$. Therefore $0 = k(1) = ig(1)$, since $D(L)$ includes functions which do not vanish at $x = 1$. This proves that $g \in \mathscr{A}'$, whence $D(L^*) \subset \mathscr{A}'$.

As another example, boundary conditions may be imposed and $D(L)$ taken to be $\{f : f \in \mathscr{A}, f(0) = f(1) = 0\}$. A similar argument proves that $D(L^*) = \mathscr{A}$.

Notice that in neither case is L equal to L^* although l is "self-adjoint" in the sense of classical differential equation theory.

In the study of the adjoint of an unbounded operator the graph (Definition 3.8.7) plays an essential part. Recall that the graph $G(L)$ and the inverse graph $G'(L)$ are respectively the subsets of $\mathscr{H} \times \mathscr{H}$ consisting of all pairs $[f, Lf]$ and $[Lf, f]$ for $f \in D(L)$. The result which makes the use of the graph effective is the following.

6.7.3 Lemma. *$G'(-L^*) = G(L)^{\perp}$, the orthogonal complement being taken in $\mathscr{H} \times \mathscr{H}$.*

Proof. From the definition (3.8.3) of the inner product in $\mathscr{H} \times \mathscr{H}$,

$$([f, Lf], [-L^*g, g]) = (f, -L^*g) + (Lf, g) \qquad (f \in D(L), g \in D(L^*)).$$

By the definition of L^* the right-hand side is zero. Therefore $G'(-L^*) \subset G(L)^{\perp}$. To prove the opposite inclusion, let $[h, g]$ be any element in $G(L)^{\perp}$. Then

$$(f, h) + (Lf, g) = ([f, Lf], [h, g]) = 0.$$

By the definition of L^* this implies that $g \in D(L^*)$ and $h = -L^*g$. That is $G(L)^{\perp} \subset G'(-L^*)$. □

As in the previous discussion of unbounded operators (Section 3.8) the concept of closedness will be important. In the following the *second adjoint* L^{**} is the operator $(L^*)^*$, and as usual $\bar{L}$ denotes the closure of L.

6.7.4 Theorem. *The adjoint is a closed operator.*

Proof. The orthogonal complement $G(L)^{\perp}$ is a closed set, and it follows from the previous lemma that $G'(-L^*)$ is closed, that is L^* is closed. □

6.7.5 Theorem. *The domain of L^* is dense if and only if L is closable. In this case $L^{**} = \bar{L}$. In particular, $L^{**} = L$ if L is closed.*

Proof. Suppose first that $D(L^*)$ is dense. If $[0, g] \in G'(-L^*)^{\perp}$, then $(g, f) = ([0, g], [-L^*f, f]) = 0$ for all $f \in D(L^*)$. Therefore $g = 0$, and it follows that $G'(-L^*)$ is a graph (Lemma 3.8.11). Further, by Lemma 6.7.3, $G(L) \perp G'(-L^*)$. Hence $\overline{G(L)} \subset G'(-L^*)^{\perp}$, and since every linear subspace of a graph is itself a graph (Problem 3.28), $\overline{G(L)}$ is a graph.

Next assume that L is closable. Then $G(\bar{L})$ is closed. But $(\bar{L})^* = L^*$ (Problem 6.26), and this implies that $G(\bar{L}) = G'(-L^*)^{\perp}$. If $D(L^*)$ is not dense there is a non-zero g with $g \perp D(L^*)$. A short calculation shows that $[0, g] \in G(-L^*)^{\perp} = G(\bar{L})$, and this contradicts the fact that $G(\bar{L})$ is a graph. Therefore $D(L^*)$ is dense.

Finally $L^{**} = \bar{L}$, since $G(L^{**}) = G(-L^*)^{\perp} = G(\bar{L})$. □

The results relating the ranges and null spaces of L and L^* are almost identical with those for bounded operators. The next two theorems are the analogues of Theorems 6.5.10 and 6.5.11.

6.7.6. Theorem. $\overline{R(L)} = N(L^*)^{\perp}$.

Proof. An exercise. □

6.7.7 Theorem. *Suppose L is closed. Then $L^{-1} \in \mathscr{L}(\mathscr{H})$ if and only if $L^{*-1} \in \mathscr{L}(\mathscr{H})$. In this case $(L^{-1})^* = L^{*-1}$.*

Proof. If $L^{-1} \in \mathscr{L}(\mathscr{H})$, by Theorem 6.5.2, $(L^{-1})^* \in \mathscr{L}(\mathscr{H})$. For any $g \in D(L^*)$ and $h \in \mathscr{H}$,

$$((L^{-1})^*L^*g, h) = (L^*g, L^{-1}h) = (g, LL^{-1}h) = (g, h).$$

Therefore $(L^{-1})^* L^*g = g$. Also $((L^{-1})^* f, Lg) = (f, L^{-1}Lg) = (f, g)$ if $f \in \mathscr{H}$ and $g \in D(L)$. It follows from the definition of the adjoint that $(L^{-1})^* f \in D(L^*)$ and that $L^*(L^{-1})^* f = f$. These two relations are exactly what is needed to prove that L^{*-1} exists and is equal to $(L^{-1})^*$.

On the other hand if $L^{*-1} \in \mathscr{L}(\mathscr{H})$, by what has just been proved, $L^{**-1} \in \mathscr{L}(\mathscr{H})$. Since $L^{**} = L$ (Theorem 6.7.5) we conclude that $L^{-1} \in \mathscr{L}(\mathscr{H})$. □

Largely because of their appearance in differential equation theory, unbounded self-adjoint operators are extremely important in applications.

6.7.8 Definition. A densely defined linear operator L from a Hilbert space into itself is said to be **self-adjoint** iff $L = L^*$. (Notice that necessarily $D(L) = D(L^*)$).

6.7.9 Example. Consider the construction of a self-adjoint operator from the formal operator $l = i\mathrm{d}/\mathrm{d}x$ on $[0, 1]$. The central problem arises in choosing boundary conditions which will ensure that $D(L) = D(L^*)$, and some insight can be gained here by using a formal argument.

Take $\mathscr{H} = \mathscr{L}_2(0, 1)$ and let $\mathscr{A}$ be the subset of $\mathscr{H}$ consisting of absolutely continuous functions with derivatives in $\mathscr{H}$. Consider provisionally operators L and M both of whose domains are subsets of $\mathscr{A}$ with $Lf = lf$ and $Mf = lf$ on $D(L)$ and $D(M)$ respectively. An integration by parts gives

$$(Lf, g) - (f, Mg) = i[f(1)\,\bar{g}(1) - f(0)\,\bar{g}(0)].$$

If $D(L)$ is fixed, M will be the adjoint of L if it has the maximal domain consistent with the vanishing of the square bracketed term for all $f \in D(L)$. Clearly the larger $D(L)$ is, the smaller $D(L^*)$ will be. For L to be self-adjoint this maximal domain must be equal to $D(L)$. From Example 6.7.2, if $D(L) = \mathscr{A}$, $D(L^*) = \mathscr{A}' = \{f : f \in \mathscr{A}, f(0) = f(1) = 0\}$; evidently $D(L)$ is too large. On the other hand if $D(L) = \mathscr{A}'(L^*) = \mathscr{A}$ and $D(L)$ is too small. It is plausible that the correct domain should lie between $\mathscr{A}$ and $\mathscr{A}'$, so let us try $D(L) = \{f : f \in \mathscr{A}, f(1) = \alpha f(0)\}$ for some $\alpha \in \mathbb{C}$. Then

$$[f(1)\,\bar{g}(1) - f(0)\,\bar{g}(0)] = f(0)[\alpha\bar{g}(1) - \bar{g}(0)],$$

and if this expression is to vanish for all $f \in D(L)$, it is necessary that $\alpha\bar{g}(1) = \bar{g}(0)$. However, this condition must be the same as the original condition $f(1) = \alpha f(0)$, which will be the case if $\alpha\bar{\alpha} = 1$. This argument suggests that a suitable domain is

$$D(L) = \{f : f \in \mathscr{A}, f(1) = \mathrm{e}^{i\theta} f(0)\} \qquad (\theta \in \mathbb{R}),$$

a conjecture which may be verified by a procedure similar to that of Example 6.7.2. There are thus an infinity of self-adjoint operators based on $i\mathrm{d}/\mathrm{d}x$ for the interval $[0, 1]$.

It is not difficult to guess an appropriate boundary condition in this simple example. However, for second order l, if the interval is infinite or l is singular, the choice of boundary conditions presents a decidedly thorny problem. Systematic methods guiding this choice will be developed in Chapter 10.

Bounded and unbounded self-adjoint operators have many properties in common, the fact that unbounded self-adjoint operators are closed (Theorem 6.7.4) being enough to compensate for the lack of continuity. The principal result concerning the spectrum is identical.

6.7.10 Theorem. *Let $\mathscr{H}$ be a complex Hilbert space, and suppose L is an (unbounded) self-adjoint operator. Then $\sigma(L)$ is real, and*

$$\|(\lambda I - L)^{-1}\| \leqslant |\operatorname{Im} \lambda|^{-1} \qquad (\operatorname{Im} \lambda \neq 0).$$

Proof. The argument is almost the same as that in Theorem 6.6.3. A repetition of the proof of Lemma 6.6.1 with f restricted to $D(L)$ yields the inequality $\|(\lambda I - L)f\| \geqslant |\operatorname{Im} \lambda| \, \|f\|$. The relation $\mathscr{H} = N(\lambda I - L) \oplus \overline{R(\bar{\lambda} I - L)}$ follows from Theorem 6.7.6. Therefore $\overline{R(\lambda I - L)} = \mathscr{H}$ if $\operatorname{Im} \lambda \neq 0$. Since L is closed, Lemma 3.8.18 is still applicable, and we conclude that $(\lambda I - L)^{-1} \in \mathscr{L}(\mathscr{H})$. □

Problems

6.1 Prove Corollary 6.2.9. (Hint: Every f in $\mathscr{M}_1 = \mathscr{M} + [g]$ can be written uniquely as $f = h + \lambda g$ ($h \in \mathscr{M}$, $\lambda \in \mathbb{C}$). Apply the Hahn–Banach Theorem to the functional f^* on $\mathscr{M}_1$ defined by $f^*(h + \lambda g) = \lambda$). Deduce Corollaries 6.2.10–6.2.12.

6.2 Suppose $g \in \mathscr{L}_p(0, 1)$ for some $p > 1$, and assume that

$$\int_0^1 f(x)\, g(x)\, dx = 0 \qquad (f \in \mathscr{C}_0^\infty([0, 1])).$$

Using Corollary 6.2.11 and Theorem 2.5.6, prove that $g = 0$ a.e.

6.3 (A geometrical version of the Hahn–Banach Theorem). Let $\mathscr{B}$ be a real Banach space, and suppose f^* is a non-zero element of $\mathscr{B}^*$. By analogy with $\mathbb{R}^3$, for real a the set $\{f : f^*(f) = a\}$ is called a hyperplane; it divides $\mathscr{B}$ into "half-spaces" $f^*(f) \geqslant a$, $f^*(f) \leqslant a$. Suppose S is the closed unit ball in $\mathscr{B}$, and assume that $f_0 \in \partial S$. f^* is said to be *tangent* to S at f_0 iff $f^*(f) \leqslant f^*(f_0)$ for all $f \in S$, and the hyperplane $f^*(f) = f^*(f_0)$ is called a *tangent hyperplane* to S at f_0. Using Corollary 6.2.10 show that there is a tangent hyperplane at every point $f_0 \in \partial S$.

6.4 Let $\mathscr{M}$ be a linear subspace which is not dense in the Banach space $\mathscr{B}$. Prove that there is a non-zero $f^* \in \mathscr{B}^*$ with $f^*(f) = 0$ ($f \in \mathscr{M}$).

6.5 Prove Lemma 6.2.20.

6.6 Show that $\{\mathcal{M}^{\perp}\}^{\perp} = \mathcal{M}$ if $\mathcal{M}$ is a closed subspace.

6.7 Suppose that $S = \{\phi_n\}$ is a countable subset of the Banach space $\mathcal{B}$. Show that $f \in [S]$—the closed linear span of S—iff $f^*(f) = 0$ for all $f^* \in \mathcal{B}^*$ such that $f^*(\phi_n) = 0$ $(n = 1, 2, \ldots)$.

6.8 In ℓ_2 construct a sequence which is weakly but not strongly convergent.

6.9 Let (f_n^*) be a sequence in $\mathcal{B}^*$. Suppose that $(f_n^*(f))$ is convergent for all $f \in \mathcal{B}$. Show that there is an $f^* \in \mathcal{B}^*$ such that $\lim f_n^*(f) = f^*(f)$ for all $f \in \mathcal{B}$. (Use Corollary 3.5.12).

6.10 Assume that S^* is a dense subset of $\mathcal{B}^*$. Suppose that the sequence (f_n) in $\mathcal{B}$ is bounded, and that $\lim f^*(f_n)$ exists for each $f^* \in S^*$. Prove that the limit exists for all $f^* \in \mathcal{B}^*$.

6.11 Let (f_n) be a sequence in Hilbert space. Prove that $f_n \to f$ if $f_n \rightharpoonup f$ and $\|f_n\| \to \|f\|$.

6.12 (Lemma 4.4.7). Let $\mathcal{B}$ and $\mathcal{C}$ be Banach spaces, let D be a convex subset of $\mathcal{B}$, and suppose that $A : D \to \mathcal{C}$ is Fréchet differentiable on D. Prove that

$$\|Af - Ag\| \leqslant \|f - g\| \sup_{h \in D} \|A'(h)\|.$$

(Prove first for real spaces: apply the Mean Value Theorem to $f^*(A[f + \theta(f - g)])$ regarded as a function of θ).

6.13 With $\mathcal{B}, \mathcal{C}, \mathcal{D}$ Banach spaces, suppose $L \in \mathcal{L}(\mathcal{B}, \mathcal{C})$, $M \in \mathcal{L}(\mathcal{C}, \mathcal{D})$. Show that $(ML)^* = L^*M^*$.

6.14 Under the natural imbedding of $\mathcal{B}, \mathcal{C}$ into $\mathcal{B}^{**}, \mathcal{C}^{**}$ respectively, an operator $L \in \mathcal{L}(\mathcal{B}, \mathcal{C})$ may be regarded as an operator with domain in $\mathcal{B}^{**}$ and range in $\mathcal{C}^{**}$. Show that L^{**} is an extension of L.

6.15 In Example 6.5.5 find the norm of $K : \mathcal{L}_1(0, 1) \to \mathcal{L}_1(0, 1)$.

6.16 Let $\mathcal{V}$ be a finite dimensional normed vector space, and suppose $L : \mathcal{V} \to \mathcal{V}$ is a linear operator. Prove that $R(L)^{\perp} = N(L^*)$ and that $R(L) = N(L^*)^{\perp}$. Deduce the relations $R(L^*) = N(L)^{\perp}$, $R(L^*)^{\perp} = N(L)$.

6.17 Let $\mathcal{B}$ be a Banach space and let $L \in \mathcal{L}(\mathcal{B})$. Show that: $R(L)^{\perp} = N(L^*)$; $R(L^*)^{\perp} = N(L)$; $\overline{R(L^*)} = N(L)^{\perp}$.

6.18 Let $\mathcal{H}$ be a complex Hilbert space. A complex valued function B on $\mathcal{H} \times \mathcal{H}$ is said to be a *bilinear form* iff for fixed g, $B(\cdot, g)$ is linear, and for fixed f, $B(f, \cdot)$ is *antilinear* (that is $\overline{B(f, \cdot)}$ is linear). B is *Hermitian* iff in addition $B(f, g) = \overline{B(g, f)}$, and *bounded* iff there is a c such that $|B(f, g)| \leqslant c\|f\|\,\|g\|$ for all $f, g \in \mathcal{H}$. Show that:

(i) (Lf, g) is a bounded Hermitian form if $L \in \mathcal{L}(\mathcal{H})$ is self-adjoint;
(ii) if B is a bounded Hermitian form, there is a self-adjoint $L \in \mathcal{L}(\mathcal{H})$ such that $(Lf, g) = B(f, g)$;

(iii) if B is Hermitian, it may be expressed in terms of the *quadratic form* $B(f) = B(f, f)$ as follows:

$$4B(f, g) = B(f + g) - B(f - g) + iB(f + ig) - iB(f - ig).$$

6.19 Assume that $L \in \mathscr{L}(\mathscr{H})$ where $\mathscr{H}$ is a complex Hilbert space. Prove that (Lf, f) is real if L is self-adjoint. Show on the other hand that L is self-adjoint if (Lf, f) is real for all $f \in \mathscr{H}$ (this need not be true if $\mathscr{H}$ is real).

6.20 Let L be a bounded self-adjoint operator. Prove that: (i) L is injective if it is surjective (false if L is not self-adjoint, see Problem 3.2); (ii) if L is injective, $R(L)$ is dense in $\mathscr{H}$.

6.21 Give an example of an operator $L \in \mathscr{L}(\mathscr{H})$ with $\|Lf\| = \|f\|$ for all $f \in \mathscr{H}$, but with $0 \in \sigma(L)$. Theorem 6.6.4 is thus false for an arbitrary bounded operator.

6.22 Show that the eigenfunctions of a self-adjoint operator corresponding to different eigenvalues are orthogonal.

6.23 Suppose $L \in \mathscr{L}(\mathscr{H})$ is self-adjoint, and assume that L is "positive", that is $(Lf, f) \geqslant 0$ $(f \in \mathscr{H})$. Prove that

$$|(Lf, g)|^2 \leqslant (Lf, f)(Lg, g) \qquad (f, g \in \mathscr{H}).$$

(Set $[f, g] = (Lf, g)$ and follow the proof of Schwartz's inequality 1.5.4).

6.24 The parallel disc condenser problem is governed by the integral equation

$$f(x) - \pi^{-1} \int_{-1}^{1} \frac{b}{b^2 + (x - y)^2} f(y)\, \mathrm{d}y = g(x) \qquad (0 \leqslant x \leqslant 1),$$

where $b > 0$. Show that the spectrum of the integral operator in $\mathscr{L}_2(-1, 1)$ is contained in $[0, 1]$. (Modify Example 6.6.8). This is not quite enough to prove that the Neumann series converges, but see Problem 7.15.

6.25 The following integral equation (Love, 1974) occurs in the calculation of the capacitance of a disc of radius 1 with a hole of radius $k < 1$ punched in it:

$$f(x) - 2\pi^{-1} \int_{0}^{k} \frac{xt}{1 - x^2 t^2} f(t)\, \mathrm{d}t = g(x) \qquad (0 \leqslant x \leqslant k).$$

Using an estimate for m_+ (Definition 6.6.5) show that the Neumann series converges in $\mathscr{L}_2(0, k)$ if

$$\frac{1}{2} \log \frac{1 + k}{1 - k} - \tan^{-1} k < \pi.$$

6.26 Assume that L is a densely defined linear operator from a Hilbert space into itself.

(i) Show that $N(L^*) = R(L)^\perp$. Deduce that $N(L) = R(L^*)^\perp$ if L is closed.
(ii) Prove that $(\bar{L})^* = L^*$ if L is closable.
(iii) If $L \subset L_1$, show that $L^* \supset L_1^*$.

6.27 Prove that Theorem 6.6.4 is valid for an unbounded self-adjoint operator.

6.28 For self-adjoint L define

$$m_- = \inf_{f \in D(L)} (Lf, f)/\|f\|^2, \qquad m_+ = \sup_{f \in D(L)} (Lf, f)/\|f\|^2,$$

the values $\pm\infty$ being allowed. Show that if $\lambda \in \sigma(L)$ then $m_- \leqslant \lambda \leqslant m_+$. Deduce that L is bounded if m_- and m_+ are finite.

6.29 If L is self-adjoint and $L^{-1} \in \mathscr{L}(\mathscr{H})$, then L^{-1} is self-adjoint.

6.30 Take $l = i d/dx$ and consider the interval $[0, 1]$. With $\mathscr{H} = \mathscr{L}_2(0, 1)$, let $\mathscr{A}$ be the subset of absolutely continuous functions with derivatives in $\mathscr{L}_2(0, 1)$. Take $D(L) = \{f : f \in \mathscr{A}, f(0) = f(1) = 0\}$ and set $Lf = lf$ on $D(L)$. Prove that $D(L^*) = \mathscr{A}$.

6.31 Show that the residual spectrum (Definition 3.73) of a self-adjoint operator is empty.

Chapter 7

Linear Compact Operators

7.1 Introduction

At the end of the last century the Swedish mathematician Ivar Fredholm proved a group of results concerning linear integral equations which was to have a profound effect on analysis. It is not appropriate here to follow through in detail the historical development, for which Bernkopf (1966), Bourbaki (1969), Steen (1973) and Monna (1973) may be consulted, but suffice it to say that Fredholm's results provided the key to the discovery of the great body of mathematics now called functional analysis. Our object in this chapter is to outline the theory of linear compact operators which is the direct generalization of Fredholm's results. This theory is extremely important in applications—as indeed is the allied theory of nonlinear compact operators, several aspects of which will be discussed later.

To motivate the direction of investigation some of the principal results for Fredholm integral equations are first recalled. Consider the equations

$$\lambda f(x) - \int_a^b k(x, y)\, f(y)\,\mathrm{d}y = g(x), \tag{7.1.1}$$

$$\lambda f(x) - \int_a^b k(x, y)\, f(y)\,\mathrm{d}y = 0, \tag{7.1.2}$$

where $\lambda \in \mathbb{C}$ and g, k are given continuous functions. These equations are known respectively as the inhomogeneous and homogeneous Fredholm integral equations of the second kind. A value of λ for which (7.1.2) has a non-zero continuous solution is called an eigenvalue, and the solution an eigenfunction. The celebrated Fredholm Alternative Theorem states the following: If $\lambda \neq 0$ and the homogeneous equation has only the zero solution, then the inhomogeneous equation has exactly one solution. On the other hand, if $\lambda \neq 0$ is an eigenvalue, the inhomogeneous equation will have a solution if and only if g is orthogonal to all the corresponding eigenfunctions of the adjoint equation (that is the equation with kernel $k(y, x)$). This result provides a powerful technique for establishing existence and uniqueness for (7.1.1). The restriction that λ is non-zero is necessary (see Example 7.4.4), but for non-zero λ the analogy with operators on a finite dimensional space is excellent, compare for example Theorem 7.3.7.

The Fredholm theory, in addition, gives detailed information about the eigenfunctions and eigenvalues which are obviously crucial in determining the properties of (7.1.1). It states that the eigenvalues form a countable set $\{\lambda_n\}$ say, that the only limit point of this set is zero, and that to each non-zero eigenvalue there corresponds a finite number of linearly independent eigenfunctions. If the kernel is Hermitian, further powerful theorems are available on the generalized Fourier expansion of an arbitrary function in terms of the eigenfunctions, and under certain circumstances it is possible to recover the analogue of the result that an n-vector may be expressed as a linear combination of the eigenvectors of a Hermitian $n \times n$ matrix.

Our aim will be to show that all these results are also true for compact linear operators on a Banach space. There are a multitude of applications of the theory. The most obvious is to a simple derivation of the principal results of Fredholm on integral equations. However, this topic is extremely well covered in the literature—see for example Riesz and Nagy (1955)—and some different illustrations are chosen here. Three of these may be mentioned in particular. First, using the theory of compact self-adjoint operators of Section 7.5, eigenfunction expansions in terms of orthogonal functions may be derived for a wide class of differential equation. Second a theory covering the numerical solution of integral equations will be outlined in Section 7.6. Lastly, in Chapter 11 we shall show how an application of compact operator theory in "Sobolev spaces" leads to an elegant direct treatment of linear elliptic partial differential equations. This approach is of special importance since it contains several of the essentials for a theory of nonlinear partial differential equations.

7.2 Examples of Compact Operators

7.2.1 Definition. Let $\mathscr{B}$ and $\mathscr{C}$ be Banach spaces, and let $T : \mathscr{B} \to \mathscr{C}$ be a linear operator. T is **compact** iff the image $T(S)$ of every bounded set S in $\mathscr{B}$

is relatively compact (in $\mathscr{C}$). The term "completely continuous" is sometimes used in the literature instead of "compact".

Obviously it is sufficient for compactness that the image of the closed unit ball should be relatively compact. Note also that since relative compactness and relative sequential compactness are equivalent (Theorem 5.2.4), T is compact iff for each bounded sequence (f_n), the sequence (Tf_n) contains a convergent subsequence.

It is easy to see that every compact operator is bounded. In finite dimensions, every bounded operator is compact, but this is not true in infinite dimensions (consider the identity). The notion of a compact operator originated in the context of Hilbert space. The original definition required that $T : \mathscr{H} \to \mathscr{H}$ should map every weakly convergent sequence into a convergent sequence. In Hilbert space (and indeed in a reflexive Banach space) the two definitions are equivalent, but this is not true in general, see Problem 7.4.

The following results assist in the characterization of compact operators.

7.2.2 Theorem. *Let $\mathscr{B}$ and $\mathscr{C}$ be Banach spaces. If (T_n) is a sequence of compact operators in $\mathscr{L}(\mathscr{B}, \mathscr{C})$ that converge uniformly to T, then T is compact, that is the set of compact operators in $\mathscr{L}(\mathscr{B}, \mathscr{C})$ is closed.*

Proof. In outline the tactics are as follows. Since total boundedness and relative compactness are equivalent (Theorem 5.2.8), if S is the closed unit ball, $T_n(S)$ is totally bounded for each n. However, $\| T_n - T\|$ is small for large n, and from this it will be deduced that $T(S)$ is also totally bounded, whence T is compact. The details are now given.

Since $T_n \to T$, given any $\varepsilon > 0$ there is an integer n such that $\| T_n f - Tf\| < \varepsilon$ for all $f \in S$. As $T_n(S)$ is totally bounded, there are a finite number of points, $f_1, \ldots, f_k$ say, such that

$$\inf_j \| T_n f - T_n f_j\| < \varepsilon \qquad (f \in S). \tag{7.2.1}$$

Therefore, for any $f \in S$,

$$\begin{aligned} \| Tf - Tf_j\| &\leqslant \| Tf - T_n f\| + \| T_n f - T_n f_j\| + \| T_n f_j - Tf_j\| \\ &\leqslant 2\varepsilon + \| T_n f - T_n f_j\|. \end{aligned}$$

We conclude from (7.2.1) that $\| Tf - Tf_j\| < 3\varepsilon$ for some j, that is $T(S)$ is totally bounded. □

7.2.3 Lemma. *Let $\mathscr{B}$ and $\mathscr{C}$ be Banach spaces, and assume that $T : \mathscr{B} \to \mathscr{C}$ is linear.*

(i) *If the dimension of $R(T)$ is finite, T is compact.*
(ii) *If T is compact and $R(T)$ is closed, its dimension is finite.*

Proof. (i) Simply note that a bounded set in a finite dimensional Banach space is relatively compact (Lemma 5.2.5).

(ii) Since $R(T)$ is closed it is a Banach space. By the Open Mapping Theorem 3.5.1, the image $T(S)$ of the open unit ball S is open in $R(T)$. Thus there is a closed ball of non-zero radius contained in $T(S)$, and as T is compact this ball is compact. But by Lemma 5.2.5 this can only be the case if $R(T)$ is finite dimensional. □

We next consider some operators that commonly occur in applications and enquire whether they are compact. The examples illustrate two standard methods of proving compactness. The first is based on Theorem 7.2.2; we look for a sequence of simple compact operators which approximate T in norm. The second exploits the Arzelà–Ascoli Theorem 5.4.4.

7.2.4 Example. An operator corresponding to an infinite matrix is first examined. If $\Sigma\Sigma|\alpha_{ij}|^2 < \infty$, by Theorem 3.4.7 the operator $L : \ell_2 \to \ell_2$ defined by

$$(Lf)_i = \sum_j \alpha_{ij} f_j \qquad (i = 1, 2, \dots)$$

is bounded. Take a sequence (P_n) of operators with

$$P_n(f_1, f_2, \dots) = (f_1, \dots, f_n, 0, 0, \dots).$$

Then $P_n L$ has finite dimensional range, and by Lemma 7.2.3 is compact. Further $P_n L \to L$, for

$$\lim_{n\to\infty} \|P_n L - L\| \leqslant \lim_{n\to\infty} \sum_{i=n+1}^{\infty} \sum_{j=1}^{\infty} |\alpha_{ij}|^2 = 0.$$

We conclude from Theorem 7.2.2 that L is compact.

7.2.5 Example. For the integral operator K where

$$Kf(x) = \int_\Omega k(x, y) f(y)\, \mathrm{d}y, \tag{7.2.2}$$

what conditions on the kernel k and on Ω ensure compactness? The following results are enough for many purposes, but Dunford and Schwartz (1958, Chap. 6) or Zabreyko *et al.* (1975) may be consulted for further information. In the first theorem the possibility that k is a Green's function is allowed.

7.2.6 Theorem. *Let Ω be a closed and bounded subset of $\mathbb{R}^n$, and equip $\mathscr{C}(\Omega)$ with the sup norm. Assume that $k(x, y)$ is a continuous function of x, y for $x, y \in \Omega$ and $x \neq y$, and suppose that there are a number m and a number $\alpha < n$ such that $|k(x, y)| \leqslant m|x - y|^{-\alpha}$. Then $K : \mathscr{C}(\Omega) \to \mathscr{C}(\Omega)$ is compact.*

Proof. The result is first established for k continuous on $\Omega \times \Omega$. The method illustrates the use of the Arzelà–Ascoli Theorem.

Let S be the closed unit ball in $\mathscr{C}(\Omega)$. It will be proved that $K(S)$, which is the set of functions g of the form

$$g(x) = \int_\Omega k(x, y) f(y)\, \mathrm{d}y \qquad (f \in S),$$

is bounded and equicontinuous. The boundedness follows from the inequality $\|Kf\| \leqslant \|K\|\,\|f\|$ as $|Kf(x)| \leqslant \|Kf\|$ for $x \in \Omega$. To prove equicontinuity note that since k is uniformly continuous, given $\varepsilon > 0$ there exists a $\delta > 0$ such that $|k(x_1, y) - k(x_2, y)| < \varepsilon$ for all $|x_1 - x_2| < \delta$ and $y \in \Omega$. Hence

$$\begin{aligned} |g(x_1) - g(x_2)| &\leqslant \int_\Omega |k(x_1, y) - k(x_2, y)|\, f(y)\, \mathrm{d}y \\ &\leqslant V(\Omega)\|f\| \sup_{y \in \Omega} |k(x_1, y) - k(x_2, y)| \\ &\leqslant \varepsilon V(\Omega)\|f\| \qquad (|x_1 - x_2| < \delta), \end{aligned}$$

where $V(\Omega)$ denotes the volume of Ω. As δ is independent of f, equicontinuity follows. Hence by the Arzelà–Ascoli Theorem 5.4.4, $K(S)$ is relatively compact. Therefore K is a compact operator.

To complete the proof, the restriction that k be continuous at points with $x = y$ must be removed. Define a sequence (k_n) of continuous kernels by setting

$$k_n(x, y) = \begin{cases} k(x, y) & (|x - y| > n^{-1}), \\ n^\beta |x - y|^\beta k(x, y) & (|x - y| \leqslant n^{-1}), \end{cases}$$

for some $\alpha < \beta < n$, and let K_n be the integral operator with kernel k_n. Then

$$\lim_{n \to \infty} \sup_{x \in \Omega} \int_\Omega |k(x, y) - k_n(x, y)|\, \mathrm{d}y = 0,$$

whence $K \in \mathscr{L}(\mathscr{C}(\Omega))$ and $\lim K_n = K$. From the first part of the proof (K_n) is a sequence of compact operators, and the result follows from Theorem 7.2.2. □

If Ω is not bounded, the above proof will not be valid since the Arzelà–Ascoli Theorem is not applicable. In fact much stronger conditions on the kernel are then required if K is to be compact. To emphasize this, suppose that $\Omega = (-\infty, \infty)$ and take $k(x, y) = k(x - y)$ where k is a continuous function in $\mathscr{L}_1(-\infty, \infty)$. Then $\|Kf\| \leqslant \|k\|_1 \|f\|$, $\|\cdot\|_1$ being the $\mathscr{L}_1$ norm, and $K : \mathscr{C}((-\infty, \infty)) \to \mathscr{C}((-\infty, \infty))$ is bounded. So far the analogy with the case Ω compact is good, the extra condition that k should be in $\mathscr{L}_1(-\infty, \infty)$ being clearly necessary to ensure the convergence of the integral. However, K is not compact.

To prove this, let h be a function in $\mathscr{C}((-\infty, \infty))$ with support $[0, 1]$, and define the bounded sequence (f_n) by setting $f_n(x) = h(x - n)$ for $x \in \mathbb{R}$. A simple calculation shows that $Kf_n(x) = g(x - n)$ where

$$g(x) = \int_0^1 k(x - y)\, h(y)\, \mathrm{d}y.$$

g tends to zero at infinity, and since (Kf_n) is a sequence of translates of g, (Kf_n) has no convergent subsequence. Therefore K is not compact.

Consider next K as an operator on $\mathscr{L}_2(\Omega)$. If k is measurable and

$$|||k|||_2 = \int_\Omega \int_\Omega |k(x, y)|^2\, \mathrm{d}x\, \mathrm{d}y < \infty,$$

then $K : \mathscr{L}_2(\Omega) \to \mathscr{L}_2(\Omega)$ is bounded and $\|K\| \leqslant |||k|||_2$ (Theorem 3.4.10). To prove that K is compact, start by assuming that Ω is compact and that k is continuous. By Weierstrass' Theorem 1.4.16, k may be approximated in the sup norm—and therefore in $|||\cdot|||_2$—by a sequence of continuous degenerate kernels, and the corresponding operators K_n are compact since their range is finite dimensional. Also $K_n \to K$, for $\|K - K_n\| \leqslant |||k - k_n|||_2$. Hence by Theorem 7.2.2, K is compact. To relax the conditions that Ω is compact and that k is continuous, simply note that by Theorem 2.5.6 the continuous functions with compact support are dense in $\mathscr{L}_2(\Omega \times \Omega)$, and again invoke Theorem 7.2.2.

An argument similar in outline may be used in $\mathscr{L}_p$ for $p \neq 2$. However, the details are not quite so readily supplied, and the following is quoted from Dunford and Schwartz (1958, p. 518).

7.2.7 Theorem. *Let Ω be a measurable subset of $\mathbb{R}^n$, suppose that k is measurable, and take any p with $1 < p < \infty$. If the number $|||k|||_p$ (defined in Theorem 3.4.10) is finite, then $K : \mathscr{L}_p(\Omega) \to \mathscr{L}_p(\Omega)$ is compact.*

In this theorem Ω need not be bounded, but for many common kernels, for example $k(x, y) = \exp(-|x - y|)$, the condition $|||k|||_p < \infty$ will not usually be satisfied if Ω is unbounded.

In the above examples, each of the compact operators may be uniformly approximated by operators of *finite rank*, that is with finite dimensional range, and it is tempting to speculate that this may be true for an arbitrary compact operator on a Banach space, that is: A compact operator is the sum of an operator of finite rank and an operator of arbitrarily small norm. Although this result is false in general (as has recently been proved), it holds for all the common Banach spaces, where it provides an illuminating characterization of compact operators, and it reinforces the analogy between a compact operator and an operator on a finite dimensional space.

7.3 The Fredholm Alternative

Consider the equation

$$(\lambda I - T)f = g \qquad (g \in \mathscr{B}), \tag{7.3.1}$$

where $T : \mathscr{B} \to \mathscr{B}$ is linear. For $\mathscr{B}$ finite dimensional, if λ is not an eigenvalue, then $(\lambda I - T)^{-1}$ exists and is bounded, and (7.3.1) has a unique solution for every g. Fredholm proved an analogous result for linear integral equations under the additional assumption that $\lambda \neq 0$. The aim of this section is the generalization of Fredholm's result to arbitrary compact linear operators on a Banach space.

Throughout $\mathscr{B}$ is a Banach space and $T : \mathscr{B} \to \mathscr{B}$ a linear operator. Theorem 6.5.10 is central to the argument, and as preparation for the application of this theorem we shall prove that T^* is compact and that $R(\lambda I - T)$ is closed.

7.3.1 Theorem. *The adjoint of a compact operator T is compact.*

Proof. Let S and S^* be the closed unit balls in $\mathscr{B}$ and $\mathscr{B}^*$ respectively. It will be shown that $T^*(S^*)$ is relatively compact. The first step is an application of the extended version of the Arzelà–Ascoli Theorem (Problem 5.11) to the linear functionals in S^* regarded as continuous complex valued functions on $T(S)$, or in other words as elements of $\mathscr{C}(T(S))$.

Since T is compact, $T(S)$ is relatively compact. Also, as $T(S)$ is a bounded set, the set of functionals in S^* is bounded on $T(S)$. Finally,

$$|f^*(g_1) - f^*(g_2)| \leqslant \|f^*\| \|g_1 - g_2\| \leqslant \|g_1 - g_2\| \quad (g_1, g_2 \in T(S), f^* \in S^*),$$

which proves equicontinuity. Therefore, from the theorem, a given sequence in S^* has a subsequence, $(f_n{}^*)$ say, convergent in $\mathscr{C}(T(S))$. The subsequence is thus Cauchy, so that

$$\lim_{m, n \to \infty} \sup_{f \in S} |f_n^*(Tf) - f_m^*(Tf)| = 0.$$

But

$$\begin{aligned}\sup_{f \in S} |f_n^*(Tf) - f_m^*(Tf)| &= \sup_{f \in S} |(T^*f_n^* - T^*f_m^*)(f)| \\ &= \|T^*f_n^* - T^*f_m^*\|.\end{aligned}$$

Hence $(T^*f_n^*)$ is Cauchy and therefore convergent as $\mathscr{B}^*$ is complete. That is $T^*(S^*)$ is relatively sequentially compact. □

7.3.2 Lemma. *Suppose that T is compact. If $\lambda \neq 0$, then $N(\lambda I - T)$ and $N(\lambda I - T^*)$ are finite dimensional.*

Proof. Since T^* is also compact, it is enough to prove the assertion for $N(\lambda I - T)$. As T is compact, every bounded sequence in $N(\lambda I - T)$ has a subsequence, (f_n) say, with (Tf_n) convergent. But since $\lambda f_n = Tf_n$, (f_n) is itself convergent. This proves that every bounded set in $N(\lambda I - T)$ is relatively sequentially compact, and the result follows from Lemma 5.25. □

This is the generalization of a familiar result in the Fredholm theory of integral equations. It states that to any non-zero eigenvalue there corresponds only a finite number of linearly independent eigenfunctions.

7.3.3 Lemma. *Assume that T is compact and that $\lambda \neq 0$. Then $R(\lambda I - T)$ and $R(\lambda I - T^*)$ are closed subspaces.*

Proof. Clearly $R(\lambda I - T)$ and $R(\lambda I - T^*)$ are linear subspaces. If it can be shown that the first is closed, as T^* is compact it will follow that so is the second. We must prove that if (f_n) is any convergent sequence in $R(\lambda I - T)$, its limit, f say, also lies in $R(\lambda I - T)$.

It is first shown that there is a bounded sequence (g_n) with $f_n = (\lambda I - T)g_n$. Since $f_n \in R(\lambda I - T)$, there is certainly a sequence (h_n) with $f_n = (\lambda I - T)h_n$, but (h_n) need not be bounded. However, if (l_n) is any sequence in $N(\lambda I - T)$ and $g_n = h_n - l_n$, then $f_n = (\lambda I - T)g_n$. Therefore if it can be proved that for some choice of the l_n the sequence $(h_n - l_n)$ is bounded, the assertion will follow.

It is evidently enough to show that $d(h_n) = \operatorname{dist}(h_n, N(\lambda I - T))$ is a bounded function of n. Suppose on the contrary that $(d(h_n))$ is not bounded. Then there is a subsequence, still denoted by (h_n), such that $d(h_n) \to \infty$ as $n \to \infty$. Set $\bar{h}_n = h_n/d(h_n)$. A short calculation shows that $d(\bar{h}_n) = 1$, and it follows that there is a sequence (k_n) in $N(\lambda I - T)$ such that $\|\bar{h}_n - k_n\| \leqslant 2$ for all n. Therefore the sequence (w_n), where $w_n = \bar{h}_n - k_n$, is bounded. Since T is compact, by replacing the original sequences by subsequences, we may assume that (Tw_n) is convergent. Also,

$$(\lambda I - T)w_n = (\lambda I - T)\bar{h}_n = (\lambda I - T)h_n/d(h_n) = f_n/d(h_n) \to 0.$$

Hence (w_n) is convergent with limit w, say. It follows that

$$\lambda w - Tw = \lim(\lambda w_n - Tw_n) = 0.$$

Thus $w \in N(\lambda I - T)$. Therefore

$$0 = d(w) = \lim d(w_n) = \lim d(\bar{h}_n) = 1.$$

This contradiction proves that $(d(h_n))$ is bounded, and establishes the existence of a bounded sequence (g_n) with $f_n = (\lambda I - T)g_n$.

The rest of the proof is easy. As T is compact, (g_n) has a subsequence, still denoted by (g_n), with (Tg_n) convergent. But (f_n) is convergent and $f_n = \lambda g_n - Tg_n$. Therefore (g_n) is convergent. If its limit is g, then

$$f = \lim f_n = \lim(\lambda g_n - Tg_n) = \lambda g - Tg.$$

That is $f \in R(\lambda I - T)$, which is what we set out to prove. □

We can now capitalize on the result of Theorem 6.5.10 relating the range of an operator to the null space of its adjoint.

7.3.4 Theorem. *If T is compact and $\lambda \neq 0$, then*

(i) $R(\lambda I - T) = N(\lambda I - T^*)^{\perp}$;
(ii) $R(\lambda I - T^*) = N(\lambda I - T)^{\perp}$.

This theorem furnishes a method for proving the existence of solutions of the equation $\lambda f - Tf = g$. For example (i) shows that the equation will have a solution for every g if $\lambda f^* = T^* f^*$ has only the zero solution. However, this result is not completely satisfactory as it does not follow from it that $\lambda I - T$ is injective and that $(\lambda I - T)^{-1}$ exists. It is in fact possible to strengthen the theorem considerably.

7.3.5 Lemma. *For compact T and non-zero λ, there is an integer $k \geqslant 1$ such that $N((\lambda I - T)^k) = N((\lambda I - T)^{k+1})$.*

Proof. Set $N_n = N((\lambda I - T)^n)$ for $n \geqslant 1$. An expansion in powers of T shows that $(\lambda I - T)^n$ is of the form $\mu I - T_1$ where $\mu \in \mathbb{C}$ and T_1 is compact. Hence N_n is finite dimensional (Lemma 7.3.2). We now argue by contradiction and suppose that $N_n \neq N_{n+1}$ for all $n \geqslant 1$. Then N_n is a proper linear subspace of N_{n+1} for each n. Thus by Problem 7.7 there is a sequence (f_n) such that $f_n \in N_{n+1}$, $\|f_n\| = 1$, and $\|f_n - f\| > \frac{1}{2}$ $(f \in N_n)$. Now for $n > m$,

$$(\lambda I - T)^n[(\lambda I - T)f_n + Tf_m] = (\lambda I - T)^{n+1}f_n + T(\lambda I - T)^n f_m = 0.$$

That is $(\lambda I - T)f_n + Tf_m \in N_n$. Therefore

$$|\lambda|^{-1}\|Tf_n - Tf_m\| = \|f_n - \lambda^{-1}[(\lambda I - T)f_n + Tf_m]\| > \tfrac{1}{2},$$

and it follows that (Tf_n) has no convergent subsequence. This contradicts the fact that T is compact. □

7.3.6 Lemma. *If T is compact and $\lambda \neq 0$, then $R(\lambda I - T) = \mathscr{B}$ if and only if $N(\lambda I - T) = 0$.*

Proof. Suppose first that $R(\lambda I - T) = \mathscr{B}$. If $N(\lambda I - T) \neq 0$, there is a non-zero f_1 in $N(\lambda I - T)$. Since $R(\lambda I - T) = \mathscr{B}$, there is a sequence (f_n) with $(\lambda I - T)f_{n+1} = f_n$ for $n \geqslant 1$, and

$$(\lambda I - T)^n f_{n+1} = f_1 \neq 0,$$
$$(\lambda I - T)^{n+1} f_{n+1} = (\lambda I - T)f_1 = 0.$$

Thus $N((\lambda I - T)^{n+1}) \neq N((\lambda I - T)^n)$ for all $n \geqslant 1$. This contradicts the previous lemma and proves that $N(\lambda I - T) = 0$.

On the other hand if $N(\lambda I - T) = 0$, by Theorem 7.3.4(ii), $R(\lambda I - T^*) = \mathscr{B}$. But as T^* is itself compact, it follows from the first part of the proof that $N(\lambda I - T^*) = 0$. We conclude from Theorem 7.3.4(i) that $R(\lambda I - T) = \mathscr{B}$. □

This lemma shows that for $\lambda \neq 0$, $(\lambda I - T)$ is injective if and only if it is surjective, in which case $(\lambda I - T)^{-1}$ is bounded (Theorem 3.5.3). It is thus a very satisfactory infinite dimensional analogue of the standard result in finite dimensions (Theorem 3.3.11). Roughly the lack of finite dimensionality of the space is compensated for by the compactness of T—a condition which ensures that the range of T is "approximately finite dimensional". The lemma is both of great theoretical and practical importance. When stated as follows in terms of the solutions of the homogeneous and inhomogeneous equations $\lambda f - Tf = 0$ and $\lambda f - Tf = g$ respectively, it is a precise generalization of the classical Fredholm alternative.

7.3.4 The Fredholm Alternative Theorem. *Suppose $\mathscr{B}$ is a Banach space. Let $T: \mathscr{B} \to \mathscr{B}$ be a linear compact operator, and assume that $\lambda \neq 0$. Then one of the two following alternatives holds:*

(i) *The homogeneous equation has only the zero solution, in which case $\lambda \in \rho(T)$, $(\lambda I - T)^{-1}$ is bounded, and the inhomogeneous equation has exactly one solution $f = (\lambda I - T)^{-1}g$ for each $g \in \mathscr{B}$.*
(ii) *The homogeneous equation has a non-zero solution, in which case the inhomogeneous equation has a solution (necessarily not unique) if and only if $\langle g, f^* \rangle = 0$ for every solution f^* of the adjoint equation $\lambda f^* = T^* f^*$.*

7.3.8 Corollary. *If for some g the inhomogeneous equation has at most one solution, then it has exactly one solution. That is, uniqueness implies existence.*

The proof of the corresponding result for integral equations provided a powerful technique for establishing the existence of solutions to boundary value problems for elliptic partial differential equations. A brief outline of the now classical argument is as follows. The equation is reformulated as a linear integral equation with the help of a Green's function. The kernel of this equation is not small enough for a Neumann series type argument to be used. However, uniqueness may be proved independently, often by invoking a maximum principle, and existence follows from Corollary 7.3.8. There are many excellent accounts of this approach, see for example Garabedian (1964), so we shall outline in Chapter 11 a line of attack that has been developed more recently, and which is more attractive from the point of view of applications. This avoids the difficult step of constructing an integral equation, the Fredholm alternative being applied directly to a certain compact operator derived directly from the partial differential equation.

7.4 The Spectrum

One of the principal consequences of the assumption that T is compact is that it is possible to characterize its spectrum $\sigma(T)$ in a particularly simple manner. Indeed it is already known from Theorem 7.3.7 that every non-zero point of $\sigma(T)$ is an eigenvalue, and it will be proved that there are only a finite number of points in $\sigma(T)$ in any region excluding a neighbourhood of the origin. Thus except near the origin the spectrum is qualitatively identical with that of an operator in finite dimensions. For the origin itself the situation is less simple, and some of the possibilities will be illustrated by examples.

7.4.1 Theorem. *Let $\mathscr{B}$ be an infinite dimensional Banach space, and suppose that $T:\mathscr{B} \to \mathscr{B}$ is a compact linear operator. Then the spectrum of T consists of the point zero together with the non-zero eigenvalues of T, and each eigenspace corresponding to a non-zero eigenvalue is finite dimensional.*

Proof. If $0 \notin \sigma(T)$, then $R(T) = \mathscr{B}$, and by Lemma 7.2.3, $\mathscr{B}$ is finite dimensional—a contradiction. The other two assertions follow from Theorem 7.3.7 and Lemma 7.3.2 respectively. □

7.4.2 Theorem. *The set of eigenvalues of a compact linear operator on a Banach space is either finite or countably infinite, and its only possible limit point is zero.*

Proof. The argument is by contradiction. Suppose that for some $\varepsilon > 0$ there is an infinite sequence (λ_n) of distinct eigenvalues with $|\lambda_n| \geqslant \varepsilon$ for each n. Let (f_n) be a sequence of eigenfunctions corresponding to the λ_n respectively, and let $\mathscr{M}_n$ be the linear span of $f_1, \ldots, f_n$; it is easily checked that the f_n are linearly independent, and it follows that $\mathscr{M}_n$ is a proper subset of $\mathscr{M}_{n+1}$.

Hence, by Problem 7.7 there is a sequence (g_n) such that for every $n > 1$, $g_n \in \mathscr{M}_n$, $\|g_n\| = 1$, and

$$\|g_n - f\| \geqslant \tfrac{1}{2} \qquad (f \in \mathscr{M}_{n-1}). \tag{7.4.1}$$

Now $g_n = \sum_1^n \alpha_j f_j$ for some $\alpha_j \in \mathbb{C}$, whence $Tg_n \in \mathscr{M}_n$. Also

$$(\lambda_n I - T)g_n = \sum_{j=1}^{n-1} \alpha_j(\lambda_n - \lambda_j)f_j,$$

and it follows that $(\lambda_n I - T)g_n \in \mathscr{M}_{n-1}$. For any integer m with $1 \leqslant m < n$ set $f = (\lambda_n I - T)g_n + Tg_m$, and note that by what has just been proved, f (and so $\lambda_n^{-1} f$) are in $\mathscr{M}_{n-1}$. Since

$$Tg_n - Tg_m = \lambda_n g_n - (\lambda_n g_n - Tg_n + Tg_m) = \lambda_n(g_n - \lambda_n^{-1} f),$$

it follows from (7.4.1) that $\|Tg_n - Tg_m\| \geqslant \frac{1}{2}|\lambda_n| \geqslant \frac{1}{2}\varepsilon$. Therefore (Tg_n) has no convergent subsequence. As (g_n) is a bounded sequence, this contradicts the fact that T is compact. □

The spectrum of a linear compact operator thus consists of eigenvalues which form either a finite set or a sequence with limit zero, together with the point zero. The point zero must be singled out for special mention, firstly because it always lies in $\sigma(T)$, and secondly because it may be in either the point, residual or continuous spectrum. Some of the possibilities are illustrated by the following examples in the Hilbert space $\mathscr{H} = \mathscr{L}_2(0, 1)$, based on the integral operator $K : \mathscr{H} \to \mathscr{H}$, where k is continuous and

$$Kf(x) = \int_0^1 k(x, y) f(y)\, \mathrm{d}y.$$

7.4.3 Example. Let $\{\phi_j\}_{j=1}^n$ be a finite set of linearly independent continuous functions, and suppose that

$$k(x, y) = \sum_1^n \phi_j(x)\, \overline{\phi_j(y)}.$$

Then

$$Kf = \sum_1^n (f, \phi_j)\phi_j \qquad (f \in \mathscr{L}_2(0, 1)),$$

and it follows that $R(K)$ is the n-dimensional linear subspace $\mathscr{M}_n$ spanned by the ϕ_j. Since $Kf = 0$ if $f \in \mathscr{M}_n^{\perp}$, there are an *infinite* number of linearly independent eigenfunctions corresponding to the eigenvalue zero.

7.4.4 Example. If k is Hermitian, K is self-adjoint. An interesting class of operator consists of those K for which in addition

$$(Kf, f) > 0 \qquad (f \neq 0); \tag{7.4.2}$$

K is then said to be non-negative—an example is the operator in Problem 7.15. It is clear from (7.4.2) that zero is not an eigenvalue of K. However, $0 \in \sigma(K)$ as K is compact, and therefore $R(K) \neq \mathscr{H}$. This confirms that the Fredholm alternative is not valid for $\lambda = 0$. The best that can be said is that $R(K)$ is dense in $\mathscr{H}$, and this follows from Problem 6.20.

7.4.5 Example. Consider the Volterra integral operator K where

$$Kf(x) = \int_0^x k(x, y)f(y)\,\mathrm{d}y.$$

The spectral radius of K is zero (Problem 3.2.4)—in other words $\sigma(K)$ consists of the single point zero. A pleasant consequence is that for $\lambda \neq 0$ the equation $\lambda f - Kf = g$ always has a unique solution, and this may be obtained by a straightforward Neumann series expansion. As in the previous example, zero is not an eigenvalue, but $R(K) \neq \mathscr{H}$.

The unpleasant properties associated with the Fredholm integral equation of the first kind

$$\int_0^1 k(x, y)f(y)\,\mathrm{d}y = g(x),$$

are directly connected with the fact that zero is in the spectrum of the integral operator. Whether or not zero is an eigenvalue, since $0 \in \sigma(K)$ the equation cannot have a solution for all $g \in H$. Usually it is not simple to characterize $R(K)$, and even if it is known that $g \in R(K)$ the solution may be difficult to find. In particular the standard approximate method based on solving a modified equation $\bar{K}\bar{f} = \bar{g}$ with $\bar{g}$ and $\bar{K}$ close to g and K respectively may not be satisfactory, for there is no guarantee that this equation has a solution, and even if it does, since K^{-1} may not exist or at best exists and is unbounded, it may not be easy to decide whether f is close to $\bar{f}$. Equations of the first kind are becoming more and more important in applications, and a number of methods for treating these equations have been developed recently, see Groetsch (1977), Hilgers (1976) and Nashed (1974).

7.5 Compact Self-adjoint Operators

The analysis of the previous section shows that the properties associated with the spectrum of a compact operator T are relatively straightforward. We now enquire what further simplification is possible if T is also self-adjoint. The theory has an obvious application to Fredholm integral equations with Hermitian kernels. Another application is to differential equations; it is less clear that the theory will be useful here for differential operators on $\mathscr{L}_2$ are not compact, but they sometimes have compact inverses, and in this case a number of interesting results may be deduced from the theory.

The central result is the Hilbert–Schmidt Theorem which states that the eigenfunctions of T form a basis for the Hilbert space. This powerful theorem illuminates the structure of the operator, and has a number of useful consequences. We shall be particularly concerned with two of these. First we shall show that an eigenfunction expansion may be written down for the solution of the equation $(\lambda I - T)f = g$; this clarifies the dependence of the solution on the parameter λ when λ lies near a point of $\sigma(T)$. Secondly we shall consider how the theorem may be used to construct orthonormal bases for $\mathscr{L}_2$ spaces.

Let $\mathscr{H}$ be a separable Hilbert space, and assume that $T : \mathscr{H} \to \mathscr{H}$ is a compact self-adjoint operator. The following convention concerning the eigenvalues and eigenfunctions of T will be adopted throughout. Let $\{\phi_n\}$ be the set consisting of all the eigenfunctions of T, and assume that this set is arranged to be orthonormal; this is always possible as eigenfunctions corresponding to different eigenvalues are orthogonal, and those corresponding to the same eigenvalue may be orthogonalized by the Gram–Schmidt Procedure. Finally the set is numbered so that the eigenvalue λ_n corresponds to ϕ_n, the λ_n being repeated according to multiplicity with $|\lambda_1| \geqslant |\lambda_2| \geqslant \ldots$.

7.5.1 The Hilbert–Schmidt Theorem. *Let T be compact and self-adjoint. Then its eigenfunctions form an orthonormal basis for $\mathscr{H}$.*

Proof. Let $\mathscr{M}$ be the closure of the linear span of $\{\phi_n\}$. By the Projection Theorem 1.5.11, $\mathscr{H} = \mathscr{M} \oplus \mathscr{M}^\perp$, and it is easy to check that $T\mathscr{M} \subset \mathscr{M}$, $T\mathscr{M}^\perp \subset \mathscr{M}^\perp$. Denote by T_0 the restriction of T to $\mathscr{M}^\perp$, and note that T_0 maps $\mathscr{M}^\perp$ into itself and that T_0 is compact and self-adjoint. If T_0 has a non-zero eigenvalue, this is an eigenvalue of T, and the corresponding eigenfunction must lie in $\mathscr{M}$. This is impossible as $\mathscr{M}$ and $\mathscr{M}^\perp$ are orthogonal. Thus as the only non-zero points of $\sigma(T_0)$ are eigenvalues, $r_\sigma(T_0) = 0$ and $T_0 = 0$ (Theorem 6.6.7). It follows that $T\mathscr{M}^\perp = 0$, which can only happen if every element of $\mathscr{M}^\perp$ is an eigenfunction of T. Therefore $\mathscr{M}^\perp \subset \mathscr{M}$, which implies that $\mathscr{M}^\perp = 0$ and that $\mathscr{M} = \mathscr{H}$. The result follows from Theorem 1.5.17. □

The next theorem asserts that a self-adjoint operator can be diagonalized—that is its "matrix" with respect to the basis $\{\phi_n\}$ is diagonal with entries λ_n—and is the generalization of a well-known result for Hermitian matrices. The theorem itself has a far-reaching generalization to unbounded self-adjoint operators which will be examined in Chapter 9. Both of the following two results are easy consequences of the Hilbert–Schmidt Theorem, and their proofs are left as exercises.

7.5.2 Theorem (Canonical form for compact self-adjoint operators). *For compact self-adjoint T,*

$$Tf = \Sigma\, \lambda_n(f, \phi_n)\phi_n \qquad (f \in \mathscr{H}).$$

7.5.3 Theorem. *Suppose that T is compact and self-adjoint.* If $\lambda \in \rho(T)$ *and* $g \in \mathscr{H}$, *the solution of the equation* $(\lambda I - T)f = g$ *is*

$$f = R(\lambda; T)g = \Sigma(\lambda - \lambda_n)^{-1}(g, \phi_n)\phi_n. \tag{7.5.1}$$

This result clarifies the structure of the resolvent $R(\lambda; T)$. It shows that $R(\lambda; T)$ is an (operator valued) analytic function of λ with simple poles at the eigenvalues, the residues at these poles yielding the eigenfunctions. From a practical point of view it provides a method for studying the solution when λ is near a non-zero eigenvalue; the analogy with the behaviour of a physical system near a mode of free oscillation may be noted.

The Hilbert–Schmidt Theorem when applied to particular operators gives a simple method for constructing bases. In $\mathscr{L}_2$ spaces most of the useful bases arise from differential operators which are not themselves compact but which have compact inverses. The following theorem shows that these operators may be dealt with just as easily. It should be remarked that a generalization of these results can be obtained if the inverse is not compact by a direct treatment of the operators, see Chapters 9 and 10.

7.5.4 Theorem. *Let L be an unbounded self-adjoint operator on the infinite dimensional separable Hilbert space* $\mathscr{H}$. *Suppose L has a compact inverse, and let* $\{\mu_n\}$ *and* $\{\phi_n\}$ *be the sets of eigenvalues and eigenfunctions of* L^{-1}. *Then the following conclusions are valid. The eigenvalues of* L^{-1} *are all non-zero, and* $\{\phi_n\}$ *is an orthonormal basis for* $\mathscr{H}$. *With* $\lambda_n = \mu_n^{-1}$, *the sequence* (λ_n) *is infinite and* $|\lambda_n| \to \infty$ *as* $n \to \infty$. $\sigma(L)$ *is the set* $\{\lambda_n\}$, *and the eigenfunctions of L are the* ϕ_n *with corresponding eigenvalues* λ_n. *Finally,*

$$(\lambda I - L)^{-1}g = \Sigma(\lambda - \lambda_n)^{-1}(g, \phi_n)\phi_n \qquad (\lambda \in \rho(L), g \in \mathscr{H}). \tag{7.5.2}$$

Proof. First note that L^{-1} is self-adjoint, and that $R(L) = D(L^{-1})$, $D(L) = R(L^{-1})$. Therefore, if $L^{-1}\phi = \mu\phi$, since $\phi \in R(L^{-1}) = D(L)$, an application of L is legitimate and yields the relation $\phi = \mu L\phi$. Similarly if $L\phi = \lambda\phi$ then $\phi = \lambda L^{-1}\phi$. It follows first that zero is not an eigenvalue of L^{-1}, and secondly that ϕ is an eigenfunction of L^{-1} (corresponding to the eigenvalue μ) if and only if it is an eigenfunction of L (corresponding to the eigenvalue $\lambda = \mu^{-1}$).

The fact that $\{\phi_n\}$ is an orthonormal basis is a consequence of the Hilbert–Schmidt Theorem.

We next show that $\sigma(L) = \{\lambda_n\}$. The inclusion $\sigma(L) \supset \{\lambda_n\}$ is an obvious consequence of the remarks above. To prove the opposite inclusion it is sufficient to show that $\lambda \in \rho(L)$ if $\lambda^{-1} \in \rho(L^{-1})$. Suppose that $\lambda^{-1} \in \rho(L^{-1})$. Then $(\lambda I - L)$ is injective, for if it is not $\lambda\phi = L\phi$ for some non-zero ϕ, and λ^{-1} is an eigenvalue of L^{-1} contrary to assumption. Also, since $R(L^{-1}) = D(L)$,

$$(\lambda I - L)\lambda^{-1}L^{-1}(L^{-1} - \lambda^{-1}I)^{-1}g = (L^{-1} - \lambda^{-1}I)(L^{-1} - \lambda^{-1}I)^{-1}g = g \tag{7.5.3}$$

for $g \in \mathscr{H}$. This proves that $(\lambda I - L)$ is surjective. Therefore it is bijective (and closed), and it follows from Theorem 3.8.16 that $\lambda \in \rho(L)$. Thus $\sigma(L) = \{\lambda_n\}$.

Finally, since $\mathscr{H}$ is infinite dimensional, from Theorem 7.4.2, $\lim|\lambda_n| = \infty$. From (7.5.3), $(\lambda I - L)^{-1} = \lambda^{-1}L^{-1}(L^{-1} - \lambda^{-1}I)^{-1}$, and (7.5.2) follows on applying Theorem 7.5.3 to L^{-1}. □

7.5.5 Example. The Fourier sine series will be derived by applying the theorem to a self-adjoint operator based on the formal operator $l = -\mathrm{d}^2/\mathrm{d}x^2$ with zero boundary conditions.

Take $\mathscr{H} = \mathscr{L}_2(0, \pi)$ and let $\mathscr{A}$ be the set of functions in $\mathscr{H}$ with absolutely continuous first derivatives and with second derivatives in $\mathscr{H}$. Set

$$D(L) = \{f : f \in \mathscr{A}, f(0) = f(1) = 0\},$$

and define $Lf = lf$ for $f \in D(L)$. We assert that L is self-adjoint. This may be proved as in Example 6.7.9, but the details are tedious, and since our assertion is a consequence of a general result (Theorem 10.5.3) that will be established later, the proof is omitted here.

The inverse of L must be compact if the theorem is to be applicable. Define the Green's function k as follows:

$$k(x, y) = \begin{cases} (\pi - y)x/\pi & (0 \leqslant x \leqslant y \leqslant \pi), \\ (\pi - x)y/\pi & (0 \leqslant y \leqslant x \leqslant \pi), \end{cases}$$

and let K be the integral operator with kernel k:

$$Kf(x) = \int_0^\pi k(x, y)f(y)\,\mathrm{d}y.$$

We shall prove that $L^{-1} = K$. Let K_0 be the restriction of K to $\mathscr{C}^2([0, \pi])$, and note that since $\mathscr{C}^2([0, \pi])$ is dense in $\mathscr{L}_2(0, \pi)$, K is the closure of K_0. Differentiation shows that for $f \in D(K_0)$, $(K_0 f)''(x) = f(x)$, and evidently $K_0 f(0) = K_0 f(\pi) = 0$. Thus $R(K_0) \subset D(L)$ and $LK_0 f = f$ for $f \in D(K_0)$. As L is injective, it follows from Lemma 3.8.19 that $L^{-1} = K$. Since K is compact

(Theorem 7.2.7), the conditions of Theorem 7.5.4 are satisfied, and we deduce that the eigenfunctions of L are a basis for $\mathscr{L}_2(0, \pi)$.

7.5.6 Theorem. *Suppose that $f \in \mathscr{L}_2(0, \pi)$. Then (almost everywhere),*

$$f(x) = 2\pi^{-1} \sum_1^\infty \sin nx \int_0^\pi f(y) \sin ny \, dy,$$

the convergence of the series being in the norm of $\mathscr{L}_2(0, \pi)$.

Theorem 7.5.4 may also be used to show that most of the standard sets of orthogonal functions are bases. The main difficulty arises from the fact that the appropriate formal differential operators are often singular (in a sense to be made precise later), in which case the construction of the self-adjoint operators presents a problem. Further discussion of this point is deferred until Chapter 10.

7.6 The Numerical Solution of Linear Integral Equations

The numerical solution of integral equations is a particularly important topic in applications, for a wide range of problems may be modelled by these equations. Nonetheless, although there is a vast literature on integral equation theory, the numerical aspects of the theory have received relatively scant attention. There are certain fundamental questions in this area which functional analysis is well suited to answer: Do the approximating equations have solutions? If so, do the approximate solutions converge to the solution of the integral equation, and what is the error in the approximation? The theory of collectively compact operators due to Anselone (1971) is explicitly designed to resolve these questions. A résumé of this theory will now be presented, and its application to a linear inhomogeneous integral equation discussed as an illustration.

Consider the integral equation

$$f(x) - \int_0^1 k(x, y) f(y) \, dy = g(x) \tag{7.6.1}$$

where k and g are given continuous functions. The most frequently used numerical method for solving this is based on the replacement of (7.6.1) by the sequence of approximating equations

$$f_n(x) - \sum_{j=1}^{n} w_j^{(n)} k(x, y_j^{(n)}) f_n(y_j^{(n)}) = g(x), \tag{7.6.2}$$

the $w_j^{(n)}$ being weights in a quadrature formula and the $y_j^{(n)}$ mesh points. Setting $x = y_i^{(n)} (i = 1, \ldots, n)$ in turn we obtain:

$$f_n(y_i^{(n)}) - \sum_{j=1}^{n} w_j^{(n)} k(y_i^{(n)}, y_j^{(n)}) f_n(y_j^{(n)}) = g(y_i^{(n)}) \quad (i = 1, \ldots, n). \tag{7.6.3}$$

For each n this is a finite system of linear algebraic equations in the unknowns $f_n(y_i^{(n)})$ and if its solution is computed, the functions f_n may be calculated by substitution in (7.6.2). Equations (7.6.2) and (7.6.3) can thus be regarded as equivalent, and we therefore concentrate on the relation between the solution f_n of (7.6.2) and the solution f of (7.6.1). To cast the problem in abstract form, take $\mathscr{B} = \mathscr{C}([0, 1])$ with the sup norm, and set

$$Kf(x) = \int_0^1 k(x, y) f(y) \, \mathrm{d}y,$$

$$K_n f(x) = \sum_{j=1}^{n} w_j^{(n)} k(x, y_j^{(n)}) f(y_j^{(n)}).$$

Then K and $K_n (n \geqslant 1)$ are bounded linear operators mapping $\mathscr{C}([0, 1])$ into itself, and (7.6.1), (7.6.2) become respectively

$$(I - K)f = g, \qquad (I - K_n) f_n = g.$$

In studying the relation between these equations, one's first thought might be to use a perturbation theoretic approach based on Theorem 3.6.3. For example, if $K_n \to K$ (recall that this means uniform convergence—Definition 3.5.7) and the sequence $(\|(I - K_n)^{-1}\|)$ is bounded, then by the theorem $(I - K_n)^{-1} \to (I - K)^{-1}$. Unfortunately this line of argument encounters a major snag, for as shown in Example 3.5.15, quadrature formulae are not uniformly convergent, and therefore certainly K_n does not tend to K uniformly. What can be counted on is the *strong* convergence (Definition 3.5.13) of K_n to K, but this is not enough to validate the above conclusion. The key to the approach to be followed is the utilization of the strong convergence together with the available compactness properties of K and (K_n) to prove that $(I - K_n)^{-1} \underset{S}{\to} (I - K)^{-1}$. This strong convergence of the inverses is not quite as satisfactory as uniform convergence, but it is enough to show that $f_n \to f$, and this result will be supplemented by the derivation of bounds for the error $\|f_n - f\|$.

The analysis will be carried out in a Banach space $\mathscr{B}$, and the operators will all map $\mathscr{B}$ into itself.

7.6.1 Definition. Let S be the closed unit ball in $\mathscr{B}$. The sequence $\mathscr{K} = (K_n)$ is said to be **collectively compact** iff the set

$$\mathscr{K}(S) = \{Kf : K \in \mathscr{K}, f \in S\}$$

is relatively compact.

7.6.2 Lemma. *Suppose $\mathscr{K} = (K_n)$ is collectively compact. Then each K_n is compact, the sequence $\mathscr{K}$ is bounded, and if $K \in \mathscr{L}(\mathscr{B})$ and $K_n \underset{S}{\rightarrow} K$, K is compact.*

Proof. The first two assertions are obvious. To prove the last, note that since $Kf = \lim K_n f$ for each $f \in S$, then $K(S) \subset \overline{\{K_n f : n \geqslant 1, f \in S\}}$. As $\mathscr{K}$ is collectively compact, this set is compact. Hence K is compact. □

Consider a sequence (K_n) of bounded operators strongly convergent to K. In finite dimensions the sequence will automatically be uniformly convergent to K. In infinite dimensions this is not of course true, but if the operators are restricted to a relatively compact subset S (which may be regarded as "nearly finite dimensional"), strong convergence (on S) does indeed imply uniform convergence (on S). An alternative point of view is obtained by noting that uniform convergence is convergence which is uniform on bounded subsets, while strong convergence is convergence which is uniform on totally bounded subsets—the same of course as relatively compact subsets. These observations are crucial for the present line of argument. For if K is compact and (K_n) is collectively compact, then with S the closed unit ball, $K(S)$ and $\mathscr{K}(S)$ are relatively compact, and so the sequences $((K_n - K)K)$ and $((K_n - K)K_n)$ converge uniformly to zero.

7.6.3 Lemma. *Suppose that (L_n) is a sequence in $\mathscr{L}(\mathscr{B}, \mathscr{C})$, and assume that $L_n \underset{S}{\rightarrow} L \in \mathscr{L}(\mathscr{B}, \mathscr{C})$. Then (L_n) converges uniformly to L on any totally bounded set S, that is,*

$$\lim_{n\to\infty} \sup_{f \in S} \|(L_n - L)f\| = 0.$$

Proof. From Definition 5.2.7 of total boundedness, given $\varepsilon > 0$ there is a finite set of points in S, $g_1, \ldots, g_j$ say, such that for every $f \in S$, $\|f - g_i\| \leqslant \varepsilon$ for some i with $1 \leqslant i \leqslant j$. Therefore,

$$\begin{aligned}\|(L_n - L)f\| &\leqslant \min_i [\|(L_n - L)(f - g_i)\| + \|(L_n - L)g_i\|], \\ &\leqslant \varepsilon(\|L_n\| + \|L\|) + \min_i \|(L_n - L)g_i\|.\end{aligned}$$

The sequence $(\|L_n\|)$ is bounded by the Principle of Uniform Boundedness 3.5.6, and the result follows as ε is arbitrary and $L_n \underset{S}{\rightarrow} L$. □

7.6.4 Corollary. *Suppose (K_n) is collectively compact and $K_n \underset{S}{\rightarrow} K$. Then $((K_n - K)K)$ and $((K_n - K)K_n)$ converge uniformly to zero.*

Before this result can be brought to bear on the convergence of the inverses $(I - K_n)^{-1}$ to $(I - K)^{-1}$, a preparatory lemma is needed. This may be compared with Theorem 3.6.3, the assumption of compactness being enough to compensate partially for a weakening of the conditions of that theorem.

7.6.5 Lemma. *Suppose that T is compact and that $L, (I - L)^{-1} \in \mathscr{L}(\mathscr{B})$. Assume that*

$$\Delta = \|(I - L)^{-1}(T - L)T\| < 1.$$

Then $(I - T)^{-1} \in \mathscr{L}(\mathscr{B})$, and for $g \in \mathscr{B}$,

$$\|(I - T)^{-1}\| \leqslant (1 - \Delta)^{-1}[1 + \|(I - L)^{-1}\| \, \|T\|], \tag{7.6.4}$$

$$\|(I - T)^{-1}g - (I - L)^{-1}g\| \leqslant (1 - \Delta)^{-1}[\|(I - L)^{-1}\| \, \|(T - L)g\| + \Delta\|(I - L)^{-1}g\|]. \tag{7.6.5}$$

Proof. The proof is based on the following idea. The condition $\Delta < 1$ should ensure that T is a reasonable approximation to L. Therefore, since $(I - L)^{-1} = I + (I - L)^{-1}L$, an approximation to $(I - T)^{-1}$ should be $I + (I - L)^{-1}T = B$, say.

By a short calculation, $B(I - T) = I - A$, where $A = (I - L)^{-1}(T - L)T$. Since $\|A\| = \Delta < 1$, $(I - A)$ is injective, and therefore so is $(I - T)$. However, T is compact, so by the Fredholm Alternative Theorem 7.3.7, $(I - T)^{-1} \in \mathscr{L}(\mathscr{B})$. (7.6.4) follows readily from the relation $B(I - T) = I - A$ when it is recalled (Theorem 3.6.1) that $\|(I - A)^{-1}\| \leqslant (1 - \|A\|)^{-1}$. Finally,

$$\begin{aligned}(I - T)^{-1} - (I - L)^{-1} &= (I - A)^{-1}B - (I - L)^{-1} \\ &= (I - A)^{-1}[B - (I - L)^{-1} + A(I - L)^{-1}] \\ &= (I - A)^{-1}[(I - L)^{-1}(T - L) + A(I - L)^{-1}],\end{aligned}$$

from which (7.6.5) follows immediately. □

7.6.6 Theorem. *Let $\mathscr{B}$ be a Banach space, let K be compact, and suppose that $(I - K)^{-1} \in \mathscr{L}(\mathscr{B})$. Assume that (K_n) is collectively compact, and that $K_n \underset{s}{\rightarrow} K$. Set*

$$\Delta_n = \|(I - K)^{-1}(K_n - K)K_n\|.$$

Then $\Delta_n \rightarrow 0$ as $n \rightarrow \infty$. If $\Delta_n < 1$, $(I - K_n)^{-1} \in \mathscr{L}(\mathscr{B})$, and for $g \in \mathscr{B}$,

$$\|(I - K_n)^{-1}\| \leqslant (1 - \Delta_n)^{-1}[1 + \|(I - K)^{-1}\| \, \|K_n\|],$$

$$\|(I - K_n)^{-1}g - (I - K)^{-1}g\| \leqslant (1 - \Delta_n)^{-1}[\|(I - K)^{-1}\| \, \|(K_n - K)g\| + \Delta_n\|(I - K)^{-1}g\|] \tag{7.6.6}$$

Proof. By Corollary 7.6.4, $\lim\|(K_n - K)K_n\| = 0$, whence $\lim \Delta_n = 0$. The result is obtained on setting $L = K$ and $T = K_n$ in Lemma 7.6.5. □

This is the central result of the theory. It proves both the existence and strong convergence of the inverses, and gives bounds for the error in the approximate solution. In practice it may sometimes be more convenient to base the analysis on estimates for $(I - K_n)^{-1}$, since these may be obtained numerically, and the reader's attention is directed to Problem 7.23.

To illustrate the application of this theorem to the numerical solution of (7.6.1), a relatively simple quadrature formula is chosen. Take

$$Q(f) = \int_0^1 f(t)\,\mathrm{d}t,$$

$$Q_n(f) = \sum_{j=1}^{n} w_j^{(n)} f(x_j^{(n)}),$$

and for each n assume that the mesh points $x_j^{(n)}$ are equally spaced, that the weights $w_j^{(n)}$ are strictly positive, and that

$$\sum_{j=1}^{n} w_j^{(n)} = 1. \tag{7.6.7}$$

These assumptions are satisfied by the trapezium rule, Simpson's rule and a number of other quadrature formulae, for which the essential property $Q_n \underset{S}{\rightarrow} Q$ holds (Example 3.5.15). The quadrature formula must now be applied to the integral equation (7.6.1), and it is convenient to define functions $k_x(\cdot)$, $k^y(\cdot)$ by requiring that $k_x(y) = k^y(x) = k(x, y)$, for then $Kf(x) = Q(k_x f)$ and $K_n f(x) = Q_n(k_x f)$.

7.6.7 Lemma. *If k is continuous and Q_n is as above, then K is compact, (K_n) is collectively compact, and $K_n \underset{S}{\rightarrow} K$.*

Proof. K is compact by Theorem 7.2.6. From (7.6.7),

$$\|K_n\| \leqslant \sup_{x \in [0,1]} \sum_{j=1}^{n} |w_j^{(n)} k(x, y_j^{(n)})| \leqslant \|k\|.$$

Therefore for $f \in S$, the closed unit ball in $\mathscr{B}$,

$$\|K_n f\| \leqslant \|K_n\| \|f\| \leqslant \|k\| \|f\|,$$

$$|K_n f(x) - K_n f(x')| \leqslant \|k_x - k_{x'}\| \|f\| \qquad (x, x' \in [0, 1]),$$

and the boundedness and equicontinuity of $\mathscr{K}(S)$ follow from the uniform continuity of k. By the Arzelà–Ascoli Theorem 5.4.4, $\mathscr{K}(S)$ is relatively compact, whence (K_n) is collectively compact.

To prove that $K_n \underset{S}{\to} K$, note that

$$\|(K_n - K)f\| = \sup_{x\in[0,1]} |K_n f(x) - Kf(x)|$$
$$= \sup_{x\in[0,1]} |Q_n(k_x f) - Q(k_x f)|. \qquad (7.6.8)$$

Now the set of functions $\{k_x f : x \in [0, 1]\}$ is bounded and equicontinuous (since k, f are uniformly continuous), and is thus relatively compact by the Arzelà–Ascoli Theorem. Therefore by Lemma 7.6.3, $Q_n(k_x f)$ converges to $Q(k_x f)$ uniformly with respect to x. Thus the right-hand side of (7.6.8) tends to zero, and the assertion follows. □

7.6.8 Theorem. *Take $\mathscr{B} = \mathscr{C}([0, 1])$ with the sup norm. Suppose that the quadrature is as described, and let k be continuous. If $(I - K)^{-1} \in \mathscr{L}(\mathscr{B})$, then for large enough n, $(I - K_n)^{-1} \in \mathscr{L}(\mathscr{B})$, and $(I - K_n)^{-1} \underset{S}{\to} (I - K)^{-1}$.*

Proof. The lemma shows that the conditions of Theorem 7.6.6 are satisfied. □

With this result we reach our first goal, for the theorem states that the matrix equations (7.6.3) have a solution if n is large enough, and that the numerical solutions $f_n = (I - K_n)^{-1}g$ of (7.6.2) converge in sup norm to the exact solution of the integral equation (7.6.1).

The final task is to find explicit error bounds. These will be based on the inequality (7.6.6), but additional restrictions must be placed on k and g in order that the error in the quadrature formula may be estimated. From the great variety of possible assumptions a simple case is selected for illustration, see Anselone (1971, Chap. 2) for further details. We quote first a standard error estimate.

7.6.9 Lemma. *Let Q_n represent the trapezium rule. If g satisfies the Lipschitz condition*

$$|g(x) - g(x')| \leqslant M|x - x'| \qquad (x, x' \in [0, 1]),$$

then

$$|Q_n(g) - Q(g)| \leqslant M/4n.$$

7.6.10 Theorem. *Let $\mathscr{C}([0, 1])$ be equipped with the sup norm. Suppose g and k satisfy the Lipschitz conditions:*

$$|g(x) - g(x')| \leqslant M|x - x'| \qquad (x, x' \in [0, 1]);$$
$$|k(x, y) - k(x', y')| \leqslant N(|x - x'| + |y - y'|) \qquad (x, x', y, y' \in [0, 1]).$$

If the trapezium rule is used, the error between the numerical solution f_n of (7.6.2) *and the solution f of* (7.6.1) *satisfies the inequality* (7.6.6), *where*

$$\|(K_n - K)g\| \leqslant (M\|k\| + N\|g\|)/4n, \qquad \|(K_n - K)K_n\| \leqslant N\|k\|/2n.$$

Proof. From the assumed Lipschitz conditions,

$$\begin{aligned}|k_x g(y) - k_x(y')\,g(y')| &= |k_x(y)[g(y) - g(y')] + [k_x(y) - k_x(y')]g(y')| \\ &\leqslant (M\|k\| + N\|g\|)\,|y - y'|.\end{aligned}$$

Therefore

$$\begin{aligned}\|K_n g - Kg\| &\leqslant \sup_{x\in[0,1]} |Q_n(k_x g) - Q(k_x g)| \\ &\leqslant (M\|k\| + N\|g\|)/4n\end{aligned}$$

from Lemma 7.6.9.

To prove the second inequality set

$$k_2(x, y) = \int_0^1 k(x, t)\,k(t, y)\,\mathrm{d}t = Q(k_x k^y),$$

$$k_2^{(n)}(x, y) = \sum_{j=1}^{n} w_j^{(n)} k(x, t_j^{(n)}) k(t_j^{(n)}, y) = Q_n(k_x k^y).$$

Then

$$K_n^2 f(x) = \sum_{j=1}^{n} w_j^{(n)} k_2^{(n)}(x, t_j^{(n)}) f(t_j^{(n)}),$$

$$KK_n f(x) = \sum_{j=1}^{n} w_j^{(n)} k_2(x, t_j^{(n)}) f(t_j^{(n)}),$$

and

$$\begin{aligned}\|(K_n - K)K_n\| &\leqslant \sup_{x\in[0,1]} \sum_{j=1}^{n} |w_j^{(n)}|\,|k_2^{(n)}(x, t_j^{(n)}) - k_2(x, t_j^{(n)})| \\ &\leqslant \sup_{x\in[0,1]} |(Q_n - Q)(k_x k^y)|.\end{aligned}$$

Using the Lipschitz condition on k, we deduce the required inequality by the argument used in the previous paragraph. □

An advantage of the approach summarized in this section is that it can be extended in a number of directions. Three possible generalizations of practical interest are to discontinuous kernels, to the eigenfunction problem, and to nonlinear integral equations, see Anselone (1971). Atkinson (1976) and Baker (1978) are useful general texts on the numerical solution of integral equations.

Problems

Throughout $\mathscr{B}$ is a Banach space, $\mathscr{H}$ is a separable Hilbert space, and T is a linear operator $\mathscr{B} \to \mathscr{B}$ or $\mathscr{H} \to \mathscr{H}$ as appropriate.

7.1 If T is compact, show that it is bounded.

7.2 If $S \in \mathscr{L}(\mathscr{B})$ and T is compact, prove that ST and TS are compact.

7.3 In Hilbert space show that T is compact iff it maps weakly convergent sequences to strongly convergent sequences.

7.4* Show that every weakly convergent sequence in ℓ_1 is strongly convergent. Deduce that the assertion in the previous problem is not true in general in Banach space.

7.5 Deduce from Problem 5.9 that if $k > j$ the imbedding of $\mathscr{C}^k([0, 1])$ in $\mathscr{C}^j([0, 1])$ is a compact operator. (An analogous result forms the basis of the treatment of elliptic equations in Chapter 11).

7.6 Let $\mathscr{V}$ be a vector space, and assume that $L : \mathscr{V} \to \mathscr{V}$ is linear. Suppose that $\lambda_1, \ldots, \lambda_n$ are eigenvalues of L no two of which are equal, and let $f_1, \ldots, f_n$ be the corresponding eigenfunctions. Prove that these eigenfunctions are linearly independent.

7.7 Suppose that $\mathscr{M}$ is a proper closed subspace of $\mathscr{B}$. Prove that given $\varepsilon > 0$, there is an $f \in \mathscr{B}$ with $\|f\| = 1$ such that $\mathrm{dist}(f, \mathscr{M}) > 1 - \varepsilon$.

7.8 Show by constructing examples that zero can be in either the point, residual or continuous spectrum of a compact operator.

7.9 If T^n is compact for some integer n, show that $\sigma(T)$ is either a finite or a countably infinite set of points.

7.10 If $T: \mathscr{H} \to \mathscr{H}$ is compact, prove that it is the uniform limit of operators of finite rank.

7.11 Suppose that $k(x, y) = k(x - y)$ where k is an even real-valued continuous function in $\mathscr{L}_1(-\infty, \infty)$. Let $K : \mathscr{C}((-\infty, \infty)) \to \mathscr{C}((-\infty, \infty))$ be the integral operator defined by 7.2.2. Show that every λ in a certain finite interval is an eigenvalue of K (substitute $e^{i\lambda x}$). This yields another proof that K is not compact (cf. Example 7.2.5).

7.12 Take a (possibly infinite) interval with end points a, b, and assume that $k \in \mathscr{L}_2((a, b) \times (a, b))$ is Hermitian ($k(x, y) = \overline{k(y, x)}$). Define the integral operator K on $\mathscr{L}_2(a, b)$ by (7.2.2). Let $\{\phi_n\}$ be the set of eigenfunctions of K. Show that

$$k(x, y) = \Sigma \lambda_n \phi_n(x) \overline{\phi_n(y)},$$

the convergence being in $\mathscr{L}_2((a, b) \times (a, b))$. For further results of this type see Riesz and Nagy (1955, p. 242 *et seq.*).

7.13 For $T: \mathscr{H} \to \mathscr{H}$ compact and self-adjoint, deduce the following from the Hilbert–Schmidt Theorem:

(i) For any polynomial p,

$$p(T)f = \Sigma p(\lambda_n)(f, \phi_n)\phi_n;$$

(ii) $R(\lambda; T)f = \Sigma(\lambda - \lambda_n)^{-1}(f, \phi_n)\phi_n \qquad (\lambda \in \rho(T)).$

If K is the integral operator on $[-1, 1]$ defined by (7.2.2) with kernel $k(x, y) = 1 + xy$, obtain $R(\lambda; T)$ explicitly in the above form.

7.14 Let $T: \mathscr{H} \to \mathscr{H}$ be compact and self-adjoint.

(i) Show that T has at least one non-zero eigenvalue if $T \neq 0$.
(ii) Prove that $\| R(\lambda; T)\| = 1/d(\lambda)$ where $d(\lambda) = \text{dist}(\lambda, \sigma(\lambda))$.

7.15 Consider the integral equation

$$\lambda f(x) - \pi^{-1} \int_{-1}^{1} \frac{b}{b^2 + (x - y)^2} f(y)\,dy = g(x)$$

of Problem 6.24 in $\mathscr{L}_2(-1, 1)$. By using the compactness of the integral operator K in addition prove that $1 \in \rho(K)$. If $|\lambda| = 1$ the convergence of the Neumann series is slow for small b. For $\lambda = -1$, prove that $\|f\| \leqslant \|g\|$, and show how the rate of convergence may be improved by using the series of Problem 3.26.

7.16 In Example 7.5.5 take $l = -d^2/dx^2$ and $D(L) = \{f : f \in \mathscr{A}, f'(0) = f'(\pi) = 0\}$. Assuming that L is self-adjoint, obtain the Fourier cosine series on $\mathscr{L}_2(0, \pi)$. (Hint: L is not injective, but consider $L + aI$ for $a > 0$).

7.17 Let $\{\phi_n\}$ be an orthonormal basis of $\mathscr{H}$.

(i) If T is compact show that $\lim T\phi_n = 0$.
(ii) Let (λ_n) be a sequence of numbers tending to zero. Set $Tf = \Sigma\lambda_n(f, \phi)\phi_n$, and prove that T is compact.

7.18 Let T be compact and self-adjoint. The equation $Tf = g$ is a generalization of the Fredholm integral equation of the first kind and presents similar difficulties. Show that $R(T)$ is the set of elements f in the closed linear span of the eigenfunctions ϕ_n corresponding to non-zero eigenvalues λ_n for which the series $\Sigma\lambda_n^{-1}(f, \phi_n)\phi_n$ is convergent.

7.19 Let T be compact and self-adjoint. If $\lambda \in \rho(T)$ but $|\lambda| < r_\sigma(T)$, the Neumann series may not be convergent, but a useful modified series can sometimes be constructed. Suppose for simplicity that the eigenvalues are non-negative, that $\lambda_1 > \lambda > \lambda_2$, and that there is exactly one eigenfunction corresponding to λ_1. Set $g_1 = g - (g, \phi_1)\phi_1$, and show that

$$(\lambda I - T)^{-1}g = (\lambda - \lambda_1)(g, \phi_1)\phi_1 + \Sigma\lambda^{-n-1}T^n g_1.$$

7.20 (Spectral Mapping Theorem). Suppose that T is compact and self-adjoint, and let g be a continuous function. Define the operator $g(T)$ by setting

$$g(T)f = \Sigma g(\lambda_n)(f, \phi_n)\phi_n.$$

Prove that $\sigma(g(T)) = g(\sigma(T))$.

The next two problems are concerned with the calculation of the eigenvalues and eigenfunctions of a compact self-adjoint operator T. For further results see Riesz and Nagy (1955, p. 237 *et seq.*), Stakgold (1968), Weinstein and Stenger (1972), or Weinberger (1974).

7.21 (Iteration). Take any g such that $Tg \neq 0$, and show that $T^n g \neq 0$ $(n \geqslant 1)$.

(i) Suppose first that T has no negative eigenvalues. For $n \geqslant 1$ set $\phi_n = T^n g/\|T^n g\|$, $r_n = \|T^n g\|/\|T^{n-1}g\|$. Prove that (r_n) and (ϕ_n) converge respectively to an eigenvalue and corresponding eigenfunction of T.

(ii) For general T, apply (i) to T^2 to obtain an eigenvalue λ^2 and eigenfunction ϕ of T^2. Prove that one of $\phi \pm \lambda^{-1}T\phi$ is an eigenfunction of T.

7.22 Arrange the eigenvalues into positive and negative sequences (μ_n^+) and (μ_n^-) respectively, the first non-increasing and the second non-decreasing. Show that

$$\mu_n^+ = \sup(Tf, f)$$

where the supremum is taken over all f with both $\|f\| = 1$ and $(f, \phi_i^+) = 0$ for $i = 1, \ldots, (n-1)$.

7.23 Prove the following result related to Theorem 7.6.6. Suppose that (K_n) is collectively compact and that $K_n \underset{S}{\rightarrow} K \in \mathscr{L}(\mathscr{B})$. Whenever $(I - K_n)^{-1} \in \mathscr{L}(\mathscr{B})$ define

$$\Gamma_n = \|(I - K_n)^{-1}\|\|(K_n - K)K\|.$$

If for some n, $(I - K_n)^{-1} \in \mathscr{L}(\mathscr{B})$ and $\Gamma_n < 1$, then $(I - K)^{-1} \in \mathscr{L}(\mathscr{B})$ and

$$\|(I - K)^{-1}\| \leqslant (1 - \Gamma_n)^{-1}[1 + \|(I - K_n)^{-1}\|\|K\|],$$

$$\|(I - K_n)^{-1}g - (I - K)^{-1}g\| \leqslant (1 - \Gamma_n)^{-1}[\|(I - K_n)^{-1}\|\|K_n g - Kg\| + \Gamma_n\|(I - K_n)^{-1}g\|].$$

Further, $(I - K_n)^{-1} \in \mathscr{L}(\mathscr{B})$ for sufficiently large n, and $\lim \Gamma_n = 0$.

Chapter 8

Nonlinear Compact Operators and Monotonicity

8.1 Introduction

The advantage of the techniques described in Chapter 4 for treating nonlinear equations is that they yield a great deal of information about the solutions. Unfortunately the class of equation to which they apply is very limited. Thus much research has been devoted to the invention of alternative methods. A major part of the effort has been directed towards compact operators. There are two main reasons for this. The first is the frequency of these operators in applications. For example, a boundary value problem for a differential equation on a bounded domain can often be posed in terms of such an operator, either by means of a Green's function, or by the use of certain special Banach spaces such as Sobolev spaces. The second reason is that, as in the linear case, compact nonlinear operators have much in common with operators on a finite dimensional space, and so are relatively tractable. A considerable amount is now known about compact nonlinear operators, and in this and later chapters some of the more important parts of their theory will be outlined.

In $\mathbb{R}$ it is obvious that a continuous operator mapping the interval $D = [-1, 1]$ into itself has a fixed point in D. In attempting to generalize this to higher dimensions it is natural to replace the interval by say a closed ball. In $\mathbb{R}^2$ the situation is illustrated by the "tea cup problem". In this we imagine that the surface of the tea is mapped continuously, by smooth stirring, into another configuration, and enquire whether any point of the surface stays in its original position. If the tea is viscous, evidently every point of the rim remains undisturbed, but if the tea is inviscid, and so does not adhere to the rim, it is far from easy to decide whether any such point exists. Thus even the generalization to two dimensions is not straightforward. The fact that the one dimensional result *does* extend to any finite number of dimensions is the conclusion of a deep and important theorem proved by Brouwer in 1910. A direct proof of this theorem is omitted since the arguments used are not needed subsequently; an easy proof using degree theory is presented later (Example 13.2.15). From a practical point of view it is interesting that constructive algorithms for finding the fixed points have recently been discovered, see Todd (1976).

8.1.1 The Brouwer Fixed Point Theorem. *Let D be a bounded closed convex subset of a finite dimensional normed vector space. If A is a continuous mapping of D into itself, A has a fixed point in D.*

The obvious next step is to enquire whether the restriction of finite dimensionality can be lifted. Unfortunately it cannot, as can be seen by constructing a counterexample (Problem 4.6). The simplest way round the difficulty is to impose a limitation on the rate of growth of the operator as in the Contraction Mapping Principle. The first result to be proved which avoids a condition of this type is the Schauder Fixed Point Theorem, in which compactness is used to bridge the gap between finite and infinite dimensions. This classical result, which is the subject of the next section, has found a multitude of applications in nonlinear analysis.

Powerful as the Schauder Fixed Point Theorem is, its use presents a number of difficulties in practice. For example, the first step in any application is to find a subset D of the domain of the operator A for which $A(D) \subset D$, and the choice of D may not be easy. This choice presents particular difficulties when A has a trivial fixed point (for example if $A0 = 0$) for since only the existence of one fixed point is asserted, if D is such that $0 \in D$ the Schauder Theorem can provide no new information. Secondly, as the theorem applies to operators which may have any number of fixed points (as in the tea cup problem), it does not give any guide on uniqueness. Finally, the nonconstructive nature of the result is a disadvantage in applications. In the last section of this chapter one group of methods which has been devised to circumvent these difficulties will be examined. In outline these methods are based on the following observations.

It is well known that positive and monotone real valued functions of a real variable are specially amenable to both theoretical and numerical treatment. For example suppose that ϕ is a monotone increasing continuous function on the interval $[u, v]$ and $\phi(u) \geqslant u$, $\phi(v) \leqslant v$ (Fig. 8.1). Then not only is the existence of a fixed point guaranteed, but also the sequence (x_n) obtained by successive substitution is increasing and converges to a fixed point $\bar{x}$. If this argument is to generalize to higher dimensions, one difficulty must first be overcome, for while there is a natural ordering of the real numbers, there is no obvious ordering of vectors in a space with dimension greater than unity. If $x, y \in \mathbb{R}^2$ and $x = (x_1, x_2)$, $y = (y_1, y_2)$, one possibility is to define $x \geqslant y$ if both $x_1 \geqslant y_1$ and $x_2 \geqslant y_2$. There is a natural extension of this notion to real valued functions by means of the relation $f(t) \geqslant g(t)$ for all t in the domain. These orderings are only partial (for example not all vectors in $\mathbb{R}^2$ are comparable in this sense), but this does not prove to be a serious disadvantage, and we can proceed to define monotone operators ($Af \geqslant Ag$ if $f \geqslant g$) and positive operators ($Af \geqslant 0$ if $f \geqslant 0$). If in addition A is compact the finite dimensional results often extend to a Banach space.

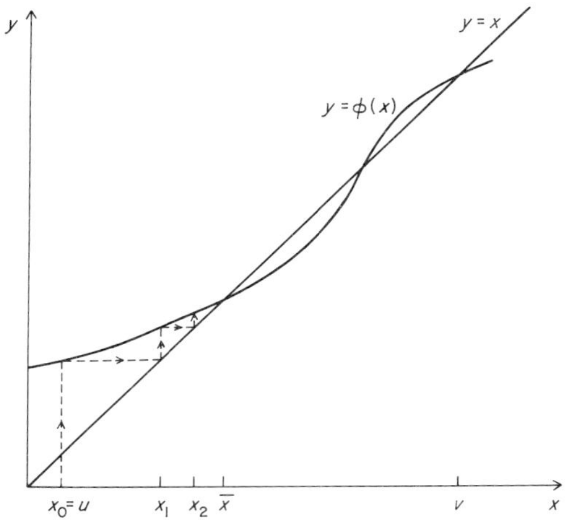

Fig. 8.1 The successive substitution method for a monotone function.

Frequently the success of these techniques in applications can be traced to a broad compatibility between the ordering in the space and some natural order implicit in the equation itself. This is the case for the following boundary value problem to which we shall pay particular attention:

$$\left.\begin{aligned} f''(x) + \psi[x, f(x)] &= 0 \qquad (0 \leqslant x \leqslant 1), \\ f(0) = f(1) &= 0, \end{aligned}\right\} \tag{8.1.1}$$

where ψ is a real valued continuous function on $[0, 1] \times (-\infty, \infty)$ and a solution in $\mathscr{C}^2([0, 1])$ is sought. First the system is rewritten as the integral equation $f = Af$ where

$$Af(x) = \int_0^1 k(x, y)\, \psi[y, f(y)]\, dy,$$

and k is the Green's function (4.2.5). It is basic to the argument which follows that k is non-negative. For then A is a positive operator if ψ is non-negative, and if $\psi[x, \cdot\,]$ is non-decreasing for each x, A is monotone. At first sight the monotonicity condition on ψ may appear severe. However, the differential equation may be rewritten as

$$\{f''(x) - \tilde{\omega} f(x)\} + \{\tilde{\omega} f(x) + \psi[x, f(x)\}] = 0,$$

and since the Green's function for $(+\mathrm{d}^2/\mathrm{d}x^2 + \tilde{\omega})$ is non-negative for any $\tilde{\omega} \geqslant 0$, the modified integral operator is monotone if for some $\tilde{\omega}, \psi[x, z] + \tilde{\omega} z$ is a non-decreasing function of z for each x. This is a much less restrictive condition which will certainly be satisfied if $\partial\psi/\partial z$ is bounded below.

These arguments may be placed in a much more general setting. The crucial condition that k should be non-negative is fulfilled as long as a maximum principle holds for the differential operator; this is the case for a wide class of elliptic partial differential operators. In combination with a technique devised by Perron which uses differential inequalities to bracket the solution, monotonicity is particularly useful in dealing with large nonlinearities in ψ. For simplicity the discussion will be restricted here to ordinary differential equations.

The classical text on operators in partially ordered spaces is Krasnoselskii (1964b). For an excellent modern survey see Amann (1976). Anselone (1964), Rall (1971), and Saaty (1967) contain a number of interesting applications.

8.2 The Schauder Fixed Point Theorem

This theorem occupies a central position in nonlinear operator theory. In its own right it is an extremely powerful and useful result. It is also of unique historical importance, providing as it did the starting point for the theory of nonlinear compact operators, which is perhaps the most effective tool in nonlinear analysis. In particular the theory of the Leray–Schauder degree in Chapter 13 is in the direct line of descent from the Schauder result.

8.2.1 Definition. Suppose that D is a subset of the Banach space $\mathscr{B}$. An operator $A : D \to \mathscr{B}$ will be said to be **compact** iff it is continuous and it maps every bounded subset of D into a relatively compact set. □

Note that in contrast with the linear case, continuity is not automatic. Unfortunately the term compact operator does not have a uniform meaning in the literature, sometimes an operator as defined above being referred to as a continuous "compact operator" and sometimes as a "completely continuous operator". Compact operators are relatively common in applications.

8.2.2 Lemma. *The operators in Lemmas 4.2.7 and 4.2.8 are compact.*

Thus as in the linear case, under mild restrictions integral operators are compact if the region of integration is finite.

8.2.3 The Schauder Fixed Point Theorem. *Let D be a non-empty closed bounded convex subset of the Banach space $\mathscr{B}$, and suppose that $A : D \to \mathscr{B}$ is compact and maps D into itself. Then A has a fixed point in D.*

The idea of the proof is to use certain approximating operators A_n to obtain a problem which is essentially finite dimensional and so can be treated by the Brouwer Fixed Point Theorem 8.1.1. In broad outline this technique is typical in compact operator theory and will reappear in more sophisticated form in the degree theoretic arguments in Chapter 13, which themselves yield a very simple alternative proof of the Schauder Theorem. The key step (Lemma 8.2.5) is the construction of the A_n.

8.2.4 Lemma. (i) *Let S be a subset of a vector space. If $f_1, \ldots, f_k \in S$ and $\alpha_1, \ldots, \alpha_k$ are non-negative numbers not all of which are zero, $\Sigma\alpha_i f_i/\Sigma\alpha_i$ is in the convex hull* co S of S.

(ii) *Let S be a subset of a Banach space. If S is relatively compact, so is* co S.

Proof. (i) follows immediately from Definition 1.2.12. For (ii) see Dunford and Schwartz (1958, p. 416). □

8.2.5 Lemma. *Suppose S is a bounded subset of the Banach space $\mathscr{B}$ and $A : S \to \mathscr{B}$ is compact. Let (ε_n) be any sequence of positive numbers tending to zero. Then there is a sequence (A_n) of continuous operators $A_n : S \to \mathscr{B}$ with the following properties.*

(i) *(A_n) converges to A in the sense that $\| A_n f - Af \| < \varepsilon_n$ for all $f \in S$.*

(ii) *For each n, the range $A_n(S)$ of A_n is finite dimensional and is contained in the convex hull of $A(S)$.*

Proof. Since $A(S)$ is relatively compact, it is totally bounded (Theorem 5.2.8). Therefore, given $\varepsilon > 0$ there is a cover of $A(S)$ by a finite number of balls $S(g_1, \varepsilon), \ldots, S(g_k, \varepsilon)$ say. Define the *Schauder Projection Operator* $P : A(S) \to \mathscr{B}$

by setting

$$Pg = \sum_{i=1}^{k} \alpha_i(g) g_i \Big/ \sum_{i=1}^{k} \alpha_i(g) \qquad (g \in A(S)), \tag{8.2.1}$$

where

$$\alpha_i(g) = \begin{cases} \varepsilon - \|g - g_i\| & (\|g - g_i\| < \varepsilon), \\ 0 & (\|g - g_i\| \geqslant \varepsilon). \end{cases}$$

Since any $g \in A(S)$ is within ε of some g_i, the denominator in (8.2.1) is strictly positive. Thus the definition of P makes sense. Also, each α_i is a continuous function $A(S) \to \mathbb{R}$, so P is continuous. In view of Lemma 8.2.4(i), it is clear from (8.2.1) that the range of P is contained in co $A(S)$. Further, for any given $g \in A(S)$, the only non-zero terms in the sum (8.2.1) are those for which $g_i \in S(g, \varepsilon)$. Hence by Lemma 8.2.4(i) again, $Pg \in \text{co } S(g, \varepsilon) = S(g, \varepsilon)$. That is $\|Pg - g\| < \varepsilon$.

It follows that if an operator $B : S \to \mathscr{B}$ is defined by setting $Bf = PAf$ for $f \in S$, then B is continuous, has finite dimensional range contained in co $A(S)$, and is close to A in the sense that

$$\|Af - Bf\| = \|Af - PAf\| < \varepsilon \qquad (f \in S).$$

Since ε is arbitrary, operators satisfying the requirements are obtained by taking $\varepsilon_n = \varepsilon$ and $A_n = B$ in turn. □

Proof of Theorem 8.2.3. With $D = S$ choose (ε_n) and (A_n) as in the lemma. Since D is convex and $A(D) \subset D$, by property (ii) $A_n(D) \subset D$. Also $A_n(D)$ is contained in some finite dimensional subspace, $\mathscr{M}_n$ say, of $\mathscr{B}$ and $D \cap \mathscr{M}_n$ is obviously convex, bounded and closed (in $\mathscr{M}_n$). Thus by the Brouwer Fixed Point Theorem 8.1.1 applied to A_n restricted to $D \cap \mathscr{M}_n$, A_n has a fixed point, f_n say, in $D \cap \mathscr{M}_n$. Therefore

$$\|Af_n - f_n\| = \|Af_n - A_n f_n\| \leqslant \varepsilon_n$$

by (i) of the previous lemma. Since A is continuous and $\overline{A(D)}$ is compact, the result follows from Theorem 5.3.5. □

A contraction condition is not required in the Schauder Fixed Point Theorem. Thus in those applications in which the operator is compact, this theorem is a much more powerful weapon for proving existence than the Contraction Mapping Principle 4.3.4. On the other hand, it is inherent in the very generality of the problems to which the theorem applies that it cannot guarantee uniqueness nor can it be used to prove convergence of the method of successive substitution. Two applications illustrate the type of situation in which the Schauder Theorem is useful.

8.2.6 Example. In the initial value problem

$$f'(t) = \psi[t, f(t)] \qquad (t \geqslant 0),$$
$$f(0) = \alpha,$$

for the $\mathbb{C}^n$ valued function f, a solution with continuous first derivative is required. Theorem 4.3.11, based on the Contraction Mapping Principle, asserts that if ψ is continuous and satisfies a Lipschitz condition, such a solution exists for all t. Because of the restrictive character of the Lipschitz condition the result is applicable to a rather limited class of equation. What then is the position if ψ is merely continuous? It is clear from Examples 4.3.8 and 4.3.9 that neither uniqueness nor global existence can be expected. In the following result (originally due to Peano) the Schauder Fixed Point Theorem is used to establish local existence.

8.2.7 Theorem. *Let $t > 0$ be given and suppose that $\psi : [0, \bar{t}] \times \mathbb{C}^n \to \mathbb{C}^n$ is continuous. Then there is a $t_0 > 0$ such that the above initial value problem has a solution in $\mathscr{C}^1([0, t_0], \mathbb{C}^n)$.*

Proof. There is no loss of generality in taking $\alpha = 0$. It is evidently enough to prove that for some $t_0 > 0$ the operator $A : \mathscr{C}([0, t_0], \mathbb{C}^n) \to \mathscr{C}([0, t_0], \mathbb{C}^n)$ defined by

$$Af(t) = \int_0^t \psi[s, f(s)]\, ds$$

has a fixed point.

Since ψ is continuous there are strictly positive numbers $t_0 \leqslant 1$ and m such that $|\psi[t, z]| \leqslant m$ for $0 \leqslant t \leqslant t_0$ and $|z| \leqslant mt_0$. Let D be the ball $\bar{S}(0, m)$ in $\mathscr{C}([0, t_0], \mathbb{C}^n)$ equipped with the sup norm. As D is bounded closed and convex, existence will follow from the Schauder Theorem 8.2.3 if it can be proved that $A(D) \subset D$ and that A is compact. By the Mean Value Theorem for integrals.

$$|Af(t)| \leqslant t_0 \sup_{0 \leqslant s \leqslant t_0} |\psi[s, f(s)]| \leqslant mt_0 \qquad (f \in D).$$

That is $\|Af\| \leqslant m$, whence $A(D) \subset D$. Lemma 8.2.2 yields the compactness. □

8.2.8 Example. Consider next the boundary value problem

$$f''(x) + \lambda\psi[x, f(x)] = g(x),$$
$$f(0) = f(1) = 0.$$

The forced oscillations of a pendulum can be modelled by taking $\psi[x, z] = \sin z$. In this case the simplest approach is to linearize and replace $\sin z$ by z, but the answers obtained in this way sometimes do not reflect the basic physics of the problem. For example, for a periodic forcing term $g(x) = \sin \omega\pi x$ the false conclusion is reached that there is no possible motion if $\lambda = (\omega\pi)^2$. This difficulty disappears if the full nonlinear equations are used.

If ψ satisfies a Lipschitz condition, the Contraction Mapping Principle shows that there is a solution for small λ which may be obtained by the method of successive substitution (Problem 4.16). However, for large λ the mapping is not a contraction and the argument breaks down. With the help of the Schauder Fixed Point Theorem it is easy to prove that if ψ is bounded the equation has a solution for *any* λ, although the theorem gives no guide as to how the solution may be constructed.

8.2.9 Theorem. *Suppose that $\psi : [0, 1] \times \mathbb{C} \to \mathbb{C}$ is continuous and bounded, and g is continuous. Then the above boundary value problem has a solution $f \in \mathscr{C}^2([0, 1])$ for any $\lambda \in \mathbb{C}$.*

Proof. Take the sup norm on $\mathscr{C}([0, 1])$. As noted in Example 4.2.2 it is enough to show that the operator A mapping $\mathscr{C}([0, 1])$ into itself has a fixed point, where

$$Af(x) = \lambda \int_0^1 k(x, y)\, \psi[y, f(y)]\, dy + h(x),$$

k is the Green's function and h is a (known) continuous function. Now A is compact (Lemma 8.2.2), therefore the Schauder Theorem will be applicable if a suitable set D can be found. Since ψ is bounded there is an m such that $|\psi[x, z]| \leqslant m$ for $0 \leqslant x \leqslant 1$ and $z \in \mathbb{C}$. Hence $\|Af - h\| \leqslant |\lambda|\, \|K\| m$, where K is the linear integral operator with kernel k. This proves that A maps the closed ball D centre h and radius $|\lambda| \|K\| m$ into itself. □

8.3 Positive and Monotone Operators in Partially Ordered Banach Spaces

Our aim will now be to enquire how compact operator theory can be strengthened when positivity and monotonicity conditions are imposed on the operator. Clearly these conditions only make sense in a space in which there is a satisfactory notion of order, and the first task will be to show how the usual ordering of the real numbers can be generalized to a Banach space setting. Since it is a property of *real* numbers that is the model, the natural assumption will be made throughout that $\mathscr{B}$ is a *real* space.

If $a, b \in \mathbb{R}$ the relation $a \geqslant b$ is equivalent to $a - b \in \mathbb{R}^+$, where $\mathbb{R}^+$ is the set of non-negative numbers. If therefore a suitable substitute for $\mathbb{R}^+$ in the Banach space can be found, there will be a natural way of defining an order on the space. In two dimensions it is reasonable to regard a vector as positive if it has positive components, and from a geometrical point of view these vectors form a cone. As a cone may be readily characterized in terms of concepts which are meaningful in an abstract Banach space, a possible substitute for $\mathbb{R}^+$ is at hand.

8.3.1 Definition. Let $\mathscr{B}$ be a real Banach space. A set $E \subset \mathscr{B}$ is a **cone** iff it satisfies the following conditions:

(i) E is closed and convex;
(ii) if $f \in E$, $tf \in E$ for every $t \geqslant 0$;
(iii) if both f and $-f$ are in E, then $f = 0$.

8.3.2 Definition. Let $E \subset \mathscr{B}$ be a cone. f is said to be **greater than or equal to** g—written $f \geqslant g$ or $g \leqslant f$—iff $f - g \in E$, with an analogous definition for the opposite inequality. With the ordering $\geqslant$, $\mathscr{B}$ is said to be **partially ordered.**

Use of the symbol $\geqslant$ in this additional sense should cause no confusion since it will be clear from the context when vectors or real numbers are being referred to. Note that $E = \{f : f \geqslant 0\}$. Thus an element of $\mathscr{B}$ may be thought of as positive if it lies in E.

8.3.3 Example. For $\mathscr{B} = \mathbb{R}^2$ a cone is shown in the figure. Note that the cone may but need not be the first quadrant, so the definition is a little more general than that suggested by the vectors with non-negative components. However, because of (iii) in the definition, θ must be less than π. If E is the first quadrant and $x = (x_1, x_2)$, $y = (y_1, y_2)$, the inequality $x \geqslant y$ is valid if and only if both $x_1 \geqslant y_1$ and $x_2 \geqslant y_2$. If $x_1 > y_1$ and $x_2 < y_2$, neither $x \leqslant y$ nor $x \geqslant y$ is true. Thus not every pair of vectors is comparable, and the order is partial only.

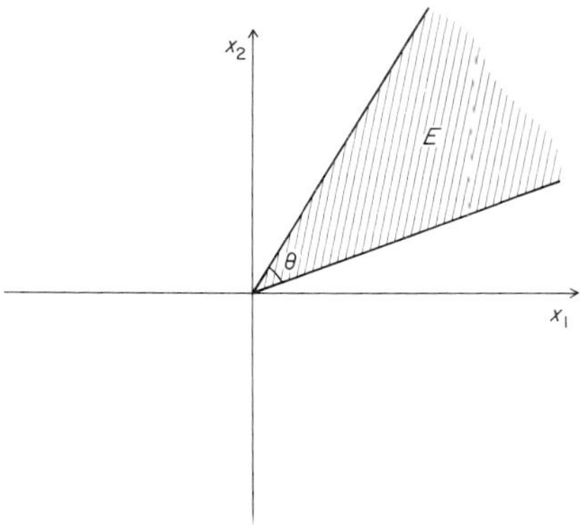

Figure 8.2.

8.3.4 Example. (a) Take $\mathscr{B} = \mathscr{C}([0, 1])$, the real valued continuous functions with the sup norm, and let E be the set of non-negative functions in $\mathscr{B}$. It is readily checked that E is closed, and the other requirements are obviously met. E is a cone. (b) Suppose that $\mathscr{B}$ is the real space $\mathscr{L}_p(0, 1)$ with $p \geqslant 1$. Again the set E of non-negative functions in $\mathscr{B}$ is a cone.

These two choices of E are by far the most useful in applications; another possibility is described in Problem 8.10, and more elaborate choices are also occasionally useful—see Krasnoselskii (1964b).

Although the ordering induced by a cone E is only partial, most of the standard relations for the ordering of real numbers hold. For example, if $f, g, h \in \mathscr{B}$,

(i) $f \geqslant g, g \geqslant h \Rightarrow f \geqslant h$,
(ii) $f \geqslant g, g \geqslant f \Rightarrow f = g$.

Some other basic properties of cones are listed in Problem 8.7.

Intervals in $\mathbb{R}$, regarded as ordered sets, often play an important role in real variable theory. The following generalization to a partially ordered space suggests itself.

8.3.5 Definition. Suppose $u, v \in \mathscr{B}$ and $u \leqslant v$. The set of elements f such that $u \leqslant f \leqslant v$ will be called an **order interval** and will be denoted by $[u, v]$.

An order interval is closed and convex. However, it need not be bounded in norm unless the cone is "normal" in the following sense.

8.3.6 Definition. A cone E is said to be **normal** iff there is an $m \in \mathbb{R}$ such that $\|f\| \leqslant m\|g\|$ for any $f, g \in \mathscr{B}$ with $0 \leqslant f \leqslant g$.

The cones of non-negative functions in $\mathscr{C}([0, 1])$ and $\mathscr{L}_p(0, 1)$ are both normal (with $m = 1$). The cone of non-negative functions in the Banach space $\mathscr{C}^1([0, 1])$ with the norm (1.3.7) is *not* normal. For if $g(x) = 1$, $f_n(x) = \sin^2 nx$, then $0 \leqslant f_n \leqslant g$ but $\lim \|f_n\|_{\mathscr{C}^1} = \infty$.

Having arranged for a suitable ordering in the space, we return to the study of compact operators. The following loose remarks set the scene for the development of the theory. In one dimension, by the Intermediate Value Theorem, the existence of an interval $[u, v]$ such that $Au \geqslant u$ and $Av \leqslant v$ is sufficient to ensure that A has a fixed point. If A is positive then $A0 \geqslant 0$, and the choice of zero is an obvious possibility for u. Also, if for large values of the argument the gradient is less than unity (this does not rule out a large negative gradient), some large v will be adequate for the other end point, whereas for bigger gradients no fixed point need exist. In higher dimensions a suitable generalization of this argument is useful in finding an order interval. However, for the existence of a fixed point of a compact operator A an additional geometric condition is needed: that A should map $[u, v]$ into itself. It can be expected that as in one dimension the monotonicity of A would be helpful in establishing this.

The following definitions are natural generalizations from one dimension. D will denote a subset of $\mathscr{B}$ and A is an operator $D \to \mathscr{B}$.

8.3.7 Definition. The operator A will be said to be **positive**† iff $f \in D$ and $f \geqslant 0 \Rightarrow Af \geqslant 0$.

8.3.8 Definition. A will be said to be **monotone**† iff for all $f, g \in D$ with $f \geqslant g$, $Af \geqslant Ag$.

Positivity and monotonicity are equivalent if the operator is linear, but not necessarily otherwise. As indicated in the introduction, it is often possible to arrange for the operator A in the integral equation $f = Af$ arising from a boundary value problem for a differential equation to be positive and monotone, and there are many other applications.

The first result is an obvious deduction from the Schauder Fixed Point Theorem 8.2.3 when it is recalled that if E is normal, an order interval is bounded closed and convex.

8.3.9 Theorem. *Let E be a cone in the real Banach space $\mathscr{B}$. Suppose E is normal, and let A be a compact operator mapping the order interval $[u, v]$ into itself. Then A has a fixed point in $[u, v]$.*

8.3.10 Example. The utility of the theorem is demonstrated in the following problem in nonlinear elasticity. The argument is due to Stuart (1975).

Consider a thin elastic sheet fixed at its edges and subjected to a normal pressure. If the deformation is large, nonlinear effects become important, and an approximate theory which takes these into account leads to the so called Föppl–Hencky equations. For a circular sheet the radial stress satisfies the differential equation

$$f''(x) + 3x^{-1}f'(x) + 2[f(x)]^{-2} = 0 \qquad (0 < x < 1) \tag{8.3.1}$$

together with the boundary conditions $f'(0) = 0$, $f(1) = a$. This is a classical problem. The following proof of the existence of a solution for all $a > 0$ is particularly simple.

† Unfortunately the terms "positive" and "monotone" are both commonly used in several senses in functional analysis. Another meaning of "positive" is encountered in Section 9.2. A class of "monotone" operators quite different from that defined above has recently been extensively studied (Brézis, 1973).

With the help of the Green's function the problem is first rewritten formally as a Hammerstein integral equation. For $g(x) = f(x) - a$ this equation is $g = Ag$ where

$$Ag(x) = \int_0^1 k(x, y)[a + g(y)]^{-2}\,\mathrm{d}y,$$

$$k(x, y) = \begin{cases} (x^{-2} - 1)y^3 & (0 \leqslant y < x \leqslant 1), \\ (y^{-2} - 1)y^3 & (0 \leqslant x < y \leqslant 1). \end{cases}$$

The setting for the analysis is the real Banach space $\mathscr{C}([0, 1])$ with the sup norm, and with the partial ordering induced by the cone E of non-negative functions. For $z \geqslant 0$ let $\psi(z) = (a + z)^{-2}$, and define an operator $\Psi : E \to \mathscr{C}([0, 1])$ by setting $\Psi g(x) = \psi(g(x))$; clearly Ψ is continuous and positive. Define

$$Lg(x) = \int_0^1 k(x, y)\, g(y)\,\mathrm{d}y \qquad (0 \leqslant x \leqslant 1).$$

It is straightforward to check that $L : \mathscr{C}([0, 1]) \to \mathscr{C}([0, 1])$ is a continuous positive linear operator. Further,

$$(Lg)'(x) = -2\int_0^x (y/x)^3 g(y)\,\mathrm{d}y \qquad (0 < x \leqslant 1),$$

$$(Lg)'(0) = 0,$$

and it follows that $\|(Lg)'\| \leqslant 2\|g\|$. Therefore L is compact (Problem 5.9). Thus the operator $A = L\Psi$ is positive and compact, and Theorem 8.3.9 will be applicable if an order interval $[u, v]$ mapped into itself by A can be found. Since A is positive, the choice $u = 0$ presents no difficulty. Also, as ψ is a monotone decreasing function, it is obvious that $Ag \leqslant A0$ for any $g \in E$. Hence $A[0, v] \subset [0, v]$ if $v = A0$. The existence of a fixed point follows from the theorem, and a routine argument shows that $f(x) = g(x) + a$ satisfies the differential equation and boundary conditions. We have proved the following.

8.3.11 Theorem. *For each $a > 0$ the Föppl–Hencky equation* (8.3.1) *with the boundary conditions $f'(0) = 0, f(1) = a$ has a solution $f \in \mathscr{C}^1([0, 1]) \cap \mathscr{C}^2((0, 1))$ such that*

$$a \leqslant f(x) \leqslant a + (2a)^{-2}(1 - x^2) \qquad (0 \leqslant x \leqslant 1).$$

This settles the existence problem, but two further questions may be asked. Is this solution unique, and how can it be constructed? Uniqueness is easy to prove (Problem 8.12), but the second question is not quite so simply resolved, and the reader is referred to the original paper for the details.

In this example the facts that A is positive and "monotone decreasing" (that is $f \geqslant g \Rightarrow Af \leqslant Ag$) makes the choice of the order interval easy. The remarks before Definition 8.3.7 suggest that a similar argument will be valid so long as A does not increase too rapidly, and this is indeed the case (Problem 8.11).

Theorem 8.3.9 suffers from two main disadvantages in practice. First, it will often be difficult to locate a suitable order interval. Second, the result, based as it is on the Schauder Fixed Point Theorem, gives no hint as to how a solution may be constructed. Guided by the one dimensional case (Fig. 8.1) we may expect that these problems will be more tractable for monotone operators. The properties of monotone sequences in partially ordered spaces are the basis of the analysis.

8.3.12 Definition. A sequence (f_n) in $\mathscr{B}$ is said to be **monotone increasing** iff $f_{n-1} \leqslant f_n$ for all n. The sequence is said to be **order bounded above** iff there is an element v, called an **order bound**, such that $f_n \leqslant v$ for all n. Similar definitions for monotone decreasing sequences are obtained by reversing the inequalities. Note that the term "bounded" will continue to mean bounded in norm.

8.3.13 Lemma. *Let E be a cone in the real Banach space $\mathscr{B}$. Let D be a subset of $\mathscr{B}$ and suppose that $A : D \to \mathscr{B}$ is monotone. Assume that there is an order interval $[u, v] \subset D$ such that $Au \geqslant u$, $Av \leqslant v$. Then*

(i) *A maps $[u, v]$ into itself;*
(ii) *the sequences (f_n), (g_n) defined by setting $f_0 = u$, $g_0 = v$ and $f_n = Af_{n-1}$, $g_n = Ag_{n-1}$ for $n \geqslant 1$ are monotone increasing and decreasing respectively.*

Proof. (i) $Af \geqslant Au \geqslant u$ if $f \geqslant u$. Similarly $Af \leqslant v$ if $f \leqslant v$.

(ii) As A is monotone, $f_n = Af_{n-1} \leqslant Af_n = f_{n+1}$ if $f_{n-1} \leqslant f_n$. Since $f_1 = Au \geqslant u$, it follows by induction that (f_n) is monotone increasing. The argument for (g_n) is similar. □

Thus if A is monotone, the choice of an order interval $[u, v]$ mapped into itself by A depends on finding u, v with $Au \geqslant u$, $Av \leqslant v$. For such A the verification of the conditions of Theorem 8.3.9 is relatively straightforward and the existence of a solution correspondingly easier to establish. In addition, if the monotone sequences (f_n) and (g_n) are convergent, their limits will be fixed points of A. However, it should be noted that neither need the fixed point be unique, nor must every fixed point in $[u, v]$ be given in this way. We next enquire whether, as in $\mathbb{R}$, order boundedness is sufficient to ensure the convergence of a monotone sequence.

8.3.14 Definition. A cone is said to be **regular** iff every monotone increasing sequence which is order-bounded above is convergent.

8.3.15 Example. In the real space $\mathscr{C}([0, 1])$ equipped with the sup norm, is the cone E of non-negative functions regular? The sequence (f_n), where $f_n(x) = 1 - x^n$, is monotone increasing and order bounded above, but does not tend in norm to a limit. Thus E is not regular.

8.3.16 Example. Consider next the cone E of non-negative functions in the real space $\mathscr{L}_p(\Omega)$ for $1 \leqslant p < \infty$ and Ω any open set in $\mathbb{R}$. Suppose that (f_n) is a monotone increasing sequence bounded above by $g \in \mathscr{L}_p(\Omega)$; we may assume that $f_n \geqslant 0$, for otherwise consider $f_n - f_1$. Then the pointwise limit, f say, of (f_n) exists almost everywhere and $f \leqslant g$. Thus $0 \leqslant (f - f_n)^p \leqslant g^p$, and

$$\|f - f_n\|_p^p = \int_\Omega |f - f_n|^p \to 0$$

by the Dominated Convergence Theorem 2.4.11. This proves that $f_n \to f$ in $\mathscr{L}_p(\Omega)$, and it follows that E is regular.

In a space ordered by a regular cone, the sequences in Lemma 8.3.13 will converge to a fixed point of A if A is continuous. Since it is not necessary that A be compact, this result can be useful in tackling problems in unbounded regions, see Example 8.3.21 below. Unfortunately, as the cone of non-negative functions in $\mathscr{C}([0, 1])$ is not regular, the most satisfactory choice of Banach space is ruled out. There are regular cones in $\mathscr{C}([0, 1])$, see Problem 8.10, but their use imposes unwelcome restrictions on the operators. On the other hand, unless the nonlinearity is mild, A will not be continuous in the $\mathscr{L}_p$ spaces. In the sequel we shall usually rely on the second condition in the following theorem, compactness being used to compensate for the lack of regularity of the cone.

8.3.17 Theorem. *Let E be a cone in the real Banach space $\mathscr{B}$. Let $A : D \to \mathscr{B}$ be continuous and monotone, and assume that there is an order interval $[u, v] \subset D$ such that $Au \geqslant u$, $Av \leqslant v$. Then A maps $[u, v]$ into itself. Further, A has a fixed point in $[u, v]$ if in addition either* (a) *E is regular, or* (b) *E is normal and A is compact.*

Define sequences $(f_n), (g_n)$ by setting $f_0 = u$, $g_0 = v$ and $f_n = Af_{n-1}$, $g_n = Ag_{n-1}$ for $n \geqslant 1$. Then (f_n) and (g_n) are monotone increasing and decreasing respectively, and each sequence converges to a fixed point of A in $[u, v]$ if either (a) *or* (b) *is satisfied.*

Proof. If (f_n) is convergent with limit f, the continuity of A ensures that f is a fixed point of A. Further, as $[u, v]$ is closed, $f \in [u, v]$. By Lemma 8.3.13, (f_n) is monotone increasing, so the assertion of the theorem will follow if the convergence of (f_n) can be proved. For condition (a) this convergence is an obvious consequence of Definition 8.3.14. Under condition (b), $[u, v]$ is bounded as E is normal, so the set $A([u, v])$ is relatively compact. Thus it is only necessary to prove that a monotone sequence in a relatively compact set is convergent.

By compactness (f_n) has a convergent subsequence (f_{n_k}) with limit f, say. Since (f_{n_k}) is monotone, $f_{n_r} - f_{n_k} \in E$ for $r \geqslant k$, and as E is closed we conclude on taking limits that $f - f_{n_k} \in E$. That is $f_{n_k} \leqslant f$ for all k. Now for all $n \geqslant n_k$, $f - f_n \leqslant f - f_{n_k}$, and it follows as E is normal that there is an m such that

$$\|f - f_n\| \leqslant m\|f - f_{n_k}\|.$$

Letting $k \to \infty$, we deduce that $f_n \to f$. □

A major advantage of this result is that it is constructive. The main problem in practice is to locate suitable u, v for the end points of the order interval, and it is to this that we next turn. In looking for a condition which will ensure the existence of an appropriate v, it is natural as in one dimension to consider restrictions on the gradient, measured by the Fréchet derivative say, of the operator. More generally it is enough to ask that the operator should be "majorized" by another operator with suitable Fréchet derivative. In many problems the obvious choice of majorant involves a linear operator, and in this case a measure of the gradient is the spectral radius; this is the situation envisaged in the following.

8.3.18 Definition. An operator V (respectively U) will be called a **majorant** (respectively a **minorant**) of $A : D \to \mathscr{B}$ iff $Af \leqslant Vf$ (respectively $Uf \leqslant Af$) for all $f \in D$. □

8.3.19 Lemma. Let E be a cone in the real Banach space $\mathscr{B}$. Suppose $A : E \to \mathscr{B}$ is continuous positive and monotone. Assume that $Vf = Lf + h$ for $f \in \mathscr{B}$, where $h \in E$ and L is a positive bounded linear operator with $r_\sigma(L) < 1$. If V is a majorant of A, then A maps the interval $[0, v]$ into itself, where v is the solution of $Vv = v$.

Proof. Since $r_\sigma(L) < 1$, by Theorem 3.7.8 the equation $f = Lf + h$ has the unique solution v, which is the limit of the sequence (v_n) where

$$v_n = h + \sum_1^n L^k h.$$

As L is positive, $L^k h \in E$ for $k \geqslant 1$. Thus v_n is a sum of vectors in E, and so $v_n \in E$. As E is closed, $v \in E$.

The result now follows from Lemma 8.3.13. For since A is positive, $A0 \geqslant 0$, while as V is a majorant of A, $Av \leqslant Vv = v$. □

8.3.20 Example. Consider the Hammerstein integral equation $f = Af$ where

$$Af(x) = \int_0^1 k(x, y)\,\psi[y, f(y)]\,\mathrm{d}y,$$

and k, ψ are real-valued, non-negative and continuous on $[0, 1] \times [0, 1]$ and $[0, 1] \times [0, \infty)$ respectively. With $\mathscr{C}([0, 1])$ the Banach space of real-valued continuous functions equipped with the sup norm, and E the cone of non-negative functions, $A : E \to \mathscr{C}([0, 1])$ is compact and positive. Suppose that

(i) $\psi[x, z]$ is non-decreasing in z for each $x \in [0, 1]$;
(ii) there are non-negative numbers m, l such that $\psi[x, z] \leqslant mz + l$ for $x \in [0, 1]$ and $z \geqslant 0$.

Then there is an obvious choice of L and h in the lemma. Set

$$Lf(x) = m \int_0^1 k(x, y) f(y)\,\mathrm{d}y,$$

$$h(x) = l \int_0^1 k(x, y)\,\mathrm{d}y.$$

Thus for $f \in E$, $0 \leqslant Af \leqslant Lf + h$, so V is a majorant of A. To apply the lemma the following further condition is needed:

(iii) $r_\sigma(L) < 1$.

With V the solution of $v = Lv + h$, A maps $[0, v]$ into itself. Hence by Theorem 8.3.17 there is a fixed point of A in $[0, v]$. Further, by setting $f_0 = 0, g_0 = v$ and $f_n = Af_{n-1}, g_n = Ag_{n-1}$ for $n \geqslant 1$, we obtain monotone iteration schemes for the solution of the integral equation.

8.3.21 Example. The argument in the previous example may not be valid if the interval is infinite, for the operator A will usually not be compact. This difficulty can sometimes be circumvented by exploiting the regularity condition (a) in Theorem 8.3.17. However, as the cone of non-negative functions in $\mathscr{C}([0, \infty))$ is not regular, a different choice of Banach space is necessary. If $\mathscr{L}_p(0, 1)$ is used the integral equation

$$f(x) = \int_0^\infty k(x, y)\,\psi[y, f(y)]\,\mathrm{d}y$$

may then be tackled as in the previous example so long as the operator is continuous.

Monotonicity methods are seen to special advantage in dealing with boundary value problems for differential equations. One possibility is to use the argument of Example 8.3.20 (see Problem 8.14), but somewhat restrictive conditions are needed on the nonlinear term, and another technique which is often more effective will be examined here. The central idea, due to Perron, is classical, but the method has been considerably developed with the aid of partially ordered space techniques; for a recent review see Amann (1976). The method is based on the utilization of "lower and upper solutions" u, v respectively for the end points of the order interval $[u, v]$ together with a rearrangement of the differential equation to produce a monotone operator.

As an illustration consider the system

$$\left.\begin{aligned} f''(x) + \psi[x, f(x)] &= 0, \\ f(0) = f(1) &= 0, \end{aligned}\right\} \tag{8.3.2}$$

and assume that ψ is real valued, continuous, and locally Lipschitz in the sense that, given any finite interval $[\alpha, \beta]$ there is a number m (which may depend on α, β) such that

$$|\psi[x, z_1] - \psi[x, z_2]| \leqslant m|z_1 - z_2| \tag{8.3.3}$$

for $x \in [0, 1]$ and $z_1, z_2 \in [\alpha, \beta]$.

8.3.22 Definition. A function $u \in \mathscr{C}^2([0, 1])$ is called a **lower solution** of (8.3.2) iff

$$-u''(x) \leqslant \psi[x, u(x)] \qquad (0 \leqslant x \leqslant 1),$$

$$u(0), u(1) \leqslant 0.$$

An **upper solution** is defined by reversing the inequalities.

Take the real Banach space $\mathscr{C}([0, 1])$ with the sup norm and with the partial ordering induced by the cone E of non-negative functions. Suppose there are lower and upper solutions u, v respectively with $u \leqslant v$, and set

$$\alpha = \inf_{x \in [0, 1]} u(x), \qquad \beta = \sup_{x \in [0, 1]} v(x).$$

Then $[\alpha, \beta]$ is a finite interval in $\mathbb{R}$, and as ψ is locally Lipschitz there is a real number ω such that for each $x \in [0, 1]$, $\psi[x, z] + \omega^2 z$ is non-decreasing in z for $z \in [\alpha, \beta]$. Since the differential equation may be rearranged as

$$-f''(x) + \omega^2 f(x) = \psi[x, f(x)] + \omega^2 f(x),$$

the system (8.3.2) is equivalent to the integral equation $f = Af$ where

$$Af(x) = \int_0^1 k(x, y)\{\psi[y, f(y)] + \omega^2 f(y)\}\, \mathrm{d}y,$$

and k is the Green's function for $-\mathrm{d}^2/\mathrm{d}x^2 + \omega^2$ with the stated boundary conditions. A is compact by Lemma 8.2.2, and as k is non-negative, A is also monotone. Thus Theorem 8.3.17 may be used if it can be proved that $u \leqslant Au$ and $v \geqslant Av$. This is shown as follows.

As u is a lower solution,

$$-u''(x) + \omega^2 u(x) \leqslant \psi[x, u(x)] + \omega^2 u(x),$$

and multiplying by k and integrating we obtain

$$u(x) \leqslant \left\{u(0)\frac{\partial k}{\partial y}(x, 0) - u(1)\frac{\partial k}{\partial y}(x, 1)\right\} + \int_0^1 k(x, y)\{\psi[y, u(y)] + \omega^2 u(y)\}\, \mathrm{d}y.$$

Since $u(0), u(1) \leqslant 0$, the first term on the right-hand side takes only negative values, and it follows that $u \leqslant Au$. A similar argument proves that $v \geqslant Av$. Theorem 8.3.17 then guarantees the existence of a solution of the integral equation in $[u, v]$. Further, if $f_0 = u$ and f_n is the solution of the linear equation

$$-f_n''(x) + \omega^2 f_n(x) = \psi[x, f_{n-1}(x)] + \omega^2 f_{n-1}(x) \qquad (n \geqslant 1)$$

under the boundary conditions $f_n(0) = f_n(1) = 0$, then (f_n) is a monotone increasing sequence which tends to a solution of (8.3.2). Evidently a similar definition for (g_n) with $g_0 = v$ yields a decreasing sequence converging to a solution. The following has been proved.

8.3.23 Theorem. *Suppose $\psi : [0, 1] \times (-\infty, \infty) \to \mathbb{R}$ is continuous and satisfies the local Lipschitz condition* (8.3.3). *Assume that there are lower and upper solutions u, v respectively with $u \leqslant v$. Then the boundary value problem* (8.3.2) *has at least one solution in $[u, v]$. Further, the sequences (f_n) and (g_n) defined above converge monotonically to solutions.*

For extensions of this result when ψ is also a function of f', see Chandra and Davis (1974).

It is frequently possible to construct upper and lower solutions by a simple geometrical argument (Problem 8.15). For example, if the equation is

$$f''(x) = \mu \sinh[f(x)] + g(x),$$

with $g \in \mathscr{C}([0, 1])$ and $\mu > 0$, u and v may be chosen to be large negative and positive constant functions respectively. The theorem is particularly useful when tackling equations such as that above with large nonlinearities. Another application is to nonlinear eigenfunction problems, where the presence of the trivial solution causes serious difficulty with most methods. In the treatment of a nonlinear oscillator which follows, the crucial step is the choice of a non-negative lower solution which is not identically zero.

8.3.24 Example. For the boundary value problem

$$f''(x) + \mu \sin[f(x)] = 0,$$
$$f(0) = f(1) = 0,$$

the linearized equation has eigenvalues $\mu_n = n^2\pi^2 (n = 1, 2, \ldots)$. In the nonlinear case there are also no nontrivial solutions if $\mu < \pi^2$, but for $\mu > \pi^2$ the position changes radically.

Assume that $\mu > \pi^2$. For a lower solution try $u(x) = \varepsilon \sin \pi x$. Then for some (small) $\varepsilon > 0$,

$$u''(x) + \mu \sin[u(x)] = \mu \sin[\varepsilon \sin \pi x] - \pi^2 \varepsilon \sin \pi x \geqslant 0.$$

As $u(0) = u(1) = 0$, u is indeed a lower solution for such ε. It is easy to see that $v(x) = \pi$ is an upper solution, and since $u \leqslant v$, Theorem 8.3.23 is applicable. We deduce that the boundary value problem has a nontrivial solution for *every* $\mu > \pi^2$. The theorem also gives a constructive method for finding solutions.

It is interesting to note that the equation

$$f''(x) + \mu \sin[f(x)] = g(x)$$

with the same boundary conditions has a solution for every continuous g and every $\mu \in \mathbb{R}$ (Theorem 8.2.9). Thus although the associated integral operator is compact, the set of eigenvalues form a continuum, and further the Fredholm alternative does not hold. It will be seen in Chapter 14 that these are typical features of nonlinear eigenfunction problems.

Problems

8.1 (Rothe). Let D be the open unit ball in the Banach space $\mathscr{B}$, and assume that $A : \bar{D} \to \mathscr{B}$ is compact. Deduce from the Schauder Fixed Point Theorem that A has a fixed point in $\bar{D}$ if $A(\partial D) \subset \bar{D}$. (Consider RA where $Rf = f (f \in \bar{D})$, $Rf = f/\|f\| (f \notin \bar{D})$).

8.2 Let $\mathscr{B}$ be a Banach space. Suppose $B : D \to \mathscr{B}$ is a contraction and set $(I - B)(D) = S$. Prove that $(I - B) : D \to S$ is a homeomorphism (Definition 4.2.9). Show also that if S is relatively compact then so is D.

8.3 (Krasnoselskii). Let D be a closed, bounded, convex subset of the Banach space $\mathscr{B}$. Suppose $A, B : D \to \mathscr{B}$ satisfy the following:

(i) $Af + Bg \in D \ (f, g \in D)$;

(ii) A is compact;
(iii) B is a contraction.

Using 8.2 and the Schauder Fixed Point Theorem, prove that $T = A + B$ has a fixed point in D.

Deduce that the following integral equation has a solution in $\mathscr{C}([0, 1])$:

$$3f(x) = x + [f(x)]^2 + \int_0^1 |x - f(y)|^{\frac{1}{2}} \, dy.$$

8.4 Let D be a closed subset of the Banach space $\mathscr{B}$. Assume that $A(D) \subset D$, that $A(D)$ is relatively compact, and that A is "contractive":

$$\|Af - Ag\| < \|f - g\| \qquad (f, g \in D, f \neq g).$$

By considering the functional $\|f - Af\|$ show that A has a unique fixed point in D. (Note that A need not be a contraction.)

8.5 Let $\mathscr{C}([0, 1])$ have the sup norm. If $k : [0, 1] \times [0, 1] \times \mathscr{C} \to \mathbb{C}$ is continuous, show that $A : \mathscr{C}([0, 1]) \to \mathscr{C}([0, 1])$ is compact, where

$$Af(x) = \int_0^1 k(x, y, f(y)) \, dy.$$

8.6 With k and A as in the previous problem, suppose there are real numbers $k_1, k_2, \alpha \geqslant 0$ and $r > 0$ such that $k_1 + k_2 r^\alpha \leqslant r$ and

$$|k(x, y, z)| \leqslant k_1 + k_2 |z|^\alpha \qquad (x, y \in [0, 1], |z| \leqslant r).$$

Prove that the Urysohn equation $f = Af$ has a solution in $\mathscr{C}([0, 1])$.

8.7 Suppose E is a cone in the real Banach space $\mathscr{B}$. Prove the following, and give geometrical interpretations for the cone in Fig. 8.2.

(i) $f, g \in E$ and $a, b \in \mathbb{R}^+ \Rightarrow af + bg \in E$.
(ii) If $-f \notin E$, $\inf_{g \in E} \|f + g\| > 0$.
(iii) Suppose $f \in E$, $g \in \mathscr{B}$. If $g \leqslant tf$ for some $t \in \mathbb{R}$, $g \leqslant sf$ for $s > t$.
(iv) Take some $g \in \mathscr{B}$, $t \in \mathbb{R}$, and assume that $g \leqslant tf$ for some non-zero $f \in E$. Then there is a smallest $s \in \mathbb{R}$ for which $g \leqslant sf$.

8.8* Prove that the cone E is normal iff there is a $\delta > 0$ such that $\|f + g\| \geqslant \delta$ for any $f, g \in E$ with $\|f\| = \|g\| = 1$.

8.9 Show that a regular cone is normal.

8.10 Let the real space $\mathscr{C}([0, 1])$ be equipped with the sup norm, and take some $m \in (0, 1)$. Let E be the set of non-negative functions in $\mathscr{C}([0, 1])$ such that

$$m \sup_{0 \leqslant x \leqslant 1} f(x) \leqslant \inf_{0 \leqslant x \leqslant 1} f(x).$$

Prove that E is a regular cone.

8.11 Let $k : [0, 1] \times [0, 1] \to \mathbb{R}$ be continuous and non-negative. Suppose that

(i) $\psi : [0, 1] \times [0, \infty) \to \mathbb{R}$ is continuous and non-negative;
(ii) $z^{-1}\psi[y, z] \to 0$ as $z \to \infty$ uniformly in y.

Prove that if $g \in \mathscr{C}([0, 1])$ is non-negative, the integral equation

$$f(x) = g(x) + \int_0^1 k(x, y)\,\psi[y, f(y)]\,\mathrm{d}y$$

has a non-negative solution in $\mathscr{C}([0, 1])$.

If (i) holds only on $[0, 1] \times (0, \infty)$, show that the conclusion is still valid if in addition $g(x) > 0$ for $0 \leqslant x \leqslant 1$. (The point is that the result applies if ψ has a large singularity at zero—for example if $\psi[x, z] = z^{-n}$).

8.12 Existence of a positive solution of the Föppl–Hencky equation was established in Example 8.3.11. Prove that

$$\int_0^1 \int_0^1 x^3 k(x, y) f(x) f(y)\,\mathrm{d}x\,\mathrm{d}y \geqslant 0 \qquad (f \in \mathscr{C}([0, 1])),$$

and deduce that there is only one strictly positive solution in $\mathscr{C}([0, 1])$.

8.13 Let E be a cone in the real Banach space $\mathscr{B}$. Let $A : D \to \mathscr{B}$ be monotone decreasing ($f \geqslant g \Rightarrow Af \leqslant Ag$). For $[u_0, v_0] \subset D$ set $u_{n+1} = Av_n$, $v_{n+1} = Au_n$ $(n \geqslant 0)$. Show that if $u_0 \leqslant u_1 \leqslant v_1 \leqslant v_0$,

$$u_0 \leqslant u_1 \leqslant u_2 \leqslant \ldots \leqslant v_2 \leqslant v_1 \leqslant v_0.$$

If further E is normal and A is compact, prove that (u_n), (v_n) converge monotonically to fixed points of A.

Apply this result to the construction of a non-negative continuous solution of the Urysohn integral equation (Collatz)

$$f(x) = \mu \int_0^1 \frac{\mathrm{d}y}{1 + x + f(y)} \qquad (\mu > 0).$$

It is instructive to note the difficulty encountered for large μ if an attempt is made to use the Contraction Mapping Principle.

8.14 Assume that $\psi : [0, 1] \times [0, \infty) \to \mathbb{R}$ is continuous. Suppose there are non-negative ω, m, l such that for $x \in [0, 1]$ and $z \geqslant 0$:

(i) $0 \leqslant \psi[x, z] + \omega z \leqslant mz + l$:
(ii) $\psi[x, z] + \omega z$ is non-decreasing in z;
(iii) $m/(\pi^2 + \omega) < 1$.

Using Lemma 8.3.19 show that

$$f''(x) + \psi[x, f(x)] = 0 \qquad (0 \leqslant x \leqslant 1)$$

has a non-negative solution with $f(0) = f(1) = 0$.

8.15 Let $\psi : [0, 1] \times \mathbb{R} \to \mathbb{R}$ be continuous, and suppose there are positive $a < \pi^2$ and z_0 such that $z\psi[x, z] \leqslant az^2$ for $|z| \geqslant z_0$ and $x \in [0, 1]$. Show that there are lower and upper solutions u, v with $u \leqslant v$ of

$$f''(x) + \psi[x, f(x)] = 0 \qquad (0 \leqslant x \leqslant 1),$$
$$f(0) = f(1) = 0.$$

(Hint: v is the solution of $v'' + av + m = 0$, $v(0) = v(1) = 0$ for some $m \in \mathbb{R}$).

8.16 Suppose g is real-valued and continuous on $[0, 1]$. Show that if $\mu > 0$

$$\left.\begin{aligned} f''(x) &= \mu \sinh[f(x)] + g(x) \qquad (0 \leqslant x \leqslant 1), \\ f(0) &= f(1) = 0, \end{aligned}\right\}$$

has a solution, and give monotone iteration schemes for constructing solutions.

8.17 The concept of a minorant is sometimes useful in eigenfunction problems. Let E be a normal cone in the real Banach space $\mathscr{B}$, and suppose that $A : E \to \mathscr{B}$ is continuous compact and montone. Assume that there exist $L, M \in \mathscr{L}(\mathscr{B})$, $h \in E$, and a $\delta > 0$ such that:

(i) L is a minorant of A on $\bar{S}(0, \delta) \cap E$, and L has an eigenfunction $u \in E$ with corresponding eigenvalue $\lambda \geqslant 1$;
(ii) M is positive and $r_\sigma(M) < 1$, and B (where $Bf = Mf + h$) is a majorant of A.

Prove that A has a nontrivial fixed point in E, and give monotone sequences converging to fixed points of A.

8.18 Using 8.17 prove that every $\mu > 0$ is an eigenvalue of

$$f''(x) + \mu[f(x)]^{\frac{1}{2}} = 0 \qquad (0 \leqslant x \leqslant 1),$$
$$f(0) = f(1) = 0.$$

8.19 Show that every $\mu > \pi^2$ is an eigenvalue of Duffing's equation

$$f''(x) + \mu\{f(x) - \tfrac{1}{6}[f(x)]^3\} = 0 \qquad (0 \leqslant x \leqslant 1)$$

under the boundary conditions $f(0) = f(1) = 0$.

8.20 Suppose $\mu > 0$. Show that the equation

$$f''(x) + \mu \sin[f(x)] = 0$$

has an odd periodic solution in $\mathscr{C}^2((-\infty, \infty))$ of period 2ω for any $\omega > (\pi^2/\mu)^{\frac{1}{2}}$.

Chapter 9

The Spectral Theorem

9.1 Introduction

In almost any physical problem which can be formulated in terms of a linear operator, the spectrum of the operator is a quantity of basic physical interest. Ample justification for this assertion is indeed provided by the common use of the term spectrum in both a physical and mathematical sense. For example in quantum mechanics the physical spectrum of the energy states of the atom and the spectrum of the governing differential equation are closely related. For a general linear operator it is often hard to obtain information even of a qualitative nature about the spectrum, but for a self-adjoint operator on Hilbert space this is a task of much less difficulty. Indeed in the latter case further investigation in the area of spectral theory shows that a rather complete description of the solution of equations involving self-adjoint operators may often be obtained. One result, the Spectral Theorem, lies at the heart of any discussion of such equations. By means of this theorem, due to Hilbert and Von Neumann, a great body of apparently disparate results are unified, and the theorem is one of the major achievements of linear operator theory.

The importance of the result which states that a Hermitian matrix (and so a Hermitian quadratic form) can be diagonalized by choice of a suitable basis is well known; an excellent account of the finite dimensional theory is given in Halmos (1948). In essence, the Spectral Theorem is a far reaching generalization of this result to self-adjoint operators on an infinite dimensional Hilbert space $\mathscr{H}$. The first step towards this generalization in which the infinite dimensionality plays a significant role is the Hilbert–Schmidt

Theorem 7.5.1 for a compact self-adjoint operator, T say. The theorem states that the set $\{\psi_n\}$ of eigenvectors of T can be arranged to form an orthonormal basis of $\mathscr{H}$. Thus an arbitrary $g \in \mathscr{H}$ can be expanded in the form

$$g = \Sigma(g, \psi_n)\psi_n, \tag{9.1.1}$$

and an immediate consequence is that

$$Tg = \Sigma\lambda_n(g, \psi_n)\psi_n, \tag{9.1.2}$$

where λ_n is the eigenvalue corresponding to ψ_n. (9.1.2) provides in an obvious sense the diagonalization of T. A further observation is that if p is a polynomial

$$p(T)g = \Sigma p(\lambda_n)(g, \psi_n)\psi_n. \tag{9.1.3}$$

The definition of a polynomial function of T of course presents no difficulty, but it turns out that (9.1.3) provides a satisfactory definition of an arbitrary continuous function of T. There are a number of reasons for interest in such functions; from the point of view of applications, exponential functions of operators are of special importance because they enter into the solution of the initial value problem for the abstract differential equation $u'(t) = Lu(t)$—cf. Example 3.5.10.

What form then should (9.1.1) and (9.1.2) take if the operator is self-adjoint but not compact? Two well known expansion formulae for differential equations provide the clue. Let L be the self-adjoint operator (Example 6.7.9) formed from $l = i\mathrm{d}/\mathrm{d}x$ by the imposition of the boundary condition $f(0) = f(2\pi)$, and let λ_n, ψ_n $(n = 1, 2, \ldots)$ be the eigenvalues and corresponding orthonormal eigenfunctions respectively of L—that is $\lambda_n = n$, $\psi_n = (2\pi)^{-\frac{1}{2}}$ e^{-inx} $(n = 1, 2, \ldots)$. The familiar Fourier expansion formula is

$$g(x) = \Sigma\psi_n(x)\int_0^{2\pi} g(y)\,\overline{\psi_n(y)}\,\mathrm{d}y,$$

and in $\mathscr{H} = \mathscr{L}_2(0, 2\pi)$ this becomes

$$g = \Sigma(g, \psi_n)\,\psi_n.$$

The domain of L is not the whole of $\mathscr{H}$, but at least for smooth g,

$$Lg = \Sigma\,\lambda_n(g, \psi_n)\,\psi_n.$$

The last two equations are the required analogues of (9.1.1) and (9.2.2) respectively.

In the last example, although L is not compact, it has a compact inverse, and the results above may be deduced from Theorem 7.5.4. For an example where this is not the case, again take $l = i\mathrm{d}/\mathrm{d}x$, but suppose now that the interval is $(-\infty, \infty)$. The Fourier transform inversion formula is

$$g(x) = \int_{-\infty}^{\infty} \psi_\lambda(x)\,\mathrm{d}\lambda \int_{-\infty}^{\infty} g(y)\,\overline{\psi_\lambda(y)}\,\mathrm{d}y,$$

where $\psi_\lambda(x) = (2\pi)^{-\frac{1}{2}}\,\mathrm{e}^{i\lambda x}$, and if $\mathscr{H} = \mathscr{L}_2(-\infty, \infty)$ and g has compact support this equation may be rewritten as

$$g(x) = \int_{-\infty}^{\infty} (g, \psi_\lambda)\,\psi_\lambda(x)\,\mathrm{d}\lambda. \tag{9.1.4}$$

Noting that the equation $lf = \lambda f$ has the solution ψ_λ for every real λ, we may tentatively interpret the spectrum as the whole real axis, and the eigenfunctions of l as ψ_λ. (9.1.4) may then be interpreted as the generalization of (9.1.1) obtained by replacing a summation by an integration over the spectrum. There are indeed a number of difficulties in making this interpretation rigorous. First it is not obvious how a self-adjoint operator L based on l should be defined. Second, the ψ_λ do not lie in $\mathscr{L}_2(-\infty, \infty)$ and thus are certainly not eigenfunctions of L itself. However, it is hard to avoid the conclusion that (9.1.4) is in some sense a "generalized eigenfunction expansion" analogous to that provided by the Hilbert–Schmidt Theorem.

Our immediate aim is to obtain an abstract version of the generalized eigenfunction expansion for an arbitrary (possibly unbounded) self-adjoint operator, and to utilize this in defining functions of operators. In the next chapter the construction of self-adjoint operators from formal ordinary differential operators will be tackled. It will then become clear that an extensive list of series expansions in orthogonal functions and transforms are nothing more than particular cases of generalized eigenfunction expansions obtained by an application of the Spectral Theorem.

9.2 Preliminaries

Some technicalities must be disposed of before the Spectral Theorem is tackled. The concept of a function of an operator is first discussed. It must be emphasized that the definition below does not provide a convenient method of calculating functions of an operator in practice; the Spectral Theorem itself will eventually furnish such a method. Throughout, $\mathscr{H}$ will denote a complex Hilbert space, and $L : \mathscr{H} \to \mathscr{H}$ will be a bounded self-adjoint operator.

If p is a polynomial, the operator $p(L)$ has an obvious interpretation. By Weierstrass' Theorem, a continuous function of the real variable λ may be approximated on compact intervals in the sup norm by a sequence of polynomials, and this suggests that continuous functions of L may be defined as the limits in the operator norm; that is in $\mathscr{L}(\mathscr{H})$, of polynomial functions of L. It is easy to show that this is indeed a sensible definition, but the restriction of continuity is too severe for the present purpose. However, a much more general class of function may be obtained by taking monotone pointwise limits of continuous functions. If this construction is to be extended to functions of operators, an appropriate interpretation of monotonicity for sequences of operators must be available. A partial ordering on the set of self-adjoint operators provides a basis for the definition of monotonicity.

9.2.1 Definition†. Let $L, M : \mathscr{H} \to \mathscr{H}$ be bounded self-adjoint operators. We shall write $L \geqslant M$ or $M \leqslant L$ iff $(Lf, f) \geqslant (Mf, f)$ for all $f \in \mathscr{H}$. L is said to be **positive** iff $L \geqslant 0$ and **strictly positive** iff $L \geqslant cI$ for some $c > 0$. A sequence (L_n) of self-adjoint operators in $\mathscr{L}(\mathscr{H})$ is said to be **monotone** (increasing) iff $L_{n-1} \leqslant L_n$ $(n = 1, 2, \ldots)$, and **order bounded above** iff there is a bounded self-adjoint L such that $L_n \leqslant L$ $(n = 1, 2, \ldots)$.

The numbers $m_{\pm}$ (Definition 6.6.5) play an important part in the analysis. Recall (Theorem 6.6.6) that the spectrum of L is contained in $[m_-, m_+]$, and note that $m_- I \leqslant L \leqslant m_+ I$. The principal properties of the ordering are listed in the following lemmas.

9.2.2 Lemma. *Let $L, M, N \in \mathscr{L}(\mathscr{H})$ be self-adjoint, and let $a, b \in \mathbb{R}$. Then the following hold:*

(i) $L \geqslant M$ *and* $M \geqslant L \Rightarrow L = M$;
(ii) $L \geqslant M \Rightarrow L + N \geqslant M + N$;
(iii) $L \geqslant M$ *and* $a \geqslant 0 \Rightarrow aL \geqslant aM$;
(iv) $L, M \geqslant 0$ *and* $a, b \geqslant 0 \Rightarrow aL + bM \leqslant \max(a, b).(L + M)$;
(v) $L \geqslant M \geqslant N \Rightarrow \|M\| \leqslant \max(\|L\|, \|N\|)$;
(vi) $L \geqslant M_1 \geqslant N, L \geqslant M_2 \geqslant N \Rightarrow \|M_1 - M_2\| \leqslant \|L - N\|$.

Proof. (i)–(iv) are obvious consequences of the definition. (v) follows from the relation

$$|(Mf, f)| \leqslant \max\left[\sup_{\|f\|=1} |(Lf, f)|, \sup_{\|f\|=1} |(Nf, f)|\right],$$

and Theorem 6.6.7. For (vi) note that $N - L \leqslant M_1 - M_2 \leqslant L - N$ and use (v). □

9.2.3 Lemma. *Let (L_n) be a sequence of bounded self-adjoint operators, and assume that the sequence is monotone increasing and order-bounded above.*

† The positive self-adjoint operators in $\mathscr{L}(\mathscr{H})$ form a cone E, and the partial ordering $\geqslant$ is that induced by E. The positive operators of Definition 8.3.7 are of course quite different objects.

Then (L_n) converges strongly to a bounded self-adjoint operator, L say. If each L_n commutes with $B \in \mathscr{L}(\mathscr{H})$, then L commutes with B.

Proof. It is obviously sufficient to establish the result for $0 \leqslant L_1 \leqslant L_2 \leqslant \ldots \leqslant I$. Take $n \geqslant m$, and note that $0 \leqslant L_n - L_m \leqslant I$, whence $\|L_n - L_m\| \leqslant 1$ (Lemma 9.2.2(v)). It then follows from Problem 6.23 and Schwartz's inequality that for all $f \in \mathscr{H}$,

$$\begin{aligned}\|[L_n - L_m]f\|^4 &\leqslant ([L_n - L_m]f, f)([L_n - L_m]^2 f, [L_n - L_m]f) \\ &\leqslant \{(L_n f, f) - (L_m f, f)\}\|f\|^2. \qquad (9.2.1)\end{aligned}$$

As the sequence $((L_n f, f))$ of real numbers is by assumption non-decreasing and bounded, it is convergent. Hence the right-hand side of (9.2.1) tends to zero as $m, n \to \infty$; $(L_n f)$ is therefore Cauchy and so convergent. By Corollary 3.5.12 there is an $L \in \mathscr{L}(\mathscr{H})$ which is the strong limit of (L_n), and L is obviously self adjoint. The commutativity follows from the relations

$$LBf = \lim L_n Bf = \lim BL_n f = BLf. \qquad \square$$

9.2.4 Lemma. *Assume that L is bounded and self-adjoint. If p, q are polynomials with real coefficients, and $p(\lambda) \geqslant q(\lambda)$ for $m_- \leqslant \lambda \leqslant m_+$, then $p(L) \geqslant q(L)$.*

Proof. Obviously $p(L)$ and $q(L)$ are self-adjoint. Suppose first that $q = 0$. Since p does not change sign on $[m_-, m_+]$, it may be factorized in the form

$$p(\lambda) = a\Pi(\lambda - a_j)\Pi(b_j - \lambda)\Pi((\lambda - c_j)^2 + d_j^2)$$

with $a > 0$, the first two products containing the simple real roots arranged with $a_j \leqslant m_-$, $b_j \geqslant m_+$, and the last containing the complex and repeated roots. Now $p(L)$ is obtained by substituting L for λ in the above expression, and since $m_- I \leqslant L \leqslant m_+ I$ each factor will be a positive operator. As the factors commute, by Problem 9.5, $p(L) \geqslant 0$. The conclusion of the lemma follows on applying this result to $p - q$. $\square$

This lemma shows that the mapping $p \to p(L)$ preserves positivity. In fact only the values taken by the polynomial $p(\lambda)$ on $[m_-, m_+]$ covering the spectrum are significant.

The preparations are now complete for the definition of more general functions of L. Suppose first that f is a real valued function continuous on $[m_-, m_+]$. By Weierstrass' Theorem there is a sequence (p_n) of polynomials with real coefficients such that (p_n) converges to f on $[m_-, m_+]$ in the sup norm. Therefore, given $\varepsilon > 0$ there is an n_0 such that

$$-\varepsilon \leqslant f(\lambda) - p_n(\lambda) \leqslant \varepsilon \qquad (n \geqslant n_0, m_- \leqslant \lambda \leqslant m_+),$$

whence

$$-2\varepsilon \leqslant p_n(\lambda) - p_m(\lambda) \leqslant 2\varepsilon \qquad (m, n \geqslant n_0, m_- \leqslant \lambda \leqslant m_+).$$

By Lemma 9.2.4,

$$-2\varepsilon I \leqslant p_n(L) - p_m(L) \leqslant 2\varepsilon I,$$

and it follows from Lemma 9.2.2(v) that $\|p_n(L) - p_m(L)\| \leqslant 2\varepsilon$. The sequence $(p_n(L))$ in $\mathscr{L}(\mathscr{H})$ is therefore Cauchy, and so convergent with a limit which we denote by $f(L)$. It is easy to check that $f(L)$ is independent of the particular sequence chosen, and the above procedure thus provides a satisfactory definition of $f(L)$. The following lemma states that the operator $f(L)$ inherits the nice properties of the real valued function f.

9.2.5 Lemma. *For every real valued function f continuous on $[m_-, m_+]$, the above procedure defines a unique bounded self-adjoint $f(L)$, and $f(L)$ commutes with every bounded operator that commutes with L. Also:*

(i) *$f(L)$ is positive iff f is non-negative on $[m_-, m_+]$;*

(ii) *the map $f \to f(L)$ is norm decreasing in the sense that*

$$\|f(L)\| \leqslant \sup_{\lambda \in [m_-, m_+]} |f(\lambda)|;$$

(iii) *if g is another function of the same type,*

$$(f \,.\, g)(L) = f(L) \,.\, g(L),$$

$$(f + g)(L) = f(L) + g(L).$$

Proof. It is first shown that

$$I \inf f(\lambda) \leqslant f(L) \leqslant I \sup f(\lambda), \tag{9.2.2}$$

where here, and throughout the proof, the infs and sups are taken over $[m_-, m_+]$. Set $\varepsilon_n = \sup|f(\lambda) - p_n(\lambda)|$; clearly $\lim \varepsilon_n = 0$. Then $p_n(\lambda) \geqslant \inf f(\lambda) - \varepsilon_n$, and by Lemma 9.2.4,

$$p_n(L) \geqslant [\inf f(\lambda) - \varepsilon_n]I.$$

Therefore for all $g \in \mathscr{H}$,

$$(p_n(L)g, g) \geqslant [\inf f(\lambda) - \varepsilon_n]\|g\|^2,$$

whence

$$\begin{aligned}(f(L)g, g) &= (p_n(L)g, g) + ([f(L) - p_n(L)]g, g) \\ &\geqslant [\inf f(\lambda) - \varepsilon_n]\|g\|^2 - \|f(L) - p_n(L)\|\,\|g\|^2 \\ &\to \|g\|^2 \inf f(\lambda) \qquad (n \to \infty),\end{aligned}$$

since both ε_n and the last term converge to zero. (9.2.2) follows on using a similar argument for the second inequality.

(i) is an immediate consequence of (9.2.2), and (ii) is obtained on applying Lemma 9.2.2(v). The remaining assertions follow readily from the definition of $f(L)$. □

The map $f \to f(L)$ may be extended to continuous complex valued functions by treating the real and imaginary parts separately—but of course $f(L)$ need not then be self-adjoint. The final step is to generalize the definition to "upper semicontinuous" functions.

9.2.6 Definition. A real valued function f defined on $[a, b]$ is said to be **upper semicontinuous** iff it is the pointwise limit of a monotone decreasing sequence of real valued continuous functions.

For the present purpose it is sufficient to note that the characteristic function of a closed interval, or of the union of a finite number of such intervals, is upper semicontinuous. An application of Lemma 9.2.3 yields the following result, for the uniqueness see Problem 9.6.

9.2.7 Theorem. *Let f be a non-negative, upper semicontinuous function on $[m_-, m_+]$, and let (f_n) be a monotone decreasing sequence of continuous functions tending pointwise to f. Then $f_n(L)$ converges strongly to a limit, $f(L)$ say, and $f(L)$ has precisely the properties stated in Lemma 9.2.5.*

Three points should be noted. First, if f is upper semicontinuous on any interval containing $[m_-, m_+]$, $f(L)$ is defined by considering the restriction of f to $[m_-, m_+]$. Again only the values of f on $[m_-, m_+]$ are relevant. Secondly, the map $f \to f(L)$ preserves both the algebraic and the order structure. Lastly, the convergence of $f_n(L)$ to $f(L)$ is strong rather than uniform.

A class of operators called projections is next examined. A projection is a generalized version of an orthogonal projection onto a linear subspace in $\mathbb{R}^3$; it is a consequence of the Projection Theorem 1.5.11 that the generalization makes sense in an infinite dimensional Hilbert space.

9.2.8 Definition. Let $\mathcal{M}$ be a closed subspace of the Hilbert space $\mathcal{H}$. For any $f \in \mathcal{H}$ let g, h be the unique vectors in $\mathcal{M}, \mathcal{M}^\perp$ respectively such that $f = g + h$, and set $Pf = g$. The operator P is called the **projection** onto $\mathcal{M}$, and g is the projection of f in $\mathcal{M}$. Two projections P_1 and P_2 are said to be **orthogonal** iff $P_1P_2 = 0$.

A projection as defined above is sometimes called an "orthogonal projection" to distinguish it from the more general projection used in Banach space theory. Since the last notion will not be used here, the extra word "orthogonal" is not included. An advantage is that any possibility of confusion with the concept of orthogonality for a pair of projections is avoided. The properties of projections are summarized in the following lemmas. The proof of the first is easy and is left as an exercise.

9.2.9 Lemma. *If P is the projection onto the closed subspace $\mathscr{M}$,*

(i) *P is a linear bounded self-adjoint operator with $P^2 = P$;*
(ii) *either $P = 0$ or $\|P\| = 1$;*
(iii) *$I - P$ is the projection onto $\mathscr{M}^\perp$;*
(iv) *$0 \leqslant P \leqslant I$.*

9.2.10 Lemma. *Let $P \in \mathscr{L}(\mathscr{H})$ be self-adjoint. P is a projection if $P^2 = P$.*

Proof. Observe first that the linear subspace $R(P)$ is closed, for if $g_n = Pf_n \to g$, then $Pg_n = P^2 f_n = Pf_n = g_n$, and $g = \lim g_n = \lim Pg_n = Pg \in R(P)$. Next, for any $f \in \mathscr{H}$ write $f = Pf + (I - P)f$. Since P is self-adjoint,

$$((I - P)f, Pf) = ((P - P^2)f, g) = 0.$$

Therefore $(I - P)f \in \mathscr{M}^\perp$, and P is the projection onto $R(P)$. □

9.2.11 Lemma. *Let P_1 and P_2 be the projections onto the closed subspaces $\mathscr{M}_1$ and $\mathscr{M}_2$ respectively. Then*

(i) *$P_1 + P_2$ is a projection $\Leftrightarrow P_1 P_2 = 0 \Leftrightarrow \mathscr{M}_1 \perp \mathscr{M}_2$;*
(ii) *$P_1 P_2 = P_2 \Leftrightarrow P_1 \geqslant P_2 \Leftrightarrow \mathscr{M}_1 \supset \mathscr{M}_2$. P_1 commutes with P_2 if any of these relations hold.*

Proof. We shall prove (ii) and leave (i) as an exercise. If $P_1 P_2 = P_2$, then $P_2 P_1 = (P_1 P_2)^* = P_2^* = P_2$. Hence P_1 and P_2 commute, and

$$(P_1 - P_2)^2 = P_1^2 - P_1 P_2 - P_2 P_1 + P_2^2 = P_1 - 2P_1 P_2 + P_2 = P_1 - P_2.$$

Since $P_1 - P_2$ is self-adjoint, it follows from Lemma 9.2.10 that $P_1 - P_2$ is a projection. Therefore by Lemma 9.2.9(iv), $P_1 \geqslant P_2$. On the other hand if $P_1 \geqslant P_2$, then $I - P_1 \leqslant I - P_2$ and

$$\begin{aligned} \|(I - P_1)P_2 f\|^2 &= ((I - P_1)P_2, (I - P_1)P_2 f) \\ &= ((I - P_1)P_2 f, P_2 f) \\ &\leqslant ((I - P_2)P_2 f, P_2 f) \\ &= 0. \end{aligned}$$

Thus $(I - P_1)P_2 = 0$, whence $P_1 P_2 = P_2$. This proves that $P_1 P_2 = P_2$ iff $P_1 \geqslant P_2$.

Assume next that $P_1P_2 = P_2$. For any $f \in \mathcal{M}_2$, $f = P_2 f = P_1P_2 f = P_1 f \in R(P_1) = \mathcal{M}_1$. Therefore $\mathcal{M}_2 \subset \mathcal{M}_1$. Suppose on the other hand that $\mathcal{M}_2 \subset \mathcal{M}_1$. For any $f \in \mathcal{H}$ write $f = g + h$ where $g \in \mathcal{M}_2$, $h \in \mathcal{M}_2^{\perp}$. Then $P_1 g = P_2 g = g$ and $P_2 h = 0$. Hence $P_1P_2 g = P_1 g = g = P_2 g$, and therefore

$$P_1P_2 f = P_1P_2 g + P_1P_2 h = P_2 g = P_2 g + P_2 h = P_2 f. \qquad \square$$

This lemma is fundamental. Notice particularly that the natural ordering by inclusion of the subspaces $\mathcal{M}_1$ and $\mathcal{M}_2$ corresponds to the ordering of the corresponding projections by the relation $\geqslant$.

Finally, in order to obtain a neat formulation of the Spectral Theorem, an integral which can cope with Banach space valued functions must be introduced. There are some rather unpleasant technicalities connected with the abstract Lebesgue integral, and the following simpler concept which is sufficient for the present purpose is preferred.

9.2.12 Definition. Let $\mathcal{B}$ be a Banach space. Let $\Omega = [a, b]$ be a finite interval in $\mathbb{R}$, and suppose that f, γ are respectively complex valued and Banach space valued functions defined on Ω. f is said to be **Riemann–Stieltjes integrable** on Ω with respect to γ iff there is a $g \in \mathcal{B}$ such that

$$\lim_{\delta \to 0} \left\| \sum_{j=1}^{n-1} f(c_j)[\gamma(a_{j+1}) - \gamma(a_j)] - g \right\| = 0$$

for any finite sets of points $a_j, b_j, c_j \in \Omega$ such that $a_1 = a, a_n = b$ and $a_j \leqslant c_j \leqslant a_{j+1}$ $(j = 1, 2, \ldots, n)$, δ being the maximum length of the subintervals $[a_j, a_{j+1}]$. g is called the Riemann–Stieltjes integral of f, and we write

$$g = \int_a^b f(\lambda)\, \mathrm{d}\gamma(\lambda).$$

If Ω is an infinite interval, and if f is integrable on each compact subinterval of Ω, we write

$$\int_{-\infty}^{\infty} f(\lambda)\, \mathrm{d}\gamma(\lambda) = \lim_{a \to -\infty} \lim_{b \to \infty} \int_a^b f(\lambda)\, \mathrm{d}\gamma(\lambda),$$

$$\int_{-\infty}^{b} f(\lambda)\, \mathrm{d}\gamma(\lambda) = \lim_{a \to -\infty} \int_a^b f(\lambda)\, \mathrm{d}\gamma(\lambda)$$

if the limits exist.

Note that if γ is constant on some subinterval Ω_0 of Ω, the terms in the defining sums vanish on Ω_0, and hence the values of f on Ω_0 do not contribute to the integral.

9.3 Background to the Spectral Theorem

The Spectral Theorem may be regarded as a generalization of the Hilbert–Schmidt Theorem to operators which are self-adjoint but not necessarily compact. However, there are difficulties in carrying out this generalization, and as a consequence the formulation of the Spectral Theorem presents some conceptual problems. Therefore before tackling the details of the analysis, we shall attempt to explain what the main difficulties are and to suggest how these should be overcome. It should be emphasized that the arguments in this section are purely formal.

Suppose first that T is compact and self-adjoint. Much of the power of the Hilbert–Schmidt Theorem rests on the fact that, since any $g \in \mathscr{H}$ may be expressed as a sum of eigenfunctions ψ_k (with corresponding eigenvalues λ_k), and as $T\psi_k$ is known explicitly, functions of T may readily be written down. In particular, if $\mathscr{M}_k$ is the eigenspace corresponding to λ_k,

$$g = \Sigma\psi_k \qquad (\psi_k \in \mathscr{M}_k), \tag{9.3.1}$$

$$Tg = \Sigma\lambda_k\psi_k. \tag{9.3.2}$$

At first sight the prospects for generalizing these relations to an operator L which is self-adjoint but not compact do not look good, for the eigenfunctions of L need not span $\mathscr{H}$—indeed L need not have any eigenfunctions at all. However, there is a hopeful sign. By Theorem 6.6.4, for any $\lambda \in \sigma(L)$ there is a sequence (ϕ_j) of vectors of norm unity such that $\lim \| L\phi_j - \lambda\phi_j \| = 0$. It is thus tempting to regard λ as an "approximate eigenvalue" and ϕ_j for large j as an "approximate eigenfunction", and to speculate that λ and ϕ_j will provide adequate substitutes for an eigenvalue and eigenfunction.

To see whether this approach is likely to be fruitful, take $\mathscr{H} = \mathscr{L}_2(0, 1)$ and consider the simple multiplication operator $M : \mathscr{H} \to \mathscr{H}$ defined by $Mg(x) = xg(x)$ $(g \in \mathscr{H})$. M is a bounded self-adjoint operator whose spectrum $[0, 1]$ is wholly continuous (Problem 9.8), and therefore M has no eigenfunctions. For given $\lambda \in [0, 1]$, take a subinterval $[\mu, \nu]$ of $[0, 1]$ containing λ. Then for any $\phi \in \mathscr{H}$ with $\|\phi\| = 1$ and with support in $[\mu, \nu]$, $\|M\phi - \lambda\phi\| \leqslant |\nu - \mu|$. Thus ϕ may be regarded as an approximate eigenfunction with $|\nu - \mu|$ providing a measure of accuracy. Next split $[0, 1]$ into n equal subintervals $[\mu_k, \mu_{k+1}]$ of "small" length δ, and let $\mathscr{M}_k$ be the set of functions with supports in these intervals respectively. The $\mathscr{M}_k$ play the role of approximate eigenspaces corresponding respectively to the approximate eigenvalues λ_k where $\mu_k \leqslant \lambda_k \leqslant \mu_{k+1}$, and a simple calculation shows that if ϕ_k is the projection of g in $\mathscr{M}_k$ (that is ϕ_k is the restriction of g to $[\mu_k, \mu_{k+1}]$),

$$g = \Sigma\phi_k \qquad (\phi_k \in \mathscr{M}_k), \tag{9.3.3}$$

$$\|Mg - \Sigma\lambda_k\phi_k\| \leqslant \delta\|g\|. \tag{9.3.4}$$

These equations may be regarded as analogues of (9.3.1) and (9.3.2), and so as a generalization of the Hilbert–Schmidt Theorem.

This argument suggests that the transition from point to continuous spectrum may be effected by considering approximate eigenspaces and using a limiting procedure. However, as it is conceptually simpler to deal with the convergence of operators rather than of subspaces, we shall rewrite our results in a somewhat different form. Recalling that ϕ_k is the projection of g in $\mathscr{M}_k$, let $P(\mu, \nu)$ be the projection onto the closed subspace of $\mathscr{H}$ consisting of functions with supports in $[\mu, \nu] \cap [0, 1]$, with the understanding that $P(\mu, \nu) = 0$ if this interval is empty. Set $P_\nu = P(-\infty, \nu)$. Then $P(\mu, \nu) = P_\nu - P_\mu$, and $P_\mu = 0$ for $\mu < 0$, $P_\mu = I$ for $\mu > 1$. The subspaces $\mathscr{M}_k$ are the ranges of the projections $P(\mu_k, \mu_{k+1})$, and (9.3.3) becomes $g = \Sigma[P_{\mu_{k+1}} - P_{\mu_k}]g$. This holds for all δ, so we may write equivalently

$$g = \lim_{\delta \to 0} \Sigma[P_{\mu_{k+1}} - P_{\mu_k}]g, \tag{9.3.5}$$

while (9.3.4) is

$$Mg = \lim_{\delta \to 0} \Sigma\lambda_k[P_{\mu_{k+1}} - P_{\mu_k}]g. \tag{9.3.6}$$

Now the right-hand sides of these equations are just the sums which appear in a Banach space valued Riemann–Stieltjes integral (Definition 9.2.12). Hence

$$g = \int_{-\infty}^{\infty} \mathrm{d}P_\lambda g, \tag{9.3.7}$$

$$Mg = \int_{-\infty}^{\infty} \lambda \, \mathrm{d}P_\lambda g. \tag{9.3.8}$$

These integrals provide a decomposition of I and M into "nice" operators P_λ, and so indirectly of $\mathscr{H}$ into the sum of subspaces $R(P(\mu, \nu))$ as in the Hilbert–Schmidt Theorem. The analogues of (9.3.1) and (9.3.2) are thus (9.3.7) and (9.3.8) respectively, and their elegance and conciseness suggest that these should be the goal of our generalization. The Spectral Theorem provides this generalization.

The properties of the P_λ in the Spectral Theorem may be expressed in a concise mathematical form, but although these properties must obviously be related to those of the eigenspaces or approximate eigenspaces, and so to the operator itself, the nature of this relationship may at first sight be somewhat obscure. Clarification may be obtained by returning to the original compact T. Number the distinct eigenvalues (λ_n) of T as an increasing sequence and let $\mathscr{M}_n$ be the corresponding eigenspaces. Let E_{λ_n} be the projection onto $\mathscr{M}_n$, and set

$$P_\lambda g = \sum_{\lambda_n \leqslant \lambda} E_{\lambda_n} g, \tag{9.3.9}$$

where the sum is interpreted as zero if it contains no terms. $\{P_\lambda\}$ is the required family. First note that the $\mathscr{M}_n$— and hence the projections E_{λ_n}—are mutually orthogonal. Therefore the P_λ are projections (Lemma 9.2.11(i)). Next regard P_λ as an operator valued function of λ. From the definition, if $\lambda > \mu$, $R(P_\lambda) \supset R(P_\mu)$, whence $P_\lambda \geqslant P_\mu$ (Lemma 9.2.11(ii)), and the P_λ's commute. In fact more can be said about commutativity. It may be proved that if $A \in \mathscr{L}(\mathscr{H})$ commutes with T then A commutes with each P_λ. The significance of this is less easy to interpret, but some insight can be obtained from a quantum mechanics model. Here the self-adjoint operators correspond to "observables", and two quantities are simultaneously observable if and only if the corresponding operators commute (Problem 9.2). Thus if A and T are simultaneously observable, so are A and P_λ for each λ. Since the Spectral Theorem may be regarded as a decomposition of T into simpler operators P_λ, this property is necessary if the decomposition is to be of practical significance.

All these properties also hold for an arbitrary self-adjoint operator. The principal difference between the compact and general cases is the behaviour of P_λ as a function of λ. For the multiplication operator M, it is obvious that P_λ is strongly continuous in the sense that $\lim_{\lambda \to \lambda_0} P_\lambda g = P_{\lambda_0} g$. In the compact case it follows from (9.3.9) that P_λ is strongly right continuous, but it is *not* continuous from the left as P_λ has a jump discontinuity at each eigenvalue and is constant in between. It is plausible, and will be proved later, that a discontinuity is associated with the point spectrum, and a smooth change with the continuous spectrum. It is therefore necessary to introduce Riemann–Stieltjes integrals as in (9.3.7) and (9.3.8) to allow for the possibility that there may be both isolated points and continuous segments in the spectrum.

The last remark concerns the method of proof to be adopted. Rather than use the construction given for obtaining P_λ for the multiplication operator M, it is easier to justify rigorously a rather different approach in the general case. For M the argument goes as follows. The family $\{P_\lambda\}$ is such that for $\lambda \in [0, 1]$,

$$(P_\lambda g)(x) = \begin{cases} g(x) & (0 \leqslant x \leqslant \lambda), \\ 0 & (x \leqslant \lambda \leqslant 1), \end{cases}$$

while $P_\lambda = 0$ $(\lambda < 0)$ and $P_\lambda = I$ $(\lambda > 1)$. For each λ define the real valued step function p_λ as follows:

$$p_\lambda(x) = \begin{cases} 1 & (x \leqslant \lambda), \\ 0 & (x > \lambda). \end{cases}$$

Now the analysis of the previous section shows that there is an operator $p_\lambda(M)$ corresponding to p_λ. For f a polynomial, $f(M)g(x) = f(x)\,g(x)$ $(0 \leqslant x \leqslant 1)$, and it follows that this relation holds for any upper semicontinuous f, and in particular for p_λ. Therefore $p_\lambda(M)$ has exactly the properties given above for P_λ. This suggests that we define $P_\lambda = p_\lambda(M)$ in the general case. From the theoretical point of view this approach has the advantage of applying to any bounded self-adjoint M. It does not provide a convenient method for constructing P_λ explicitly in anything but the simplest examples, but an alternative method for this will be given later.

9.4 The Spectral Theorem for Bounded Self-adjoint Operators

The conclusions of the heuristic arguments of the previous section will now be justified for an arbitrary bounded self-adjoint operator L. Throughout $\mathscr{H}$ is a complex Hilbert space.

The first stage is the definition of the family $\{P_\lambda\}$ of projections. For each $\lambda \in \mathbb{R}$ set

$$p_\lambda(t) = \begin{cases} 1 & (t \leqslant \lambda), \\ 0 & (t > \lambda). \end{cases}$$

p_λ is a step function and is obviously upper semicontinuous. Thus by Theorem 9.2.7 there is a corresponding self-adjoint operator $p_\lambda(L)$. For each t, $[p_\lambda(t)]^2 = p_\lambda(t)$, and it follows from the same theorem that $[p_\lambda(L)]^2 = p_\lambda(L)$. Hence $p_\lambda(L)$ is a projection (Lemma 9.2.10). The required family $\{P_\lambda\}$ of projections is obtained by taking $P_\lambda = p_\lambda(L)$. The following lemmas show that the behaviour of P_λ is as asserted in the previous section.

9.4.1 Lemma. *$\{P_\lambda\}$ is a family of self-adjoint projections with the following properties:*

(i) *P_λ is a strongly right continuous function of λ. That is $P_\mu \underset{S}{\rightarrow} P_\lambda$ as $\mu \rightarrow \lambda +$.*

(ii) *The P_λ commute with each other and with any bounded operator that commutes with L.*

(iii) *$P_\lambda = 0$ $(\lambda < m_-)$, $P_\lambda = I$ $(\lambda \geqslant m_+)$.*

(iv) *If $\lambda \geqslant \mu$, $P_\lambda \geqslant P_\mu$—or equivalently $P_\lambda P_\mu = P_\mu$.*

Proof. (i) Let (ε_n) be a sequence of positive numbers tending to zero. Fix λ, and let (p_n) be a sequence of continuous functions which decrease monotonically to p_λ on some interval including $[m_-, m_+]$, and which are such that $p_n(t) \geqslant p_{\lambda+\varepsilon_n}(t)$ for $t \in \mathbb{R}$. Then by Theorem 9.2.7,

$$p_n(L) \geqslant p_{\lambda+\varepsilon_n}(L) \geqslant p_\lambda(L).$$

But $p_n(L) \underset{S}{\to} p_\lambda(L)$, whence $p_{\lambda+\varepsilon_n}(L) \underset{S}{\to} p_\lambda(L)$ as $n \to \infty$. Since further $p_\mu(L)$ is a monotone decreasing function of μ, this implies that $p_{\lambda+\varepsilon}(L) \underset{S}{\to} p_\lambda(L)$ as $\varepsilon \to 0+$.

(ii) is a consequence of Theorem 9.2.7. (iii) and (iv) follow from the corresponding properties of the real valued function p_λ on application of the same theorem. □

The following inequality will be useful in the sequel:

$$\mu(P_\nu - P_\mu) \leqslant L(P_\nu - P_\mu) \leqslant \nu(P_\nu - P_\mu) \qquad (\mu \leqslant \nu). \tag{9.4.1}$$

To establish (9.4.1) note that $p_\nu - p_\mu$ is the characteristic function of $(\mu, \nu]$, and hence

$$\mu[p_\nu(t) - p_\mu(t)] \leqslant t[p_\nu(t) - p_\mu(t)] \leqslant \nu[p_\nu(t) - p_\mu(t)] \tag{9.4.2}$$

since all three terms are zero except when $t \in (\mu, \nu]$. Replacing t by L in (9.4.2) and recalling that $P_\lambda = p_\lambda(L)$, we deduce (9.4.1) from Theorem 9.2.7.

9.4.2 Lemma. *If $\lambda_0 \in \rho(L)$, there is a neighbourhood of λ_0 on which P_λ is constant.*

Proof. By Theorem 6.6.4, there is an $m > 0$ such that $\|(L - \lambda_0 I)f\| \geqslant m\|f\|$ for all $f \in \mathscr{H}$. Choose $\varepsilon < m$, and suppose that P_λ is not constant on $[\lambda_0 - \varepsilon, \lambda_0 + \varepsilon]$. Then $P = P_{\lambda_0+\varepsilon} - P_{\lambda_0-\varepsilon} \neq 0$, and there is an h with $\|h\| = 1$ and $Ph = h$. It is an obvious consequence of (9.4.1) that $-\varepsilon P \leqslant (L - \lambda_0 I)P \leqslant \varepsilon P$, whence $\|(L - \lambda_0 I)P\| \leqslant \varepsilon$ (Lemma 9.2.2(v)). Therefore, as $Ph = h$, $\|(L - \lambda_0 I)h\| \leqslant \varepsilon\|h\|$. This contradicts the first assertion in the proof. □

9.4.3 Lemma. *Take any $\mu_0, \ldots, \mu_n$ and $\lambda_1, \ldots, \lambda_n$ such that $\mu_0 < m_-$, $\mu_n = m_+$ and $\mu_0 \leqslant \lambda_1 \leqslant \mu_1 \leqslant \lambda_2 \leqslant \mu_2 \leqslant \ldots \leqslant \mu_n$. Let δ be the maximum length of the subintervals $[\mu_k, \mu_{k+1}]$ for $k = 0, \ldots, (n-1)$. Then with the limit understood in the sense of uniform operator convergence* (*that is in $\mathscr{L}(\mathscr{H})$*),

$$L = \lim_{\delta \to 0} \sum_{k=1}^{n} \lambda_k [P_{\mu_k} - P_{\mu_{k-1}}].$$

Proof. It follows on setting $\nu = \mu_k$, $\mu = \mu_{k-1}$ in turn in (9.4.1) and adding that

$$\Sigma\mu_{k-1}[P_{\mu_k} - P_{\mu_{k-1}}] \leqslant L\Sigma[P_{\mu_k} - P_{\mu_{k-1}}] \leqslant \Sigma\mu_k[P_{\mu_k} - P_{\mu_{k-1}}], \tag{9.4.3}$$

and evidently,

$$\Sigma\mu_{k-1}[P_{\mu_k} - P_{\mu_{k-1}}] \leqslant \Sigma\lambda_k[P_{\mu_k} - P_{\mu_{k-1}}] \leqslant \Sigma\mu_k[P_{\mu_k} - P_{\mu_{k-1}}]. \tag{9.4.4}$$

Now

$$L\Sigma[P_{\mu_k} - P_{\mu_{k-1}}] = L[P_{\mu_n} - P_{\mu_0}] = L, \tag{9.4.5}$$

since $P_{\mu_0} = 0$, $P_{\mu_n} = I$ (Lemma 9.4.1(iii)). Also the difference between the extreme terms in (9.4.3) and (9.4.4) is

$$\Sigma(\mu_k - \mu_{k-1})[P_{\mu_k} - P_{\mu_{k-1}}] \leqslant \delta\Sigma[P_{\mu_k} - P_{\mu_{k-1}}] = \delta I.$$

Hence the norm of this difference is not greater than δ. Therefore the middle terms in (9.4.3) and (9.4.4) differ in norm by δ at most (Lemma 9.2.2(vi)). The result follows from (9.4.5). □

9.4.4 Corollary. *For any complex-valued continuous function f defined on* $\mathbb{R}$,

$$f(L) = \lim_{\delta\to 0} \Sigma f(\lambda_k)[P_{\mu_k} - P_{\mu_{k-1}}] \tag{9.4.6}$$

Proof. Since f may be decomposed into real and imaginary parts it is only necessary to establish the result for real-valued f. By Lemma 9.4.1(iv),

$$[P_{\mu_k} - P_{\mu_{k-1}}][P_{\mu_j} - P_{\mu_{j-1}}] = 0 \qquad (j \neq k),$$

and straightforward algebraic manipulation shows that

$$\Sigma\lambda_k^r[P_{\mu_k} - P_{\mu_{k-1}}] = \{\Sigma\lambda_k[P_{\mu_k} - P_{\mu_{k-1}}]\}^r.$$

The result for $f(L) = L^r$ is obtained on letting $n \to \infty$ and using the argument in the proof of the previous lemma. The extension to f a polynomial is obvious.

Suppose finally that f is an arbitrary continuous function. Then given $\varepsilon > 0$ there is a polynomial, p say, such that $|f(\lambda) - p(\lambda)| \leqslant \varepsilon$ for $\lambda \in [m_-, m_+]$. Hence by Theorem 9.2.5(ii), $\|f(L) - p(L)\| \leqslant \varepsilon$. If

$$S(f) = \Sigma f(\lambda_k)[P_{\mu_k} - P_{\mu_{k-1}}],$$

it is easily seen that $\|S(f) - S(p)\| \leqslant \varepsilon$. Since for a sufficiently fine subdivision $\|p(L) - S(p)\| \leqslant \varepsilon$, the result follows from the inequality

$$\|f(L) - S(f)\| \leqslant \|f(L) - p(L)\| + \|p(L) - S(p)\| + \|S(p) - S(f)\| \leqslant 3\varepsilon. \quad □$$

9.4.5 The Spectral Theorem. *Let $\mathscr{H}$ be a complex Hilbert space, and let $L: \mathscr{H} \to \mathscr{H}$ be a bounded self-adjoint operator. Then there is a unique family $\{P_\lambda\}$ of self-adjoint projections with the following properties:*

(i) *P_λ is a strongly right continuous function of λ;*
(ii) *the P_λ commute with each other and with any bounded operator which commutes with L;*
(iii) $P_\lambda = 0$ $(\lambda < m_-)$, $P_\lambda = I$ $(\lambda \geqslant m_+)$;
(iv) $P_\lambda \geqslant P_\mu$ if $\lambda \geqslant \mu$;
(v) *for any $f : \mathbb{R} \to \mathbb{C}$ which is continuous on an open set containing the spectrum, and with the integrals being understood in the Riemann–*

Stieltjes sense (*Definition 9.2.12*),

$$f(L) = \int_{-\infty}^{\infty} f(\lambda)\, \mathrm{d}P_\lambda, \tag{9.4.7}$$

$$f(L)g = \int_{-\infty}^{\infty} f(\lambda)\, \mathrm{d}P_\lambda g \qquad (g \in \mathscr{H}), \tag{9.4.8}$$

$$(f(L)g, h) = \int_{-\infty}^{\infty} f(\lambda)\, \mathrm{d}(P_\lambda g, h) \qquad (g, h \in \mathscr{H}). \tag{9.4.9}$$

Proof. The existence of a family $\{P_\lambda\}$ with the properties (i)–(iv) is just the assertion of Lemma 9.4.1. To prove (9.4.7) simply note that P_λ is constant on $\rho(L)$ (Lemma 9.4.2). Hence from the definition of the integral, there are no contributions to the integral when $\lambda \in \rho(L)$. (9.4.7) then follows from (9.4.6), and (9.4.8) and (9.4.9) are obtained in a similar manner. For a proof of uniqueness see Riesz and Nagy (1955, p. 276). □

In the theorem the limits of integration have been taken as $\pm\infty$ to simplify the notation. In fact P_λ is constant if $\lambda \notin [m_-, m_+)$ and is right continuous. It would thus be equivalent to take the upper limit to be m_+ and the lower limit to be $m_- - \varepsilon$ for some $\varepsilon > 0$ (to allow for a possible discontinuity in P_λ at m_-) or indeed to integrate only over an open set Ω containing the spectrum; obviously the values of f outside Ω are irrelevant to the value of the integral. The integrals in (9.4.7) and (9.4.8) may also be treated in terms of Banach space valued measures, but some extra technicalities are involved in such an approach.

9.4.6 Definition. The family of projections with the properties stated in the theorem is known as the **spectral family** of L (or the **resolution of the identity** for L), and each P_λ is called a **spectral projection**.

The Spectral Theorem is the required generalization of the Hilbert–Schmidt Theorem to arbitrary bounded self-adjoint operators. In practice, in order to apply the Spectral Theorem a convenient method for constructing the spectral family must be available. Such a method will be described in the next section.

The importance of functions of operators has been pointed out in the introduction. The Spectral Theorem provides a practicable method for calculating such functions. In the next theorem the emphasis is on the manipulation of these functions; in essence the theorem asserts that this may be carried out merely by considering the corresponding operations for complex valued functions of a real variable. The theorem is principally a restatement of previous results, and its proof is omitted.

9.4.7 The Spectral Calculus. *With L and* $\mathscr{H}$ *as in the previous theorem, let f, g be complex valued functions continuous on an open set containing* $\sigma(L)$. *Then the following hold*:

(i) *the laws of combination of functions of a real variable carry over to functions of operators, and thus*
$(f+g)(L) = f(L) + g(L)$, $(\alpha f)(L) = \alpha f(L)$, $(f \cdot g)(L) = f(L) \cdot g(L)$;

(ii) $f(L)$ *is self-adjoint iff f is real-valued, and positive iff f is non-negative*;

(iii) (a) $\|f(L)\| = \sup_{\lambda \in \sigma(L)} |f(\lambda)|$,

(b) $\|f(L)h\|^2 = \int_{-\infty}^{\infty} |f(\lambda)|^2 \, \mathrm{d}(P_\lambda h, h) \quad (h \in \mathscr{H})$;

(iv) *the Spectral Mapping Theorem*: $f(\sigma(L)) = \sigma(f(L))$.

9.5 The Spectrum and the Resolvent

Before the power of the Spectral Theorem can be fully exploited, a closer examination of the spectral family must be carried out. The first question to be resolved is the relationship between the behaviour of P_λ as a function of λ and the spectrum. Confirming the tentative conclusions reached in Section 9.3, we shall prove that P_λ is constant on $\rho(L)$, P_λ is continuous but not constant on a neighbourhood of λ if λ lies in the continuous spectrum, and P_λ has a discontinuity at λ if λ is an eigenvalue.

The starting point for the analysis is a formula for the resolvent $R(\lambda; L)$ in terms of P_λ. Let λ_0 be any complex number not in $\sigma(L)$. Then $f(\lambda) = (\lambda_0 - \lambda)^{-1}$ is continuous on an open set containing $\sigma(L)$, and by the Spectral Theorem 9.4.5 (v),

$$f(L) = \int_{-\infty}^{\infty} (\lambda_0 - \lambda)^{-1} \, \mathrm{d}P_\lambda.$$

Further, by Theorem 9.4.7(i),

$$(\lambda_0 I - L) f(L) = f(L)(\lambda_0 I - L) = \int_{-\infty}^{\infty} f(\lambda)(\lambda_0 - \lambda) \, \mathrm{d}P_\lambda = \int_{-\infty}^{\infty} \mathrm{d}P_\lambda = I, \tag{9.5.1}$$

whence $R(\lambda_0; L) = f(L)$ and

$$R(\lambda_0; L) = \int_{-\infty}^{\infty} (\lambda_0 - \lambda)^{-1} \, \mathrm{d}P_\lambda \qquad (\lambda_0 \in \rho(L)). \tag{9.5.2}$$

Suppose next that λ_0 is a real number such that P_λ is constant on a neighbourhood $U = [\lambda_0 - \varepsilon, \lambda_0 + \varepsilon]$ of λ_0. Let f_0 be any function which is continuous on $\mathbb{R}$ and which is equal to $(\lambda_0 - \lambda)^{-1}$ for $\lambda \notin U$. Since P_λ is constant on U, there is no contribution to the integral from U, and it is legitimate to replace f by f_0 in (9.5.1). It follows that $(\lambda_0 I - L)$ has an inverse, that is $\lambda_0 \in \rho(L)$. Since it is already known from Lemma 9.4.2 that P_λ is constant on a neighbourhood of any $\lambda_0 \in \rho(L)$, the following result has been proved.

9.5.1 Theorem. *Let $\mathscr{H}$ be a complex Hilbert space, and let $L: \mathscr{H} \to \mathscr{H}$ be bounded and self-adjoint. A real number $\lambda_0 \in \rho(L)$ if and only if there is a neighbourhood of λ_0 on which P_λ is constant.*

Define P_{λ_0-} to be the strong limit of P_λ as $\lambda \to \lambda_0-$; the existence of the limit is guaranteed by Lemma 9.2.3.

9.5.2 Theorem. *Let L and $\mathscr{H}$ be as in the previous theorem. The real number λ_0 is in the point spectrum of L if and only if $P_{\lambda_0} \neq P_{\lambda_0-}$. The eigenspace $N(\lambda_0 I - L)$ corresponding to λ_0 is equal to the range of $P_{\lambda_0} - P_{\lambda_0-}$. The continuous spectrum of L consists of those points at which P_λ is continuous, but which do not have any neighbourhood on which P_λ is constant.*

Proof. Suppose that $\varepsilon, \eta \geqslant 0$, and set $P_{\varepsilon,\eta} = P_{\lambda_0+\eta} - P_{\lambda_0-\varepsilon}$. From (9.4.1),

$$(\lambda_0 - \varepsilon)P_{\varepsilon,\eta} \leqslant LP_{\varepsilon,\eta} \leqslant (\lambda_0 + \eta)P_{\varepsilon,\eta}, \tag{9.5.3}$$

whence

$$\|(L - \lambda_0 I)P_{\varepsilon,\eta}\| \leqslant \max(\varepsilon, \eta). \tag{9.5.4}$$

Assume that $P_{\lambda_0} \neq P_{\lambda_0-}$. Then P_λ is not constant on $[\lambda_0 - \varepsilon, \lambda_0]$ for any $\varepsilon > 0$. Choose any $h \in R(P_{\lambda_0} - P_{\lambda_0-})$ with $\|h\| = 1$; evidently $P_{\varepsilon,0}h = h$ for any $\varepsilon > 0$. It follows from (9.5.4) that $\|(L - \lambda_0 I)h\| = \|(L - \lambda_0 I)P_{\varepsilon,0}h\| = 0$. Therefore $R(P_{\lambda_0} - P_{\lambda_0-}) \subset N(\lambda_0 I - L)$.

Suppose on the other hand that $h \in N(\lambda_0 I - L)$, and arrange that $\|h\| = 1$. Take any $\nu > m_+$ and any $\eta > 0$ such that $\lambda_0 + \eta < \nu$. By (9.5.3),

$$(\lambda_0 + \eta)(P_\nu - P_{\lambda_0+\eta}) \leqslant L(P_\nu - P_{\lambda_0+\eta}),$$

and since each P_λ commutes with L it follows that

$$((\lambda_0 + \eta)(P_\nu - P_{\lambda_0+\eta})h, h) \leqslant ((P_\nu - P_{\lambda_0+\eta})Lh, h) = \lambda_0((P_\nu - P_{\lambda_0+\eta})h, h).$$

But $P_\nu - P_{\lambda_0+\eta} \geqslant 0$, therefore $((P_\nu - P_{\lambda_0+\eta})h, h) = 0$. Recalling that $P_\nu = I$, we deduce that $P_{\lambda_0+\eta}h = h$. By a similar argument $P_{\lambda_0-\varepsilon}h = 0$, and letting $\varepsilon, \eta \to 0$ and using the right continuity of P_λ, we obtain the relation $(P_{\lambda_0} - P_{\lambda_0-})h = h$. That is $N(\lambda_0 I - L) \subset R(P_{\lambda_0} - P_{\lambda_0-})$. In combination with the conclusion reached in the previous paragraph, this proves the first two statements of the theorem. The final statement now follows from Theorem 9.5.1. □

A formula providing a convenient method of calculating the spectral family $\{P_\lambda\}$ will next be derived. For $\lambda_1 < \lambda_2$ define $P(\lambda_1, \lambda_2) = P_{\lambda_2-} - P_{\lambda_1}$ ($P(\lambda_1, \lambda_2)$ is the projection corresponding to the characteristic function of the open interval (λ_1, λ_2)). Since $P_\lambda = 0$ for $\lambda < m_-$, the family $\{P_\lambda\}$ may be found if $P(\lambda_1, \lambda_2)$ is known for all λ_1, λ_2. $P(\lambda_1, \lambda_2)$ is obtained by solving the "integral equation" (9.5.2).

9.5.3 Theorem. *Let $\mathscr{H}$ be a complex Hilbert space, and assume that $L: \mathscr{H} \to \mathscr{H}$ is bounded and self-adjoint. For $g, h \in \mathscr{H}$,*

$$(P(\lambda_1, \lambda_2)g, h) = \lim_{\delta\to 0+} \lim_{\varepsilon\to 0+} \frac{1}{2\pi i} \int_{\lambda_1+\delta}^{\lambda_2-\delta} ([R(\lambda - i\varepsilon; L) - R(\lambda + i\varepsilon; L)]g, h)\, d\lambda. \tag{9.5.5}$$

Proof. Observe first that

$$(P_{\lambda_0}g, g) = \int_{-\infty}^{\lambda_0} d(P_\lambda g, g),$$

this formula being obtained simply by writing out the partial sums in the definition of the integral. Therefore, since P_λ is right continuous,

$$(P(\lambda_1, \lambda_2)g, g) = \lim_{\delta\to 0+} \int_{\lambda_1+\delta}^{\lambda_2-\delta} d(P_\lambda g, g). \tag{9.5.6}$$

It is sufficient to prove the result with $g = h$, for (9.5.5) will follow on using the representation for a bilinear form in terms of a quadratic form (Problem 6.18). The advantage is that $(P_\lambda g, g)$ is a non-decreasing real-valued function of λ, hence it gives rise to a Lebesgue–Stieltjes measure (Example 2.2.15), and the theory of Chapter 2 is valid.

For $\varepsilon > 0$ set $q(\lambda, \mu, \varepsilon) = (\lambda - \mu - i\varepsilon)^{-1} - (\lambda - \mu + i\varepsilon)^{-1}$. An elementary calculation shows that for $\delta > 0$,

$$\begin{aligned} p(\mu, \delta, \varepsilon) &= \frac{1}{2\pi i} \int_{\lambda_1+\delta}^{\lambda_2-\delta} q(\lambda, \mu, \varepsilon)\, d\lambda \\ &= \frac{1}{\pi}\left[\tan^{-1}\left(\frac{\lambda_2 - \delta - \mu}{\varepsilon}\right) - \tan^{-1}\left(\frac{\lambda_1 + \delta - \mu}{\varepsilon}\right)\right]. \end{aligned}$$

Therefore $|p(\mu, \delta, \varepsilon)| < 1$, and $\lim_{\varepsilon\to 0+} p(\mu, \delta, \varepsilon)$ is 1 in the interior of the interval $[\lambda_1 + \delta, \lambda_2 - \delta]$, $\frac{1}{2}$ at its end points, and 0 outside it. The Dominated Convergence Theorem 2.4.11 gives

$$\lim_{\varepsilon \to 0+} \int_{-\infty}^{\infty} p(\mu, \delta, \varepsilon)\, \mathrm{d}(P_\mu g, g) = \int_{-\infty}^{\infty} \lim_{\varepsilon \to 0+} p(\mu, \delta, \varepsilon)\, \mathrm{d}(P_\mu g, g).$$

If $0 < \delta' < \delta < \delta''$ we have

$$\int_{\lambda_1+\delta''}^{\lambda_2-\delta''} \mathrm{d}(P_\mu g, g) \leqslant \int_{-\infty}^{\infty} \lim_{\varepsilon \to 0+} p(\mu, \delta, \varepsilon)\, \mathrm{d}(P_\mu g, g) \leqslant \int_{\lambda_1+\delta'}^{\lambda_2-\delta'} \mathrm{d}(P_\mu g, g),$$

whence from (9.5.6)

$$\lim_{\delta \to 0+} \lim_{\varepsilon \to 0+} \int_{-\infty}^{\infty} p(\mu, \delta, \varepsilon)\, \mathrm{d}(P_\mu g, g) = (P(\lambda_1, \lambda_2)g, g).$$

Since $q/2i \geqslant 0$ and q is measurable, by Theorem 2.4.17(i) we may interchange the order of integration to find

$$\begin{aligned}\int_{-\infty}^{\infty} p(\mu, \delta, \varepsilon)\, \mathrm{d}(P_\mu g, g) &= \frac{1}{2\pi i} \int_{\lambda_1+\delta}^{\lambda_2-\delta} \int_{-\infty}^{\infty} q(\lambda, \mu, \varepsilon)\, \mathrm{d}(P_\mu g, g)\, \mathrm{d}\lambda \\ &= \frac{1}{2\pi i} \int_{\lambda_1+\delta}^{\lambda_2-\delta} ([R(\lambda - i\varepsilon; L) - R(\lambda + i\varepsilon; L)]g, g)\, \mathrm{d}\lambda,\end{aligned}$$

and the conclusion follows. □

9.6 Unbounded Self-adjoint Operators

The restriction that the operator should be bounded is too severe for most applications, and in particular is not satisfied by differential operators. However, the results of this chapter can be extended to unbounded self-adjoint operators if some rather natural modifications are made.

The most significant change arises from the fact that the spectrum is not bounded. Therefore the spectral family $\{P_\lambda\}$ will vary over an infinite range of λ, and the relations $P_\lambda = 0\ (\lambda < m_-)$, $P_\lambda = I\ (\lambda \geqslant m_+)$ become

$$P_\lambda \underset{S}{\rightarrow} 0 \qquad (\lambda \to -\infty), \qquad P_\lambda \underset{S}{\rightarrow} I \qquad (\lambda \to \infty).$$

The support of the integrand in the various integrals in the Spectral Theorem is no longer compact, and extra care is needed in dealing with the convergence of the integrals. With the proviso that $g \in D(L)$, the integrals for Lg and (Lg, h) (equations (9.4.8) and (9.4.9) respectively) may be interpreted as (improper) Riemann–Stieltjes integrals over $(-\infty, \infty)$. However, it is less easy to attach a meaning to $\int \lambda\, \mathrm{d}P_\lambda$, and a discussion of such integrals is omitted.

Another change will be mentioned for the sake of completeness although it will have no direct consequence in the subsequent analysis. If $L, B \in \mathscr{L}(\mathscr{H})$, the definition of commutativity is simply that $BL = LB$. This definition is not appropriate if L is unbounded. For if $L^{-1} \in \mathscr{L}(\mathscr{H})$, LL^{-1} is defined on all of $\mathscr{H}$, but $L^{-1}L$ only makes sense on $D(L)$, and it follows that L would not commute with its inverse. To avoid this difficulty L is said to *commute* with $B \in \mathscr{L}(\mathscr{H})$ iff $BL \subset LB$.

A proof of the Spectral Theorem may be supplied by taking the limit of a sequence of bounded operators, but the technicalities are rather unpleasant, and the reader is referred to Riesz and Nagy (1955, Chap. 8) for the details.

9.6.1 The Spectral Theorem for Unbounded Operators. *Suppose that L is a self-adjoint (unbounded) operator on a complex Hilbert space $\mathscr{H}$. Then there exists a unique spectral family $\{P_\lambda\}$ of self-adjoint projections with the following properties:*

(i) *P_λ is a strongly right continuous function of λ;*
(ii) *the P_λ commute with each other and with any bounded operator that commutes with L;*
(iii) $P_\lambda \underset{s}{\to} 0 \ (\lambda \to -\infty)$, $P_\lambda \underset{s}{\to} I \ (\lambda \to \infty)$;
(iv) $P_\lambda \geqslant P_\mu$ if $\lambda \geqslant \mu$;
(v) *with the convergence of the vector valued integrals being in $\mathscr{H}$,*

$$g = \int_{-\infty}^{\infty} \mathrm{d}P_\lambda g, \qquad (g, h) = \int_{-\infty}^{\infty} \mathrm{d}(P_\lambda g, h) \qquad (g, h \in \mathscr{H}),$$

$$Lg = \int_{-\infty}^{\infty} \lambda \, \mathrm{d}P_\lambda g, \qquad (Lg, h) = \int_{-\infty}^{\infty} \lambda \, \mathrm{d}(P_\lambda g, h) \quad (g \in D(L),\ h \in \mathscr{H}),$$

and $D(L)$ is the set of g such that $\int_{-\infty}^{\infty} \lambda^2 \, \mathrm{d}(P_\lambda g, g) < \infty$.

If L is self-adjoint and bounded there are a number of methods for defining the operator valued function $f(L)$. One possibility is to use a power series expansion; an extension to unbounded L presents obvious difficulties. A second method is the construction of Section 9.2. Yet another method is to define

$$f(L)g = \int_{-\infty}^{\infty} f(\lambda) \, \mathrm{d}P_\lambda g \qquad (g \in \mathscr{H}).$$

For unbounded L the last method is the one that presents fewest problems, and which at the same time is convenient for calculation. Motivation for the definition of $D(f(L))$ is provided by Theorem 9.4.7(iii)(b).

9.6.2 Definition. Let L be self-adjoint, and assume that f is continuous on an open set containing $\sigma(L)$. Let $D(f(L))$ be the set of $g \in \mathscr{H}$ such that

$$\int_{-\infty}^{\infty} |f(\lambda)|^2 \, d(P_\lambda g, g) < \infty,$$

and define the operator $f(L)$ by requiring that

$$f(L)g = \int_{-\infty}^{\infty} f(\lambda) \, dP_\lambda g \qquad (g \in D(f(L))). \qquad \square$$

9.6.3 Theorem. *Suppose that L is an unbounded self-adjoint operator on a complex Hilbert space $\mathscr{H}$, and let f, g be complex valued functions continuous on an open set containing $\sigma(L)$. Then:*

(i) $(f + g)(L) = f(L) + g(L)$, $(\alpha f)(L) = \alpha f(L)$, *and if g is bounded on* $\sigma(L)$, *then* $D(f(L) \,.\, g(L)) = D((f \,.\, g)(L))$ *and* $(f \,.\, g)(L) = f(L) \,.\, g(L)$;
(ii) $f(L)$ *is self-adjoint if f is real-valued, and positive if f is non-negative*;
(iii) (a) $\|f(L)\| = \sup_{\lambda \in \sigma(L)} |f(\lambda)| \quad$ (*f bounded on* $\sigma(L)$),
(b) $\|f(L)h\|^2 = \int_{-\infty}^{\infty} |f(\lambda)|^2 \, d(P_\lambda h, h) \quad (h \in D(f(L)))$;
(iv) *The Spectral Mapping Theorem*: $f(\sigma(L)) = \sigma(f(L))$.

9.6.4 Theorem. *Theorems 9.5.1, 9.5.2 and 9.5.3 are also valid for unbounded self-adjoint L. In particular, for g, h* $\in \mathscr{H}$,

$$(P(\lambda_1, \lambda_2)g, h) = \lim_{\delta \to 0+} \lim_{\varepsilon \to 0+} \frac{1}{2\pi i} \int_{\lambda_1 + \delta}^{\lambda_2 - \delta} ([R(\lambda - i\varepsilon; L) - R(\lambda + i\varepsilon; L)]g, h) \, d\lambda.$$

9.7 The Solution of an Evolution Equation

The abstract initial value problem:

$$\left.\begin{aligned} u'(t) &= Lu(t) \qquad (t > 0), \\ u(0) &= u_0, \end{aligned}\right\} \tag{9.7.1}$$

where L is a bounded linear operator on a Banach space, has been discussed previously (Example 3.5.10). Our aim now is to extend the analysis in two directions, first by considering unbounded L, and second by using the Spectral Theorem to provide a qualitative description of the solution in terms of the spectrum of L.

The purpose of studying unbounded L is that partial differential equations may be included as special cases of (9.7.1). For example, a standard mixed problem for the diffusion equation is to solve

$$\frac{\partial u}{\partial t} = \frac{\partial^2 u}{\partial x^2} \qquad (t > 0, 0 \leqslant x \leqslant a)$$

with $u(x, 0)$ given and, say, $u(0, t) = u(a, t) = 0$. This system may be put in the form (9.7.1) by letting $\mathscr{H} = \mathscr{L}_2(0, a)$ and taking L to be the self-adjoint operator derived from $l = \mathrm{d}^2/\mathrm{d}x^2$ with boundary conditions $u(0) = u(a) = 0$ (more general l are considered in the next chapter). u is then regarded as an $\mathscr{L}_2(0, a)$-valued function of t.

9.7.1 Theorem. *Let L be a (possibly unbounded) self-adjoint operator on the complex Hilbert space $\mathscr{H}$, and assume that $\sigma(L) \subset (-\infty, -m]$ for some $m > 0$. Take fixed $u_0 \in \mathscr{H}$, and set*

$$u(t) = \mathrm{e}^{tL}u_0 = \int_{-\infty}^{-m} \mathrm{e}^{t\lambda}\, \mathrm{d}P_\lambda u_0 \qquad (t \geqslant 0).$$

Then u is continuous on $[0, \infty)$, for $t > 0$ $u(t)$ lies in $D(L)$ and is differentiable, and u is a solution of the initial value problem (9.7.1). *Also*:

(i) $\|u(t)\| \leqslant \mathrm{e}^{-mt}\|u_0\| \quad (t > 0)$;
(ii) *the semigroup property holds*: $\mathrm{e}^{(t_1+t_2)L}u_0 = \mathrm{e}^{t_1L}\mathrm{e}^{t_2L}u_0 \quad (t_1, t_2 \geqslant 0)$.

Proof. Since

$$\int_{-\infty}^{-m} |\lambda\, \mathrm{e}^{t\lambda}|^2\, \mathrm{d}(P_\lambda g, g) < \infty \qquad (g \in \mathscr{H}),$$

from Definition 9.6.2 and Theorem 9.6.3(i), $\mathrm{e}^{tL}u_0 \in D(L)$ for $t > 0$. By Theorem 9.6.3(iii)(a),

$$h^{-1}\|\mathrm{e}^{(t+h)L} - \mathrm{e}^{tL} - hL\,\mathrm{e}^{tL}\| = \sup_{\lambda\in\sigma(L)} h^{-1}|\mathrm{e}^{(t+h)\lambda} - \mathrm{e}^{t\lambda} - h\lambda\,\mathrm{e}^{t\lambda}|,$$

and a routine calculation based on the Mean Value Theorem shows that the right-hand side tends to zero as $h \to 0$ for any $t > 0$. This proves that $\mathrm{e}^{tL}u_0$ is differentiable and that it satisfies the differential equation. The continuity of u is obtained by a similar argument. Finally (i) and (ii) are easily proved by use of the Spectral Calculus Theorem 9.6.3(iii)(a) and (i) respectively. □

This theorem emphasizes the strong analogy between the properties of the solution of the abstract initial value problem and of the corresponding equation in one dimension in which L is a real number. In one dimension the solution $\mathrm{e}^{tL}u_0$ is asymptotically stable if L is a negative number, in the sense that $\mathrm{e}^{tL}u_0 \to 0$ as $t \to \infty$. In the general case, the fact that the spectrum of L is negative means that L may sensibly be regarded as a negative operator, the solution being again asymptotically stable (by (i) of the theorem). The difficulties connected with the "backward" diffusion equation are well known—obviously stability depends critically on the spectrum.

The semigroup property (ii) is, as noted in Example 3.5.10, typical of autonomous equations. This characteristic property has been heavily exploited in tackling more general versions of the differential equation, and has led to the development of a sophisticated theory of semi-groups of operators. A good introduction to the extensive body of theory on evolution equations is Ladas and Lakshmikantham (1972).

Problems

Throughout $\mathscr{H}$ will be a complex Hilbert space.

9.1 The physical significance of self-adjointness is clarified by the following examples. In both assume that L is self-adjoint.

(i) Consider the "abstract wave equation"

$$u''(t) + Lu(t) = 0$$

for the $\mathscr{H}$-valued function u of t. By analogy with a dynamical system, $-Lf$ may be regarded as the force, and $V = \frac{1}{2}(u, Lu)$, $K = \frac{1}{2}(u', u')$, $E = V + K$ as the potential, kinetic and total energies respectively. Show formally that

$$E' = \tfrac{1}{2}(u', u'' + Lu) + \tfrac{1}{2}(u'' + Lu, u'),$$

and deduce the principle of *conservation of energy*: $E =$ constant.

(ii) For the "abstract Schrödinger equation"

$$u'(t) + iLu(t) = 0,$$

where L is the "Hamiltonian operator", show formally that $\|u(t)\|$ is constant. The significance of this in quantum mechanics is that the state u must have norm unity for all time.

9.2 *The Heisenberg Uncertainty Principle.* Let L, M be unbounded self-adjoint operators on $\mathscr{H}$ with $D = D(LM) \cap D(ML)$ dense in $\mathscr{H}$. Take fixed $f \in D$ and set

$$m(L) = (Lf, f), \qquad \lambda(L) = \|(L - m(L)I)f\|.$$

Show that

$$\lambda(L)\lambda(M) \geqslant \tfrac{1}{2}|m(LM - ML)|.$$

In quantum mechanics, $\lambda(L)$ gives the uncertainty in the measurement of the observable corresponding to L. Unless the commutator $ML - LM$ is zero there is a limit on the accuracy of simultaneous observations. The best known example is when L is the momentum operator arising from $-i\hbar\,\mathrm{d}/\mathrm{d}x$ and M is the position operator corresponding to multiplication by x.

9.3 Construct an example of self-adjoint L, M on $\mathbb{R}^2$ for which neither $L \geqslant M$ nor $L \leqslant M$.

9.4 Prove that every positive bounded self-adjoint L has a unique positive square root $L^{\frac{1}{2}}$. Show that $L^{\frac{1}{2}}$ can be represented as the strong limit of polynomials in L, and so commutes with every operator that commutes with L. (Hint: Arrange that $0 \leqslant L \leqslant I$, set $L_0 = 0$, $L_{n+1} = L_n + \frac{1}{2}(L - L_n^2)$ for $n \geqslant 1$, and use Lemma 9.2.3).

9.5 Suppose $L, M, N \in \mathscr{L}(\mathscr{H})$ are self-adjoint. Show using 9.4 that

(i) LM is positive and self-adjoint if L, M are positive and commute.
(ii) $L \geqslant M \Rightarrow LN \geqslant MN$ if N is positive and commutes with L, M.

9.6 In Theorem 9.2.7 show that $f(L)$ is independent of the particular sequence chosen. (Use Dini's theorem: If a sequence of continuous functions on a compact set converges monotonically to a continuous function, the convergence is uniform).

9.7 Let P_1, P_2 be the projections onto the closed subspaces $\mathscr{M}_1, \mathscr{M}_2$ respectively. Prove the following:

(i) P_1 and P_2 commute iff P_1P_2 is a projection, and then $R(P_1P_2) = \mathscr{M}_1 \cap \mathscr{M}_2$;
(ii) $P_1 + P_2$ is a projection $\Leftrightarrow P_1P_2 = 0 \Leftrightarrow \mathscr{M}_1 \perp \mathscr{M}_2$;
(iii) $P_1 - P_2$ is a projection $\Leftrightarrow \mathscr{M}_1 \supset \mathscr{M}_2$;
(iv)(*) $\mathscr{M}_1$ and $\mathscr{M}_2$ are isometrically isomorphic if $\|P_1 - P_2\| < 1$.

9.8 Consider $M : \mathscr{L}_2(0, 1) \to \mathscr{L}_2(0, 1)$ where $Mg(x) = xg(x)$ for $x \in [0, 1]$. Show that $\sigma(M) = [0, 1]$, and that every point of $\sigma(M)$ is in the continuous spectrum.

9.9 If $L \in \mathscr{L}(\mathscr{H})$, show that $L^{-1} \in \mathscr{L}(\mathscr{H})$ iff L^*L has a strictly positive lower bound m_-.

9.10 Let E be the projection onto an eigenspace of a compact self-adjoint operator T. Show that $A \in \mathscr{L}(\mathscr{H})$ commutes with E if A commutes with T.

9.11 Let L be bounded and self-adjoint, and let f be a complex valued continuous function. Show that

$$\|f(L)\| = \sup_{\lambda \in \sigma(L)} |f(\lambda)|.$$

9.12 Let μ be a measurable bounded real-valued function, and define $M : \mathscr{L}_2(0, 1) \to \mathscr{L}_2(0, 1)$ by setting $Mg(x) = \mu(x)g(x)$ for all $g \in \mathscr{L}_2(0, 1)$. Find the projections $P(\lambda_1, \lambda_2)$ by using Theorem 9.5.3.

9.13(*) Let k be an even real-valued continuous function in $\mathscr{L}_1(-\infty, \infty)$. Define $K : \mathscr{L}_2(-\infty, \infty) \to \mathscr{L}_2(-\infty, \infty)$ by setting

$$Kg(x) = \int_{-\infty}^{\infty} k(x - y)\, g(y)\, dy.$$

Using the Fourier transform find the resolvent, and deduce a formula for $P(\lambda_1, \lambda_2)$. (Taylor, 1958; p. 356).

Chapter 10

Generalized Eigenfunction Expansions Associated with Ordinary Differential Equations

10.1 Introduction

As a reason for the study of the Spectral Theorem, it was claimed that a wide class of superficially dissimilar expansion formulae involving series or integrals can be obtained by an application of this theorem to self-adjoint operators associated with certain formal ordinary differential operators l. This claim will now be substantiated. To emphasize their close relationship these formulae will be called "generalized eigenfunction expansions". This terminology is justified by noting that the expansions are all sums or integrals, or perhaps a combination of both, of solutions of $lf = \lambda f$, the word "generalized" being included as a reminder that the solutions need not be eigenfunctions of the self-adjoint operator itself.

Before the Spectral Theorem can be invoked, an appropriate self-adjoint operator L associated with l must evidently first be found. Sometimes, as in Example 6.7.9, L may be constructed simply by guesswork, but this is not always possible.

10.1.1 Example. As in Example 6.7.9 take $l = i\,\mathrm{d}/\mathrm{d}x$, but now suppose that the interval is $[0, \infty)$. To illustrate the difficulty it is sufficient to use a formal argument, so assume throughout that the domains of all operators include only smooth functions in $\mathscr{L}_2(0, \infty)$. If L, M are any operators with $Lf = lf$, $Mf = lf$ on $D(L)$, $D(M)$ respectively, an integration by parts yields

$$(Lf, g) - (f, Mg) = i[\lim_{x\to\infty} f(x)\,\bar{g}(x) - f(0)\,\bar{g}(0)] = -if(0)\,\bar{g}(0)$$

if the assumption (not unreasonable as $f, g \in \mathscr{L}_2(0, \infty)$) is made that the limit is zero. For given $D(L)$, M will be the adjoint of L if $D(M)$ is the maximal domain consistent with the vanishing of $f(0)\,\bar{g}(0)$ for all $f \in D(L)$. It follows that if no conditions (other than the smoothness requirement mentioned above) are placed on $D(L)$, then necessarily $g(0) = 0$ if g is to lie in the domain of L^*; in this case L^* is a proper restriction of L, and L is not self-adjoint. On the other hand, if we suppose that $f(0) = 0$ for all $f \in D(L)$, the only restriction required of an element g of $D(L^*)$ is that it should be sufficiently smooth; in this case L^* is a proper extension of L, and again L is not self-adjoint. There are no further obvious possibilities likely to produce self-adjoint L, and the tentative conclusion (confirmed in Example 10.5.5) must be that no such operator exists.

Thus even for simple l, difficulties can arise in constructing an associated self-adjoint operator, and there are even doubts as to the existence of such an operator. For higher order l, perhaps on an infinite interval or with leading coefficient vanishing at an end point, the difficulties are severe, and most of the technicalities in this chapter arise in overcoming these difficulties. Evidently the first objective must be to devise a systematic procedure for deciding on the existence of a self-adjoint operator in a given case, and for constructing such an operator.

A sensible first step in this direction is to start with an operator L_0 with domain so severely restricted as to be contained in the domains of all likely candidates for self-adjoint operators, and to attempt to construct these operators by expanding the initial domain—in other words by considering self-adjoint extensions of L_0. For example if the interval is (a, b) a reasonable choice of $D(L_0)$ might be $\mathscr{C}_0^\infty((a, b))$. L_0 is not self-adjoint, but satisfies the weaker condition of symmetry. A detailed study of extensions of symmetric operators is therefore first undertaken with the above objective in view. The theory is then used to decide whether L_0 has self-adjoint extensions, and to construct these extensions when they exist. The final step of deriving the generalized eigenfunction expansions by an application of the Spectral Theorem is itself relatively straightforward.

Standard references are the voluminous Chapter 13 of Dunford and Schwartz (1963), and Naimark (1968). A different approach based on classical analysis is followed in Titchmarsh (1962), where a wealth of examples is given. For a wide range of applications see Sneddon (1972). A qualitative description of the spectrum is of interest in many applications; this topic is covered extensively in the first three references above, and is discussed from the point of view of quantum mechanics by Reed and Simon (1972, 1975).

For the perturbation theoretic viewpoint see Kato (1966). Aleksandrjan *et al.* (1976) survey the spectral theory of partial differential equations.

10.2 Extensions of Symmetric Operators

The aims of this section are first to provide a systematic method of deciding whether a symmetric operator has self-adjoint extensions, and second to find a characterization of these extensions (when they exist) suitable for the ordinary differential equation case.

Throughout $\mathscr{H}$ will be a complex Hilbert space, and all operators will be from $\mathscr{H}$ into itself. The operators are usually unbounded.

10.2.1 Definition. A linear operator L_0 is said to be **symmetric** iff its domain is dense in $\mathscr{H}$ and one of the following (obviously equivalent) conditions hold:

(i) $(L_0 f, g) = (f, L_0 g) \qquad (f, g \in D(L_0))$,
(ii) $L_0 \subset L_0^*$.

Clearly symmetry and self-adjointness are equivalent conditions for a bounded operator, but for an unbounded operator the situation is different. For while a self-adjoint L_0 is certainly symmetric, if L_0 is symmetric it may happen that $D(L_0) \neq D(L_0^*)$, in which case L_0 is not self-adjoint (compare Example 6.7.2). Indeed a symmetric operator need not even be closed. However, next best, it is closable, and this is enough for the present purpose.

10.2.2 Lemma. *Let L_0 be a symmetric operator. Then it is closable with closure $\bar{L}_0 = L_0^{**}$, $\bar{L}_0$ is symmetric, and $\bar{L}_0^* = L_0^*$. If L_1 is a symmetric extension of L_0, $L_0 \subset L_1 \subset L_1^* \subset L_0^*$.*

Proof. Since $L_0^* \supset L_0$, L_0^* is densely defined. Hence by Theorem 6.7.5, L_0 is closable and $\bar{L}_0 = L_0^{**}$. Now L_0^* is closed (Theorem 6.7.4) and $L_0 \subset L_0^*$, therefore $\bar{L}_0 \subset L_0^*$. From the definition of closure, $G(\bar{L}_0) \subset \overline{G(L_0)}$, where G denotes the graph (Definition 3.8.7). Hence from Lemma 6.7.3, $L_0^* = \bar{L}_0^*$. This proves that $\bar{L}_0 \subset \bar{L}_0^*$, whence $\bar{L}_0$ is symmetric. The final statement follows from Lemma 6.7.3. □

In the following analysis it is convenient to simplify the notation by setting $\bar{L}_0 = L$. Then by the lemma, $L_0^* = \bar{L}_0^* = L^*$.

10.2.3 Lemma. *Let L be closed and symmetric, and assume that* $\operatorname{Im}\lambda \neq 0$. *Then $N(L - \lambda I) = 0$, and*

$$\mathscr{H} = R(L - \lambda I) \oplus N(L^* - \bar{\lambda} I) = R(L - \bar{\lambda} I) \oplus N(L^* - \lambda I).$$

$R(L - \lambda I)$ and $N(L^* - \bar{\lambda} I)$ are closed orthogonal subspaces.

Proof. Repetition of the proof of Lemma 6.6.1 under the restriction $f \in D(L)$ yields the inequality $\|(L - \lambda I)f\| \geqslant v\|f\|$ where $v = |\operatorname{Im}\lambda|$. It follows from Lemma 3.8.18(i) that $R(L - \lambda I)$ is closed. Theorem 6.7.6 then gives the result. □

Lemma 10.2.2 narrows down the area to which the search for self-adjoint extensions need be directed. Because a self-adjoint operator is closed, and $\bar{L}_0$ is by definition the minimal closed extension of L_0, if L_0 has a self-adjoint extension M, then M is symmetric, an extension of $\bar{L}_0$, and a restriction of L_0^*. For formal ordinary differential operators it is easy to find a suitable symmetric L_0, but although L_0 is closable, the calculation of its closure is in general by no means an easy task. However, it is not difficult to find L_0^*, and it is therefore natural to proceed by restricting L_0^* rather than by extending $\bar{L}_0$; in fact this approach leads incidentally to a characterization of $\bar{L}_0$. Since $D(M)$ is a linear subspace the problem may be formulated as follows. Do there exist linear subspaces $\mathscr{M}$ with $D(\bar{L}_0) \subset \mathscr{M} \subset D(L_0^*)$ such that the restrictions of L_0^* to $\mathscr{M}$ are self-adjoint? If there are, how may these subspaces be characterized?

Lemma 10.2.3 indicates how we should continue. If $\bar{L}_0$ is self-adjoint, its spectrum is real, and therefore $R(\bar{L}_0 - \lambda I) = \mathscr{H}$ if $\operatorname{Im}\lambda \neq 0$. Thus for symmetric L_0, the dimensions of $N(L_0^* - \lambda I)$ and $N(L_0^* - \bar{\lambda} I)$ are a measure of the extent by which $\bar{L}_0$ fails to be self-adjoint. A careful study of these spaces will furnish the answers to the questions above. In fact the dimension of each of the spaces is constant as λ varies over the upper half plane $\operatorname{Im}\lambda > 0$ (Dunford and Schwarz, 1963; p. 1232), and it is enough to consider $\lambda = i$ only.

10.2.4 Definition. The closed subspaces

$$N_+ = N(L_0^* - iI), \qquad N_- = N(L_0^* + iI)$$

are known as the **deficiency spaces** of L_0 and their dimensions, $n_\pm$ respectively, as the **deficiency indices**. We shall also say for brevity that L_0 has deficiency indices (n_+, n_-).

10.2.5 Lemma. *If L_0 is symmetric and $L = \bar{L}_0$, $D(L^*) = D(L) \oplus N_+ \oplus N_-$.*

Proof. Take any $f \in D(L^*)$. By Lemma 10.2.3,

$$L^*f - if = (Lg - ig) + h_-$$

for some $g \in D(L)$ and $h_- \in N_-$. Since $g \in D(L) \subset D(L^*)$, this equation can be rewritten as

$$\begin{aligned} L^*(f - g - \tfrac{1}{2}ih_-) &= i(f - g) + h_- + L^*(-\tfrac{1}{2}ih_-) \\ &= i(f - g) + h_- - \tfrac{1}{2}h_- \qquad (\text{as } L^*h_- = -ih_-) \\ &= i(f - g - \tfrac{1}{2}ih_-). \end{aligned}$$

Thus $f - g - \frac{1}{2}ih_- \in N_+$, and with $\phi_- = \frac{1}{2}ih_-$, $\phi_+ = f - g - \frac{1}{2}ih_-$, we have proved that

$$f = g + \phi_+ + \phi_- \qquad (g \in D(L), \phi_\pm \in N_\pm). \tag{10.2.1}$$

Since $D(L)$ and $N_\pm$ are contained in the linear subspace $D(L^*)$, the result will follow if it can be shown that the representation (10.2.1) is unique, or equivalently that $f = 0 \Rightarrow g = \phi_+ = \phi_- = 0$. Now an application of $L^* - iI$ to (10.2.1) with $f = 0$ yields

$$(L - iI)g - 2i\phi_- = 0,$$

and since $\phi_- \in N_-$ the terms on the left-hand side are orthogonal (Lemma 10.2.3). Thus each term is zero and $\phi_- = 0$. Similarly $\phi_+ = 0$, whence also $g = 0$. That is $g = \phi_+ = \phi_- = 0$. □

10.2.6 Corollary. *Suppose L is closed, and L_1 is a linear operator with $L \subset L_1 \subset L^*$. Then there is a unique linear subspace N of $N_+ \oplus N_-$ such that $D(L_1) = D(L) \oplus N$.*

Proof. Let N be the set consisting of all $n \in N_+ \oplus N_-$ such that there is an $f_1 \in D(L_1)$, $f \in D(L)$ with $f_1 = f + n$. $D(L)$ and $D(L_1)$ are linear subspaces, and it follows easily that so also is N. But $N \cap D(L) = 0$, therefore $D(L_1) = D(L) \oplus N$ (Lemma 1.2.9). The uniqueness of N is obvious. □

Thus the domain of any symmetric extension (and so of any self-adjoint extension) of $\bar{L}_0$ is obtained by "adding" to $D(\bar{L}_0)$ a linear subspace N of $N_+ \oplus N_-$. The properties of N_+ and N_- are further exploited in the following analysis.

10.2.7 Definition. For symmetric L_0 define

$$\langle f, g \rangle = (L_0^* f, g) - (f, L_0^* g) \qquad (f, g \in D(L_0^*)).$$

Thus $\langle \cdot, \cdot \rangle$ is a bilinear form on $D(L_0^*) \times D(L_0^*)$, and $\langle f, g \rangle = -\overline{\langle g, f \rangle}$.

10.2.8 Example. Consider the interval $[0, 1]$ and let $l = i\,d/dx$. Recall from Example 6.7.2 that if $\mathscr{A}$ denotes the set of absolutely continuous functions with $f' \in \mathscr{L}_2(0, 1)$, and $Lf = lf$ on $D(L)$ where

$$D(L) = \{f : f \in \mathscr{A}, f(0) = f(1) = 0\},$$

then $D(L^*) = \mathscr{A}$. For $g \in \mathscr{A}$,

$$\langle f, g \rangle = (L^*f, g) - (f, L^*g) = i[f(1)\bar{g}(1) - f(0)\,\bar{g}(0)]. \tag{10.2.2}$$

The analysis in Example 6.7.9 shows that $\langle \cdot , \cdot \rangle$ is of central importance in a discussion of self-adjointness. The domain of a self-adjoint extension of L is obtained by restricting $D(L^*)$ by a boundary condition of the form $\beta f(1) - \alpha f(0) = 0$. It is clear from (10.2.2) that this may alternatively be written $\langle f, g \rangle = 0$ if $\bar{g}(1) = \beta$, $\bar{g}(0) = \alpha$. Thus in the following abstract theory such a condition may be regarded as a generalized boundary condition, and the aim will be to find conditions of this type on $D(L^*)$ which yield the domain of a self-adjoint extension of L.

10.2.9 Lemma. *Let L_0 be symmetric. A vector f in $D(L_0^*)$ lies in $D(\bar{L}_0)$ if and only if one of the following equivalent conditions holds:*

(i) $\langle f, g \rangle = 0 \qquad (g \in D(L_0^*))$; (10.2.3)

(ii) $\langle f, \phi \rangle = 0 \qquad (\phi \in N_+ \oplus N_-)$. (10.2.4)

If $N_+ \oplus N_-$ is finite dimensional with basis $\phi_1, \ldots, \phi_n$, these conditions are equivalent to

(iii) $\langle f, \phi_i \rangle = 0 \qquad (i = 1, \ldots, n)$ (10.2.5)

Proof. Set $\bar{L}_0 = L$. By Lemma 10.2.2, $L_0^* = L^*$, and $D(L^{**}) = D(L) \subset D(L^*)$. Hence from the definition of the adjoint, an f in $D(L^*)$ lies in $D(L^{**}) = D(L)$ if and only if $\langle f, g \rangle = 0$ for all $g \in D(L^*)$. This proves that (i) is necessary and sufficient, and the fact that (ii) is necessary is an obvious consequence. To prove sufficiency note that by Lemma 10.2.5 every $g \in D(L^*)$ can be written as $g = g_0 + \phi$ where $g_0 \in D(L)$ and $\phi \in N_+ \oplus N_-$. Then

$$\langle f, g \rangle = \langle f, g_0 \rangle + \langle f, \phi \rangle.$$

By (i), $\langle f, g_0 \rangle = 0$, and $\langle f, \phi \rangle = 0$ by assumption. Thus $\langle f, g \rangle = 0$ for all $g \in D(L^*)$, and the result follows from (i). (iii) is an obvious consequence when linearity is used. □

This result provides a description of the domain of the closure of a symmetric operator. For our present purpose the most significant point is that $\langle \cdot , g \rangle$ vanishes on $D(\bar{L}_0)$. The following additional terminology will be useful in the analysis, but will not appear in the final result.

10.2.10 Definition. Let N be a linear subspace of $N_+ \oplus N_-$. The **adjoint** N^* of N is the linear subspace consisting of all vectors f in $N_+ \oplus N_-$ such that $\langle f, g\rangle = 0$ $(g \in N)$. N is **symmetric** iff $N \subset N^*$ (equivalently iff $\langle f, g\rangle = 0$ for all $f, g \in N$), and **self-adjoint** iff $N = N^*$.

10.2.11 Lemma. *Let L_0 be symmetric, and let L_1 be a linear operator such that $\bar{L}_0 \subset L_1 \subset L_0^*$. Then $D(L_1) = D(\bar{L}_0) \oplus N$ where N is a linear subspace of $N_+ \oplus N_-$, and $D(L_1^*) = D(\bar{L}_0) \oplus N^*$. L_1 is symmetric iff N is symmetric, and self-adjoint iff N is self-adjoint.*

Proof. Set $L = \bar{L}_0$. For any $f \in D(L_1)$, $g \in D(L^*)$,

$$\langle f, g\rangle = (L^*f, g) - (f, L^*g) = (L_1 f, g) - (f, L^*g).$$

Since $D(L) \subset D(L_1^*) \subset D(L^*)$, from the definition of L_1^* a necessary and sufficient condition for any $g \in D(L^*)$ to be in $D(L_1^*)$ is that $\langle f, g\rangle = 0$ for all $f \in D(L_1)$.

By Corollary 10.2.6, $D(L_1) = D(L) \oplus N$, $D(L_1^*) = D(L) \oplus P$ where N, P are linear subspaces of $N_+ \oplus N_-$. Now Lemma 10.2.5 shows that for any $f \in D(L_1)$, $g \in D(L^*)$, $f = f_0 + \phi$, $g = g_0 + \psi$ for some $f_0, g_0 \in D(L)$, $\phi, \psi \in N_+ \oplus N_-$, and it is an easy deduction from Lemma 10.2.9 that $\langle f, g\rangle = 0$ iff $\langle \phi, \psi\rangle = 0$. Thus in view of what was proved in the previous paragraph, $g \in D(L_1^*)$ iff $\langle \phi, \psi\rangle = 0$ for all $\phi \in N$, and by the definition of N^* this happens iff $\psi \in N^*$. Therefore $P = N^*$. The final statement of the lemma follows from the definitions of symmetry and self-adjointness. □

Questions about the symmetry and self-adjointness of extensions of L_0 are thus equivalent to the same questions about linear subspaces of $N_+ \oplus N_-$. This is a striking simplification if L_0 has finite deficiency indices, for the problem may be resolved by elementary finite dimensional methods. We shall see that for symmetric operators arising from formal ordinary differential operators, $N_\pm$ consist respectively of the $\mathscr{L}_2$ solutions of $lf = \pm if$, and $N_\pm$ are thus certainly finite dimensional. The theory will therefore proceed on this assumption. A simple criterion for deciding whether a given symmetric operator has self-adjoint extensions will next be derived.

10.2.12 Lemma. *Suppose $\phi, \psi \in N_+ \oplus N_-$, and let $\phi = \phi_+ + \phi_-$, $\psi = \psi_+ + \psi_-$ where $\phi_\pm, \psi_\pm \in N_\pm$ respectively. Then*

$$\langle \phi, \psi\rangle = 2i[(\phi_+, \psi_+) - (\phi_-, \psi_-)], \tag{10.2.6}$$

$$\langle \phi, \phi\rangle = 2i[\|\phi_+\|^2 - \|\phi_-\|^2]. \tag{10.2.7}$$

Proof. By definition $L^*\phi_\pm = \pm i\phi_\pm$, $L^*\psi_\pm = \pm i\psi_\pm$. Hence

$$\begin{aligned}\langle\phi, \psi\rangle &= (L^*\phi, \psi) - (\phi, L^*\psi)\\ &= (L^*[\phi_+ + \phi_-], \psi_+ + \psi_-) - (\phi_+ + \phi_-, L^*[\psi_+ + \psi_-])\\ &= i(\phi_+ - \phi_-, \psi_+ + \psi_-) + i(\phi_+ + \phi_-, \psi_+ - \psi_-)\\ &= 2i[(\phi_+, \psi_+) - (\phi_-, \psi_-)].\end{aligned}$$ □

10.2.13 Corollary. *Let $N \subset N_+ \oplus N_-$ be symmetric, and for any $\phi \in N$ set $\phi = \phi_+ + \phi_-$ where $\phi_\pm \in N_\pm$ respectively. Then $\|\phi_+\| = \|\phi_-\|$.*

Proof. From Definition 10.2.10, $\langle\phi, \phi\rangle = 0$. The result follows from (10.2.7). □

10.2.14 Lemma. *Suppose L_0 has finite deficiency indices $n_\pm$. If N is a linear subspace of $N_+ \oplus N_-$ of dimension n, N^* has dimension $n_+ + n_- - n$.*

Proof. Suppose $\phi \in N_+ \oplus N_-$ is such that $\langle f, \phi\rangle = 0$ for all $f \in N_+ \oplus N_-$. Set $\phi = \phi_+ + \phi_-$ where $\phi_\pm \in N_\pm$ respectively. Then by taking $f = \phi_+, \phi_-$ in turn and using (10.2.6) it follows that $\phi_+ = \phi_- = 0$. That is $\phi = 0$.

Now let $\phi_1, \ldots, \phi_n$ be a basis of N, and define linear functionals $\phi_1^*, \ldots, \phi_n^*$ on $N_+ \oplus N_-$ by setting $\phi_i^*(f) = \langle f, \phi_i\rangle$. Then

$$(\Sigma\alpha_i\phi_i^*)(f) = \langle f, \Sigma\bar{\alpha}_i\phi_i\rangle \qquad (f \in N_+ \oplus N_-, \alpha_i \in \mathbb{C}),$$

and by the first result of the proof the ϕ_i^* are linearly independent. But N^* is the orthogonal complement of the $\{\phi_i^*\}$, and the result follows from a standard theorem on finite dimensional vector spaces (Halmos, 1948; p. 26). □

10.2.15 Theorem. *A symmetric operator with finite deficiency indices has a self-adjoint extension if and only if its deficiency indices are equal.*

Proof. In view of Lemma 10.2.11 it is sufficient to show that there is a self-adjoint linear subspace of $N_+ \oplus N_-$ iff $n_+ = n_-$. The proof is based on an examination of the relation between a symmetric subspace N of $N_+ \oplus N_-$ and N_+, N_-. The clue is provided by Corollary 10.2.13 which shows that every non-zero element in N is the sum of a pair of elements, *neither of which can be zero*, of N_+ and N_-. This suggests the form which must be taken by elements of a basis of N.

Suppose first that $n_+ = n_- = n$, say. Let $\{\phi_{i\pm}\}_1^n$ be orthonormal bases of $N_\pm$ respectively, and define $\phi_i = \phi_{i+} + \phi_{i-}$. Let N be the linear span of the (obviously linearly independent) vectors $\phi_1, \ldots, \phi_n$. For any $f, g \in N$

we can write $f = \Sigma\alpha_i\phi_i$, $g = \Sigma\beta_i\phi_i$ for some $\alpha_i, \beta_i \in \mathbb{C}$, and from (10.2.6),

$$\begin{aligned}\langle f, g\rangle &= \left\langle \sum_i \alpha_i[\phi_{i+} + \phi_{i-}], \sum_j \beta_j[\phi_{j+} + \phi_{j-}]\right\rangle \\ &= 2i\sum_i\sum_j \alpha_i\bar{\beta}_j[(\phi_{i+}, \phi_{j+}) - (\phi_{i-}, \phi_{j-})] \\ &= 0\end{aligned}$$

by orthonormality. Hence $\langle f, g\rangle = 0$, N is symmetric, and $N \subset N^*$. By Lemma 10.2.14, $\dim N^* = n_+ + n_- - n = n = \dim N$. Therefore $N = N^*$ and N is self-adjoint.

Suppose on the other hand that N is self-adjoint. Let $\phi_1, \ldots, \phi_k$ be a basis of N, and write $\phi_i = \phi_{i+} + \phi_{i-}$ where $\phi_{i\pm} \in N_\pm$ respectively. The set $\{\phi_{i+}\}$ is linearly independent. For if not there are scalars $\alpha_1, \ldots, \alpha_k$ not all zero such that $\Sigma\alpha_i\phi_{i+} = 0$, in which case $\Sigma\alpha_i\phi_i = \Sigma\alpha_i\phi_{i-}$. Since the left-hand side is in the symmetric linear subspace N, $\Sigma\alpha_i\phi_{i-} = 0$ by Corollary 10.2.13. Therefore $\Sigma\alpha_i\phi_i = 0$, and the ϕ_i are linearly dependent contrary to assumption. Similarly $\{\phi_{i-}\}$ is linearly independent. It follows that $k \leqslant \min(n_+, n_-)$. However, as N is self-adjoint $2k = n_+ + n_-$ (Lemma 10.2.14), and this is inconsistent with the previous inequality unless $n_+ = n_-$. □

This theorem neatly settles the question of the existence of self-adjoint extensions. The next step is to enquire how these extensions may be found. The construction most suitable for applications to ordinary differential equations avoids as far as possible the use of the detailed properties of the functions in N_+ and N_-, and is most conveniently stated in terms of the following concept.

10.2.16 Definition. A set of functions $f_1, \ldots, f_k$ in $D(L_0^*)$ will be said to be **linearly independent relative** to $D(L_0)$ iff in the representation $f_j = h_j + \phi_j$ where $h_j \in D(\bar{L}_0)$, $\phi_j \in N_+ \oplus N_-$, the ϕ_j are linearly independent.

10.2.17 Lemma. *The functions $f_1, \ldots, f_k$ in $D(L_0^*)$ are linearly independent relative to $D(L_0)$ if and only if there exist $g_1, \ldots, g_k$ in $D(L_0^*)$ such that* $\det(\langle f_i, g_j\rangle) \neq 0$. *Then $k \leqslant n_+ + n_-$.*

Proof. With the notation of the definition define linear functionals ϕ_j^* on $D(L_0^*)$ by setting $\phi_j^*(f) = \langle f, \phi_j\rangle$; as noted in the proof of Lemma 10.2.14 the ϕ_j^* are linearly independent if and only if the same is true of the ϕ_j. The necessity of the condition is obvious on choosing a dual basis (Halmos, 1948; p. 22) for the linear span of the ϕ_j^* and recalling that $\langle\,\cdot\,, g\rangle$ vanishes on $D(\bar{L}_0)$. The sufficiency is left as an exercise. □

10.2.18 Theorem. *Let L_0 be symmetric and have finite deficiency indices $n_+ = n_- = n$, say. If $n = 0$ the closure of L_0 is its unique self-adjoint extension. If $n \neq 0$ suppose that $f_1, \ldots, f_n \in D(L_0^*)$ are linearly independent relative to $D(L_0)$ and satisfy*

$$\langle f_i, f_j \rangle = 0 \qquad (i, j = 1, \ldots, n). \tag{10.2.8}$$

Let $\mathcal{M}$ be the linear subspace of $\mathcal{H}$ consisting of all $f \in D(L_0^*)$ such that

$$\langle f, f_i \rangle = 0 \qquad (i = 1, \ldots, n). \tag{10.2.9}$$

Then $\mathcal{M}$ is the domain of a self-adjoint extension M of L_0 given by $Mf = L_0^* f$ for $f \in \mathcal{M}$.

Conversely, let M be a self-adjoint extension of L_0, and suppose that $h_1, \ldots, h_{2n} \in D(L_0^*)$ are linearly independent relative to $D(L_0)$. Then there are vectors $f_1, \ldots, f_n$ in the linear span of the h_i which satisfy (10.2.8) and are such that the linear subspace $\mathcal{M}$ defined by (10.2.9) is the domain of M.

Proof. Set $f_i = g_i + \phi_i$ where $g_i \in D(\bar{L}_0)$, $\phi_i \in N_+ \oplus N_-$. The conditions (10.2.8) ensure that the linear span, N say, of the ϕ_i is symmetric, so that $N \subset N^*$. By Lemma 10.2.14 the dimensions of N and N^* are equal. It follows that $N = N^*$ and N is self-adjoint. As $\mathcal{M} = D(\bar{L}_0) \oplus N$, the restriction M of L_0^* to $\mathcal{M}$ is self-adjoint (Lemma 10.2.11). Conversely if M is a self-adjoint extension of L_0, $D(M) = D(\bar{L}_0) \oplus N$ where N is self-adjoint, and the result follows on choosing a basis for N. □

Theorems 10.2.15 and 10.2.18 together with the simple criterion for linear independence relative to $D(L_0)$ of Lemma 10.2.17 provide a convenient theoretical basis for answering questions concerning the existence and construction of self-adjoint extensions. Before these results can be applied to differential operators, a concrete representation of L_0, L_0^* and the bilinear form $\langle \cdot, \cdot \rangle$ must be calculated.

10.3 Formal Ordinary Differential Operators: Preliminaries

For simplicity the analysis is restricted to the most useful case, that of second order operators. Consider the formal ordinary differential operator l, where

$$lf = (pf')' + qf,$$

acting on functions defined on an interval Ω with end points a, b $(a < b)$ either or both of which may be finite or infinite; the question of whether a and b are contained in Ω will shortly be answered. It will be assumed throughout that $p, q \in \mathscr{C}^{\infty}((a, b))$ and that $p(x) \neq 0$ for $x \in (a, b)$. The finiteness of a and b and the vanishing or otherwise of p at these points will have a critical effect on the analysis.

10.3.1 Definition. The point a is said to be a **regular end point** iff a is finite, $p(a) \neq 0$, and $p, q \in \mathscr{C}^{\infty}([a, b))$; a is **singular** iff it is not regular. The terminology for b is analogous. The formal operator l is called **regular** iff both a and b are regular, otherwise it is said to be **singular**. The interval Ω with end points a, b is defined by the requirement that it should contain a iff a is regular, and b iff b is regular. Thus Ω can be any one of $[a, b]$, $[a, b)$, $(a, b]$ or (a, b). For example $\Omega = [a, b)$ if a is regular and b is singular.

In order to obtain from l operators in $\mathscr{L}_2(\Omega)$ which are symmetric, the coefficients of l will be required to satisfy an additional condition.

10.3.2 Definition. The formal operator l^* where

$$l^*g = (\bar{p}g')' + \bar{q}g$$

is called the **formal adjoint** of l. l is said to be **formally self-adjoint** iff $l = l^*$.

It is obvious that l is formally self-adjoint if and only if p and q are real-valued. There is again an unfortunate clash with the classical terminology in which "formal self-adjointness" is just "self-adjointness". It will be seen shortly that the symmetry is essentially a consequence of the following lemma obtained by elementary manipulation.

10.3.3 Lemma. *If l is formally self-adjoint, for all f, g for which the following make sense,*

$$\bar{g}lf - f\overline{lg} = \frac{\mathrm{d}}{\mathrm{d}x}[f, g], \tag{10.3.1}$$

where

$$[f, g] = p(f'\bar{g} - f\bar{g}'). \tag{10.3.2}$$

If $[\alpha, \beta]$ is a finite subinterval of Ω, Green's formula holds:

$$\int_{\alpha}^{\beta} (\bar{g}lf - f\overline{lg})\,\mathrm{d}x = [f, g]_{\alpha}^{\beta}, \tag{10.3.3}$$

where

$$[f, g]_{\alpha}^{\beta} = [f, g](\beta) - [f, g](\alpha). \tag{10.3.4}$$

The conditions on the formal operator l which will be dealt with throughout are collected together as follows; the generality allowed for l is sufficient for almost all applications.

10.3.4 Definition. Let Ω be as in Definition 10.3.1, and suppose that p, q are real-valued functions in $\mathscr{C}^{\infty}(\Omega)$ with $p(x) \neq 0$ for $x \in \Omega$. Let l be the formal operator defined by

$$lf = (pf')' + qf. \tag{10.3.5}$$

For a proof of the basic existence theorem see Dunford and Schwartz (1963, p. 1281).

10.3.5 Theorem. *Assume that $g \in \mathscr{L}_1^{\text{loc}}(\Omega)$, take any $x_0 \in \Omega$, and let c_0, c_1, λ be arbitrary complex numbers. Then there exists a unique function f on Ω with f' absolutely continuous and with*

$$lf(x) + \lambda f(x) = g(x) \qquad \text{(a.e.)},$$

$$f(x_0) = c_0, \qquad f'(x_0) = c_1.$$

If $g \in \mathscr{C}^j(\Omega)$, $f \in \mathscr{C}^{j+2}(\Omega)$.

10.4 Symmetric Operators Associated with Formal Ordinary Differential Operators

Before the theory of Section 10.2 can be applied, three quantities must be calculated: a symmetric L_0 based on the formal operator l, the adjoint of L_0, and a reasonably simple expression for the bilinear form $\langle \cdot, \cdot \rangle$. For purposes of illustration the easier case of l regular is first tackled. The simplifying factor here is that no difficulties connected with the end points of Ω arise.

The analysis will throughout be in the complex Hilbert space $\mathscr{L}_2(\Omega)$. The definitions which follow are modelled on the first order case treated in Example 6.7.2. We first define $\mathscr{A}$ to be the largest set on which lf makes sense and lies in $\mathscr{L}_2(\Omega)$; if l is regular the condition $lf \in \mathscr{L}_2(\Omega)$ is equivalent to $f'' \in \mathscr{L}_2(\Omega)$, but for singular l, $\mathscr{A}$ depends on the coefficients of l. Next operators L and L' are defined and it is proved that L is symmetric and that L' is the adjoint of L.

10.4.1 Definition. Let l be the second order formal operator of Definition 10.3.4. Denote by $\mathscr{A}$ the set of functions f in $\mathscr{L}_2(\Omega)$ with f' absolutely continuous and with $lf \in \mathscr{L}_2(\Omega)$.

10.4.2 Definition. For l regular set

$$D(L) = \{f : f \in \mathscr{A}, f(a) = f'(a) = f(b) = f'(b) = 0\},$$

$$D(L') = \mathscr{A},$$

and let $Lf = lf$, $L'f = lf$ on $D(L)$, $D(L')$ respectively. It will occasionally be necessary to consider a subinterval Δ of Ω, and L_Δ, L'_Δ will denote the corresponding operators for Δ.

10.4.3 Lemma. *If l is regular, then $D(L)$ is dense in $\mathscr{L}_2(\Omega)$, $L \subset L'$, and*

$$(L'f, g) - (f, L'g) = [f, g]_a^b \qquad (f, g \in D(L')), \tag{10.4.1}$$

$$(Lf, g) = (f, L'g) \qquad (f \in D(L), g \in D(L')). \tag{10.4.2}$$

Proof. Since $D(L) \supset \mathscr{C}_0^\infty(\Omega)$ the first statement follows from Theorem 2.5.6. Equation (10.4.1) is just Green's formula (10.3.4). If $f \in D(L)$, f and f' vanish at a and b, whence $[f, g]_a^b = 0$. This yields (10.4.2). □

Relations (10.4.1) and (10.4.2) are direct consequences of the formal self-adjointness of l, and supply an essential connecting link between the formal operator l and the operators L, L' on $\mathscr{L}_2(\Omega)$. The symmetry of L will follow from (10.4.2), and (10.4.1) will provide a simple representation for the bilinear form $\langle \cdot, \cdot \rangle$. Two technical results are needed as preparation for the main theorem.

10.4.4 Lemma. *If l is regular, $N(L')^\perp = R(L)$ and $\mathscr{L}_2(\Omega) = R(L) \oplus N(L')$.*

Proof. From the Existence Theorem 10.3.5, if $g \in \mathscr{L}_2(\Omega)$ there is a unique solution, f say, of $lf = g$ with f' absolutely continuous and $f(a) = f'(a) = 0$; $f \in \mathscr{A}$ since $lf = g \in \mathscr{L}_2(\Omega)$, which proves that $L'f = g$ has a solution satisfying the above boundary conditions. Theorem 10.3.5 also shows that $N(L')$ has a basis $\{k_1, k_2\}$ such that $k_1(b) = 1$, $k_1'(b) = 0$, $k_2(b) = 0$, $k_2'(b) = 1$. Noting the boundary conditions on f and using (10.4.1) we deduce that for $j = 1, 2$,

$$(g, k_j) = (L'f, k_j) = [f, k_j]_a^b = (-1)^{j+1} p(b) f^{(2-j)}(b).$$

Since l is regular, $p(b) \neq 0$. Therefore $(g, k_j) = 0$ for $j = 1, 2$ iff $f(b) = f'(b) = 0$. In view of the definition of $D(L)$, this is equivalent to the statement that $f \in D(L)$ iff $g \in N(L')^\perp$. That is $N(L')^\perp = R(L)$. The final result follows as $N(L')$ is closed. □

10.4.5 Lemma. *Take any $\alpha, \beta, \gamma, \delta \in \mathbb{C}$. Then there is an $f \in \mathscr{C}^\infty([a, b])$ such that $f(a) = \alpha$, $f'(a) = \beta$, $f(b) = \gamma$, $f'(b) = \delta$.*

Proof. A polynomial with these properties may obviously be constructed. □

10.4.6 Theorem. *Suppose that l is regular. Then L is a closed symmetric operator with adjoint $L^* = L'$, and $L'^* = L$.*

Proof. Since $L \subset L'$, the symmetry of L is a consequence of (10.4.2), and this equation then shows also that $L^* \supset L'$. It will follow that $L' = L^*$ if the opposite inclusion can be proved. Choose any $g \in D(L^*)$ and set $L^*g = h$. By Theorem 10.3.5 there is a solution f of $L'f = h$, and for any $k \in D(L)$, $(k, h) = (k, L'f) = (Lk, f)$ by (10.4.2). But $(k, h) = (k, L^*g) = (Lk, g)$ from the definition of the adjoint, and subtraction gives $(Lk, f - g) = 0$. Therefore $f - g \in R(L)^{\perp}$, whence $f - g \in N(L')$ (Lemma 10.4.4). Since $f \in D(L')$, also $g \in D(L')$, and as g is an arbitrary element of $D(L^*)$ this proves that $D(L^*) \subset D(L')$ and that $L'g = L'f = h = L^*g$. That is $L^* \subset L'$. Hence $L' = L^*$.

If it can be proved that $L'^* = L$, it will follow from Theorem 6.7.4 that L is closed. Now $\bar{L} = L^{**}$ by Lemma 10.2.2, and since $L^* = L'$, $\bar{L} = L'^*$ and $L'^* \supset L$. Thus it is enough to prove that $L'^* \subset L$. Since $L \subset L'$, $L' = L^* \supset L'^*$ and

$$(L'g, f) = (g, L'^*f) = (g, L'f) \qquad (f \in D(L'^*), g \in D(L')).$$

Hence from (10.4.1), $[g, f]_a^b = 0$. But by Lemma 10.4.5 there is a $g \in D(L')$ with $g'(a) = 1$ and $g(a) = g(b) = g'(b) = 0$, and it follows on substitution in the last relation that $f(a) = 0$. Similarly also $f'(a) = f(b) = f'(b) = 0$. Thus $f \in D(L)$. □

We have thus found a closed symmetric operator L and its adjoint $L^* = L'$, and the bilinear form $\langle \cdot, \cdot \rangle$ of Definition 10.2.7 is known from Lemma 10.4.3 to be $[\cdot, \cdot]_a^b$. The preparations for the application of the theory of symmetric extensions for regular l are therefore complete.

For singular l a direct generalization of the argument above proves to be awkward. The principal reason for this is that the domain of the basic closed symmetric operator arising from l turns out to depend both on the coefficients of l and on Ω and cannot be readily specified as in Definition 10.4.2. The difficulties are connected with the end points of Ω, and the tactics will be to choose the domain of a symmetric L_0 to consist only of functions which are zero on neighbourhoods of each end point of Ω. L_0 is not closed, but the theory of Section 10.2 allow for just this contingency, and $\bar{L}_0$ and its symmetric extensions are obtained by restricting $D(L_0^*)$ by conditions analogous to boundary conditions.

10.4.7 Definition. Let l be regular or singular. A function f on Ω belongs to the set $\mathscr{A}_0$ iff both the following hold:

(i) f' is absolutely continuous and $f'' \in \mathscr{L}_2(\Omega)$;
(ii) f has compact support contained in the interior of Ω.

Evidently $\mathscr{A}_0 \subset \mathscr{A}$. A simple but useful observation is that if $f \in \mathscr{A}_0$, there exist α, β with $a < \alpha < \beta < b$ such that the support of f is contained in $[\alpha, \beta]$, and hence with $f(\alpha) = f'(\alpha) = f(\beta) = f'(\beta) = 0$.

10.4.8 Definition. For l regular or singular set $D(L_0) = \mathscr{A}_0$, $D(L') = \mathscr{A}$, and let $L_0 f = lf$, $L'f = lf$ on $D(L_0)$, $D(L')$ respectively.

Note that in contrast with the regular case, $D(L')$ depends on l if l is singular. An analogue of Lemma 10.4.3 connecting l with L_0 and L' is needed. A little more care is now necessary because of difficulties with the interpretation of $[f, g]_a^b$, and we start with that part of the result which does not involve this quantity.

10.4.9 Lemma. *Let l be regular or singular. Then $L_0 \subset L'$ and*

$$(L_0 f, g) = (f, L'g) \qquad (f \in D(L_0),\, g \in D(L')). \tag{10.4.3}$$

L_0 is symmetric.

Proof. $D(L_0)$ contains $\mathscr{C}_0^\infty(\Omega)$ and is therefore dense (Theorem 2.5.6). Suppose $f \in D(L_0)$ and let $[\alpha, \beta]$ be a compact interval containing the support of f and contained in the interior of Ω. Then f, f' are zero at α, β, and the term $[f, g]_\alpha^\beta$ in Green's formula (10.3.3) is zero. Equation (10.4.3) follows, and the symmetry of L_0 is an easy consequence. □

10.4.10 Definition. For l regular or singular set $L = \bar{L}_0$.

That this is consistent with Definition 10.4.2 is a consequence of the next result, since in each case L is closed with adjoint L'.

10.4.11 Theorem. *Let l be a regular or singular formally self-adjoint operator satisfying the conditions of Definition 10.3.4. Then L_0 is symmetric. L is a (closed) symmetric operator with adjoint $L^* = L_0^* = L'$, and $L'^* = L = \bar{L}_0$.*

Proof. Since L is by definition the closure of L_0, only the assertion $L_0^* = L'$ requires proof. From (10.4.3) and the definition of the adjoint, $L' \subset L_0^*$, so it is enough to show that $L' \supset L_0^*$. The idea of the proof is to note that l is regular on any compact interval Δ contained in the interior of Ω, and to use the results for the regular case. Let then L_Δ, L'_Δ be the operators of Definition 10.4.2 on the Hilbert space $\mathscr{L}_2(\Delta)$ with inner product $(\cdot, \cdot)_\Delta$. For f defined on Ω, f_Δ will denote the restriction of f to Δ.

Take any $f \in D(L_0^*)$. By the definition of the adjoint there is a $g \in \mathscr{L}_2(\Omega)$ such that

$$(L_0 h, f) = (h, g) \qquad (h \in D(L_0)). \tag{10.4.4}$$

Choose any fixed Δ and let h have support in Δ. Then h, h' are zero at the end points of Δ, whence $h \in D(L_\Delta)$, and (10.4.4) becomes

$$(L_\Delta h, f_\Delta)_\Delta = (h, g_\Delta)_\Delta.$$

By the definition of the adjoint, $f_\Delta \in D(L_\Delta^*)$ and $L_\Delta^* f_\Delta = g_\Delta$. But $L_\Delta^* = L'_\Delta$ (Theorem 10.4.6). Therefore $f_\Delta \in D(L'_\Delta)$, and as Δ is arbitrary f' is absolutely continuous on Ω. Also $L'_\Delta f_\Delta = g_\Delta$ in $\mathscr{L}_2(\Delta)$ and so almost everywhere. Hence $(lf)_\Delta = g_\Delta$ a.e., and as Δ is arbitrary $lf = g$ a.e. That is $lf = g \in \mathscr{L}_2(\Omega)$, and so $f \in \mathscr{A} = D(L')$. □

The last requirement in the singular case is an explicit expression for the bilinear form $\langle \cdot\,, \cdot \rangle$ of Definition 10.2.7, and this is obtained by generalizing (10.4.1). That the next definition is reasonable is a consequence of the lemma which follows it.

10.4.12 Definition. Let l be regular or singular. For $f, g \in D(L')$ set

$$[f, g]^b = \lim_{x \to b_-} [f, g](x),$$

$$[f, g]_a = \lim_{x \to a_+} [f, g](x),$$

$$[f, g]_a^b = [f, g]^b - [f, g]_a.$$

10.4.13 Lemma. *For l regular or singular the above limits exist and are finite if $f, g \in D(L')$, and then*

$$\langle f, g \rangle = (L'f, g) - (f, L'g) = [f, g]_a^b.$$

Proof. If $a < \alpha < x < b$, by Green's formula (10.4.3)

$$\int_\alpha^x (\bar{g} lf - f l\bar{g})\, dy = [f, g](x) - [f, g](\alpha).$$

From the definition of $D(L')$, each of $f, g, lf, lg \in \mathscr{L}_2(\Omega)$. Hence the limit as $x \to b_-$ of the integral exists and is finite. □

10.4.14 Definition. Let l be regular or singular. Suppose that $g \in D(L')$ is given. For $f \in D(L')$ the relation $[f, g]_a^b = 0$ will be called a **boundary condition**. This terminology is natural since the boundary conditions are obviously of the usual type when l is regular.

10.4.15 Lemma. *Let l be the formally self-adjoint operator of Definition 10.3.4. If l is regular, the deficiency indices of L_0 are* (2, 2). *If l is singular the deficiency indices are* (m, m) *where* $m = 0, 1$, *or* 2.

Let $\phi_1, \ldots, \phi_{2m}$ be any basis of $N_+ \oplus N_-$. A function f in $D(L')$ lies in $D(L)$ iff it satisfies the boundary conditions

$$[f, \phi_i]_a^b = 0 \ (i = 1, \ldots, 2m).$$

Proof. Since the coefficients of l are real, $l\phi = i\phi$ iff $l\bar{\phi} = -i\bar{\phi}$. Hence $n_+ = n_-$. As l is second order there are at most two linearly independent solutions of $l\phi = i\phi$. If l is regular both solutions are in $\mathscr{L}_2(\Omega)$. $D(L)$ is deduced from Lemmas 10.4.13 and 10.2.9. □

Examples will be constructed later of singular l where $m = 0, 1$, or 2. Thus the number of boundary conditions needed to define $D(L)$ depends on the detailed behaviour of the coefficients of l and on Ω, and the nature of these conditions is more complex. Of course if l is regular the lemma yields the four boundary conditions of Definition 10.4.2.

10.5 The Construction of Self-adjoint Extensions

The preparations are now complete for the application of the theory to obtain self-adjoint extensions for formally self-adjoint second order l. The final result of the previous section asserts that such extensions always exist, and from Theorem 10.2.18 their domain is obtained by restricting $D(L')$ by exactly m conditions where (m, m) are the deficiency indices of L_0; by Lemma 10.4.13 these conditions are the boundary conditions of Definition 10.4.14. If l is regular the conditions involve linear combinations of the function and its first derivative evaluated at the end points and are of familiar form. If l is singular but has one regular end point, a say, in some cases the boundary conditions need only be applied at a and are no more complicated than when l is regular. However, in general conditions at singular end points must be considered, and these will involve limits rather than function values.

The technicalities which arise in applying Theorem 10.2.18 are much reduced if a simple choice of the functions $f_1, \ldots, f_n$ is made, and it is at this point that Lemma 10.2.17 is useful. The following simple example illustrates the method for a first order operator.

10.5.1 Example. Take $\Omega = [0, 1]$ and $l = i\,d/dx$, and let $\mathscr{A}$ be the set of absolutely continuous functions with derivatives in $\mathscr{L}_2(0, 1)$. If $D(L) = \{f : f \in \mathscr{A}, f(0) = f(1) = 0\}$ and $Lf = lf$ on $D(L)$, then $D(L^*) = \mathscr{A}$. The solutions of $lf = \pm if$ are $\exp(\pm x)$ respectively and are both in $D(L^*)$. Therefore the deficiency indices are (1, 1) and L has self-adjoint extensions.

The construction of Theorem 10.2.18 is used to find every such extension, M say.

Since $n = 1$, $D(M)$ is the set of functions f in $D(L^*)$ satisfying $\langle f, f_1 \rangle = 0$ where f_1 is some function linearly independent relative to $D(L)$ with $\langle f_1, f_1 \rangle = 0$, while on the other hand every such choice of f_1 leads to a self-adjoint M. From the expression for $\langle \cdot, \cdot \rangle$ given in Example 10.2.8, these conditions are respectively

$$f(1)\bar{f}_1(1) - f(0)\bar{f}_1(0) = 0, \qquad f_1(1)\bar{f}_1(1) - f_1(0)\bar{f}_1(0) = 0. \tag{10.5.1}$$

A simple choice for h_1, h_2 is $h_1(x) = x$, $h_2(x) = 1 - x$, since $h_1(0) = 0$ and $h_2(1) = 0$; a brief calculation gives $\langle h_1, h_1 \rangle = -\langle h_2, h_2 \rangle = -1$ and $\langle h_1, h_2 \rangle = 0$, whence from Lemma 10.2.17 h_1 and h_2 are linearly independent relative to $D(L)$. Therefore $f_1 = \bar{\alpha} h_1 + \bar{\beta} h_2$ for some $\alpha, \beta \in \mathbb{C}$. Since f_1 must satisfy (10.5.1), $|\alpha| = |\beta|$. Thus $\alpha/\beta = e^{i\theta}$ for some $\theta \in \mathbb{R}$, and the boundary condition yielding a self-adjoint extension must be of the form $f(1) = e^{i\theta} f(0)$ for some $\theta \in \mathbb{R}$. This confirms the conclusions reached formally in Example 6.7.9, and proves further that *every* self-adjoint extension is obtained in this manner.

The basic result on the construction of self-adjoint extensions for second order l is as follows, its proof being obtained simply by setting $\langle f, g \rangle = [f, g]_a^b$ in Theorem 10.2.18.

10.5.2 Theorem. *Let l be the regular or singular formally self-adjoint second order operator of Definition 10.3.4. Then L_0 has equal deficiency indices (m, m) with $m \leqslant 2$, and so L_0 has self-adjoint extensions.*

If $m = 0$, $\bar{L}_0 = L = L'$, and the closure of L_0 is its unique self-adjoint extension. For $m \neq 0$, let $f_1, \ldots, f_m$ be vectors in $D(L')$ which are linearly independent relative to $D(L_0)$ and which satisfy the conditions

$$[f_i, f_j]_a^b = 0 \qquad (i, j = 1, \ldots, m). \tag{10.5.2}$$

Let $\mathcal{M}$ be the linear subspace consisting of all $f \in D(L')$ such that

$$[f, f_i]_a^b = 0 \qquad (i = 1, \ldots, m). \tag{10.5.3}$$

Then $\mathcal{M}$ is the domain of a self-adjoint extension of L_0. Conversely every self-adjoint extension of L_0 is obtained in this manner, and the f_i may be chosen without loss of generality from among the linear combinations of any set of $2m$ functions in $D(L')$ which are linearly independent relative to $D(L)$.

The fact that L_0 has self-adjoint extensions is a consequence of the form assumed for l; there are formal operators (Example 10.5.5) for which no such extension exists. The regular case may now readily be disposed of.

10.5.3 Theorem. *Assume that l in Definition 10.3.4 is regular. For $i = 1, 2$ suppose that $\alpha_i, \alpha'_i, \beta_i, \beta'_i$ satisfy the following:*

(i) *The 4-vectors $(\alpha_i, \alpha'_i, \beta_i, \beta'_i)$ are linearly independent.*
(ii) $p(b)(\bar{\beta}'_i\beta_j - \bar{\beta}_i\beta'_j) - p(a)(\bar{\alpha}'_i\alpha_j - \bar{\alpha}_i\alpha'_j) = 0 \qquad (i, j = 1, 2). \qquad (10.5.4)$

Let $\mathcal{M}$ be the linear subspace of all functions $D(L')$ which satisfy the boundary conditions

$$p(b)(\beta'_i f'(b) - \beta_i f(b)) - p(a)(\alpha'_i f'(a) - \alpha_i f(a)) = 0 \qquad (i = 1, 2).$$

Then $\mathcal{M}$ is the domain of a self-adjoint extension of L. Conversely every self-adjoint extension of L may be obtained in this manner.

Proof. By Lemma 10.4.5 there are functions f_1, f_2 in $D(L')$ with $f_i(a) = \bar{\alpha}'_i$, $f'_i(a) = \bar{\alpha}_i$, $f_i(b) = \bar{\beta}'_i$, $f'_i(b) = \bar{\beta}_i$ for $i = 1, 2$. An application of Theorem 10.5.2 yields the result, for (i) gives the linear independence while (ii) is just (10.5.2). □

10.5.4 Example. Let l be regular. The possible boundary conditions are conveniently divided into two classes. The simpler class consists of *separated conditions* obtained when each condition involves only the values of f and f' at one end point. The conditions are otherwise said to be *mixed*.

Separated Conditions. Choose $\alpha_1 = \alpha'_1 = \beta_2 = \beta'_2 = 0$. Since l is regular, $p(b) \neq 0$. Hence (10.5.4) with $i = j = 1$ requires that $\bar{\beta}'_1\beta_1$ be real, and by (i) of the theorem both β_1 and β'_1 cannot be zero. A similar argument applies for α_2, α'_2, while (10.5.4) is identically satisfied if $i \neq j$. Thus the most general separated conditions are

$$f(b)\cos\theta - f'(b)\sin\theta = 0, \qquad f(a)\cos\phi - f'(a)\sin\phi = 0$$

for some real θ, ϕ.

Mixed Conditions. Consider for example conditions of the form $f(a) = \gamma f(b)$, $f'(a) = \delta f'(b)$ obtained by choosing $\alpha_1 = \alpha'_2 = 1$, $\beta_1 = \gamma p(a)/p(b)$, $\beta'_2 = \delta p(a)/p(b)$, $\alpha'_1 = \alpha_2 = \beta'_1 = \beta_2 = 0$. The only condition of the theorem that is not automatically satisfied is (ii) with $i \neq j$, and calculation shows that this holds iff $\gamma\bar{\delta} = p(b)/p(a)$. A familiar example is $l = -\mathrm{d}^2/\mathrm{d}x^2$ with periodic boundary conditions $f(a) = f(b)$, $f'(a) = f'(b)$.

The discussion of the more interesting singular case starts with a look at two simple examples.

10.5.5 Example. Take $l = i\mathrm{d}/\mathrm{d}x$; the modification of the theory needed for first order l will be obvious to the reader. The question was raised in Example 10.1.1 as to whether any self-adjoint operator can be constructed from l if $\Omega =$

$[0, \infty)$. An answer may now readily be supplied by checking on the dimensions of the deficiency spaces $N_{\pm}$. The solutions of $lf = \pm if$ are $\exp(\pm x)$ respectively, and since $\exp(-x)$ is in $\mathscr{L}_2(0, \infty)$ but $\exp(x)$ is not, the deficiency indices are (0, 1). It can immediately be concluded that L_0 has *no* self-adjoint extensions.

If $\Omega = (-\infty, \infty)$, $D(L')$ is the set of all absolutely continuous f with $f, f' \in \mathscr{L}_2(-\infty, \infty)$. Since neither of $\exp(\pm x)$ is in $\mathscr{L}_2(-\infty, \infty)$, the deficiency indices are (0, 0). Thus $\bar{L}_0 = L = L'$, and the closure of L_0 is itself self-adjoint—and indeed is the only self-adjoint extension of L_0. The classical Fourier transform will be derived later by an application of the Spectral Theorem to L'.

10.5.6 Example. The easiest second order formal operator is $l = -\mathrm{d}^2/\mathrm{d}x^2$. Since it is known from Theorem 10.5.2 that the deficiency indices are equal, it is sufficient to consider $-f'' = if$. This has solutions $\exp(\lambda_1 x)$, $\exp(\lambda_2 x)$ where $\lambda_1 = \exp(-i\pi/4)$, $\lambda_2 = \exp(-5i\pi/4)$.

Suppose first that $\Omega = (-\infty, \infty)$. Then neither of these solutions is in $\mathscr{L}_2(-\infty, \infty)$ and the deficiency indices are (0, 0). Thus the only self-adjoint extension of L_0 is its closure, and L' is itself self-adjoint.

If $\Omega = [0, \infty)$ the second solution is in $\mathscr{L}_2(0, \infty)$ but not the first. The deficiency indices are (1, 1), and self-adjoint extensions will be obtained by imposing one boundary condition.

It is clear that the situation is easiest if the deficiency indices are (0, 0), for then L' is itself self-adjoint and no boundary conditions have to be considered. In a number of important cases (of which $\Omega = [0, \infty)$ above is an example) one end point is regular. The following investigation shows that here also there will sometimes be simplifying factors.

10.5.7 Lemma. *Suppose a is a regular end point of the formal operator l in Definition 10.3.4. Then L_0 has deficiency indices* (1, 1) *or* (2, 2).

Proof. Take any c such that $a < c < b$, and let h_1, h_2 be real-valued functions in $\mathscr{C}^\infty(\Omega)$ with supports in $[a, c]$ and with $h_1(a) = 1$, $h_1'(a) = 0$, $h_2(a) = 0$, $h_2'(a) = 1$. Since a is regular, $h_1, h_2 \in D(L')$. As h_1, h_2 are zero on (c, b) it is easy to check that

$$\det([h_i, h_j]_a^b) = -p^2(a) \neq 0.$$

Thus by Lemma 10.2.17, h_1 and h_2 are linearly independent relative to $D(L_0)$ and $n_+ + n_- \geqslant 2$. □

This is *Weyl's Alternative*. The possibilities (1, 1) and (2, 2) for the deficiency indices are known as the *limit point* and *limit circle* cases respectively; this terminology arises from the technique used in the classical theory. In the limit point case the boundary condition may be treated very simply.

10.5.8 Lemma. *With l as above, suppose that a is a regular end point and assume that L_0 has deficiency indices (1, 1). Let M be a self-adjoint extension of L_0. Then there is a real number θ such that $D(M)$ is the subset of $D(L')$ of all f such that*

$$f(a)\cos\theta - f'(a)\sin\theta = 0. \tag{10.5.5}$$

Conversely every such restriction of $D(L')$ yields the domain of a self-adjoint extension of L_0.

Proof. Since L_0 has deficiency indices (1, 1), the domain of any self-adjoint extension is determined by one boundary condition, $[f, f_1]_a^b = 0$ say, and by Theorem 10.5.2 there is no loss of generality in taking $f_1 = \alpha_1 h_1 + \alpha_2 h_2$ $(\alpha_1, \alpha_2 \in \mathbb{C})$ where h_1, h_2 are the functions defined in the proof of the last lemma. As h_1, h_2 are zero on (c, b), this condition is $\bar{\alpha}_1 f'(a) - \bar{\alpha}_2 f(a) = 0$. However, f_1 must satisfy (10.5.2), that is $[f_1, f_1]_a^b = 0$, and an easy calculation shows that this holds iff $\alpha_1\bar{\alpha}_2$ is real. (10.5.5) follows. □

10.5.9 Example. We saw in Example 10.5.6 that for $l = -\mathrm{d}^2/\mathrm{d}x^2$ and $\Omega = [0, \infty)$ the domain of a self-adjoint extension is specified by one boundary condition, the deficiency indices being (1, 1). Since a is a regular end point, the lemma shows that the condition is of the form

$$f(0)\cos\theta - f'(0)\sin\theta = 0$$

for some real θ. Applications of the Spectral Theorem will yield the cosine transform if $\theta = \pi/2$ and the sine transform if $\theta = 0$, while other values of θ lead to less familiar transforms—see Example 10.6.4.

10.5.10 Example. Suppose l is the Bessel operator:

$$lf(x) = -f''(x) + x^{-2}(v^2 - \tfrac{1}{4})f(x),$$

and let Ω be (0, 1]. The notation and properties of Bessel functions are taken from Whittaker and Watson (1927). The solutions of $lf = if$ are $x^{\frac{1}{2}}J_v(x\lambda^{\frac{1}{2}})$, $x^{\frac{1}{2}}Y_v(x\lambda^{\frac{1}{2}})$, where $\lambda^{\frac{1}{2}} = \exp(i\pi/4)$ and J_v and Y_v are the Bessel functions of the first and second kinds respectively. The first solution is bounded on (0, 1] for all $v \geqslant 0$, and near the origin the second is proportional to $x^{\frac{1}{2}-v}$ (or $x^{\frac{1}{2}}\log x$ if $v = 0$). Thus for $v \geqslant 1$ only one solution lies in $\mathscr{L}_2(0, 1)$ and the deficiency indices are (1, 1), while for $0 \leqslant v < 1$ both solutions are in $\mathscr{L}_2(0, 1)$ and the deficiency indices are (2, 2). In the first case Lemma 10.5.8 shows that self-adjoint extensions are obtained by the imposition of the single condition

$$f(1)\cos\theta - f'(1)\sin\theta = 0 \qquad (\theta \in \mathbb{R}). \tag{10.5.6}$$

For $\theta = 0$ the Spectral Theorem yields the familiar Fourier–Bessel series, while other choices of 0 give rise to what are sometimes called Fourier–Dini series.

In the (2, 2) case an additional boundary condition is needed, and by analogy with regular l it is likely that this will involve the singular end point. In order to treat this case and also examples in which both end points are singular, boundary conditions at a singular end point must be discussed. In general the analysis will be less straightforward, but the complications are somewhat reduced if the conditions are separated, this being a generalization of the corresponding concept when l is regular. In almost all applications the conditions are separated.

10.5.11 Definition. Let g be a given function in $D(L')$. The boundary condition $[f, g]_a^b = 0$ is said to be a **condition at** a iff $[f, g]^b = 0$ for all $f \in D(L')$, with a similar definition for the other end point; note that if g is zero on (c, b) a condition at a is obtained. The condition is said to be **real** iff $[\bar{f}, g]_a^b = \overline{[f, g]_a^b}$ for all $f \in D(L')$. A **separated** set of boundary conditions is one in which each condition is either a condition at a or a condition at b. □

10.5.12 Example. The procedure for applying a boundary condition at a singular end point is illustrated by returning to the Bessel operator of Example 10.5.10 for $0 < \nu < 1$; in outline the treatment when $\nu = 0$ is the same, but the details are different because of the logarithmic singularity in the second solution.

The solutions of $lf = if$ are $x^{\frac{1}{2}}J_{\pm\nu}(x\lambda^{\frac{1}{2}})$ where $\lambda^{\frac{1}{2}} = \exp(i\pi/4)$, and near the origin these are proportional to

$$x^{\pm\nu+\frac{1}{2}}[1 + O(x^2)]$$

respectively. Let h_1, h_2 be real-valued functions in $\mathscr{C}^\infty((0, 1])$ with support in $[0, \frac{1}{2}]$, and which on $(0, \frac{1}{4})$ are equal to $x^{\pm\nu+\frac{1}{2}}$ respectively. A simple calculation shows that $lh_1 = lh_2 = 0$ on $(0, \frac{1}{4})$, and it follows that $h_1, h_2 \in D(L')$. Further $[h_1, h_1]_0^1 = [h_2, h_2]_0^1 = 0$ and

$$[h_1, h_2]_0^1 = [h_1, h_2]_0 = \lim_{x\to 0+} \{h_1(x)h_2'(x) - h_1'(x)h_2(x)\} = 2\nu. \tag{10.5.7}$$

For the other end point take h_3, h_4 to be real-valued functions in $\mathscr{C}^\infty((0, 1])$ with support in $[\frac{1}{2}, 1]$ and with $h_3(1) = 1, h_3'(1) = 0, h_4(1) = 0, h_4'(1) = 1$. The most general choice of f_1, f_2 leading to separated conditions is evidently

$f_1 = \alpha_1 h_1 + \alpha_2 h_2, f_2 = \alpha_3 h_3 + \alpha_4 h_4$, and it is easy to check using (10.5.7) and Lemma 10.2.17 that f_1 and f_2 are linearly independent relative to $D(L_0)$ if at least one of α_1, α_2 and one of α_3, α_4 is non-zero. Condition (10.5.2) is satisfied if and only if $\alpha_1 \bar{\alpha}_2$ and $\alpha_3 \bar{\alpha}_4$ are real, and general separated conditions are thus $[f, f_1]_0 = 0, [f, f_2]^1 = 0$ where $\alpha_i \in \mathbb{R}$. The condition at 1 is just (10.5.6) as before. For the other end point, as an example the choice $\alpha_1 = 1, \alpha_2 = 0$ yields the condition at zero:

$$\lim_{x \to 0+} \{x^{\nu + \frac{1}{2}} f'(x) - (\nu + \tfrac{1}{2}) x^{\nu - \frac{1}{2}} f(x)\} = 0. \tag{10.5.8}$$

Suppose finally that both end points are singular. A direct calculation of the deficiency indices will often involve tedious manipulations of special functions since the behaviour of a solution of $lf = if$ at both end points must be known. The following result, see Dunford and Schwartz (1963, p. 1302), reduces the problem to a discussion of operators with one regular end point and much simplifies the difficulties.

10.5.13 Lemma. *Let l be the second order formal operator of Definition 10.3.4. Suppose $a < c < b$, and let L_{0a}, L_{0b} be the symmetric operators associated with l on $(a, c]$ and $[c, b)$ respectively. Let $n_{\pm}, n_{\pm a}, n_{\pm b}$ be the deficiency indices of L, L_{0a}, L_{0b} respectively. Then $n_{\pm} = n_{\pm a} + n_{\pm b} - 2$ respectively.*

10.5.14 Example. Again consider the Bessel operator of Example 10.5.10, but this time with $\Omega = (0, \infty)$. Both end points are now singular. The deficiency indices are first calculated. With the notation of the lemma, it is evident from the previous analysis that $n_{+0} = 1\ (\nu \geqslant 1), n_{+0} = 2\ (0 \leqslant \nu < 1)$. From the asymptotic formulae for the Bessel functions, $x^{\frac{1}{2}}[J_\nu(x\lambda^{\frac{1}{2}}) \pm iY_\nu(x\lambda^{\frac{1}{2}})]$ are proportional respectively to $\exp(\pm ix\lambda^{\frac{1}{2}})$ as $x \to \infty$. Hence $n_{+\infty} = 1$. It follows from the lemma that the deficiency indices are (0, 0) for $\nu \geqslant 1$ and (1, 1) for $0 \leqslant \nu < 1$.

For $\nu \geqslant 1$ the only self-adjoint extension of L_0 is its closure and no boundary conditions are required. For $0 < \nu < 1$ one boundary condition is needed. That this is a condition at zero can be seen from Example 10.5.12 by noting that h_1, h_2 are linearly independent relative to $D(L_0)$, so every self-adjoint extension is given by taking a linear combination of these functions, and the most general such extension is thus obtained by the imposition of the single condition $[f, f_1]_0 = 0$ where $f_1 = \alpha_1 h_1 + \alpha_2 h_2\ (\alpha_1, \alpha_2 \in \mathbb{R})$. If $\alpha_1 = 1, \alpha_2 = 0$ this condition is just (10.5.8).

The deficiency indices are clearly of central importance in the construction of self-adjoint extensions. These may be found easily if explicit solutions of $lf = if$ are known, but unless the coefficients of l are simple, the calculations will involve difficult special functions. For an interesting qualitative theory which has been developed to deal with this problem, see Dunford and Schwartz (1963, Section 13.6) or Naimark (1968; Chap. 7).

10.6 Generalized Eigenfunction Expansions

The preliminaries are finally complete, and may now show how the generalized eigenfunction expansions are obtained from formal ordinary differential operators by an application of the Spectral Theorem to the associated self-adjoint operator. The analysis rests on the following formulae (Theorems 9.6.1 and 9.6.4) for the spectral family $P(\lambda_1, \lambda_2)$ of the self-adjoint operator M:

$$\lim_{\lambda \to \infty} P(-\lambda, \lambda)f = f \qquad (f \in \mathscr{H}), \tag{10.6.1}$$

$$(P(\lambda_1, \lambda_2)f, g) = \lim_{\delta \to 0+} \lim_{\varepsilon \to 0+} \frac{1}{2\pi i} \int_{\lambda_1 + \delta}^{\lambda_2 - \delta} ([R(\lambda - i\varepsilon; M) - R(\lambda + i\varepsilon; M)]f, g)\, \mathrm{d}\lambda, \tag{10.6.2}$$

together with an explicit expression for the resolvent $R(\lambda; M)$ when $\operatorname{Im} \lambda > 0$ as an integral operator with kernel the Green's function. Since M is self-adjoint, $\sigma(M)$ is real, and $R(\lambda; M)$ is an analytic function of λ in the upper and lower half planes and may be extended by analytic continuation to those parts of the real axis which lie in $\rho(M)$. Therefore $P(\lambda_1, \lambda_2) = 0$ unless (λ_1, λ_2) includes a point of $\sigma(M)$. In other words contributions to the integral in (10.6.2) come only from points of $\sigma(M)$. These points fall into two main classes. The first consists of points where $R(\lambda; M)$ has isolated poles; in this case the integral is evaluated by the Residue Theorem. Secondly $R(\lambda; M)$ may have a discontinuity across some section of the real axis, and a limiting procedure is then used to calculate the integral. The two classes correspond to the point and continuous spectra respectively.

The boundary conditions encountered in applications are almost always separated, and only self-adjoint operators determined by separated conditions will be considered. The advantage of this limitation is that a simple formula for the resolvent can be given. In the following theorem, quoted from Dunford and Schwartz (1963, p. 1329), uniqueness is to be interpreted as uniqueness up to a multiplicative constant. Also if no boundary condition at a is required to determine M, ϕ is simply the unique solution in $\mathscr{L}_2(a, c)$ of $l\phi = \lambda\phi$, with a similar proviso for ψ.

10.6.1 Theorem. *Let l be the second order, formally self-adjoint operator of Definition* 10.3.4. *Let M be a self-adjoint extension of L_0 obtained from a real* *separated set of boundary conditions. Suppose* $\text{Im}\lambda \neq 0$, *and take any* c *with* $a < c < b$. *Then there is exactly one solution* ϕ *of* $l\phi = \lambda\phi$ *which lies in* $\mathscr{L}_2(a, c)$ *and which satisfies any boundary condition at* a, *and exactly one solution* ψ *of* $l\psi = \lambda\psi$ *which lies in* $\mathscr{L}_2(c, b)$ *and which satisfies any boundary condition at* b. *Also*

$$R(\lambda; M)g = \int_a^b k(x, y; \lambda)\, g(y)\, \mathrm{d}y \qquad (g \in \mathscr{L}_2(\Omega)),$$

with

$$k(x, y; \lambda) = \begin{cases} \gamma\phi(x, \lambda)\, \psi(y, \lambda) & (x < y), \\ \gamma\phi(y, \lambda)\, \psi(x, \lambda) & (x > y), \end{cases}$$

where $\gamma = -1/[p(x)W]$ *and* W *is the Wronskian of* ϕ *and* ψ. *The kernel* k *is symmetric and*

$$\overline{k(x, y; \lambda)} = k(x, y; \bar{\lambda}). \qquad (10.6.3)$$

If L_0 *has deficiency indices* $(2, 2)$, $R(\lambda; M)$ *is compact.*

If the deficiency indices are (2, 2) the expansion formula may be obtained directly from the theory of compact self-adjoint operators of Section 7.5. For by the theorem $(\lambda I - M)^{-1}$ is compact for every λ with $\text{Im}\,\lambda \neq 0$, and it may readily be deduced that this is also the case for *some* real λ. An application of Theorem 7.5.4 then shows that the eigenfunctions of $(\lambda I - M)$, and so if M, are a basis for $\mathscr{L}_2(\Omega)$. This result treats all regular l (compare Example 7.5.5), but much more significantly, although most of the standard sets of orthogonal functions arise from singular l, their completeness may be deduced directly from Theorem 7.5.4. As its applicability is limited to cases where the deficiency indices are (2, 2), this approach is not pursued further here, and the generalized eigenfunction expansions are derived by means of the Spectral Theorem. Two examples chosen to illustrate the procedure with the minimum of complexity are first considered.

10.6.2 Example. (*The Fourier sine series*). Take $l = -\mathrm{d}^2/\mathrm{d}x^2$ and $\Omega = [0, \pi]$. Let M be the self-adjoint extension of L_0 determined by the boundary conditions $f(0) = f(\pi) = 0$. By Theorem 10.6.1, for $\text{Im}\,\lambda > 0$,

$$k(x, y; \lambda) = \frac{\sin \lambda^{\frac{1}{2}}(x - \pi) \sin \lambda^{\frac{1}{2}} y}{\lambda^{\frac{1}{2}} \sin \lambda^{\frac{1}{2}}\pi} \qquad (x > y),$$

while k is given for $x < y$ by symmetry and for $\text{Im}\,\lambda < 0$ by (10.6.3). Obviously k may be extended by analytic continuation to the whole λ plane excluding the zeros of $\sin \lambda^{\frac{1}{2}}\pi$ where there are poles.

The next step is write out (10.6.2). A change of order of integration, legitimate since Ω is bounded, yields

$$(P(\lambda_1, \lambda_2)f, g) = \int_0^{\pi} \bar{g}(x)\,dx \int_0^{\pi} f(y)\,dy$$

$$\times \lim_{\delta\to 0+} \lim_{\varepsilon\to 0+} \frac{1}{2\pi i} \int_{\lambda_1+\delta}^{\lambda_2+\delta} [k(x, y; \lambda - i\varepsilon) - k(x, y; \lambda + i\varepsilon)]\,d\lambda. \tag{10.6.4}$$

If Γ is the rectangle with vertices $(\lambda_2 - \delta - i\varepsilon), (\lambda_2 - \delta + i\varepsilon), (\lambda_1 + \delta + i\varepsilon), (\lambda_1 + \delta - i\varepsilon)$,

$$\lim_{\delta\to 0+} \lim_{\varepsilon\to 0+} \int_{\lambda_1+\delta}^{\lambda_2-\delta} [k(x, y; \lambda - i\varepsilon) - k(x, y; \lambda + i\varepsilon)]\,d\lambda = \lim_{\delta\to 0+} \lim_{\varepsilon\to 0+} \int_{\Gamma} k(x, y; \lambda)\,d\lambda,$$

since the contributions from the "short" sides of Γ tend to zero as $\varepsilon \to 0$, and a straightforward calculation based on the Residue Theorem then gives

$$\lim_{\delta\to 0+} \lim_{\varepsilon\to 0+} \frac{1}{2\pi i} \int_{\Gamma} k(x, y; \lambda)\,d\lambda = (2/\pi) \sum_{\lambda_1 < n^2 < \lambda_2} \sin nx \sin ny.$$

For any $f \in \mathscr{L}_2(\Omega)$ set

$$\hat{f}(n) = (2/\pi)^{\frac{1}{2}} \int_0^{\pi} \sin ny\, f(y)\,dy \qquad (n = 1, 2, \ldots),$$

and let $\hat{f}$ be the vector with components $\hat{f}(1), \hat{f}(2), \ldots$. In this notation (10.6.4) becomes

$$(P(\lambda_1, \lambda_2)f, g) = \sum_{\lambda_1 < n^2 < \lambda_2} \hat{f}(n)\,\overline{\hat{g}(n)}. \tag{10.6.5}$$

Hence from (10.6.1), $\hat{f} \in \ell_2$, and if $\|[\cdot]\|$ and $[\cdot, \cdot]$ denote the inner product and norm of ℓ_2,

$$\|f\|^2 = \Sigma\,|\hat{f}(n)|^2 = \|[\hat{f}]\|^2, \qquad (f, g) = \Sigma \hat{f}(n)\,\overline{\hat{g}(n)} = [\hat{f}, \hat{g}]. \tag{10.6.6}$$

These relations are known as Parseval's formula (cf. (1.5.9)) and Plancherel's formula respectively.

It is now easy to obtain the Fourier series itself. For with $\phi_n(x) = (2/\pi)^{\frac{1}{2}} \sin nx$, (10.6.5) is

$$(P(\lambda_1, \lambda_2)f, g) = \left(\sum_{\lambda_1 < n^2 < \lambda_2} \hat{f}(n)\,\phi_n, g \right),$$

and since this holds for any $g \in \mathscr{L}_2(\Omega)$, it follows that

$$P(\lambda_1, \lambda_2)f = \sum_{\lambda_1 < n^2 < \lambda_2} \hat{f}(n)\,\phi_n.$$

Hence by (10.6.1),

$$f(x) = (2/\pi)^{\frac{1}{2}} \sum_{1}^{\infty} \hat{f}(n) \sin nx,$$

the convergence being in $\mathscr{L}_2(\Omega)$.

In this example the resolvent is compact, the resolvent kernel is a meromorphic function of λ, the functions $(2/\pi)^{\frac{1}{2}} \sin nx$ are normalized eigenfunctions of M, and the $\hat{f}(n)$ are the Fourier sine coefficients. It is easy to show that the mapping $f \to \hat{f}$ of $\mathscr{L}_2(\Omega)$ into ℓ_2 is surjective, and Plancherel's formula thus expresses the fact that the mapping is an isometric isomorphism.

10.6.3 Example. (*The Fourier Transform*). An operator with a continuous rather than a discrete spectrum is next considered. With $l = i\mathrm{d}/\mathrm{d}x$ and $\Omega = (-\infty, \infty)$ the only self-adjoint extension of L_0 is its closure M (Example 10.5.5). The resolvent is easy to calculate, and we find that

$$R(\lambda; M)\, g(x) = \begin{cases} -i \displaystyle\int_x^{\infty} e^{i\lambda(y-x)} g(y)\,\mathrm{d}y & (\operatorname{Im} \lambda > 0), \\ i \displaystyle\int_{-\infty}^{x} e^{i\lambda(y-x)}\, g(y)\,\mathrm{d}y & (\operatorname{Im} \lambda < 0). \end{cases}$$

There are now some extra difficulties because the ranges of integration are infinite. The tactics will be to work initially in the space $\mathscr{L}_{20}(\Omega)$ consisting of those functions in $\mathscr{L}_2(\Omega)$ with compact support, for there will then be no problems with the exchange of orders of integration. The results obtained in this way may readily be extended by continuity to the whole of $\mathscr{L}_2(\Omega)$ since $\mathscr{L}_{20}(\Omega)$ is dense in this space. From (10.6.2), for $f, g \in \mathscr{L}_{20}(\Omega)$,

$$\begin{aligned} (P(\lambda_1, \lambda_2) f, g) &= \lim_{\delta \to 0+} \lim_{\varepsilon \to 0+} \frac{1}{2\pi i} \int_{\lambda_1+\delta}^{\lambda_2-\delta} \mathrm{d}\lambda \int_{-\infty}^{\infty} \bar{g}(x)\,\mathrm{d}x \\ &\quad \times \left\{ i \int_{-\infty}^{x} e^{i(\lambda - i\varepsilon)(y-x)}\, f(y)\,\mathrm{d}y + i \int_{x}^{\infty} e^{i(\lambda + i\varepsilon)(y-x)} f(y)\,\mathrm{d}y \right\} \\ &= \frac{1}{2\pi} \int_{\lambda_1}^{\lambda_2} \mathrm{d}\lambda \int_{-\infty}^{\infty} \bar{g}(x)\,\mathrm{d}x \int_{-\infty}^{\infty} e^{i\lambda(y-x)} f(y)\,\mathrm{d}y, \end{aligned}$$

by the Dominated Convergence Theorem 2.4.11. That is

$$(P(\lambda_1, \lambda_2) f, g) = \int_{\lambda_1}^{\lambda_2} \hat{f}(\lambda)\, \bar{\hat{g}}(\lambda)\,\mathrm{d}\lambda,$$

where

$$\hat{f}(\lambda) = (1/2\pi)^{\frac{1}{2}} \int_{-\infty}^{\infty} e^{i\lambda y} f(y)\,\mathrm{d}y.$$

Thus

$$(f, g) = \int_{-\infty}^{\infty} \hat{f}(\lambda)\,\bar{\hat{g}}(\lambda)\,\mathrm{d}\lambda \tag{10.6.7}$$

by (10.6.1), and it follows on taking $f = g$ that $\hat{f} \in \mathscr{L}_2(-\infty, \infty)$. The mapping $f \to \hat{f}$ is thus an isometric isomorphism of $\mathscr{L}_{20}(\Omega)$ into $\mathscr{L}_2(-\infty, \infty)$. Since $\mathscr{L}_{20}(\Omega)$ is dense in $\mathscr{L}_2(\Omega)$ a continuity argument shows that these results hold for all of $\mathscr{L}_2(\Omega)$, and that in particular

$$\lim_{A \to \infty} \int_{-A}^{A} \mathrm{e}^{i\lambda x} f(x)\,\mathrm{d}x$$

exists in the norm of $\mathscr{L}_2(\Omega)$. The inversion formula is obtained as in the previous example by using (10.6.1). Finally it is obvious from symmetry that the mapping $f \to \hat{f}$ is surjective, and so is an isometric isomorphism of $\mathscr{L}_2(\Omega)$ *onto* $\mathscr{L}_2(-\infty,\infty)$. We thus have obtained a proof of the standard Fourier Transform Theorem 2.6.1, (10.6.7) being Plancherel's formula.

The two previous examples illustrate the analysis underlying the construction of generalized eigenfunction expansions. With the understanding that the details may be supplied in a similar manner, only the formal argument is given in the remaining examples.

10.6.4 Example. Here $l = -\mathrm{d}^2/\mathrm{d}x^2$ and $\Omega = [0, \infty)$; the choice of boundary conditions leads to a somewhat unfamiliar transform (for an application see Problem 10.11). Self-adjoint extensions are obtained by imposing a boundary condition at zero. Consider in particular $f(0) + \alpha f'(0) = 0$ for some $\alpha > 0$. The resolvent is found from Theorem 10.6.1 by choosing the solution ϕ of $l\phi = \lambda\phi$ which satisfies this boundary condition, and the solution ψ in $\mathscr{L}_2(c, \infty)$. Thus for $\operatorname{Im}\lambda > 0$

$$\phi(x, \lambda) = \sin(x\lambda^{\frac{1}{2}}) - \alpha\lambda^{\frac{1}{2}}\cos(x\lambda^{\frac{1}{2}}),$$

$$\psi(x, \lambda) = \exp(ix\lambda^{\frac{1}{2}}),$$

and

$$k(x, y; \lambda) = -\frac{1}{\lambda^{\frac{1}{2}}(1 + i\alpha\lambda^{\frac{1}{2}})}\phi(y, \lambda)\,\psi(x, \lambda) \qquad (x > y),$$

with k given for $x < y$ by symmetry, and for $\operatorname{Im}\lambda < 0$ by $\overline{k(x, y; \lambda)} = k(x, y; \bar{\lambda})$. k may be extended by analytic continuation across the negative real axis excluding the point $\lambda = 1/\alpha^2$. The occurrence of a pole at this point reflects the fact that the point is in the point spectrum of M; the corresponding eigenfunction is $\exp(-x/\alpha)$. k has a discontinuity across the positive real axis which is therefore in the continuous spectrum of M. The spectrum is thus a mixture of point and continuous parts. The evaluation of the integral defining $P(\lambda_1, \lambda_2)$ is carried out by a combination of the methods of the two previous examples, and the following is obtained:

$$(P(\lambda_1, \lambda_2)f, g) = 2\alpha^{-1}\int_0^\infty \exp(-y/\alpha) f(y)\,dy \int_0^\infty \exp(-x/\alpha)\,\bar{g}(x)\,dx$$
$$+ \int_{\lambda_1}^{\lambda_2} \frac{d\lambda}{\lambda^{\frac{1}{2}}(1+\alpha^2\lambda)} \int_0^\infty \phi(y,\lambda) f(y)\,dy \int_0^\infty \phi(x,\lambda)\,\bar{g}(x)\,dx, \tag{10.6.8}$$

where the first term is omitted if $-1/\alpha^2 \notin (\lambda_1, \lambda_2)$, and the second term is omitted if $\lambda_2 \leqslant 0$. With

$$\tau(x,\lambda) = [\lambda^{\frac{1}{2}}(1+\alpha^2\lambda)]^{-\frac{1}{2}}\,\phi(x,\lambda),$$

(10.6.8) gives on taking limits and using (10.6.1),

$$(f, g) = 2\alpha^{-1}\int_0^\infty \exp(-y/\alpha) f(y)\,dy \int_0^\infty \exp(-x/\alpha)\,\bar{g}(x)\,dx$$
$$+ \int_0^\infty d\lambda \int_0^\infty \tau(y,\lambda) f(y)\,dy \int_0^\infty \tau(x,\lambda)\,\bar{g}(x)\,dx.$$

In order to express the result in a form analogous to that of the usual transform formula, define

$$\hat{f}_1 = (2/\alpha)^{\frac{1}{2}}\int_0^\infty \exp(-y/\alpha) f(y)\,dy,$$

$$\hat{f}_2(\lambda) = \lim_{A\to\infty}\int_0^A \tau(y,\lambda) f(y)\,dy.$$

Then Plancherel's formula is

$$(f, g) = \hat{f}_1\bar{\hat{g}}_1 + \int_0^\infty \hat{f}_2(\lambda)\,\bar{\hat{g}}_2(\lambda)\,d\lambda,$$

and the inversion formula is, with convergence in $\mathscr{L}_2(\Omega)$,

$$f(x) = \hat{f}_1 \exp(-x/\alpha) + \lim_{A\to\infty}\int_0^A \tau(x,\lambda)\hat{f}_2(\lambda)\,d\lambda.$$

This situation where the spectrum has both a point and a continuous part is not uncommon. A Bessel function example is given by Titchmarsh (1962, p. 90).

10.6.5 Example. (*The Hankel Transform*). Let l be the Bessel operator of Example 10.5.10 and take $\Omega = (0, \infty)$. From Example 10.5.14 the deficiency indices are (1, 1) for $0 \leqslant \nu < 1$ and (0, 0) for $\nu \geqslant 1$. To avoid unilluminating detail assume that ν is greater than 1 and is not an integer. For $\operatorname{Im} \lambda > 0$, ϕ, ψ in Theorem 10.6.1 are

$$\phi(x, \lambda) = x^{\frac{1}{2}} J_\nu(x\lambda^{\frac{1}{2}}),$$

$$\psi(x, \lambda) = x^{\frac{1}{2}}[\mathrm{e}^{-\nu\pi i} J_\nu(x\lambda^{\frac{1}{2}}) - J_{-\nu}(x\lambda^{\frac{1}{2}})],$$

and

$$k(x, y; \lambda) = (\pi/2 \sin \nu\pi)\, \psi(x, \lambda)\, \phi(y, \lambda) \qquad (x > y),$$

with k defined for other values of x, y, λ by symmetry and by taking complex conjugates. Since $\exp(\nu\pi i)$ is the only term that is not real as λ approaches a point of the positive real axis, k has a discontinuity there. Next consider the negative real axis. Recalling that

$$J_\nu(iz) = \exp(\nu\pi i/2)\, I_\nu(z),$$

we see that as $\lambda \to -\mu$ $(\mu > 0)$,

$$\phi(x, \lambda)\, \psi(y, \lambda) \to (xy)^{\frac{1}{2}}\, \mathrm{e}^{\nu\pi i/2} I_\nu(x\mu^{\frac{1}{2}})[\mathrm{e}^{-\nu\pi i}\, \mathrm{e}^{\nu\pi i/2} I_\nu(y\mu^{\frac{1}{2}}) - \mathrm{e}^{-\nu\pi i/2} I_{-\nu}(y\mu^{\frac{1}{2}})].$$

Since $I_{\pm\nu}(y\mu^{\frac{1}{2}})$ are both real, it follows that k is continuous on the negative real axis which thus lies in $\rho(M)$. The rest of the analysis is as before, and the inversion formula is

$$f(x) = \frac{1}{2} \int_0^\infty x^{\frac{1}{2}} J_\nu(x\lambda^{\frac{1}{2}})\, H_\nu f(\lambda)\, \mathrm{d}\lambda,$$

where

$$H_\nu f(\lambda) = \int_0^\infty y^{\frac{1}{2}} J_\nu(y\lambda^{\frac{1}{2}}) f(y)\, \mathrm{d}y,$$

the convergence of each of the integrals being in $\mathscr{L}_2(0, \infty)$.

These examples indicate how the Spectral Theorem is used in obtaining generalized eigenfunction expansions. The method is a powerful one and may be applied to a wide range of second order operators to yield most of the standard expansions, and some that are less familiar. It may be noted in conclusion that there is a more general approach based on the Titchmarsh–Weyl–Kodeira Theorem, see Dunford and Schwartz (1963, p. 1364 *et seq.*). However, the technicalities connected with this theorem are rather formidable.

Problems

Throughout $\mathscr{H}$ is a complex Hilbert space, and L, L_0 are densely defined linear operators from $\mathscr{H}$ into itself.

10.1 Prove that L_0 is symmetric iff $(L_0 f, f)$ is real for all $f \in D(L_0)$.

10.2 Suppose that L is closed and symmetric. If $\operatorname{Im} \lambda \neq 0$ show that λ is either in $\rho(L)$ or in the residual spectrum of L.

10.3 Take $l = -\mathrm{d}^2/\mathrm{d}x^2$ and $\Omega = [0, \infty)$. Let M be the self-adjoint operator obtained from the boundary condition $f(0) = 0$. Deduce from Problem 6.28 that $\sigma(M)$ is contained in the positive real axis. Does this result remain true for the boundary condition $f'(0) + \alpha f(0) = 0$ $(\alpha \in \mathbb{R})$?

10.4 Suppose $p(\cdot)$ is a polynomial of degree m with real constant coefficients. Take $l = p(i\mathrm{d}/\mathrm{d}x)$ and $\Omega = [0, \infty)$. Show that the associated operator L_0 with domain $\mathscr{C}_0^\infty(\Omega)$ is symmetric. Prove that if p has only even powers the deficiency indices $n_\pm$ of L_0 are equal, whereas if m is odd, $n_+ \neq n_-$.

10.5 If L is symmetric and $R(L) = \mathscr{H}$, prove that L is self-adjoint.

10.6 Let L be closed and symmetric with finite and equal deficiency indices. Let L_1 be a closed symmetric extension of L. Does L_1 necessarily have self-adjoint extensions?

10.7 Let M be a self-adjoint (unbounded) operator from $\mathscr{H}$ into itself, and assume that $(\lambda I - M)^{-1}$ is compact for each λ with $\operatorname{Im} \lambda \neq 0$. Prove that:

(i) There is a real λ for which $(\lambda I - M)^{-1} \in \mathscr{L}(\mathscr{H})$;
(ii) For such λ, $(\lambda I - M)^{-1}$ is compact.

10.8 Consider the Bessel operator of Example 10.5.10 on $(0, 1]$ for $0 < \nu < 1$. Show that a self-adjoint extension is determined by the boundary conditions $f(1) = 0$ and

$$\lim_{x \to 0+} [x^{\nu + \frac{1}{2}} f'(x) - (\nu + \tfrac{1}{2}) x^{\nu - \frac{1}{2}} f(x)] = 0.$$

Using the result of the previous problem and Theorem 7.5.4 establish the validity of the Fourier–Bessel series in $\mathscr{L}_2(0, 1)$:

$$f(x) = \sum_{n=1}^{\infty} c_n x^{\frac{1}{2}} J_\nu(\alpha_n x) \int_0^1 y^{\frac{1}{2}} J_\nu(\alpha_n y) f(y)\, \mathrm{d}y,$$

where α_n are the zeros of J_ν and c_n are normalizing constants. (If this result is obtained from the Spectral Theorem, the c_n appear explicitly.)

10.9 Take $\Omega = (-1, 1)$ and $lf(x) = [(1 - x^2) f'(x)]'$. Find the deficiency indices of L_0. Give separated boundary conditions for self-adjoint extensions, and derive the standard Legendre series expansion.

10.10 Obtain the Hankel transform formula (Example 10.6.5) for $\nu = 0$.

10.11 The "wave maker" problem of linearized water wave theory in deep water is as follows. With y measured vertically downwards from the surface a solution of $\nabla^2 \phi = 0$ is required satisfying the boundary conditions: (i) $\partial^2 \phi / \partial t^2 - g \partial \phi / \partial y = 0$ on $y = 0$. (ii) $\partial \phi / \partial x = u(y) \sin \omega t$ at the wave maker $x = 0$. (iii) $\phi \to 0$ as $y \to \infty$. (iv) A radiation condition that as $x \to \infty$, ϕ behaves like an outgoing wave. With $\phi(x, y, t) = \operatorname{Im} \mathrm{e}^{i\omega t} \phi(x, y)$, a standard boundary value problem is obtained, the boundary condition at $y = 0$ becoming $\partial\phi/\partial y(x, 0) + k\phi(x, 0) = 0$ where $k = \omega^2/g$. The generalized eigenfunction expansion of Example 10.6.4 is exactly fitted to this problem. Show that if u is a continuous function with compact support, a solution is

$$\phi(x, y, t) = 2\cos(kx - \omega t)\,\mathrm{e}^{-ky}\int_0^\infty \mathrm{e}^{-ks}\,u(s)\,\mathrm{d}s$$

$$-(2/\pi)\sin\omega t\int_0^\infty \mathrm{e}^{-\mu x}(\mu\cos\mu y - k\sin\mu y)\,A(\mu)\,\mathrm{d}\mu,$$

where

$$A(\mu) = [\mu(k^2 + \mu^2)]^{-1}\int_0^\infty (\mu\cos\mu s - k\sin\mu s)\,u(s)\,\mathrm{d}s.$$

Chapter 11

Linear Elliptic Partial Differential Equations

11.1 Introduction

A standard classical argument for tackling the boundary value problem for elliptic partial differential equations is based on reformulating the equation as an integral equation by using a Green's function and then invoking the theory of integral equations. While this argument has achieved considerable success, it is somewhat artificial to base the theory on the integral equation rather than on the differential equation itself, and recent theoretical developments show that a direct attack on the differential equation often yields more information, and at the same time avoids the tedious technical problems associated with the construction of the integral equation. The numerical solution is one context in which the advantages may be clearly seen, for the use of an integral equation does not fit in with standard numerical procedures, and is clearly unnatural. The purpose of this chapter is the introduction of the direct approach.

The starting point for the analysis is the replacement of the original boundary value problem by a certain weak analogue. For illustration consider Poisson's equation

$$\nabla^2 f - g = 0 \tag{11.1.1}$$

on a bounded open region Ω, a solution $f \in \mathscr{C}^2(\Omega) \cap \mathscr{C}(\overline{\Omega})$ vanishing on the boundary $\partial\Omega$ being required. Multiplication of this equation by any $\phi \in \mathscr{C}_0^\infty(\Omega)$ and integration gives

$$\int_\Omega (\nabla^2 f - g)\, \phi \, \mathrm{d}x = 0 \qquad (\phi \in \mathscr{C}_0^\infty(\Omega)), \tag{11.1.2}$$

and since $\mathscr{C}_0^\infty(\Omega)$ is dense in $\mathscr{L}_2(\Omega)$, (11.1.1) and (11.1.2) are equivalent. Integration by parts yields (since ϕ and its derivatives vanish on $\partial\Omega$)

$$\int_\Omega f\nabla^2\phi \, dx = \int_\Omega g\phi \, dx. \tag{11.1.3}$$

Therefore (11.1.3) and (11.1.1) are equivalent for smooth f. However, (11.1.3) makes perfectly good sense for any f in $\mathscr{L}_2(\Omega)$, although (11.1.1) cannot be recovered directly, and indeed does not have any obvious interpretation. This heuristic argument leads to the following weak version of the original problem: for given g find a function f with $f = 0$ on $\partial\Omega$ which satisfies (11.1.3) for all $\phi \in \mathscr{C}_0^\infty(\Omega)$.

There are two attractive features of the weak problem which are immediately apparent. First, it makes sense for a wide class of right-hand sides, g, certainly for any $g \in \mathscr{L}_2(\Omega)$. Second, since derivatives of f do not appear in (11.1.3), the smoothness of f is, at least initially, not such a pressing problem as it would be if the original equation were to be considered. In fact the discussion in the previous paragraph is rather imprecise, the statement $f = 0$ on $\partial\Omega$ being meaningless for a general $f \in \mathscr{L}_2(\Omega)$, and in order to make sense of the boundary condition some smoothness must be required of f, although not as much as in the original formulation.

The key to success in tackling the weak boundary value problem (known as the generalized Dirichlet problem) is the choice of space. The problem fits naturally in a certain Sobolev space of functions satisfying relatively mild smoothness conditions, the fact that this space is a Hilbert space significantly simplifying the analysis. Sobolev spaces are now a basic tool in partial differential equation theory. Thus although for simplicity only a relatively simple example is considered, the homogeneous Dirichlet problem for a linear elliptic equation, the Sobolev space theory described has applications to both linear and nonlinear equations and equations of elliptic and evolution type.

The advantages of posing the problem in a Sobolev space become apparent when existence and uniqueness for a general elliptic equation is tackled, for the problem may be formulated in terms of a bounded linear operator whose properties are relatively simple to study. In certain cases, for example for Poisson's equation on a bounded domain, it is easy to show that the operator has a bounded inverse, and it follows that the generalized Dirichlet problem has exactly one solution for every reasonable right-hand side g. However, in general the homogeneous equation may have non-trivial solutions, and then existence for arbitrary g cannot be expected. The most that can be hoped for is existence and uniqueness if the homogeneous equation has only the zero solution, in other words a theorem analogous to the Fredholm alternative. Under certain conditions, the most important of which is the boundedness of the domain, a certain related operator has a compact inverse and the required Alternative Theorem is readily deduced. It is interesting to note that compact operator theory, originally devised to treat the integral equation arising in the Green's function approach, is still the main tool needed in the direct method, where however it is applied to a "Green's operator" whose precise form need not be calculated. After a preparatory discussion of Sobolev spaces in Section 3, the proof of this Alternative Theorem will be the main business of this chapter, and will be tackled in Sections 4 and 5.

The above argument yields criteria for the existence and uniqueness of solutions in a Sobolev space for an elliptic operator of order $2m$. However, these solutions need not be in $\mathscr{C}^{2m}(\Omega)$, and cannot therefore be regarded as solutions in the classical sense. A separate investigation is needed to determine when these solutions have their classical meaning, and some results in this area are given in Section 6.

Agmon (1965) and Friedman (1969) are standard texts on linear partial differential equations, and the recent books by Folland (1976), Schechter (1977), Showalter (1977) and Treves (1975) are other useful references. Much of the known theory of linear and nonlinear second-order elliptic equations is contained in Gilbarg and Trudinger (1977).

11.2 Notation

The complexity of the notation presents something of a difficulty, and for convenience of reference some of the main conventions to be used are summarized below.

We shall deal with partial differential equations on subsets Ω of $\mathbb{R}^n$. Ω will always be an *open* set, $\bar{\Omega}$ will be its closure, and $\partial\Omega = \bar{\Omega}\backslash\Omega$ its boundary.

Several spaces of functions will be needed. Since for the most part the functions will be defined on the standard domain Ω, in order to simplify the rather cumbersome notation, in this chapter the symbol Ω is omitted from the expressions for these spaces unless the domain is different from Ω. Thus $\mathscr{L}_2$ will denote $\mathscr{L}_2(\Omega)$. The functions will always be complex valued unless stated to the contrary.

In addition to the spaces $\mathscr{C}^m$ and $\mathscr{C}^m(\bar{\Omega})$ of m times differentiable functions mapping Ω and $\bar{\Omega}$ respectively into $\mathbb{C}$, frequent use is made of the space $\mathscr{C}_0^m$ (Definition 1.3.23) of functions in $\mathscr{C}^m$ with bounded support in Ω. Since Ω is open and a support is closed, it is easy to prove that given $f \in \mathscr{C}_0^m$ there is an $\varepsilon > 0$ such that for every $x \in \partial\Omega$, $f = 0$ on $S(x, \varepsilon)$, the open ball centre x and radius ε. Thus f vanishes at all points within a distance ε of the boundary—

that is in a strip at the boundary. Note that $\mathscr{C}_0^m$ is dense in $\mathscr{L}_2$ for $0 \leqslant m \leqslant \infty$ (Theorem 2.5.6).

For a general partial differential equation in n dimensions the classical notation is extremely awkward. The complexity is considerably reduced by the use of multi-indices.

11.2.1 Definition. A **multi-index** α is an n-tuple $(\alpha_1, \ldots, \alpha_n)$ of non-negative integers. We write $|\alpha| = \alpha_1 + \ldots + \alpha_n$; this conflicts with the notation for the Euclidean distance in $\mathbb{R}^n$, but the meaning will always be clear from the context. Multi-indices will be denoted by α and β.

A point in $\mathbb{R}^n$ will be $x = (x_1, \ldots, x_n)$ with $|x|^2 = \Sigma x_j^2$, and $x^\alpha = x_1^{\alpha_1} \ldots x_n^{\alpha_n}$. We write $D_j = \partial/\partial x_j$ and $D^\alpha = D_1^{\alpha_1} \ldots D_n^{\alpha_n}$. With these conventions the notation for a partial differential equation may be simplified by writing

$$\sum_{|\alpha| \leqslant m} p_\alpha D^\alpha = \sum_{j=0}^{m} \sum_{\alpha_1 + \ldots + \alpha_n = j} p_{\alpha_1, \ldots, \alpha_n} D_1^{\alpha_1} \ldots D_n^{\alpha_n}.$$

Although the theory may be carried through under more general conditions, here the simplifying assumption that the coefficients are smooth will be made, and the operator will usually be written in divergence form.

11.2.2 Definition. Assume that for some α, β with $|\alpha| = |\beta| = m$, $p_{\alpha\beta} \neq 0$, and that for all α, β, $p_{\alpha\beta} \in \mathscr{C}^\infty(\bar{\Omega})$. The $p_{\alpha\beta}$ are thus complex valued variable coefficients. For $\phi \in \mathscr{C}^{2m}$ define

$$l\phi = \sum_{|\alpha|, |\beta| \leqslant m} (-1)^{|\alpha|} D^\alpha (p_{\alpha\beta} D^\beta \phi), \tag{11.2.1}$$

$$l_P\phi = (-1)^m \sum_{|\alpha| = |\beta| = m} D^\alpha (p_{\alpha\beta} D^\beta \phi).$$

l is called a **formal partial differential operator** of order $2m$. l_P is known as the **principal part** of l.

As for formal ordinary differential operators the idea of a formal adjoint is needed. Suppose for the moment that $n = 1$ and $\Omega = (-1, 1)$. For $\phi, \psi \in \mathscr{C}_0^\infty$ an integration by parts gives

$$\int_{-1}^{1} \phi \,.\, D_1\bar{\psi} \,\mathrm{d}x = -\int_{-1}^{1} D_1\phi \,.\, \bar{\psi} \,\mathrm{d}x.$$

Note that the integrated term vanishes because ϕ, ψ are zero near $\partial\Omega$, this being itself a consequence of the assumption that $\phi, \psi \in \mathscr{C}_0^\infty$. Analogously, for the general case, repeated integration by parts gives for $\phi, \psi \in \mathscr{C}_0^\infty$,

$$
\begin{aligned}
(l\phi, \psi)_0 &= \sum_{|\alpha|,|\beta| \leqslant m} (-1)^{|\alpha|} \int_\Omega D^\alpha(p_{\alpha\beta} D^\beta \phi) . \bar{\psi}\, \mathrm{d}x \\
&= \sum_{|\alpha|,|\beta| \leqslant m} \int_\Omega p_{\alpha\beta} D^\beta \phi . D^\alpha \bar{\psi}\, \mathrm{d}x \\
&= \sum_{|\alpha|,|\beta| \leqslant m} (-1)^{|\beta|} \int_\Omega \phi . \overline{D^\beta(\bar{p}_{\alpha\beta} D^\alpha \psi)}\, \mathrm{d}x \\
&= \sum_{|\alpha|,|\beta| \leqslant m} (-1)^{|\alpha|} \int_\Omega \phi . \overline{D^\alpha(\bar{p}_{\beta\alpha} D^\beta \psi)}\, \mathrm{d}x \\
&= (\phi, l^*\psi)_0
\end{aligned}
$$

say, where $(. , .)_0$ is the inner product of $\mathscr{L}_2$. (The suffix zero will be used in this chapter for reasons which will soon become apparent).

11.2.3 Definition. The operator l^*, where

$$l^*\phi = \sum_{|\alpha|,|\beta| \leqslant m} (-1)^{|\alpha|} D^\alpha(\bar{p}_{\beta\alpha} D^\beta \phi)\ (\phi \in \mathscr{C}^{2m}),$$

is called the **formal adjoint** of l. l is said to be **formally self-adjoint** iff $l = l^*$.

In order to obtain a well-posed boundary value problem for l, some condition of ellipticity must be imposed, and the following will be used here.

11.2.4 Definition. l will be said to be **strongly elliptic** (on Ω) iff there is a $c > 0$ such that

$$\mathrm{Re}(-1)^m l_p(\xi) = \mathrm{Re} \sum_{|\alpha| = |\beta| = m} \xi^\alpha p_{\alpha\beta}(x) \xi^\beta \geqslant c|\xi|^{2m} \qquad (x \in \bar{\Omega})$$

for all $\xi \in \mathbb{R}^n$, where $|\xi|^2 = \xi_1^2 + \ldots + \xi_n^2$. The term *uniformly strongly elliptic* is sometimes used for this condition in the literature.

Evidently if l is strongly elliptic, so is l^*. It is also worth noting that the condition of strong ellipticity is invariant under a change of coordinates with non-vanishing Jacobian (Problem 11.1).

11.2.5 Example. For $l = -\nabla^2 = -\sum_1^n D_i^2$, $-l_p(\xi) = \sum_1^n \xi_i^2$. Thus (minus) the Laplacian is strongly elliptic. If $l = -(x_1 D_1^2 + D_2^2)$ in $\mathbb{R}^2$, l is strongly elliptic on $\{(x_1, x_2) : x_1 \geqslant d\}$ for $d > 0$, but not for $d = 0$. If $l = D_1 - D_2^2$, $(-1)^m l_p(\xi) = \xi_2^2$, but $\xi^2 = \xi_1^2 + \xi_2^2$ and the condition does not hold. The heat conduction equation is therefore not strongly elliptic.

11.3 Weak Derivatives and Sobolev Spaces

In the classical theory of differential equations, it is usual to regard a function f as a solution only if all derivatives of f appearing in the equation exist and are continuous. This prompts the following definition.

11.3.1. Definition. Let Ω be bounded and have a $\mathscr{C}^\infty$ boundary†, and let l be the formal operator of Definition 11.2.2. For given $g \in \mathscr{C}$, a function f is said to be a **classical solution** of $lf = g$ iff $f \in \mathscr{C}^{2m}$ and

$$lf = g \tag{11.3.1}$$

in Ω. f is said to be a classical solution of the **homogeneous Dirichlet problem** iff in addition $f \in \mathscr{C}^{m-1}(\overline{\Omega})$ and

$$\partial^j f/\partial v^j = 0 \qquad (j = 0, 1, \ldots, (m-1)) \tag{11.3.2}$$

on $\partial\Omega$, where $\partial/\partial v$ denotes differentiation in the direction normal to the boundary.

The corresponding inhomogeneous Dirichlet problem may be expressed in the above form under mild restrictions on the boundary data, see Friedman (1969, p. 38). Following the tactics outlined in the introduction, the concept of a weak solution is now introduced. In view of Definition 11.2.3 of the formal adjoint l^*, a generalization of the argument leading to (11.1.3) yields the following.

11.3.2 Definition. For given $g \in \mathscr{L}_2$, a function $f \in \mathscr{L}_2$ is said to be a **weak solution** of $lf = g$ iff

$$(f, l^*\phi)_0 = (g, \phi)_0 \tag{11.3.3}$$

for all $\phi \in \mathscr{C}_0^\infty$, and we then write

$$lf \overset{w}{=} g. \tag{11.3.4}$$

No boundary conditions have been imposed on weak solutions, and these are therefore analogues of classical solutions of $lf = g$ rather than of classical solutions of the Dirichlet problem. We shall return to this point later, but first let us consider (11.3.4) further.

Observe that it is not legitimate to interpret the left-hand side of (11.3.4) as a sum of ordinary derivatives since the existence in the usual sense of any

† For the purposes of this chapter an intuitive understanding of this as a "very smooth" boundary will suffice. A precise definition would be on the following lines. For each point P of $\partial\Omega$, there are open sets $S_1 \subset \mathbb{R}^{n-1}$, $S_2 \subset \partial\Omega$ with $0 \in S_1$, $P \in S_2$, and a bijection $\phi: S_1 \to S_2$ such that $\phi \in \mathscr{C}^\infty(S_1)$, and such that the rank of the Jacobian of ϕ is $(n-1)$ at all points of S_1.

of the terms which would appear there is not guaranteed. With the eventual aim of assigning a meaning to these terms, we first show that it is sometimes possible to give a sensible interpretation to the derivatives of a function which is not smooth in the conventional sense.

11.3.3. Definition. A function f in $\mathscr{L}_2^{\text{loc}}$ (Definition 2.5.2) is said to have an αth **weak derivative** iff there is a $g \in \mathscr{L}_0^{\text{loc}}$ such that

$$\int_\Omega g\phi \,\mathrm{d}x = (-1)^{|\alpha|} \int_\Omega f.(D^\alpha \phi)\,\mathrm{d}x$$

for all $\phi \in \mathscr{C}_0^\infty$. g is called the αth **weak derivative** of f, and we write $D^\alpha f = g$.

The following remarks are intended to clarify the idea of a weak derivative.

(i) Since $\phi, D^\alpha\phi$ have compact support, both integrals in the definition exist.

(ii) Weak derivatives are essentially an $\mathscr{L}_2$ notion, and as usual in this context functions equal almost everywhere are identified. With this understanding weak derivatives are unique. For if g_1 and g_2 are both weak αth derivatives of f,

$$\int_\Omega (g_1 - g_2)\phi \,\mathrm{d}x = 0 \qquad (\phi \in \mathscr{C}_0^\infty).$$

Now for any compact set $S \subset \Omega$, $g_1 - g_2 \in \mathscr{L}_2(S)$, and $\mathscr{C}_0^\infty(S)$ is dense in $\mathscr{L}_2(S)$. It follows that $g_1 = g_2$ a.e. in S, and as S is arbitrary, a.e. in Ω.

(iii) If a function has an αth derivative g in the ordinary sense lying in $\mathscr{L}_2^{\text{loc}}$, then g is the weak αth derivative of f. To see this simply consider the left-hand side in the definition and integrate by parts.

(iv) The weak derivatives may be thought of as averaging out the discontinuities in f. However, note that f may have an ordinary derivative almost everywhere without having a weak derivative. For example if $f(x) = 1$ $(x > 0)$, $f(x) = 0$ $(x < 0)$, then

$$\int_{-1}^{1} f\phi' \,\mathrm{d}x = \int_0^1 \phi' \,\mathrm{d}x = -\phi(0),$$

and since there is no $g \in \mathscr{L}_2^{\text{loc}}$ such that $\phi(0) = \int_{-1}^{1} g\phi \,\mathrm{d}x$ for all $\phi \in \mathscr{C}_0^\infty$, f does not have a weak derivative.

(v) In one dimension there is a simple characterization of functions with weak first derivative. They are just those functions which are absolutely continuous and have first derivative in $\mathscr{L}_2^{\text{loc}}$ (Problem 11.6).

(vi) The averaging property of the weak derivative has the nice consequence that an exchange of order of differentiation is always permitted. For $D_iD_j\phi =$

$D_jD_i\phi$ when $\phi \in \mathscr{C}_0^\infty$, and if f has weak derivative D_iD_jf, then

$$\int_\Omega D_iD_jf \,.\, \phi \,\mathrm{d}x = \int_\Omega f \,.\, D_jD_i\,\phi \,\mathrm{d}x = \int_\Omega f \,.\, D_iD_j\phi \,\mathrm{d}x = \int_\Omega D_jD_if \,.\, \phi \,\mathrm{d}x,$$

whence $D_iD_jf = D_jD_if$.

(vii) If f is a solution of $lf \overset{w}{=} g$ we cannot immediately conclude that each term in lf can be interpreted as a weak derivative. Certainly this will follow from the definition if l has just one term, but the argument breaks down for general l. We shall see in Section 6 that for strongly elliptic l the conclusion is broadly correct but non-trivial to prove.

The Hilbert spaces in which the analysis will be carried out are now introduced. The norm may be regarded as measuring the average value of the weak derivatives.

11.3.4 Definition. Let m be a non-negative integer. Denote by $\mathscr{H}^m$ (or $\mathscr{H}^m(\Omega)$ if the domain requires emphasis) the set of functions f such that for $0 \leqslant |\alpha| \leqslant m$ all the weak derivatives $D^\alpha f$ exist and are in $\mathscr{L}_2$, and equip $\mathscr{H}^m$ with an inner product and norm as follows:

$$(f, g)_m = \sum_{|\alpha| \leqslant m} \int_\Omega D^\alpha f \,.\, \overline{D^\alpha g} \,\mathrm{d}x,$$

$$\|f\|_m^2 = (f, f)_m = \sum_{|\alpha| \leqslant m} \int_\Omega |D^\alpha f|^2 \,\mathrm{d}x.$$

$\mathscr{H}^m$ is known as a **Sobolev space** of order m.

$\mathscr{H}^m$ is a proper subset of the set of functions with mth weak derivatives, for the $D^\alpha f$ are required to be in $\mathscr{L}_2$—and not just in $\mathscr{L}_2^{\mathrm{loc}}$. Evidently $\mathscr{H}^0 = \mathscr{L}_2$ and $(\cdot, \cdot)_0$, $\|\cdot\|_0$ are the inner product and norm of $\mathscr{L}_2$; for consistency with the higher order Sobolev spaces, in the notation in this chapter the suffix zero is retained. An obvious relation is

$$\mathscr{C}^\infty \subset \ldots \subset \mathscr{H}^{m+1} \subset \mathscr{H}^m \subset \ldots \subset \mathscr{H}^0 = \mathscr{L}_2.$$

11.3.5 Theorem. *$\mathscr{H}^m$ is a Hilbert space.*

Proof. It is easy to check that $\mathscr{H}^m$ is pre-Hilbert. To prove completeness, let (f_j) be a Cauchy sequence in $\mathscr{H}^m$. Then for $|\alpha| \leqslant m$,

$$\|D^\alpha f_j - D^\alpha f_k\|_0^2 = \int_\Omega |D^\alpha f_j - D^\alpha f_k|^2 \,\mathrm{d}x \leqslant \sum_{|\alpha| \leqslant m} \int_\Omega |D^\alpha f_j - D^\alpha f_k|^2 \,\mathrm{d}x$$
$$= \|f_j - f_k\|_m^2.$$

Hence $(D^\alpha f_j)$ is Cauchy, and so convergent, in $\mathscr{L}_2$ with limit $f^{(\alpha)}$ say. Thus for $\phi \in \mathscr{C}_0^\infty$

$$(f^{(0)}, D^\alpha\phi)_0 = \lim(f_j, D^\alpha\phi)_0 = (-1)^{|\alpha|}\lim(D^\alpha f_j, \phi)_0 = (-1)^{|\alpha|}(f^{(\alpha)}, \phi)_0.$$

Therefore $f^{(0)}$ has weak derivatives $D^\alpha f^{(0)} = f^{(\alpha)}$ for $|\alpha| \leqslant m$, and so is in $\mathscr{H}^m$. Further

$$\|f_j - f^{(0)}\|_m^2 = \sum_{|\alpha| \leqslant m} \int_\Omega |D^\alpha f_j - D^\alpha f^{(0)}|^2 \, dx,$$

whence $f_j \to f^{(0)}$ in $\|\cdot\|_m$. This proves that (f_j) has a limit in $\mathscr{H}^m$ and establishes completeness. □

It is standard that the set $\mathscr{C}^\infty$ of smooth functions is dense in $\mathscr{L}_2$. In other words, if $m = 0$ the closure of $\mathscr{C}^\infty$ in $\mathscr{H}^m$ is $\mathscr{H}^m$ itself. The following theorem, see Friedman (1969, p. 15), states that this is also true for arbitrary m, and gives an alternative description of a function in $\mathscr{H}^m$ as a limit of a sequence of smooth functions.

11.3.6 Theorem. *The closure of $\mathscr{C}^\infty$ in $\|\cdot\|_m$ is $\mathscr{H}^m$.*

The existence of weak solutions locally in $\mathscr{H}^{2m}$ and so having $2m$ weak derivatives may be proved under minimal assumptions on l, see Agmon (1965, p. 49). For such a solution the left-hand side of $lf \overset{w}{=} g$ may be written as a sum of weak derivatives. However, $\mathscr{C}^{2m}(\bar{\Omega}) \subset \mathscr{H}^{2m}$ if Ω is bounded, and it is therefore evident that even smooth functions in $\mathscr{H}^{2m}$ do not vanish on the boundary as is required in the classical homogenous Dirichlet problem. To obtain a weak analogue of this problem a modification of $\mathscr{H}^m$ is needed.

11.3.7 Definition. Let $\mathscr{H}_0^m$ be the closure in $\mathscr{H}^m$ of $\mathscr{C}_0^\infty$. $\mathscr{H}_0^m$ is also called a **Sobolev space** of order m.

As before there is a chain of inclusions

$$\mathscr{C}_0^\infty \subset \ldots \subset \mathscr{H}_0^{m+1} \subset \mathscr{H}_0^m \subset \ldots \subset \mathscr{H}_0^0 = \mathscr{L}_2.$$

Since functions in $\mathscr{C}_0^\infty$ vanish near $\partial\Omega$, functions in their closure $\mathscr{H}_0^m$ may be expected to behave at $\partial\Omega$ in a manner which reflects this fact.

11.3.8 Example. As for $\mathscr{H}^1$, in one dimension a simple description of $\mathscr{H}_0^1$ may be given. For $\Omega = (-1, 1)$, by remark (v) above every $f \in \mathscr{H}_0^1 (\subset \mathscr{H}^1)$ is absolutely continuous and has first derivative in $\mathscr{L}_2$. By the definition, there is a sequence (ϕ_j) in $\mathscr{C}_0^\infty$ such that $\lim\|\phi_j - f\|_1 = 0$. Since ϕ_j has support in Ω,

$$\varphi_j(x) = \int_{-1}^{x} \phi_j'(t)\,dt,$$

and from Schwarz's inequality

$$\left| \phi_j(x) - \int_{-1}^{x} f'(t)\,dt \right| = \left| \int_{-1}^{x} [\phi_j'(t) - f'(t)]\,dt \right| \leqslant 2\|\phi_j' - f'\|_0$$
$$\leqslant 2\|\phi_j - f\|_1.$$

Hence $\phi_j(x) \to \int_{-1}^{x} f'(t)\,dt$ in sup norm, and since $\phi_j \to f$ in $\|\cdot\|_1$ and so in $\mathscr{L}_2$, it follows that $f(x) = \int_{-1}^{x} f'(t)\,dt$ a.e. Thus $f(-1) = 0$, and by a similar argument also $f(1) = 0$. The result in the opposite direction is easy (Problem 11.8), and we conclude that $f \in \mathscr{H}_0^1$ if and only if it is absolutely continuous with $f' \in \mathscr{L}_2$ and f vanishes at ± 1.

11.3.9 Lemma. *Suppose that Ω is bounded and has smooth boundary. If $f \in \mathscr{H}_0^m \cap \mathscr{C}^{m-1}(\overline{\Omega})$, then $\partial^j f/\partial \nu^j = 0$ on $\partial\Omega$ for $0 \leqslant j \leqslant m-1$. On the other hand, if $f \in \mathscr{C}^m(\overline{\Omega})$ and the normal derivatives above vanish on $\partial\Omega$, then $f \in \mathscr{H}_0^m$.*

Proof. See Friedman (1969, p. 3[illegible]). □

If $m = 0$ or $\Omega = \mathbb{R}^n$ then $\mathscr{H}_0^m = \mathscr{H}^m$, but it is obvious from the lemma that in general these spaces are not equal. Except in one dimension a simple characterization of functions in $\mathscr{H}_0^m$ is not usually possible. However, the lemma shows that a smooth function in $\mathscr{H}_0^m$ vanishes together with its first $(m-1)$ normal derivatives on $\partial\Omega$. Since this property is just what is required of a solution of the classical homogeneous Dirichlet problem, a weak analogue of this problem may reasonably be posed as follows: find a solution of $lf \overset{w}{=} g$ in $\mathscr{H}_0^m$. This is called the *generalized Dirichlet problem*, and its study is the primary object of this chapter. The resolution of this problem depends on the properties of Sobolev spaces which are established below.

Suppose f is defined on Ω. For open $\Omega' \supset \Omega$ define an extension of f to Ω', which will also be denoted by f, by requiring that $f = 0$ on $\Omega' \backslash \Omega$. We shall say that f is *extended to Ω' by zero.*

In general an $f \in \mathscr{H}^m(\Omega)$ extended by zero to Ω' will not be in $\mathscr{H}^m(\Omega')$. In fact if such an extension of the characteristic function of Ω is considered, it is obvious that all that can be said is that it is in $\mathscr{H}^0(\Omega')$. However, in view of the above remarks more may be expected of functions in $\mathscr{H}_0^m(\Omega)$.

11.3.10 Lemma. *Suppose that Ω, Ω' are open and $\Omega' \supset \Omega$.*

(i) *If $f \in \mathscr{H}_0^m(\Omega)$ is extended by zero to Ω', then $f \in \mathscr{H}_0^m(\Omega')$. In this sense $\mathscr{H}_0^m(\Omega) \subset \mathscr{H}_0^m(\Omega')$.*

(ii) *If* $f \in \mathscr{H}_0^m(\Omega')$ *and* f *has compact support in* Ω, *then* (*the restriction to* Ω *of*) $f \in \mathscr{H}_0^m(\Omega)$.

Proof. (i) If (ϕ_j) is a sequence in $\mathscr{C}_0^\infty(\Omega)$ with $\phi_j \to f$ in $\|\cdot\|_m$, the extension by zero of each ϕ_j to Ω' is in $\mathscr{C}_0^\infty(\Omega')$, and (ϕ_j) is a Cauchy sequence in $\mathscr{H}_0^m(\Omega')$. The result follows immediately.

(ii) is readily proved using a sequence of mollifiers and is left as an exercise (Problem 11.11). □

Although the functions considered are usually defined on a proper subset of $\mathbb{R}^n$, the properties of the Fourier transform (Theorem 2.6.1) may be usefully exploited if the functions are first extended to $\mathbb{R}^n$ by zero. The following states that if $f \in \mathscr{H}_0^m(\Omega)$, $\hat{f}$ tends rapidly to zero at infinity. This is of course not in general true if merely $f \in \mathscr{H}^m(\Omega)$.

11.3.11 Lemma. *Suppose that* $f \in \mathscr{H}_0^m(\Omega)$ *and extend* f *to* $\mathbb{R}^n$ *by zero. Then for* $|\alpha| \leqslant m$,

$$\widehat{(D^\alpha f)}(\xi) = (-i\xi)^\alpha \hat{f}(\xi), \tag{11.3.5}$$

$\xi^\alpha \hat{f}(\xi) \in \mathscr{L}_2(\mathbb{R}^n)$, *and*

$$\|f\|_m^2 = \int_{\mathbb{R}^n} \sum_{|\alpha| \leqslant m} |\xi^\alpha|^2 \, |\hat{f}(\xi)|^2 \, d\xi. \tag{11.3.6}$$

Proof. For $f \in \mathscr{C}_0^\infty(\Omega)$ integration by parts gives (11.3.5). Now by Lemma 11.3.10 the extension of a general f is in $\mathscr{H}_0^m(\mathbb{R}^n)$, and therefore $D^\alpha f \in \mathscr{L}_2(\mathbb{R}^n)$. Hence for all $\phi \in \mathscr{C}_0^\infty(\mathbb{R}^n)$,

$$(\widehat{D^\alpha f}, \hat{\phi})_0 = (D^\alpha f, \phi)_0 = (-1)^{|\alpha|}(f, D^\alpha \phi)_0 = (-1)^{|\alpha|}(\hat{f}, \widehat{D^\alpha \phi})_0$$

$$= \int_{\mathbb{R}^n} (-i)^{|\alpha|} \xi^\alpha \hat{f}(\xi) \overline{\hat{\phi}}(\xi) \, d\xi.$$

(11.3.5) follows since $\mathscr{C}_0^\infty(\mathbb{R}^n)$ is dense in $\mathscr{L}_2(\mathbb{R}^n)$, and hence so is its image under the isometric isomorphism $f \to \hat{f}$. (11.3.6) is an immediate consequence of the definition of $\|\cdot\|_m$ and the Parseval formula. □

The following theorem is the key to the resolution of the generalized Dirichlet problem, as it will enable us to show that the basic operator in the theory is compact. The restriction that Ω is bounded cannot in general be lifted, but under certain conditions on $\partial\Omega$, the result is also true for $\mathscr{H}^m$, see Agmon (1965, p. 30).

11.3.12 The Rellich Imbedding Theorem. *Suppose that* Ω *is open and bounded, and let* m, k *be non-negative integers with* $m > k$. *Then the imbedding of* $\mathscr{H}_0^m$ *in* $\mathscr{H}_0^k$ *is compact.*

Proof. We must show that any sequence (f_j) in the closed unit ball of $\mathscr{H}_0^m$ has a subsequence convergent in $\|\cdot\|_k$. From the weak sequential compactness of the closed unit ball in Hilbert space (Theorem 6.4.3), there is a subsequence, denoted also by (f_j), weakly convergent in $\mathscr{H}_0^m$ to an f such that $\|f\|_m \leqslant 1$. Now for $0 \leqslant k \leqslant m$ the imbedding of $\mathscr{H}_0^m$ in $\mathscr{H}_0^k$ is continuous, so every continuous linear functional on $\mathscr{H}_0^k$ may be regarded as a continuous linear functional on $\mathscr{H}_0^m$, and it follows that $f_j \rightharpoonup f$ in $\mathscr{H}_0^k$. It will be shown that if $k < m$, then $\|f_j\|_k \to \|f\|_k$, from which the result readily follows (Problem 6.11).

Extend f_j, f to $\mathbb{R}^n$ by zero. Since $f_j \rightharpoonup f$ in $\mathscr{H}_0^0 = \mathscr{L}_2(\Omega)$, from the definition of the Fourier transform, $\hat{f}_j \to \hat{f}$ pointwise, and so

$$\sum_{|\alpha| \leqslant m} |\xi^\alpha|^2 |\hat{f}_j(\xi)|^2 \to \sum_{|\alpha| \leqslant m} |\xi^\alpha|^2 |\hat{f}(\xi)|^2 \tag{11.3.7}$$

for each ξ. By Schwarz's inequality $|\hat{f}_j(\xi)| < c$ for some c dependent only on Ω. Therefore, for any $\tau < \infty$ each term in (11.3.7) is dominated for $|\xi| \leqslant \tau$ by a constant, and so from the Dominated Convergence Theorem 2.4.11,

$$\lim_{j \to \infty} \int_{|\xi| \leqslant \tau} \sum_{|\alpha| \leqslant k} |\xi^\alpha|^2 |\hat{f}_j(\xi)|^2 \, d\xi = \int_{|\xi| \leqslant \tau} \sum_{|\alpha| \leqslant k} |\xi^\alpha|^2 |\hat{f}(\xi)|^2 \, d\xi. \tag{11.3.8}$$

In view of equation (11.3.6) this is almost what is needed. If it can be shown that the contribution from $|\xi| > \tau$ to the integrals is small, the proof will be complete. Take any $\varepsilon > 0$. Since $m > k$ there is a $\tau > 0$ such that

$$\sum_{|\alpha| \leqslant k} |\xi^\alpha|^2 \Big/ \sum_{|\alpha| \leqslant m} |\xi^\alpha|^2 < \varepsilon \qquad (|\xi| \geqslant \tau). \tag{11.3.9}$$

By Lemmas 11.3.10 and 11.3.11, $\xi^\alpha \hat{f}(\xi) \in \mathscr{L}_2(\mathbb{R}^n)$ for $|\alpha| \leqslant m$. Thus

$$\begin{aligned}
\int_{|\xi| > \tau} \sum_{|\alpha| \leqslant k} |\xi^\alpha|^2 |\hat{f}(\xi)|^2 \, d\xi &= \int_{|\xi| > \tau} \sum_{|\alpha| \leqslant m} |\xi^\alpha|^2 \left\{ \sum_{|\alpha| \leqslant k} |\xi^\alpha|^2 \Big/ \sum_{|\alpha| \leqslant m} |\xi^\alpha|^2 \right\} |\hat{f}(\xi)|^2 \, d\xi \\
&\leqslant \varepsilon \|f\|_m^2 \\
&\leqslant \varepsilon,
\end{aligned} \tag{11.3.10}$$

where the first inequality is obtained by using (11.3.9) and (11.3.6). A similar argument shows that (11.3.10) holds with $\hat{f}$ replaced by $\hat{f}_j$. Now from (11.3.8) there is an n_0 such that for $j > n_0$

$$\left| \int_{|\xi| \leqslant \tau} \sum_{|\alpha| \leqslant k} |\xi^\alpha|^2 \{ |\hat{f}_j(\xi)|^2 - |\hat{f}(\xi)|^2 \} \, d\xi \right| < \varepsilon.$$

Thus for $j > n_0$,

$$\big|\|f_j\|_k^2 - \|f\|_k^2\big| = \left|\int_{\mathbb{R}^n} \sum_{|\alpha| \leqslant k} |\xi^\alpha|^2 \{|\hat{f}_j(\xi)|^2 - |\hat{f}(\xi)|^2\}\, d\xi\right|$$
$$\leqslant 3\varepsilon,$$

where $\mathbb{R}^n$ has been split into the interior and exterior of the ball $\bar{S}(0, \tau)$ and the last inequality and (11.3.10) have been used. This shows that $\lim \|f_j\|_k = \|f\|_k$ as required. □

The next result will be used to determine when solutions of the generalized Dirichlet problem are smooth. The bounded domain Ω will be said to have the *cone property* iff there are positive numbers θ, h such that for every $x \in \Omega$ there is a right circular cone with vertex x, angle θ and height h contained in Ω.

11.3.13 Lemma. *Suppose that Ω is open and bounded, and let $m > \frac{1}{2}n$ be an integer. Assume that either $f \in \mathscr{C}_0^\infty$, or that $f \in \mathscr{C}^\infty(\bar{\Omega})$ and Ω has the cone property. Then there is a real number c (depending only on n, m, Ω) such that*

$$\sup_{x \in \bar{\Omega}} |f(x)| \leqslant c\|f\|_m. \tag{11.3.11}$$

Proof. We prove the result for $f \in \mathscr{C}_0^\infty$, and refer to Friedman (1969, p. 22) for the other case. Let ρ be any number greater than the diameter of Ω. Take an arbitrary point P in Ω as centre of coordinates and let r be the distance from P. Extend f to $\mathbb{R}^n$ by zero. Then

$$f(P) = -\int_0^\rho \frac{\partial f}{\partial r}\, dr = b\int_0^\rho r^{m-1} \frac{\partial^m f}{\partial r^m}\, dr,$$

on integrating by parts $(m - 1)$ times, b being a constant depending only on m. Integration over the angle of the ball $S(P, \rho)$ gives

$$f(P) = b' \int_{S(P,\rho)} r^{m-n} \frac{\partial^m f}{\partial r^m}\, d\tau,$$

where τ is the element of volume and b' depends on n, m only. By Schwarz's inequality

$$|f(P)|^2 \leqslant b'^2 \int_{S(P,\rho)} r^{2(m-n)}\, d\tau \int_{S(P,\rho)} \left|\frac{\partial^m f}{\partial r^m}\right|^2 d\tau.$$

The first integral is finite as $m > \frac{1}{2}n$, and depends only on m, n, Ω, while the second is not greater than $\|f\|_m^2$. The result follows as P is arbitrary. □

11.3.14 Sobolev Imbedding Theorem. *Suppose that Ω is open and bounded, and let k be an integer less than $m - \frac{1}{2}n$. If $f \in \mathscr{H}_0^m$, or $f \in \mathscr{H}^m$ and Ω has the* *cone property, then f is equal a.e. to a function in $\mathscr{C}^k(\bar{\Omega})$, and the imbeddings of $\mathscr{H}_0^m$ and $\mathscr{H}^m$ in $\mathscr{C}^k(\bar{\Omega})$ are continuous with norms determined only by n, m, Ω.*

Proof. (11.3.11) applied to $D^\alpha f$ for $|\alpha| \leqslant k$ shows that the imbedding of $\mathscr{H}_0^m$ in $\mathscr{C}^k(\bar{\Omega})$ is continuous on the dense subspace $\mathscr{C}_0^\infty$ of $\mathscr{H}_0^m$. The result follows on extending the imbedding by continuity (Theorem 3.4.4). The proof in the second case is similar. □

11.4 The Generalized Dirichlet Problem

We may now proceed towards our primary goal, the solution of the weak analogue of the classical homogeneous Dirichlet problem. Recall that this is the generalized Dirichlet problem, which consists of finding solutions in $\mathscr{H}_0^m$ of $lf \overset{w}{=} g$, the boundary conditions being modelled by means of the restriction that only solutions in $\mathscr{H}_0^m$ should be allowed. In fact this formulation is rather awkward, for since the order of the Sobolev space is only m, it is not possible to interpret the terms in lf of order greater than m as weak derivatives. A more convenient formulation is suggested by considering Poisson's equation again.

If $f \in \mathscr{H}_0^1$ is a solution of the generalized Dirichlet problem for $l = -\nabla^2$,

$$-\nabla^2 f \overset{w}{=} g, \tag{11.4.1}$$

which from Definition 11.3.2 is equivalent to

$$(f, -\nabla^2\phi)_0 = (g, \phi)_0 \qquad (\phi \in \mathscr{C}_0^\infty).$$

Since $f \in \mathscr{H}_0^1$, it has first order weak partial derivatives, and *one* integration by parts is legitimate. With the components of ∇ interpreted as weak derivatives we obtain

$$\int_\Omega \nabla f \,.\, \nabla\phi \,\mathrm{d}x = \int_\Omega g\bar{\phi}\,\mathrm{d}x. \tag{11.4.2}$$

Now if

$$B[f, \phi] = \int_\Omega \nabla f \,.\, \nabla\bar{\phi}\,\mathrm{d}x,$$

B is a bilinear form (Problem 6.18) on $\mathscr{H}_0^1 \times \mathscr{C}_0^\infty$ and

$$\begin{aligned}
|B[f, \phi]| &= \left|\int_\Omega \sum_{i=1}^n D_i f \,.\, D_i\bar{\phi}\,\mathrm{d}x\right| \\
&\leqslant \sum_{i=1}^n |(D_i f, D_i\phi)_0| \\
&\leqslant \sum_{i=1}^n \|D_i f\|_0 \|D_i\phi\|_0 \\
&\leqslant c\|f\|_1 \|\phi\|_1
\end{aligned}$$

for some c depending only on n. Thus B is bounded, and as $\mathscr{H}_0^1$ is the closure of $\mathscr{C}_0^\infty$ in $\|\cdot\|_1$, B may be extended by continuity to a bounded form on $\mathscr{H}_0^1 \times \mathscr{H}_0^1$, and from (11.4.2)

$$B[f, \phi] = (g, \phi)_0 \qquad (\phi \in \mathscr{H}_0^1). \tag{11.4.3}$$

The problem of finding an $f \in \mathscr{H}_0^1$ satisfying this equation is thus equivalent to (11.4.1), and (11.4.3) is the required alternative form of the generalized Dirichlet problem. It is a rather natural formulation in the setting of $\mathscr{H}_0^1$, for each derivative in $B[f, \phi]$ is defined in the weak sense, whereas in the original problem (11.4.1) the terms in $\nabla^2 f$ cannot be so interpreted.

It is interesting to note that the last formulation is essentially variational. To see this assume for simplicity that the functions are real valued, and let Q be the quadratic functional defined by

$$Q(f) = \int_\Omega |\nabla f|^2 \,\mathrm{d}x - 2\int_\Omega gf \,\mathrm{d}x.$$

The formal Fréchet derivative of Q at ϕ is

$$2\int_\Omega \nabla f \,.\, \nabla\phi \,\mathrm{d}x - 2\int_\Omega g\phi,$$

and the condition that this should vanish is just (11.4.2). Thus the generalized Dirichlet problem may be regarded as the Euler equation for Q. Further, it is easy to check that if $\tilde{f}$ is a solution of (11.4.2), then $Q(\tilde{f} + \phi) \geqslant Q(\tilde{f})$ for any $\phi \in \mathscr{H}_0^1$, which shows that Q takes a minimum at a solution of (11.4.2). This approach may be pursued to prove the existence and uniqueness of a function in $\mathscr{H}_0^1$ minimizing Q, and thus to yield a proof of existence for the generalized Dirichlet problem. This is the direct method of the calculus of variations, which has its historical roots in an argument of Riemann for the Dirichlet integral (see Problem 11.17). A somewhat different line of attack will be followed here.

We return to the generalized Dirichlet problem for arbitrary l, and follow the argument leading to (11.4.3).

11.4.1 Definition. With l as in Definition 11.2.2, for all $f, \phi \in \mathscr{H}_0^m$ set

$$B[f, \phi] = \sum_{|\alpha|, |\beta| \leqslant m} (p_{\alpha\beta} D^\alpha f, D^\beta \phi)_0.$$

B will be called the **bilinear form associated with** l.

11.4.2 Lemma. *B is a bounded bilinear form on $\mathscr{H}_0^m \times \mathscr{H}_0^m$. If l is formally self-adjoint, B is Hermitian.*

Proof. By assumption $p_{\alpha\beta} \in \mathscr{C}(\overline{\Omega})$. Hence

$$\begin{aligned} |B[f, \phi]| &\leqslant \sum_{|\alpha|, |\beta| \leqslant m} \| p_{\alpha\beta} D^\alpha f \|_0 \| D^\beta \phi \|_0 \\ &\leqslant c \sum_{|\alpha|, |\beta| \leqslant m} \| D^\alpha f \|_0 \| D^\beta \phi \|_0 \\ &\leqslant c' \| f \|_m \| \phi \|_m \end{aligned}$$

where c, c' are constants depending only on the $p_{\alpha\beta}$ and m, n. This shows that B is bounded. The remaining assertion follows from Definition 11.2.3. □

11.4.3 Definition. Let l be the formal operator of order $2m$ of Definition 11.2.2. Given $g \in \mathscr{L}_2$ the problem of finding $f \in \mathscr{H}_0^m$ such that

$$B[f, \phi] = (g, \phi)_0 \tag{11.4.4}$$

for all $\phi \in \mathscr{H}_0^m$, is called the **generalized Dirichlet problem.**

This is the final version of the weak analogue of the classical homogeneous Dirichlet problem.

As a first step towards a solution, the problem is recast in abstract form in the single Hilbert space $\mathscr{H}_0^m$. By Schwarz's inequality, $|(g, \phi)_0| \leqslant \|g\|_0 \|\phi\|_0$, therefore, since $\|\phi\|_0 \leqslant \|\phi\|_m$, $|(g, \phi)_0| \leqslant \|g\|_0 \|\phi\|_m$. Thus g^* defined by $g^*(\phi) = (g, \phi)_0$ is a continuous linear functional on $\mathscr{H}_0^m$, and it follows from the Riesz Representation Theorem 6.4.1 that there is a unique $h \in \mathscr{H}_0^m$ such that $(g, \phi)_0 = g^*(\phi) = (h, \phi)_m$. The generalized Dirichlet problem is then to find, for such an $h \in \mathscr{H}_0^m$ an $f \in \mathscr{H}_0^m$ satisfying

$$B[f, \phi] = (h, \phi)_m \qquad (\phi \in \mathscr{H}_0^m). \tag{11.4.5}$$

The method by which this is solved may be clarified if initially the additional assumption is made that B is Hermitian. As B is bounded, by Problem 6.18 there is a unique bounded self-adjoint operator, L say, such that

$$B[f, \phi] = (Lf, \phi)_m \qquad (f, \phi \in \mathscr{H}_0^m),$$

and (11.4.5) becomes $(Lf, \phi)_m = (h, \phi)_m$ for all $\phi \in \mathscr{H}_0^m$, or equivalently $Lf = h$. If $L^{-1} \in \mathscr{L}(\mathscr{H}_0^m)$, this equation, and so the generalized Dirichlet problem, has the unique solution $f = L^{-1}h$. Since this provides a complete answer to existence and uniqueness, we look for a condition on B which will ensure that L has this property. The simplest such condition is that B is strictly positive, that is there exists a $c > 0$ such that $(Lf, f)_m = B[f, f] \geqslant c\|f\|_m^2$, for then by Theorem 6.6.6, $0 \in \rho(L)$. If B is not Hermitian (and so not real-valued), this condition is not appropriate, but a suitable generalization is readily found. It is the following.

11.4.4. Definition. B will be said to be **coercive** iff there is a real number $c > 0$ such that for all $f \in \mathscr{H}_0^m$

$$\operatorname{Re} B[f,f] \geqslant c\|f\|_m^2.$$

11.4.5 The Lax–Milgram Lemma. *Let B be a bounded bilinear form on $\mathscr{H}_0^m \times \mathscr{H}_0^m$. Then there is a unique $L \in \mathscr{L}(\mathscr{H}_0^m)$ such that $B[f, \phi] = (Lf, \phi)_m$ for all $f, \phi \in \mathscr{H}_0^m$. If B is coercive, $L^{-1} \in \mathscr{L}(\mathscr{H}_0^m)$.*

Proof. Since B is bounded, for fixed f, $B[f, \cdot\,] = B^*(\cdot)$, say, is a continuous antilinear functional on $\mathscr{H}_0^m$. Hence by the Riesz Representation Theorem 6.4.1, there is a unique $k \in \mathscr{H}_0^m$ such that $B[f, \phi] = (k, \phi)_m$ for all $\phi \in \mathscr{H}_0^m$, and $\|k\|_m = \|B^*\|$. The relation $k = Lf$ for each $f \in \mathscr{H}_0^m$ then defines an operator L which is evidently linear, which is bounded since $\|B^*\| \leqslant d\|f\|_m$ for some $d \in \mathbb{R}$, and which is such that $B[f, \phi] = (Lf, \phi)_m$.

For B coercive, there is a $c > 0$ such that

$$\|Lf\|_m \|f\|_m \geqslant |B[f,f]| \geqslant \operatorname{Re} B[f,f] \geqslant c\|f\|_m^2.$$

Hence $\|Lf\|_m \geqslant c\|f\|_m$. Also $\overline{R(L)} = \mathscr{H}_0^m$, for if $h \in R(L)^\perp$, $B[h, h] = (Lh, h)_0 = 0$, whence $h = 0$ by coercivity. The result follows from Lemma 3.8.18. □

11.4.6 Theorem. *Let l be the formal operator of order $2m$ of Definition 11.2.2, and let B be the associated bilinear form. Then if B is coercive there is exactly one solution (in $\mathscr{H}_0^m$) of the generalized Dirichlet problem for any given right-hand side $g \in \mathscr{L}_2$.*

Proof. Apply the Lax–Milgram Lemma to (11.4.5). □

11.4.7 Example. Take $l = -\nabla^2 + k + p$ with $k \in \mathbb{R}$ and $p \in \mathscr{C}^\infty(\overline{\Omega})$. Then

$$B[f, \phi] = \int_\Omega [\nabla f . \nabla \bar{\phi} + (k + p) f \bar{\phi}] \, dx,$$

and

$$\begin{aligned} \operatorname{Re} B[f,f] &= \|f\|_1^2 - \|f\|_0^2 + \operatorname{Re}((k + p)f, f)_0 \\ &\geqslant \|f\|_1^2 + (k - 1 + p_0)\|f\|_0^2, \end{aligned}$$

where $p_0 = \inf_{x \in \Omega} \operatorname{Re} p(x)$. Thus if $k \geqslant 1 - p_0$ the form is coercive, and the generalized Dirichlet problem has a unique solution by Theorem 11.4.6. In general more cannot be asserted, but in the important case when Ω is bounded, the condition $k \geqslant 1 - p_0$ can be considerably weakened by using Poincare's inequality (Problem 11.15). The next lemma shows how the argument goes in one case, and proves in particular existence and uniqueness for Laplace's equation.

On the other hand if $p = 0$ and k is a large negative number, B is not coercive and the argument breaks down. This is natural, for since the Laplacian has negative eigenvalues (corresponding to values of k for which $lf = 0$ has non-trivial solutions), existence must fail for some negative k. The best that can be hoped for is a Fredholm alternative. This will be the subject of the next section.

11.4.8 Lemma. *Let Ω be bounded. Suppose that l is homogeneous of degree m with constant coefficients:*

$$lf = \sum_{|\alpha|, |\beta| = m} p_{\alpha\beta} D^{\alpha+\beta} f.$$

Then the associated form is coercive if l is strongly elliptic.

Proof. It is evidently enough to consider functions in $\mathscr{C}_0^\infty$ since this is dense in $\mathscr{H}_0^m$. Take any $\phi \in \mathscr{C}_0^\infty$ and extend ϕ to $\mathbb{R}^n$ by zero. Then

$$\begin{aligned}
\operatorname{Re} B[\phi, \phi] &= \operatorname{Re} \int_{\mathbb{R}^n} \sum_{|\alpha| = |\beta| = m} p_{\alpha\beta} \xi^{\alpha+\beta} |\hat{\phi}(\xi)|^2 \, d\xi \\
&\geqslant c \int_{\mathbb{R}^n} |\xi|^{2m} |\hat{\phi}(\xi)|^2 \, d\xi \qquad \text{(by ellipticity)} \\
&\geqslant c' \int_{\mathbb{R}^n} \sum_{|\alpha| = m} |\xi^\alpha|^2 |\hat{\phi}(\xi)|^2 \, d\xi \\
&= c' \int_{\mathbb{R}^n} \sum_{|\alpha| = m} |D^\alpha \phi|^2 \, dx,
\end{aligned} \tag{11.4.6}$$

c, c' being strictly positive numbers independent of ϕ. Now it is an immediate consequence of Poincare's inequality (Problem 11.15) that there is an $a > 0$ such that for all $\phi \in \mathscr{C}_0^\infty$,

$$(1 + a) \int_{\mathbb{R}^n} \sum_{|\alpha| = m} |D^\alpha \phi|^2 \, dx \geqslant \int_{\mathbb{R}^n} \sum_{|\alpha| \leqslant m} |D^\alpha \phi|^2 \, dx = \|\phi\|_m^2.$$

Combining this with (11.4.6) we obtain

$$\operatorname{Re} B[\phi, \phi] \geqslant c'(1 + a)^{-1} \|\phi\|_m^2. \qquad \square$$

When B is coercive, existence and uniqueness for the generalized Dirichlet problem are guaranteed by Theorem 11.4.6. It will be seen below that the solution f depends continuously on the right-hand side g in the sense that $f = Gg$ where $G : \mathscr{L}_2 \to \mathscr{H}_0^m$ is a bounded linear operator.

Since $\mathscr{H}_0^m \subset \mathscr{H}_0^0 = \mathscr{L}_2$, any $\phi \in \mathscr{H}_0^m$ can also be regarded as an element of $\mathscr{L}_2$, and often no confusion arises if we do this. However, it is now necessary to distinguish between these two possibilities, and to this end the following definition is introduced.

11.4.9 Definition. Let $K: \mathcal{H}_0^m \to \mathcal{L}_2$ be the imbedding of $\mathcal{H}_0^m$ in $\mathcal{L}_2$, and let its adjoint be $K^*: \mathcal{L}_2 \to \mathcal{H}_0^m$.

Obviously K and K^* are bounded linear operators with norms not greater than unity. One advantage of defining K explicitly is that the properties of its adjoint may be exploited. From Definition 6.5.6, if $g \in \mathcal{L}_2$ and $\phi \in \mathcal{H}_0^m$,

$$(g, \phi)_0 = (g, K\phi)_0 = (K^*g, \phi)_m. \tag{11.4.7}$$

Now from the Lax–Milgram Lemma 11.4.5 and equation (11.4.5), the solution f of the generalized Dirichlet problem is $L^{-1}h$, where h is related to the right-hand side g by $(h, \phi)_m = (g, \phi)_0$ for all $\phi \in \mathcal{H}_0^m$. Therefore from (11.4.7), $h = K^*g$ and $f = L^{-1}K^*g$. This proves that f depends continuously on g, for $L^{-1}K^*$ is bounded.

11.4.10 Definition. The operators $G = L^{-1}K^*: \mathcal{L}_2 \to \mathcal{H}_0^m$ and $\tilde{G} = KL^{-1}K^*: \mathcal{L}_2 \to \mathcal{L}_2$ will be called **Green's operators.**

The term Green's operator is used because G plays the role here taken in the classical theory by an integral operator with kernel the Green's function. An advantage of the present method is that all the properties of G required for proofs of existence and uniqueness may be obtained without a detailed examination of the Green's function. The results of this section are summarized as follows.

11.4.11 Theorem. *Suppose that the form associated with l is coercive. Then for any right-hand side $g \in \mathcal{L}_2$, the solution of the generalized Dirichlet problem is $f = Gg$, where $G = L^{-1}K^*$ is a bounded linear operator from $\mathcal{L}_2$ into $\mathcal{H}_0^m$.*

11.4.12 Corollary. *Assume that in addition l is formally self-adjoint. Then $\tilde{G} = KL^{-1}K^*$ is a bounded self-adjoint operator $\mathcal{L}_2 \to \mathcal{L}_2$.*

Proof. Since B is Hermitian, L is self-adjoint, whence so also is $\tilde{G}$. □

11.5 A Fredholm Alternative for the Generalized Dirichlet Problem

Theorem 11.4.6 settles existence and uniqueness for the generalized Dirichlet problem when the form B associated with l is coercive. However, it is clear from elementary examples (such as the operator $-\nabla^2 + k$ of Example 11.4.7) that if B is not coercive, the corresponding homogeneous equation (that is with right-hand side $g = 0$) may have non-trivial solutions. These examples also suggest that uniqueness and existence for arbitrary g can be expected if and only if there are no such solutions. We shall now prove that this powerful analogue of the Fredholm Alternative is in fact true for all strongly elliptic operators on a bounded domain. The proof is based on the compactness of the Green's operator, and is reminiscent of the argument used in the classical integral equation approach.

The method to be used is suggested by the tactics for the simpler equation

$$Mf = g, \tag{11.5.1}$$

where $\mathscr{H}$ is a Hilbert space and $M \in \mathscr{L}(\mathscr{H})$ is self-adjoint. If M is strictly positive then $0 \in \rho(M)$ and (11.5.1) has the unique solution $f = M^{-1}g$. However, suppose that instead of positivity only the weaker condition $(Lf, f) \geqslant -b\|f\|^2$ $(b > 0)$ holds. Then for $a > b$, still $M_a = M + aI$ is strictly positive, $0 \in \rho(M_a)$, and M_a has a bounded inverse. A rearrangement of (11.5.1) gives $M_a f = g + af$, and then

$$f = aM_a^{-1}f + \bar{g} \qquad (\bar{g} = M_a^{-1}g). \tag{11.5.2}$$

Existence and uniqueness for (11.5.1) may thus be settled by considering (11.5.2). If M_a^{-1} is compact, the advantage of using (11.5.2) is apparent, for the Fredholm Alternative for compact operators may be applied.

In adapting this argument to the generalized Dirichlet problem the fact that two Hilbert spaces $\mathscr{H}_0^m$ and $\mathscr{L}_2 = \mathscr{H}_0^0$ appear and that coercive forms rather than strictly positive operators must be dealt with slightly complicates matters, but the technique is similar. We show that by adding a suitable term to B a new form is obtained which is coercive (and so has a bounded Green's operator). That this is the case is a consequence of the fundamental Gårding's inequality. For the somewhat complicated proof of this result see Friedman (1969, p. 34).

11.5.1 Theorem (Gårding's Inequality). *Suppose that Ω is bounded, and assume that the formal operator l of Definition 11.2.2 is strongly elliptic. Then there are real numbers $c > 0$ and a such that for all $\phi \in \mathscr{H}_0^m$,*

$$\operatorname{Re} B[\phi, \phi] \geqslant c\|\phi\|_m^2 - a\|\phi\|_0^2.$$

11.5.2 Corollary. *With the assumptions of the theorem, the form B_a defined by*

$$B_a[f, \phi] = B[f, \phi] + a(f, \phi)_0$$

is coercive.

Proof. By Gårding's inequality, for all $\phi \in \mathscr{H}^m_0$,

$$\begin{aligned}\operatorname{Re} B_a[\phi, \phi] &= \operatorname{Re} B[\phi, \phi] + a\|\phi\|_0^2 \\ &\geqslant c\|\phi\|_m^2 - a\|\phi\|_0^2 + a\|\phi\|_0^2 \\ &= c\|\phi\|_m^2.\end{aligned}$$

□

The symbols L_a, G_a, $\tilde{G}_a$ will henceforth denote the operators derived from B_a in the same manner as L, G, $\tilde{G}$ respectively are derived from B.

The Rellich Imbedding Theorem is now invoked in order to prove the compactness of the Green's operator. As remarked previously the condition that Ω is bounded cannot in general be dispensed with.

11.5.3 Theorem. *If Ω is bounded, the Green's operator $\tilde{G}_a : \mathscr{L}_2 \to \mathscr{L}_2$ corresponding to a coercive form B_a is compact.*

Proof. By Theorem 11.3.12 the imbedding K of $\mathscr{H}^m_0$ in $\mathscr{H}^0_0 = \mathscr{L}_2$ is compact. The compactness of $\tilde{G}_a = KL_a^{-1}K^*$ follows since L_a^{-1} and K^* are continuous.

□

11.5.4 Definition. A complex number λ will be called an **eigenvalue** of the generalized Dirichlet problem iff there is a non-zero $f \in \mathscr{H}^m_0$ such that

$$B[f, \phi] = \lambda(f, \phi)_0 \qquad (\phi \in \mathscr{H}^m_0),$$

and f will be known as an **eigenfunction.**

11.5.5 Theorem (Fredholm Alternative). *Suppose that Ω is open and bounded, and let the formal operator l of Definition 11.2.2 be strongly elliptic. Then either the generalized Dirichlet problem has exactly one solution for any $g \in \mathscr{L}_2$, or zero is an eigenvalue.*

Proof. By Corollary 11.5.2 and Theorem 11.5.3, for some $a \in \mathbb{R}$ the Green's operator $\tilde{G}_a$ is compact. We first show that the generalized Dirichlet problem is equivalent to the equation

$$f = a\tilde{G}_a\tilde{f} + \tilde{g}, \tag{11.5.3}$$

where $\tilde{g} = \tilde{G}_a g$ and $\tilde{f} = Kf$. (f and $\tilde{f}$ are the same function, but a distinction is made in the notation in the proof to emphasize that they are being regarded as elements of $\mathscr{H}^m_0$ and $\mathscr{L}_2$ respectively. To be absolutely precise, we should write the generalized Dirichlet problem as $B[f, \phi] = (g, K\phi)_0$, since the first appearance of ϕ is as an element of $\mathscr{H}^m_0$, the second as an element of $\mathscr{L}_2$. However, here as elsewhere this somewhat pedantic notation is avoided).

First, addition of $a(\tilde{f}, \phi)_0$ to each side of $B[f, \phi] = (g, \phi)_0$ yields

$$B_a[f, \phi] = (a\tilde{f} + g, \phi)_0,$$

and from Theorem 11.4.11, $\tilde{f} = a\tilde{G}_a\tilde{f} + \tilde{G}_a g$. Suppose on the other hand that $\tilde{f} \in \mathscr{L}_2$ satisfies (11.5.3), and set $f = G_a(a\tilde{f} + g)$. Then $f \in \mathscr{H}_0^m$ and

$$Kf = KG_a(a\tilde{f} + g) = \tilde{G}_a(a\tilde{f} + g) = \tilde{f}.$$

Thus for all $\phi \in \mathscr{H}_0^m$,

$$\begin{aligned} B_a[f, \phi] &= (L_a f, \phi)_m \\ &= (L_a G_a(a\tilde{f} + g), \phi)_m \\ &= (K^*(a\tilde{f} + g), \phi)_m \\ &= (a\tilde{f} + g, K\phi)_0 \\ &= (a\tilde{f} + g, \phi)_0. \end{aligned}$$

It follows on subtracting $(a\tilde{f}, \phi)_0$ from each side of this equation that f is a solution of the generalized Dirichlet problem.

To complete the proof recall that $\tilde{G}_a$ is compact and apply the Fredholm Alternative Theorem 7.3.7 to (11.5.3). □

The compactness of $\tilde{G}_a$ may be further exploited to show that if zero is an eigenvalue, the generalized Dirichlet problem has a solution if and only if g is orthogonal to all the (finite number of) corresponding eigenfunctions.

We conclude with two results which emphasize the analogy between the spectral properties of ordinary and partial differential equations.

11.5.6 Theorem. *Assume that the conditions of the last theorem hold. Then either the generalized Dirichlet problem for $l - \lambda$ has exactly one solution for any $g \in \mathscr{L}_2$, or λ is an eigenvalue. The eigenvalues have no finite limit point and to each eigenvalue there corresponds only a finite number of linearly independent eigenfunctions.*

Proof. Apply Theorem 11.5.5 to $l - \lambda$ and use Theorems 7.4.1 and 7.4.2. □

11.5.7 Theorem. *Assume that the conditions of Theorem 11.5.5 hold, and suppose in addition that l is formally self-adjoint. Then the eigenfunctions form a basis for $\mathscr{L}_2$.*

Proof. By the argument used in the proof of Theorem 11.5.5, to a solution of $\mu_n\tilde{f} = a\tilde{G}_a\tilde{f}$ there corresponds an eigenvector f and eigenvalue $\lambda_n = a(\mu_n^{-1} - 1)$. Since $\tilde{G}_a$ is self-adjoint (Corollary 11.4.12), the result will follow from the Hilbert–Schmidt Theorem 7.5.1 if it can be shown that zero is not an eigenvalue of $\tilde{G}_a$. To prove this suppose that $\tilde{G}_a \tilde{f} = 0$. Then

$$0 = (KL_a^{-1}K^* \tilde{f}, \tilde{f})_0 = (L_a^{-1}K^* \tilde{f}, K^* \tilde{f})_m = B_a[K^* \tilde{f}, K^* \tilde{f}],$$

and since B_a is coercive, $K^* \tilde{f} = 0$. Therefore, for all $\phi \in \mathscr{H}_0^m$,

$$0 = (K^* \tilde{f}, \phi)_m = (\tilde{f}, K\phi)_0 = (\tilde{f}, \phi)_0,$$

whence $\tilde{f} = 0$ as $\mathscr{H}_0^m$ is dense in $\mathscr{L}_2$. □

11.6 Smoothness of Weak Solutions

The analysis of the previous sections provides natural criteria for the existence and uniqueness of solutions of the generalized Dirichlet problem. In applications it is sometimes relevant to ask further whether such a solution is smooth enough to be a classical solution of the Dirichlet problem. If this is to be the case, first f should have $2m$ continuous derivatives in Ω, and second f should be smooth enough for the boundary conditions to be meaningful—that is f should have $m - 1$ continuous derivatives in the closure $\bar{\Omega}$. These properties are called "interior regularity" and "regularity up to the boundary" respectively, and the position is broadly that f has both these properties if the right-hand side g and the boundary $\partial\Omega$ are reasonably smooth. The proof of this assertion presents quite considerable technical difficulties, and since our primary interest is in the generalized Dirichlet problem we shall quote the main result and refer the reader to one of the cited texts for a proof. Nonetheless in order to give the flavour of the type of argument that is used, we shall sketch a proof of interior regularity in one relatively simple but important case—when the principal part of l has constant coefficients. In outline the method is to show that the order of the Sobolev space to which f belongs can be raised step by step to $2m + k$ if $g \in \mathscr{H}^k$. The continuity of the derivatives then follows from the Sobolev Imbedding Theorem 11.3.14.

In tackling interior regularity, rather than use the divergence form of Definition 11.2.2 it is more convenient to take l as follows.

11.6.1 Definition. Assume that $p_\alpha \in \mathscr{C}^\infty(\bar{\Omega})$ for $|\alpha| \leqslant 2m$. The formal operator l of order $2m$, its adjoint l^*, and its principal part l_p are defined as follows:

$$lf = \sum_{|\alpha| \leqslant 2m} p_\alpha D^\alpha f,$$

$$l^* f = \sum_{|\alpha| \leqslant 2m} (-1)^{|\alpha|} D^\alpha(\bar{p}_\alpha f),$$

$$l_p f = \sum_{|\alpha| = 2m} p_\alpha D^\alpha f.$$

l is said to be **strongly elliptic** iff there is a $c > 0$ such that for all real ξ, and all $x \in \overline{\Omega}$,

$$(-1)^m \operatorname{Re} l_p(\xi) = (-1)^m \operatorname{Re} \sum_{|\alpha| = 2m} p_\alpha \xi^\alpha \geqslant c|\xi|^{2m}.$$

If $l, l_1, \ldots, l_r$ are formal operators, the relation

$$lf \overset{w}{=} l_1 g_1 + \ldots + l_r g_r$$

on Ω will mean that for all $\phi \in \mathscr{C}_0^\infty$,

$$(f, l^*\phi)_0 = (g_1, l_1^*\phi) \square \ldots + (g_r, l_r^*\phi)_0.$$

Interior regularity is a local property. f is differentiable on the open set Ω if it is differentiable on a neighbourhood of each point of Ω. It follows intuitively that the boundary of Ω should not have any significance in the discussion. In particular it should not matter whether f lies in $\mathscr{H}_0^m$ or $\mathscr{H}^m$, and indeed interior regularity will be established for all solutions of $lf \overset{w}{=} g$ if g is smooth enough. In order to exploit the fact that only a local result is sought, for a point P in Ω an equation for ψf is formulated where $\psi = 1$ on a neighbourhood S of P and $\psi \in \mathscr{C}_0^\infty(\Omega)$. Then if $f \in \mathscr{H}^k$, it follows that $\psi f \in \mathscr{H}_0^k$, and by Lemma 11.3.10 ψf may be extended by zero to $\mathbb{R}^n$, and $\psi f \in \mathscr{H}_0^k(\mathbb{R}^n)$. The advantage of this approach is that it is quite easy to tackle regularity on $\mathbb{R}^n$ since the Fourier transform may be used. The behaviour of f on S itself is readily deduced since $\psi = 1$ on S. We start then with two results in $\mathbb{R}^n$.

11.6.2 Lemma. *Suppose that $f \in \mathscr{L}_2(\mathbb{R}^n)$. Then $f \in \mathscr{H}^k(\mathbb{R}^n)$ if and only if $(1 + |\xi|)^k \hat{f}(\xi) \in \mathscr{L}_2(\mathbb{R}^n)$.*

Proof. Recall that $\mathscr{H}_0^k(\mathbb{R}^n) = \mathscr{H}^k(\mathbb{R}^n)$ and proceed as in the proof of Lemma 11.3.11. □

11.6.3 Lemma. *Assume that l_P has constant coefficients and is strongly elliptic. Suppose that on $\mathbb{R}^n$*

$$(1 + l_P) f \overset{w}{=} \sum_{|\alpha| < 2m} D^\alpha g_\alpha,$$

where $g_\alpha \in \mathscr{H}^{k_\alpha}(\mathbb{R}^n)$. Then $f \in \mathscr{H}^j(\mathbb{R}^n)$ where

$$j = \min_{|\alpha| < 2m} (2m + k_\alpha - |\alpha|).$$

Proof. It is given that for all $\phi \in \mathscr{C}_0^\infty(\mathbb{R}^n)$,

$$(f, (1 + l_P^*)\phi)_0 = \sum_{|\alpha| < 2m} (g_\alpha, D^\alpha \phi)_0,$$

and since $\mathscr{C}_0^\infty(\mathbb{R}^n)$ is dense in $\mathscr{L}_2(\mathbb{R}^n)$, it follows from the Plancherel formula that

$$[1 + l_p(\xi)]\hat{f}(\xi) = \sum_{|\alpha| < 2m} (-i)^{|\alpha|} \xi^\alpha \hat{g}_\alpha(\xi).$$

Using the ellipticity, we deduce after rearranging that for some $c > 0$,

$$c(1 + |\xi|)^j |\hat{f}(\xi)| \leqslant \sum_{|\alpha| < 2m} (1 + |\xi|)^{j-(2m+k_\alpha-|\alpha|)} \{(1 + |\xi|)^{k_\alpha} |\hat{g}_\alpha(\xi)|\}.$$

By Lemma 11.6.2 each term in the curly brackets lies in $\mathscr{L}_2(\mathbb{R}^n)$, while the factors multiplying these terms are bounded for the stated values of j. Hence each term on the right-hand side, and therefore the left-hand side, is in $\mathscr{L}_2(\mathbb{R}^n)$, and the result follows from Lemma 11.6.2. □

Following the tactics outlined above, to recast the equation $lf \overset{w}{=} g$ in a form to which the last lemma can be applied, a new equation for ψf is formulated. This requires nothing more in principle than liberal use of Leibnitz's Theorem for weak derivatives (Problem 11.12). We spare the reader the tedious details.

11.6.4 Lemma. *Let S be an open subset of Ω, and assume that $\psi \in \mathscr{C}_0^\infty(S)$. Let $g \in \mathscr{H}^k(\Omega)$, and suppose that $f \in \mathscr{H}^t(S)$ satisfies $lf \overset{w}{=} g$ on S. Then there are functions $\psi_\alpha \in \mathscr{C}_0^\infty(S)$ such that on $\mathbb{R}^n$,*

$$[1 + l_p)(\psi f) \overset{w}{=} \psi g + \sum_{|\alpha| < 2m} D^\alpha(\psi_\alpha f), \tag{11.6.1}$$

and (the extensions by zero of) $\psi g \in \mathscr{H}^k(\mathbb{R}^n)$, $\psi_\alpha f \in \mathscr{H}^t(\mathbb{R}^n)$.

The highest derivative on the right-hand side of (11.6.1) is of order $(2m - 1)$. Therefore, from Lemma 11.6.3, $\psi f \in \mathscr{H}^j(\mathbb{R}^n)$ where $j = \min(2m + k, t + 1)$. This proves that unless already $t = 2m + k$, the order of the Sobolev space to which ψf belongs can be raised from t to $t + 1$. Now by Problem 11.10, for any bounded open S' with $\bar{S}' \subset S$ there is a $\psi \in \mathscr{C}_0^\infty(S)$ with $\psi = 1$ on S'. Hence $f \in \mathscr{H}^{t+1}(S')$. An inductive application of this argument proves that $f \in \mathscr{H}^{2m+k}(\Omega')$ for any bounded open Ω' such that $\overline{\Omega}' \subset \Omega$. Finally the Sobolev Imbedding Theorem 11.3.14 is applied to $\psi f \in \mathscr{H}_0^{2m+k}(\Omega)$ to prove continuity of the derivatives. This establishes interior regularity.

11.6.5 Theorem. *Let Ω be an open set in $\mathbb{R}^n$. Assume that the formal operator l of order $2m$ in Definition 11.6.1 is strongly elliptic and that the principal part of l has constant coefficients. Suppose that $g \in \mathscr{H}^k(\Omega)$ and that f is a solution in $\mathscr{L}_2$ of $lf \overset{w}{=} g$. Then $f \in \mathscr{H}^{2m+k}(\Omega')$ for any bounded open set Ω' such that $\overline{\Omega}' \subset \Omega$. Further $f \in \mathscr{C}^s(\Omega')$ for integers $s < 2m + k - \frac{1}{2}n$, and $f \in \mathscr{C}^\infty(\Omega')$ if $g \in \mathscr{C}^\infty(\Omega)$.*

The assumption that l_p has constant coefficients simplifies the technicalities. For a proof in the general case see Friedman [1969, Section 1.15].

For regularity up to the boundary some smoothness is required of $\partial\Omega$. For the somewhat complicated proof of the following and further discussion see Friedman (1969, Section 1.17).

11.6.6 Theorem. *Let Ω be a bounded domain with $\mathscr{C}^\infty$ boundary, and let the formal operator l of Definition 11.2.2 be strongly elliptic. Assume that f is a solution of the generalized Dirichlet problem with right-hand side $g \in \mathscr{H}^k(\Omega)$. Then $f \in \mathscr{H}^{2m+k}(\Omega)$. Also $f \in \mathscr{C}^s(\overline{\Omega})$ for integers $s < 2m + k - \frac{1}{2}n$.*

Problems

Throughout Ω will be an open subset of $\mathbb{R}^n$.

11.1 Suppose that l in Definition 11.2.2 is strongly elliptic on the bounded domain Ω. Make the change of variable $y = \psi(x)$ where $\psi: \overline{\Omega} \to \mathbb{R}^n$ is in $\mathscr{C}^\infty(\overline{\Omega}, \mathbb{R}^n)$. If ψ has non-vanishing Jacobian determinant in $\overline{\Omega}$, show that the new operator is also strongly elliptic.

11.2 Show that $(-1)^k (\nabla^2)^k$ is strongly elliptic.

11.3 The operator l in Definition 11.6.1 is said to be *elliptic* if

$$\sum_{|\alpha|=2m} p_\alpha \xi^\alpha \neq 0 \qquad (\xi \neq 0)$$

for any real ξ and any $x \in \overline{\Omega}$. If complex coefficients are allowed, ellipticity is not enough for the Dirichlet problem to be well posed. For take Ω to be the open unit ball in $\mathbb{R}^2$ and consider $lf = f_{xx} + 2if_{xy} - f_{yy}$. Prove that l is elliptic but not strongly elliptic. With $z = x + iy$ show that $f(x, y) = (1 - |z|^2)u(z)$ is a solution of $lf = 0$ for any analytic function u, but that f vanishes on $\partial\Omega$.

11.4 Let Ω' be a compact subset of Ω. Prove that there is an $\varepsilon > 0$ such that $S(x, \varepsilon) \cap \Omega' = \varnothing$ for every $x \in \partial\Omega$.

11.5 Take $\Omega = (-1, 1)$ and let $\tau \in \mathscr{C}_0^\infty$ be such that $\int_\Omega \tau \, dx = 1$. If $\psi \in \mathscr{C}_0^\infty$ show that also $\phi \in \mathscr{C}_0^\infty$, where

$$\phi(x) = \int_{-1}^{x} \psi(t)\,dt - \int_{-1}^{1} \psi(t)\,dt \, . \int_{-1}^{x} \tau(s)\,ds.$$

Deduce that if $h \in \mathscr{L}_2^{\text{loc}}$ and $\int_\Omega h\phi' \, dx = 0$ for all $\phi \in \mathscr{C}_0^\infty$, then h is (equal a.e. to) a constant.

11.6 With $\Omega = (-1, 1)$ prove that an $f \in \mathscr{L}_2^{\text{loc}}$ has weak first derivative iff f is absolutely continuous and $f' \in \mathscr{L}_2^{\text{loc}}$.

11.7 For $\Omega = (0, 1)$ prove that the closure of $\mathscr{C}^\infty(\bar{\Omega})$ in $\|\cdot\|_1$ is $\mathscr{H}^1$ itself (cf. Theorem 11.3.6). One possibility is to take $\phi(x) = \cos n\pi x$ $(n = 0, 1, \ldots)$ in

$$(f, \phi)_1 = \int_0^1 f(\bar{\phi} - \bar{\phi}'')\,dx + \bar{\phi}'(1)f(1) - \bar{\varphi}'(0)f(0).$$

11.8 Complete the proof of Example 11.3.8.

11.9 Let $\{j_\varepsilon\}$ be a family of mollifiers (Definition 2.6.6). Show that if $f \in \mathscr{H}^m(\Omega)$, and Ω_0 is a compact subset of Ω, then as $\varepsilon \to 0$, $j_\varepsilon * f \to f$ in $\mathscr{H}^m(\Omega_0)$.

11.10 Assume that $\Omega_0 \subset \Omega$ is compact. Show that there is a real-valued $g \in \mathscr{C}_0^\infty$ such that $g(x) = 1$ $(x \in \Omega_0)$ and $0 \leqslant g(x) \leqslant 1$ $(x \in \Omega)$. (Hint: use mollifiers).

11.11 Prove that if $f \in \mathscr{H}^m(\Omega)$ has support in a compact set contained in Ω, then $f \in \mathscr{H}_0^m(\Omega)$.

11.12 Let $\phi \in \mathscr{C}_0^\infty$. If f has weak derivatives up to order m, show that so also does ϕf, and prove the n dimensional version of Leibnitz's Theorem:

$$D^\alpha(\phi f) = \sum_{\beta \leqslant \alpha} \binom{\alpha}{\beta} D^\beta f\, D^{\alpha-\beta}\phi,$$

where $\beta \leqslant \alpha$ means $\beta_i \leqslant \alpha_i$ for $1 \leqslant i \leqslant n$, and where

$$\binom{\alpha}{\beta} = \binom{\alpha_1}{\beta_1} \cdots \binom{\alpha_n}{\beta_n}.$$

11.13 Suppose $\psi \in \mathscr{C}_0^\infty(\mathbb{R}^n)$ is equal to 1 on $\bar{S}(0, 1)$ and has support in $\bar{S}(0, 2)$, and set $\psi_j(x) = \psi(x/j)$. If $f \in \mathscr{H}^m(\mathbb{R}^n)$ show that as $j \to \infty$, $\psi_j f \to f$ in $\|\cdot\|_m$. Deduce that $\mathscr{H}_0^m(\mathbb{R}^n) = \mathscr{H}^m(\mathbb{R}^n)$.

11.14 Show that if $f \in \mathscr{H}_0^1$,

$$\int_0^1 |f|^2\,dx \leqslant b \int_0^1 |f'|^2\,dx$$

where $b = \pi^{-2}$, and prove that this value of b is best possible.

Show that the bilinear form $B[f, \phi] = \int_0^1 (f'\bar{\phi}' - kf\bar{\phi})\,dx$ associated with $l = -d^2/dx^2 - k$ $(k \in \mathbb{R})$ is coercive iff $k < \pi^2$. Interpret this in terms of the homogeneous Dirichlet problem for $-d^2/dx^2$.

11.15 Suppose that $\Omega \subset \mathbb{R}^n$ is bounded and let $d = \sup_{x, y \in \Omega} |x - y|$ be its diameter.

(i) If $f \in \mathscr{C}_0^\infty$,

$$f(x_1, \ldots, x_n) = \int_x^{x_1} \frac{\partial f}{\partial t}(t, x_2, \ldots, x_n)\,dt,$$

where $(x, x_2, \ldots, x_n)$ lies in Ω but outside the support of f. Deduce that

$$\|f\|_0 \leqslant 2^{-\frac{1}{2}} d\, \|\partial f/\partial x_1\|_0.$$

Hence prove that for $f \in \mathscr{H}_0^1$, $\|f\|_0 \leqslant (2n)^{-\frac{1}{2}} d |f|_1$, where

$$|f|_m^2 = \int_\Omega \sum_{|\alpha| = m} |D^\alpha f|^2\,dx.$$

Show that the bilinear form $\int_\Omega(\nabla f.\nabla\bar{\phi} - kf\bar{\phi})\,dx$ associated with $-\nabla^2 - k\,(k \notin \mathbb{R})$ is coercive on $\mathscr{H}_0^1$ if $k < d^2/2n$. Deduce a lower bound for the smallest eigenvalue of the generalized Dirichlet problem for $-\nabla^2$.

(ii) *Poincaré's Inequality*. Prove that there is a constant a dependent only on n, m such that for all $f \in \mathscr{H}_0^m$,

$$|f|_j \leqslant ad^{m-j}|f|_m \qquad (0 \leqslant j \leqslant m-1).$$

11.16 For k a positive integer, show that there are coercive forms associated with $(-1)^k(\nabla^2)^k + 1$ and $(1 - \nabla^2)^k$.

11.17 Suppose that l (Definition 11.2.2) is formally self-adjoint and the associated form B is coercive. Define $(f, \phi)_E = B[f, \phi]$ and $\|f\|_E^2 = (f,f)_E$ for all $f, \phi \in \mathscr{H}_0^m$. Show that $\|\cdot\|_E$ and $\|\cdot\|_m$ are equivalent norms and deduce that $\mathscr{H}_0^m$ is a Hilbert space $\mathscr{H}_E$ with the new inner product $(\cdot,\cdot)_E$ and norm $\|\cdot\|_E$.

Take given $g \in \mathscr{L}_2$, and for all $f \in \mathscr{H}_E$ define the "total energy"

$$Q(f) = \|f\|_E^2 - (g, f)_0 - (f, g)_0.$$

Let $\mathscr{M}$ be a closed subspace of $\mathscr{H}_E$. Show that there is exactly one $\tilde{f} \in \mathscr{M}$ such that

$$Q(\tilde{f}) = \inf_{\phi \in \mathscr{M}} Q(\phi),$$

and deduce that the generalized Dirichlet problem has a unique solution.

Historical note. This is an existence proof by the direct method of the calculus of variations. Its history starts with the Dirichlet Principle from which important results in a number of fields were deduced by Riemann and others. The original argument on which the principle was based was simply that since a certain integral related to Q has a lower bound, there must be a function for which the minimum is attained. Because of the evident fallacy in this and other difficulties with the class of admissible functions, the method fell into disrepute until rescued by Hilbert. For details of its interesting history see Courant (1950) and Monna (1975).

11.18 (Continuous dependence of the solution on the right-hand side). With the notation of Problem 11.15(i), take $k < a$, and set $a = 2nd^{-2}$. If $g \in \mathscr{L}_2$ show that the solution f of the generalized Dirichlet problem satisfies the relations

$$\|f\|_0 \leqslant (a-k)^{-1}\|g\|_0, \qquad \|f\|_1 \leqslant c\,\|g\|_0,$$

where $c^2 = (a+1)/(a-k)^2$ for $-1 < k < a$, $c^2 = 1/(a-k)$ for $k \leqslant -1$.

11.19 Extend Theorem 11.5.5 as follows. If zero is an eigenvalue, show that the generalized Dirichlet problem has a solution for $g \in \mathscr{L}_2$ iff $(g, \psi_i)_0 = 0$ for all the eigenvectors ψ_i of the adjoint problem $B[\phi, \psi_i] = 0\ (\phi \in \mathscr{H}_0^m)$.

Chapter 12

The Finite Element Method

12.1 Introduction

The first technique to be used for the numerical solution of elliptic partial differential equations was the finite difference method. The later study of complicated and irregular boundaries arising in engineering problems led to the invention of the finite element method, in which the domain was replaced by a set of "elements" of fairly simple shape on which the unknown function was approximated. Physical arguments of equilibrium for each element then produced a set of linear simultaneous equations whose solution could be computed. A major practical advantage of this approach is flexibility, it being possible to match the elements to the geometry and physical characteristics of the expected solution, and the finite element method soon became the preferred approach in a wide variety of problems.

In the early stages, numerical techniques for the solution of partial differential equations could be adequately justified on an intuitive basis. However, the very complexity of the problems which could be treated by the finite element method, and the wide range of possible elements and approximations to the unknown function sometimes raised difficulties which could not be resolved in this manner. It was therefore natural to look for a mathematical framework in which the validity and accuracy of the method could be examined. Unfortunately there was no theory at first available which was adequate for this difficult task. The turning point in the mathematical development was probably the realization that the method was essentially variational. It was shown that the simultaneous equations could be obtained by choice of simple test functions approximating the unknown function and by the use of an energy minimization argument—that is by a Ritz procedure. It could then be seen that the theoretical analysis of the finite element method fitted naturally into the variational Sobolev space approach to partial differential equations. As a result an extensive theory has been developed capable of answering many of the questions of most practical interest.

In order to introduce the theory without overburdening the reader with detail, a relatively simple situation will be considered, the homogeneous Dirichlet problem for the formal operator $l = -\nabla^2 + p$ in two dimensions. This may be treated by the Ritz method; for general elliptic operators a more elaborate technique such as the Galerkin method must be used. The principal object of the investigation is to obtain global bounds for the error.

A good general reference for the finite element method is Strang and Fix (1973), and Prenter (1975) and Mitchell and Wait (1977) are also useful. Whiteman (1973, 1977) covers a wide variety of topics including time dependent and nonlinear problems, and also contains an interesting review of the historical background by Zienkiewicz. For a deeper discussion of the theory see Aziz (1972), Aubin (1972), Oden and Reddy (1976), or Temam (1970).

12.2 The Ritz Method

The generalized Dirichlet problem for $l = -\nabla^2 + p$ on Ω will be tackled under the following conditions:

(i) $\Omega \subset \mathbb{R}^2$ is bounded and open and $\partial\Omega$ is smooth;
(ii) p is a real valued non-negative function in $\mathscr{C}^\infty(\bar{\Omega})$;
(iii) $g \in \mathscr{L}_2$.

For $f, \phi \in \mathscr{H}^1_0$ define

$$B[f, \phi] = \int_\Omega (\nabla f \cdot \nabla\bar{\phi} + pf\bar{\phi})\,\mathrm{d}x.$$

12.2.1 Lemma. *B is Hermitian and coercive. The generalized Dirichlet problem has exactly one solution f in $\mathscr{H}^1_0$. Further $f \in \mathscr{H}^2$ and $f \in \mathscr{C}(\bar{\Omega})$.*

Proof. The coercivity follows from Lemma 11.4.8, for since p is non-negative, the additional term $(pf, f) \geqslant 0$. Theorem 11.4.6 proves existence and uniqueness, and $f \in \mathscr{H}^2 \cap \mathscr{C}(\bar{\Omega})$ by Theorem 11.6.6. □

The first step is to show by an application of the Ritz Theory that questions concerning the validity and accuracy of the finite element method for the above equation can be recast as problems in approximation theory in $\mathscr{H}_0^1$. The basic question then becomes the following: How closely can the solution be approximated in $\mathscr{H}_0^1$ by an element of $\mathscr{M}_k$—a k-dimensional linear subspace of simple trial functions? This approach leads naturally to a rather stringent global error test, for the norm measures the average value of the square of the moduli of the error and of its first derivative over the whole of Ω.

The technical details of the analysis are much simplified by exploiting the special properties of the bilinear form B associated with l. For since B is coercive and Hermitian, B is actually an inner product on $\mathscr{H}_0^1$. As the corresponding norm is equivalent to $\|\cdot\|_1$, the previous results on $\mathscr{H}_0^1$, in particular its completeness, are still applicable. However, the advantage of dealing with inner products rather than bilinear forms throughout is considerable. In addition the physical background is emphasized, since the square of the norm has a natural interpretation in terms of energy (a factor 2 is added for convenience in manipulation); in the finite element method it is often called the strain energy because of the roots of the technique in structural engineering.

12.2.2 Definition. For all $f, \phi \in \mathscr{H}_0^1$ set

$$(f, \phi)_E = B[f, \phi] = \int_\Omega (\nabla f \cdot \nabla \bar{\phi} + p f \bar{\phi})\, dx,$$

$$\|f\|_E^2 = B[f, f] = \int_\Omega (|\nabla f|^2 + p|f|^2)\, dx.$$

$\|\cdot\|_E$ will be known as the **energy norm**. The set $\mathscr{H}_0^1$ equipped with the inner product $(\cdot, \cdot)_E$ and norm $\|\cdot\|_E$ is denoted by $\mathscr{H}_E$.

12.2.3 Lemma. *If* (i) *and* (ii) *above hold,* $\|\cdot\|_1$ *and* $\|\cdot\|_E$ *are equivalent norms, and* $\mathscr{H}_E$ *is a Hilbert space.*

Proof. Since $\|f\|_E^2 = B[f, f]$, by Lemmas 12.2.1 and 11.4.2 respectively there are real numbers $c, c' > 0$ such that

$$c\|f\|_1 \leqslant \|f\|_E \leqslant c'\|f\|_1 \qquad (f \in \mathscr{H}_0^1).$$

This proves the equivalence of the norms, and hence the completeness of $\mathscr{H}_E$. $(\cdot, \cdot)_E$ is an inner product, for it is a Hermitian bilinear form and $(f, f)_E = \|f\|_E^2 = 0 \Rightarrow f = 0$. □

Since $B[f, \phi] = (f, \phi)_E$, the generalized Dirichlet problem takes a very simple form in $\mathscr{H}_E$: find the solution f in $\mathscr{H}_E$ of

$$(f, \phi)_E = (g, \phi)_0 \qquad (\varphi \in \mathscr{H}_E). \tag{12.2.1}$$

12.2.4 Lemma. *Suppose that assumptions* (i)–(iii) *above hold. Let $\mathscr{M}$ be a closed subspace of $\mathscr{H}_E$. Then given any $f \in \mathscr{H}_E$, there is exactly one $\tilde{f} \in \mathscr{M}$ such that*

$$\|f - \tilde{f}\|_E = \inf_{\phi \in \mathscr{M}} \|f - \phi\|_E, \tag{12.2.2}$$

and $\tilde{f}$ is the projection of f onto $\mathscr{M}$. Further, if f is the solution of the generalized Dirichlet problem, $\tilde{f}$ is the solution—unique in $\mathscr{M}$—of

$$(\tilde{f}, \phi)_E = (g, \phi)_0 \qquad (\phi \in \mathscr{M}). \tag{12.2.3}$$

Proof. The first assertion is just the Projection Theorem 1.5.11, from which it follows also that $f = \tilde{f} + (f - \tilde{f})$ with $\tilde{f} \in \mathscr{M}$ and $f - \tilde{f} \in \mathscr{M}^\perp$. On substituting for f in 12.2.1, we obtain for all $\phi \in \mathscr{M}$,

$$(g, \phi)_0 = (\tilde{f} + f - \tilde{f}, \phi)_E = (\tilde{f}, \phi)_E + (f - \tilde{f}, \phi)_E = (\tilde{f}, \phi)_E,$$

which is (12.2.3). Lastly, if $\tilde{f}_1$ is another solution of (12.2.3) in $\mathscr{M}$, then $(\tilde{f} - \tilde{f}_1, \phi)_E = 0$ for all $\phi \in \mathscr{M}$. Therefore $\tilde{f} - \tilde{f}_1 \in \mathscr{M}^\perp$. Since $\mathscr{M}$ is a linear subspace this can only happen if $\tilde{f} = \tilde{f}_1$, whence the uniqueness. □

12.2.5 Definition. A k-dimensional linear subspace $\mathscr{M}_k$ of $\mathscr{H}_E$ will be known as a **trial function space**, and any element of $\mathscr{M}_k$ as a **trial function**. Given $f \in \mathscr{H}_E$, the element $\tilde{f}_k$ in $\mathscr{M}_k$ closest to f in $\|\cdot\|_E$ will be called the **Ritz approximation** (in $\mathscr{M}_k$) to f.

12.2.6 Theorem (Ritz). *Assume that* (i)–(iii) *above hold. Let $\phi_1, \ldots, \phi_k$ be a basis for $\mathscr{M}_k$. Then the simultaneous equations*

$$\sum_{i=1}^{k} c_i(\phi_i, \phi_j)_E = (g, \phi_j)_0 \qquad (j = 1, \ldots, k) \tag{12.2.4}$$

have a unique solution, and the Ritz approximation $\tilde{f}_k$ in $\mathscr{M}_k$ to the solution of the generalized Dirichlet problem is given by

$$\tilde{f}_k = \sum_{i=1}^{k} c_i \phi_i. \tag{12.2.5}$$

The error $f - \tilde{f}_k$ satisfies the inequality

$$\|f - \tilde{f}_k\|_E \leqslant \|f - \phi\|_E \qquad (\phi \in \mathscr{M}_k). \tag{12.2.6}$$

Proof. Set $\tilde{f}_k = \Sigma c_i \phi_i$. By Lemma 12.2.4, $\tilde{f}_k$ satisfies (12.2.3), and if we substitute in this equation taking $\phi = \phi_j$ in turn, we obtain (12.2.4). On the other hand suppose that $c_1, \ldots, c_k$ is a solution of (12.2.4). Multiplying these equations in turn by $\alpha_1, \ldots, \alpha_k \in \mathbb{C}$ and adding we find that

$$(\Sigma c_i \phi_i, \Sigma \bar{\alpha}_i \phi_i)_E = (g, \Sigma \bar{\alpha}_i \phi_i)_0.$$

Since $\{\phi_i\}$ is a basis of $\mathscr{M}_k$, it follows that

$$(\Sigma c_i\phi_i, \phi)_E = (g, \phi)_0 \qquad (\phi \in \mathscr{M}_k),$$

so by Lemma 12.2.4, $\tilde{f}_k = \Sigma c_i\phi_i$. This also proves uniqueness. Finally (12.2.6) is an immediate consequence of the definition of the Ritz approximation. □

The original theoretical treatment of the finite element method was based on an energy minimization argument. The connection between this and the last results is that $\tilde{f}_k$ is the function in $\mathscr{M}_k$ which minimizes the "total energy"—see Problem 11.17.

This theorem provides the theoretical background against which the finite element method will be studied. In practice typical tactics for the numerical method are as follows. First the domain Ω is subdivided into a finite number of subdomains of simple shape—assumed here to be triangular—together with a thin skin along the boundary if necessary; the vertices of the triangles are usually known as *nodes* and will be denoted by x_i. A set of trial functions $\{\phi_i\}$ is then chosen, these being usually low order polynomials in each triangle with some continuity condition at the edges, and with ϕ_i zero except in those triangles with a vertex at x_i. The coefficients $(\phi_i, \phi_j)_E$ and the $(g, \phi_j)_0$ in (12.2.4) are then calculated, in a simple case explicitly, but in general by numerical integration. The solution of the simultaneous equations (12.2.4) are next obtained by computation (the fact that a solution exists and is unique is guaranteed by the Ritz Theorem above), and the approximate solution $\tilde{f}_k$ of the differential equation is then deduced from (12.2.5). The final step, which will be our principal concern, is the error analysis, and this is based on (12.2.6).

From a practical point of view the following points are of obvious importance.

(i) It should be reasonably simple to calculate $(\phi_i, \phi_j)_E$. This means that the trial functions must not be too complicated.

(ii) The matrix should be fairly sparse. This will be achieved in practice if, as in the above case, the supports of the ϕ_i do not overlap too much.

(iii) The simultaneous equations (12.2.4) should be well behaved from a computational angle. This point is certainly important, but as it belongs naturally to the field of matrix algebra, we shall not discuss it here and refer the reader to Strang and Fix (1973) or Mitchell and Wait (1977).

(iv) $\tilde{f}_k$ should be a good approximation to f for a reasonable number k of simultaneous equations.

There is an obvious conflict between the first two requirements and the last, and it may be difficult to decide which trial space gives a satisfactory compromise. This raises two fundamental questions, answers to which are provided by the theory. First, what choice of trial space is legitimate? The answer is easy, any finite dimensional linear subspace of $\mathscr{H}_E$ ($= \mathscr{H}_0^1$) is allowed. Second, how accurate is the approximate solution?

12.2.7 Example. Before coming to grips with the technicalities needed to cope with the partial differential equation, an easy example in one dimension is tackled to clarify ideas. Take $lf = -f'' + f$ and smooth g, and consider the generalized Dirichlet problem on $\Omega = (0, 1)$. The assumption that g is smooth ensures that this is equivalent to the classical problem of finding $f \in \mathscr{C}^2(\overline{\Omega})$ such that $-f'' - f = g$ on $(0, 1)$ and $f(0) = f(1) = 0$. Divide Ω into equal subintervals (x_i, x_{i+1}) for $i = 0, 1, \ldots, k$ of length $h = (k + 1)^{-1}$.

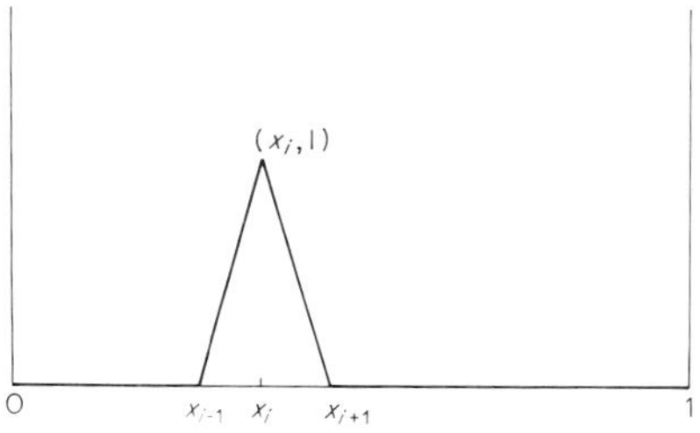

Fig. 12.1 A hat function.

Since the equation is of second order, we might intuitively require that the trial functions be twice differentiable. However, the theory tells us that all that is required is that they should lie in $\mathscr{H}_E$, and for this it is sufficient if they are absolutely continuous and vanish at the end points (Example 11.3.8). This extra flexibility allowed in the variational approach is extremely valuable in practice, as it is much easier to satisfy the smoothness condition at the nodes. The simplest choice of the trial functions would be piecewise constant functions equal to unity on the ith interval and zero elsewhere, but these do not lie in $\mathscr{H}_E$ and must be ruled out. Another easy choice is the set of "hat" functions as in Fig. 12.1; although these do not have even weak second derivatives, they lie in $\mathscr{H}_E$ and so are a legitimate choice—which we shall adopt here. Each trial space will then consist of the linear span of these hat functions, and every $\phi \in \mathscr{M}_k$ will be linear on each subinterval and continuous on $\overline{\Omega}$. Obviously for any $\phi \in \mathscr{M}_k$,

$$\phi(x) = \sum_1^k \phi(x_i)\phi_i(x).$$

The matrix associated with the simultaneous equations (12.2.4) for the coefficients of the Ritz approximation is first found. The entries are

$$(\phi_i, \phi_j)_E = \int_0^1 (\phi_i'\bar{\phi}_j' + \phi_i\bar{\phi}_j)\,dx,$$

and by elementary integration

$$(\phi_i, \phi_i)_E = \tfrac{2}{3}h^{-2}[3 + h^3],$$

$$(\phi_i, \phi_{i+1})_E = -\tfrac{1}{3}h^{-2}[3 - h^3],$$

while the other entries vanish since each ϕ_i is zero except on (x_{i-1}, x_{i+1}). The matrix is thus of simple tridiagonal form. The first two requirements above are certainly met, and we turn to the error analysis.

Theorem (12.2.6) states that the error $\|f - \tilde{f}_k\|_E$ is the distance of the solution f from the subspace $\mathscr{M}_k$, and it follows that the Ritz approximation $\tilde{f}_k$ is better than *any* other approximation in $\mathscr{M}_k$. This observation enables us to avoid the awkward problem of working with $\tilde{f}_k$ itself, for an adequate upper bound for the error may be obtained by using any $\phi \in \mathscr{M}_k$ which is reasonably close to f. The obvious choice is the linear interpolate, $\hat{f}_k$ say, of f, which takes the values $f(x_i)$ at x_i and is linear in between. The crucial step, and the one which gives considerable trouble in the partial differential equation case, is that of finding a bound for $\|f - \hat{f}_k\|_E$. In the present example nothing more than elementary calculus and Schwarz's inequality is needed (Problem 12.2) to prove that there are real numbers c_1, c_2 such that for any $f \in \mathscr{C}^2(\bar{\Omega})$,

$$\|f - \hat{f}_k\|_0 \leqslant c_1 h^2 \|f''\|_0, \tag{12.2.7}$$

$$\|f' - \hat{f}_k'\|_0 \leqslant c_2 h \|f''\|_0, \tag{12.2.8}$$

whence for some constant c,

$$\|f - \hat{f}_k\|_E \leqslant ch\|f''\|_0. \tag{12.2.9}$$

Therefore by (12.2.6), if f is the solution of the generalized Dirichlet problem,

$$\|f - \tilde{f}_k\|_E \leqslant ch\|f''\|_0, \tag{12.2.10}$$

c being a constant independent of f and h. This proves that $\tilde{f}_k \to f$ in $\|\cdot\|_E$ as $h \to 0$ (and so as $k \to \infty$), and gives the required estimate for the convergence rate.

For most purposes this is enough. However, the bound (12.2.10) for the error is not completely satisfactory as it is expressed in terms of the unknown function itself. This deficiency may be repaired by finding an *a priori* bound for f involving only the given right-hand side g. In one dimension this is easy, for as f is the solution of the generalized Dirichlet problem, $\|f\|_E^2 = (f, f)_E = (g, f)_0$. Hence

$$\|f\|_0^2 \leqslant \|f\|_E^2 = (g, f)_0 \leqslant \|g\|_0 \|f\|_0,$$

and it follows that $\|f\|_0 \leqslant \|g\|_0$. But $f'' = -g + f$, therefore $\|f''\|_0 \leqslant \|g\|_0 + \|f\|_0$, whence

$$\|f''\|_0 \leqslant 2\|g\|_0. \tag{12.2.11}$$

Thus from (12.2.10),

$$\|f - \tilde{f}_k\|_E \leqslant ch\|g\|_0, \tag{12.2.12}$$

which is the required estimate.

One final point may be noted. In practice the right-hand side $(g, \phi_j)_0$ of the simultaneous equations must usually be evaluated by numerical integration. How does this affect the error? In fact if g is replaced by its linear interpolate, the difference between the corresponding Ritz approximation and $\tilde{f}_k$ is of order h^2 in $\|\cdot\|_E$. The proof of this can be obtained by a minor modification of the above argument and presents little difficulty—see Problem 12.3. This amendment obviously does not affect the previous estimate (12.2.10).

12.3 The Rate of Convergence of the Finite Element Method

We now return to the generalized Dirichlet problem for $-\nabla^2 + p$ in two dimensions and obtain bounds for the error in the finite element solution. We continue to assume that conditions (i)–(iii) of the previous section hold.

The simplest and most usual way of subdividing Ω is as follows. In the kth subdivision $\partial\Omega$ is approximated by the boundary of an inscribed polygon Ω_k, and Ω_k is subdivided into triangles, T_i say, whose maximum diameter is h. The nodes (that is the vertices of the triangles) are denoted by x_i. We envisage a situation where the triangulation is successively refined, and then $h \to 0$ as $k \to \infty$.

12.3.1 Definition. The triangulations will be said to be **admissible** iff

(i) no vector of a triangle lies part way along the side of another triangle;
(ii) there is a number $\theta_0 > 0$ such that every angle in every triangle in each triangulation (that is for all k) is not less than θ_0.

The last condition will be essential in the derivation given here of the error bounds; for a detailed discussion see Strang and Fix (1973, p. 139). In fact the condition can be weakened, although not without bringing in extra technicalities—see Babuška and Aziz (1976).

The simplest possible choice of trial space will be adopted. For the kth subdivision $\mathscr{M}_k$ is taken to be the set of functions vanishing on $\partial\Omega_k$ and on the boundary "skin" $\Omega\backslash\Omega_k$, each function being continuous on $\overline{\Omega}$ and a first degree polynomial in each triangle. The surface described by such a function will thus be a union of flat "slabs". For a basis $\{\phi_i\}$ "pyramid functions" (generalizing the hat functions in $\mathbb{R}$) are used. Corresponding to each internal node x_i, let ϕ_i be the function in $\mathscr{M}_k$ which has height 1 at x_i, and which has support the polygon with vertices the nodes adjacent to x_i. Thus in the figure, ϕ_5 has support $x_1 x_2 x_3 x_4$ and vanishes on the edge of this polygon. Then for every $\phi \in \mathscr{M}_k$,

$$\phi(x) = \Sigma\phi(x_i)\,\phi_i(x) \qquad (x \in \bar{\Omega}).$$

Our aim in the error analysis is to obtain an analogue of the bound (12.2.10) of the one dimensional case. In outline the argument is the same.

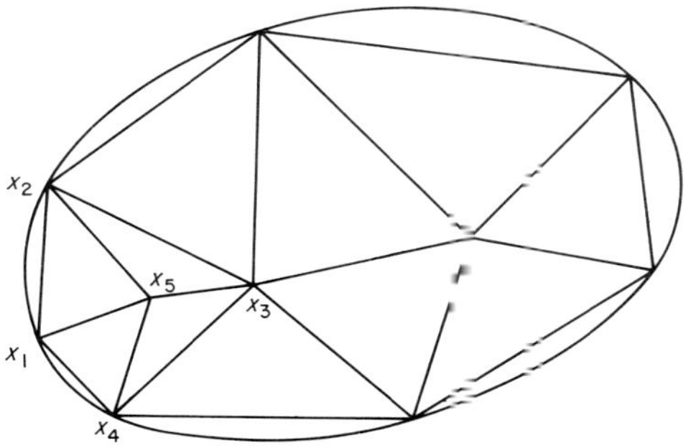

Fig. 12.2 A subdivision of Ω.

A bound for $\|f - \hat{f}_k\|_E$, where $\hat{f}_k$ is the linear interpolate of f, is first calculated, and then (12.2.6) is used. However, it is by no means easy to obtain analogues of (12.2.7) and (12.2.8), and a preparatory result whose proof is rather technical is needed.

For $f \in \mathscr{H}^j$ set

$$|f|_j = \int_\Omega \sum_{|\alpha| = j} |D^\alpha f|^2 \, dx.$$

Where necessary we shall write $|f|_{j,T}$, $\|f\|_{j,T}$ and so on to indicate that the integration is over the domain T.

12.3.2 The Bramble–Hilbert Lemma. *Let $\theta_0 > 0$ be given. There exist real numbers c_0, c_1, d such that for any $h > 0$, for any closed triangle T with diameter h and with angles bounded below by θ_0, and for any $f \in \mathscr{H}^2(T)$, there is a first degree polynomial q with*

$$|f - q|_{j,T} \leqslant c_j h^{2-j} |f|_{2,T} \qquad (j = 0, 1), \tag{12.3.1}$$

$$\sup_{x \in T} |f(x) - q(x)| \leqslant d\,h\,|f|_{2,T}. \tag{12.3.2}$$

Proof. We start by proving (12.3.1). It is enough to do this for a triangle with $h = 1$, for the factor h^{2-j} will appear as the result of a shrinkage of the independent variables. Further, it is sufficient to consider a single fixed triangle T. For there is a linear transformation of the independent variables mapping T onto any other triangle, T_1 say, and the ratio of the constants which appear on the right-hand sides of the inequalities corresponding to T_1 and T respectively depends only on the minimum angle of T_1, this ratio being bounded if T_1 satisfies the minimum angle restriction. Finally, since T is fixed throughout the proof, the suffix will be dropped from the notation in the proof.

The first step is to reformulate the problem—essentially by removing q from the left-hand side of (12.3.1).

12.3.3 Proposition. *Given $f \in \mathscr{H}^2(T)$ there is a unique first degree polynomial q such that*

$$\int_T D^\alpha(f - q)\,\mathrm{d}x = 0 \qquad (|\alpha| \leqslant 1). \tag{12.3.3}$$

Proof. There are three restrictions imposed by (12.3.3), and there are three arbitrary constants in a first degree polynomial. □

Returning to the proof of the lemma, if we take q satisfying (12.3.3) and set $u = f - q$, we need to show that

$$|u|_j \leqslant c_j |u|_2 \qquad (j = 0, 1)$$

for all u such that

$$\int_T D^\alpha u\,\mathrm{d}x = 0 \qquad (|\alpha| \leqslant 1).$$

The argument is by contradiction. If the result is false, then for either $j = 0$ or $j = 1$ there is a sequence, (u_i) say, in $\mathscr{H}^2$ such that

$$\|u_i\|_2 = 1, \tag{12.3.4}$$

$$|u_i|_j \geqslant i|u_i|_2, \tag{12.3.5}$$

$$\int_T D^\alpha u_i\,\mathrm{d}x = 0. \tag{12.3.6}$$

Now the Rellich Imbedding Theorem 11.3.12 also holds for $\mathscr{H}^2(T)$, see Agmon (1965, p. 30). Hence there is a subsequence, still denoted by (u_i), and a $u \in \mathscr{H}^1$ such that $u_i \to u$ in $\|\cdot\|_1$. Since $|u_i|_j \leqslant \|u_i\|_2$, from (12.3.4) and (12.3.5), $|u_i|_2 \to 0$. Therefore (u_i) is Cauchy in $\mathscr{H}^2$ and so convergent, with limit obviously u. This proves that $\|u\|_2 = 1$ by (12.3.4). It also shows that $D^\alpha u = 0$ for $|\alpha| = 2$, whence u is a first degree polynomial. But from (12.3.6),

$$\int_T D^\alpha u \, dx = 0 \qquad (|\alpha| \leqslant 1).$$

Hence by Proposition 12.3.3, $u = 0$. This contradicts the fact that $\|u\|_2 = 1$.

To prove (12.3.2) it is again enough to take $h = 1$ and change variables. We have

$$\sup_{x \in T} |f(x) - q(x)| \leqslant d_1 \|f - q\|_{2,T} \leqslant d_2 |f|_{2,T},$$

the inequalities being deduced from the Sobolev Imbedding Theorem 11.3.14 and equation (12.3.1) respectively. □

We are finally in a position to prove the main result. In the thin skin $\Omega \backslash \Omega_k$ not covered by the triangulation, all functions in $\mathscr{M}_k$ vanish. The contribution to the error from this region will thus be $\|f\|_{E,\Omega\backslash\Omega_k}$, and our task will be to find a bound for the error in Ω_k.

12.3.4 Theorem. *Let $\Omega \subset \mathbb{R}^2$ be a bounded open set with smooth boundary. Suppose that $p \in \mathscr{C}^\infty(\bar{\Omega})$ is real-valued and non-negative. Then there is a real number c such that*

$$\|f - \tilde{f}_k\|_{E,\Omega} \leqslant ch|f|_{2,\Omega_k} + \|f\|_{E,\Omega\backslash\Omega_k} \tag{12.3.7}$$

for any admissible triangulation (Definition 12.3.1) and any solution f of the generalized Dirichlet problem for $-\nabla^2 + p$ with right-hand side $g \in \mathscr{L}_2$.

Proof. Throughout $d_1, d_2, \ldots$ will denote fixed real numbers independent of f and the subdivision. As in $\mathbb{R}$, it is enough to prove (12.3.7) with $\tilde{f}_k$ replaced by $\hat{f}_k$ the linear interpolate of f (the function which is a first degree polynomial on each triangle and which is equal to f at the nodes), and to use (12.2.6). Also if the result can be established for a single triangle, (12.3.7) will follow on summation over the triangulation. A single triangle T is thus considered throughout, and the suffix T is omitted from the notation. The linear interpolate of any continuous function, say u, defined on T is denoted by $\hat{u}$.

By Lemma 12.3.2 there is a first degree polynomial q such that the error $\delta = f - q$ satisfies

$$|\delta|_j \leqslant c_j h^{2-j} |f|_2 \qquad (0 \leqslant j \leqslant 2). \tag{12.3.8}$$

Since q is of first degree, $q = \hat{q}$, and so

$$\hat{f} = \hat{q} + \hat{\delta} = q + \hat{\delta}.$$

Hence $f - \hat{f} = \delta - \hat{\delta}$, and if we can show that (12.3.8) holds with δ replaced by $\hat{\delta}$, we can obtain an estimate for $|f - \hat{f}|_j$ from which the result may readily be deduced.

From the definition of the pyramid functions ϕ_i,

$$\hat{\delta}(x) = \Sigma\hat{\delta}(x_i)\,\phi_i(x) = \Sigma\delta(x_i)\,\phi_i(x),$$

the summation being over the vertices of T. Hence for $|\alpha| \leqslant 1$,

$$|D^\alpha\hat{\delta}(x)| \leqslant 3\max_i|\delta(x_i)|\,.\,\max_i|D^\alpha\phi_i(x)|$$

$$= 3h^{-|\alpha|}\max_i|\delta(x_i)|\,.\,\max_i|h^{|\alpha|}D^\alpha\phi_i(x)|. \qquad (12.3.9)$$

Now from (12.3.2),

$$\max_i|\delta(x_i)| \leqslant d_1 h|f|_2.$$

Also, the last factor in (12.3.9) is bounded by unity if $\alpha = 0$ (since by definition $|\phi_i(x)| \leqslant 1$ for $x \in T$), while if $|\alpha| = 1$ the first derivatives of ϕ_i are dominated by h^{-1} times a constant (because of the restriction on the angles of T). We deduce that

$$|D^\alpha\hat{\delta}(x)| \leqslant d_2 h^{1-|\alpha|}|f|_2.$$

Now square, integrate, sum over α, and take the square root to obtain

$$|\hat{\delta}|_j \leqslant d_3 h^{2-j}|f|_2 \qquad (0 \leqslant j \leqslant 2).$$

This is the required estimate for $\hat{\delta}$. The bound for $|f - \hat{f}|_j$ follows on recalling that $f - \hat{f} = \delta - \hat{\delta}$ and using (12.3.8). Summation over j gives

$$\|f - \hat{f}\|_1 \leqslant d_4 h|f|_2,$$

and (12.3.7) is deduced from the equivalence of the norms $\|\cdot\|_E$ and $\|\cdot\|_1$. □

This theorem furnishes the required estimate for the global rate of convergence of the finite element method in the energy norm. We conclude with some general remarks.

The estimate (12.3.7) for the error involves the unknown function f. In the one dimensional case (Example 12.2.7) we have seen that a bound may easily be deduced in terms of the $\mathscr{L}_2$ norm of the known right-hand side g by means of an *a priori* inequality. Similar inequalities are available for partial differential equations, but they are considerably harder to establish, and the reader is referred to Friedman (1969, Section 1.17) for example for further details.

The method of proof described is non-constructive and does not give a numerical figure for the constant c in the error estimate (12.3.7), which must therefore be regarded as asymptotic. In the one dimensional case (Example 12.2.6) a bound for c is easily found, but in general there are much greater difficulties—see for example Barnhill and Whiteman (1973).

The Ritz method used in this chapter is applicable to general partial differential equations of order $2m$ so long as l is formally self-adjoint and the bilinear form is coercive; otherwise a Galerkin type method will be needed. Of course for higher order equations it may be necessary to use more complicated trial functions. This will also be the case if increased accuracy is sought without refinement of the subdivision. Almost always trial functions are polynomial on each element with appropriate continuity conditions at the edges. The central problem is the "convergence" of the trial spaces to the basic Sobolev space $\mathscr{H}_0^m$, which is therefore the problem we have concentrated on in this introduction. There is an extensive literature on this topic. In addition to the general references already cited, see Bramble and Zlámal (1970) where piecewise polynomial trial functions on triangles are discussed in detail.

One intriguing point may be noted finally. We have seen that although the equation is second order, it is enough to use trial functions which have only piecewise continuous first derivatives and not even weak second derivatives. In fact this is perfectly natural in the Sobolev space approach. However, especially for higher order equations, the conditions at the edges of the elements can be tedious to apply in practice, and trial functions have sometimes been used which do *not* belong to the basic Sobolev space. These "non-conforming" elements do not always work, but sometimes give excellent results, see Strang and Fix (1973, p. 174). The patch test (Mitchell and Wait, 1977; p. 167) has been devised to examine the eligibility of these trial functions.

Problems

12.1 Suggest a trial space for the finite element method for d^4/dx^4 on (0, 1) with homogeneous Dirichlet conditions.

12.2 Take $\Omega = (0, 1)$. Let $f \in \mathscr{H}^2$ and let $\hat{f}$ be linear on Ω with $\hat{f}(0) = f(0)$, $\hat{f}(1) = f(1)$. Find explicitly c_1, c_2 such that

$$|f - \hat{f}|_j \leqslant c_j |f|_2 \qquad (j = 0, 1),$$

and deduce (12.2.7) and (12.2.8).

12.3 In Example 12.2.7 the right-hand side $(g, \varphi_j)_0$ must be evaluated by numerical integration. With $\mathscr{M}_k$ as defined there, let f_k^* be the Ritz approximation to the solution of the equation with right-hand side the linear interpolate of g. Show that there is a c such that for any $g \in \mathscr{H}^2$,

$$\|\tilde{f}_k - f_k^*\|_E \leqslant ch^2 \|g''\|_0.$$

12.4 Error estimates other than those in energy norm are also of interest. For example the average pointwise error satisfies

$$\|f - \tilde{f}_k\|_0 \leqslant ch^2 \|g\|_0.$$

At first sight this seems to follow from (12.2.7) and (12.2.11). Why is this false?

The following method is known as the *Nitsche trick*. Set $\delta = f - \tilde{f}_k$, let u be the solution of the generalized Dirichlet problem with right-hand side δ, and let $\tilde{u}_k$ be its Ritz approximation. Show that

$$(u - \tilde{u}_k, \delta)_E = \|\delta\|_0^2.$$

Deduce the above estimate using the inequalities in Example 12.2.7.

Chapter 13

Introduction to Degree Theory

13.1 Introduction

For a number of important nonlinear equations arising in applications, general constructive techniques of solution are not known, and at the present time it is necessary to fall back on methods which yield only qualitative information about the behaviour of the system. It is to one of the most effective of these, the degree theory of Leray–Schauder, that we now turn our attention.

Some simple situations are first examined in order to clarify the argument to be used, the discussion being initially purely formal. Let D be the bounded open interval (a, b), and suppose that $\phi : \bar{D} \to \mathbb{R}$ is continuous. With the aim of using as little information as possible about ϕ, we enquire what can be deduced about the existence of a solution of the equation $\phi(x) = p$ from the values of ϕ on the boundary ∂D alone. One possible argument goes as follows. If ϕ is the identity I, the equation is just $x = p$, and this has a solution in D if $p \in D$. For general ϕ it is geometrically obvious that if the curve $y = \phi(x)$ can be deformed continuously into the straight line $y = Ix$ without either end point crossing the line $y = p$, then $\phi(x) = p$ will also have a solution when $p \in D$ (see Fig. 13.1). As an existence principle this assertion is potentially extremely powerful, for since only the values of ϕ on ∂D are used, there is no restriction on the slope of ϕ nor on its size. Also this principle may at least be meaningfully stated in higher dimensions, although it is far from easy to offer even a formal geometrical argument for its validity. On the other hand, it must be recognized that as an inevitable consequence of

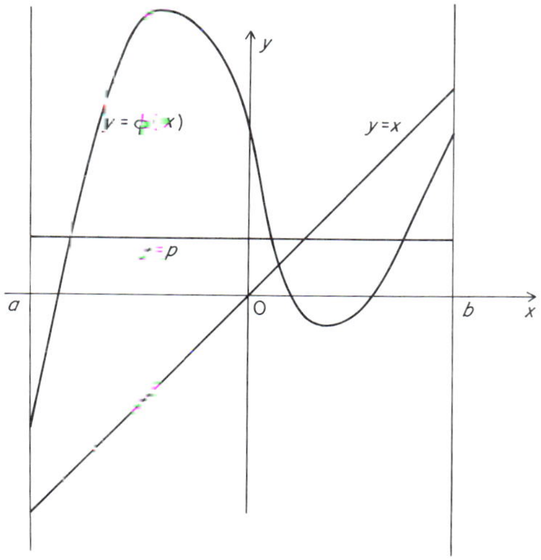

Fig. 13.1

its generality, the method has certain shortcomings, for it is not constructive, nor does it lend itself naturally to a proof of uniqueness, while for certain boundary values (as in Fig. 13.2) the argument is not applicable and no conclusion on existence can be reached.

The existence principle above was arrived at by a perturbation type argument, and for purposes of comparison it is illuminating to attempt to recover the result by using a standard continuity method based on the Implicit Function Theorem 4.4.9. Let $\{h_t\}$ be a family of smooth functions depending continuously on the parameter t with $h_1 = \phi, h_0 = I$. If $p \in D$, the theorem shows that under certain conditions on $\partial h_t/\partial x$, the existence of

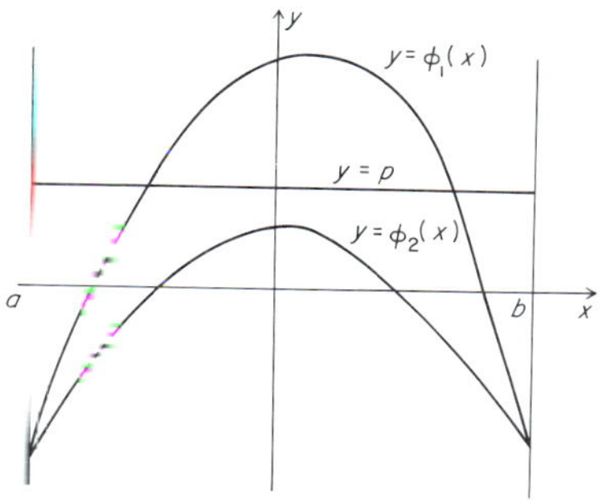

Fig. 13.2 ϕ_1 and ϕ_2 have the same boundary values, but the equations $\phi_1(x) = p$ and $\phi_2(x) = p$ have no solution and two solutions respectively.

a solution of $h_t(x) = p$ for sufficiently small t follows from the fact that $h_0(x) = p$ has a solution; this procedure could be repeated to enlarge the range of t. However, the Implicit Function Theorem is a local result, and there is no guarantee that the value $t = 1$ will be reached. In addition the procedure will break down if $\partial h_t/\partial x$ vanishes. In contrast the argument of the previous paragraph suffers from neither of these disadvantages, and provides the basis of a theory of large perturbations considerably more powerful than that derived from the continuity method.

In order to combine the above observations into a systematic theory, it is natural to enquire whether there is an integer $d(\phi, p, D)$—which may be regarded as a count in some sense of the solutions of $\phi(x) = p$—having the following properties.

(i) The *degree* $d(\phi, p, D)$ is defined for any bounded open set D and any function ϕ continuous on $\bar{D}$ such that $\phi(x) \neq p$ for $x \in \partial D$, and depends only on p and the values of ϕ on ∂D.

(ii) The degree is invariant under any continuous perturbation of the function, so long as a solution does not escape across the boundary. This can be stated precisely as follows. Let $h_t(x)$ be continuous in t and x for $x \in \bar{D}$ and $t \in [0, 1]$, and assume that $h_t(x) \neq p$ for any $x \in \partial D$ and $t \in [0, 1]$. Then $d(h_t, p, D)$ is independent of t. The family $\{h_t\}$ will be called a *homotopy*, and the above property will be known as *homotopy invariance*.

(iii) $d(I, p, D) = 0$ if $p \notin \bar{D}$, $d(I, p, D) = 1$ if $p \in D$.

(iv) If $d(\phi, p, D) \neq 0$, the equation $\phi(x) = p$ has a solution in D.

Assuming that such an integer does exist, the previous result for the function in Fig. 13.1 can be recovered easily. For by a combination of (ii) with $h_r = t\phi + (1 - t)I$ and (iii), $d(\phi, p, D) = 1$, and the existence of a solution follows from (iv). Note that (iv) says nothing about the solutions if $d(\phi, p, D) = 0$; as can be seen in Fig. 13.2, $\phi(x) = p$ may or may not have solutions.

Consider next how an explicit definition of degree could be given. In Fig. 13.1, ϕ and I are homotopic yet $\phi(x) = p$ has three solutions while $Ix = p$ has only one. Therefore if (ii) is to hold, it is obvious that $d(\phi, p, D)$ cannot be a straight count of solutions. A more promising approach is to attach a sign to each solution, depending on the direction in which the graph crosses the line $y = p$, and to define the degree as the sum of the "signed solutions". Then in the above example two solutions are lost, one positive and one negative, as ϕ changes into I, but the degree is unaltered. Set therefore

$$d(\phi, p, D) = \sum_j \operatorname{sgn} \phi'(x_j), \tag{13.1.1}$$

where the summation runs over all the solutions x_j in D of $\phi(x) = p$; if there is no solution in D, define $d(\phi, p, D) = 0$. This definition only makes sense if

ϕ is differentiable, $\phi'(x_j) \neq 0$ for any j, and there are only a finite number of solutions, but we shall ignore these technical difficulties for the moment, and simply note that the degree defined in this way satisfies (i)–(iv) above if ϕ is well behaved.

In one dimension the geometry is easily visualized, and it is readily seen that (13.1.1) yields an integer with the required properties. In higher dimensions the conceptual difficulties are much greater, and it is not even possible to fall back on a formal generalization of (13.1.1) since it is not clear how the right-hand side should be interpreted. Fortunately in two dimensions there is a result from complex variable theory which suggests what the generalization ought to be.

Let D be a bounded open domain in $\mathbb{R}^2$ whose boundary is a simple closed contour (in the sense usual in complex variable theory), and let $\phi : \bar{D} \to \mathbb{R}^2$ be the mapping $(\xi, \eta) \to (u, v)$. Suppose first that u, v are the real and imaginary parts respectively of the analytic function ϕ of $x = \xi + i\eta$, assume that $\phi(x) \neq p$ on ∂D, and let $x_1, \ldots, x_k$ be the solutions of $\phi(x) = p$ in D. Define

$$d(\phi, p, D) = \frac{1}{2\pi i} \int_{\partial D} \frac{\phi'(x)}{\phi(x) - p} \, \mathrm{d}x. \tag{13.1.2}$$

Then by Cauchy's Theorem

$$d(\phi, p, D) = \sum_{j=1}^{k} \frac{1}{2\pi i} \int_{C_j} \frac{\phi'(x)}{\phi(x) - p} \, \mathrm{d}x, \tag{13.1.3}$$

where for each j, C_j is a small circle centre x_j containing no solution other than x_j. The Residue Theorem shows that $d(\phi, p, D)$ is a non-negative integer which is the sum of the multiplicities of the solutions, and indeed it is easily checked by standard complex variable methods that all of (i)–(iv) above are satisfied. Thus this line of attack may be continued with some confidence. However, if the definition of degree is not to be restricted to analytic functions, (13.1.3) must be modified.

A purely formal argument goes as follows. First rewrite (13.1.3) as

$$d(\phi, p, D) = \sum_{j=1}^{k} \frac{1}{2\pi} \int_{C_j} \mathrm{d}\left\{\tan^{-1}\left(\frac{v - \beta}{u - \alpha}\right)\right\}, \tag{13.1.4}$$

where $p = \alpha + i\beta$. Now that analyticity has been dropped the integrals are no longer positive. However, as each integral is obviously the number of rotations of the vector from $p = (\alpha, \beta)$ to $\phi = (u, v)$ as (ξ, η) makes one circuit round a solution, an easy linearization argument enables us to calculate $d(\phi, p, D)$, at least when the zeros of $\phi(x) - p$ are simple. For by Taylor's Theorem, in the neighbourhood of x_j

$$\phi(x) - p = \phi'(x_j)(x - x_j) + r,$$

where $\phi'(x_j)$ is the Jacobian matrix, assumed non-singular, of ϕ at x_j and r is a small remainder. $\phi'(x_j)$ is a linear bijection of $\mathbb{R}^2$ onto itself, and therefore as x makes one circuit of C_j, its image $\phi'(x_j)(x - x_j)$ describes a circuit of an ellipse, C'_j say, the orientation of C'_j being determined by the sign of the Jacobian determinant $J_\phi(x_j)$. Hence the value of the jth integral in (13.1.4) is $\operatorname{sgn} J_\phi(x_j)$. The following definition of degree is obtained by this method:

$$d(\phi, p, D) = \sum_{j=1}^{k} \operatorname{sgn} J_\phi(x_j) \tag{13.1.5}$$

if there are no solutions in D set $d(\phi, p, D) = 0$. Compared with the one dimensional definition (13.1.1) this is a much more promising formula, for the Jacobian has a clear meaning in any finite number of dimensions.

The initial object in this chapter will be to show that a degree with the properties (i)–(iv) above may be defined for certain classes of operator on a Banach space. Initially the degree is defined for continuous operators in finite dimensions, the starting point being (13.1.5); it is here that the greatest technical difficulties are encountered. As an indication of the power of the theory a simple proof of the Brouwer Fixed Point Theorem is given. The next stage is the extension to infinite dimensions. It turns out that a direct generalization for arbitrary continuous operators is not possible. Operators of the form $I + A$ with A compact form the best known and most useful class for which the degree may be defined, a limiting procedure being used to extend the finite dimensional theory. The conclusions so obtained may be described as a theory of (large) compact perturbations of the identity. The result of which most use will be made is the celebrated Leray–Schauder Fixed Point Theorem 13.3.6(iii), the analogue in infinite dimensions of property (iv) above with ϕ replaced by $I + A$.

Two general points are worth noting. First, although (13.1.5) is the starting point for the definition, this formula will not be used for the calculation of degree in any application. Indeed in (13.1.5) it is assumed that the solutions are known, whereas it is the object of the investigation to find these. When the theoretical development is complete, the formula plays no further part, the calculation of degree being based on the property of homotopy invariance (ii) above. Second, from the point of view of applications alone, it is not usually essential to know the rather technical details of the analysis underlying the theory. Indeed the results of the theory summarized in Theorem 13.3.6 are elegant and simple, and their use is straightforward—at least in principle.

Degree theory has over the years found a multitude of applications, and is now a standard technique in nonlinear analysis. It is extremely useful in tackling nonlinear partial differential equations, such as the Navier–Stokes equations, for which it was originally devised. However, for such equations the preparatory analysis is rather difficult, and we shall confine ourselves here to problems for which the preliminaries are less taxing—essentially to those in which the basic equations may readily be reduced to an integral equation. An application is given in Section 4 to the Chandresekhar H-equation of radiative transfer, which in a simple context emphasizes the main points of the argument. The theory will also play an essential part in the later discussion of nonlinear eigenfunction problems.

The original development of degree theory depended on deep results in algebraic topology. While this approach may better illuminate the underlying geometrical structure, it requires considerable theoretical background, and here the analytical method as described by Berger and Berger (1968) and Schwartz (1969) is followed. On applications Krasnoselskii (1964a) for integral equations, Cronin (1964), Berger (1977) and a recent survey by Serrin (1976) for differential equations, are recommended. Ladyzenskaya (1969) is a standard text on the Navier–Stokes equations, and Sattinger (1973) gives a relatively elementary introduction.

13.2 The Degree in Finite Dimensions

The heuristic remarks above will now be developed into a rigorous theory of degree in a *real* finite dimensional Banach space. Since any such space may be identified with $\mathbb{R}^n$ by a suitable choice of basis, it is sufficient to restrict the analysis to $\mathbb{R}^n$ itself.

The following notation is used. $\|\cdot\|$ is the Euclidean norm in $\mathbb{R}^n$. $D \subset \mathbb{R}^n$ is a *bounded* open set with closure $\bar{D}$ and boundary ∂D. Mappings $\phi : \bar{D} \to \mathbb{R}^n$ always in $\mathscr{C}(\bar{D}, \mathbb{R}^n)$ and sometimes in $\mathscr{C}^1(\bar{D}, \mathbb{R}^n)$ will be considered. These spaces are equipped with the sup norm (denoted for emphasis in this section by $\|\cdot\|_{\mathscr{C}}$) and the norm $\|\cdot\|_{\mathscr{C}^1}$ respectively in which they are Banach spaces (Corollary 1.4.9 and Lemma 1.4.10). As usual $\phi(S)$ denotes the image of the set S under ϕ.

Let $p \in \mathbb{R}^n$ be given. The degree $d(\phi, p, D)$ will be defined only if $\phi(x) = p$ has no solution on ∂D, in other words if $p \notin \phi(\partial D)$. For such ϕ, since ∂D is compact and ϕ is continuous, by Theorem 5.3.3, $\text{dist}(p, \phi(\partial D)) > 0$. Straightforward applications of the triangle inequality are enough to prove the remainder of the following.

13.2.1 Lemma. *For D bounded and open, let $\phi, \psi \in \mathscr{C}(\bar{D}, \mathbb{R}^n)$, and assume that $p \notin \phi(\partial D)$. Then $r = \text{dist}(p, \phi(\partial D)) > 0$, and p, p_0 are in the same component*† *of $\mathbb{R}^n \backslash \phi(\partial D)$ if $\|p - p_0\| < \frac{1}{2}r$. If $\|\phi - \psi\|_{\mathscr{C}} < \frac{1}{2}r$, then $p \notin \psi(\partial D)$.*

† A *component* of an open set $S \subset \mathbb{R}^n$ is a maximal subset of S any two points of which may be joined by a continuous curve lying wholly in S (and thus not crossing ∂S).

13.2.2 Definition. For $\phi \in \mathscr{C}^1(\bar{D}, \mathbb{R}^n)$, the Jacobian matrix (the Fréchet derivative) of ϕ at x will be denoted by $\phi'(x)$, and the corresponding determinant by $J_\phi(x)$.

A **critical point** is any $x \in D$ such that $J_\phi(x) = 0$. Z will be used for the set of all critical points.

The aim is to define the degree for D, p, ϕ satisfying the conditions:

(i) D is a bounded open set in $\mathbb{R}^n$, and $\phi \in \mathscr{C}(\bar{D}, \mathbb{R}^n)$;

(ii) no solution of $\phi(x) = p$ lies on ∂D, that is $p \notin \phi(\partial D)$.

However, the starting point for the definition is (13.1.5), and initially the following further restrictions must be imposed if this formula is to be meaningful:

(iii) $\phi \in \mathscr{C}^1(\bar{D}, \mathbb{R}^n)$;

(iv) no solution of $\phi(x) = p$ is a critical point, that is $p \notin \phi(Z)$.

Under these conditions the set of solutions $\phi^{-1}(p) = \{x : x \in D, \phi(x) = p\}$ is finite. For if not, as $\bar{D}$ is compact, there is a sequence (x_j) of distinct points in $\phi^{-1}(p)$ converging to some $x \in \bar{D}$. As ϕ is continuous, $\phi(x) = p$, and from (ii), $x \in D$. Thus there are an infinite number of solutions in any neighbourhood of x, a possibility ruled out by the Implicit Function Theorem 4.4.9, since by (iv), $J_\phi(x) \neq 0$. The following definition, which it should be noted is independent of the basis, therefore makes sense. We set

$$d(\phi, p, D) = \begin{cases} 0 & (p \notin \phi(D)), \\ \displaystyle\sum_{x \in \phi^{-1}(p)} \operatorname{sgn} J_\phi(x) & (p \in \phi(D)). \end{cases} \tag{13.2.1}$$

In outline the plan is to relax successively conditions (iv) and (iii) as follows. First, if $p \in \phi(Z)$ the fact that there is a $p_0 \notin \phi(Z)$ arbitrarily close to p is used, and we define $d(\phi, p, D) = d(\phi, p_0, D)$. Second, since $\mathscr{C}^1(\bar{D}, \mathbb{R}^n)$ is dense in $\mathscr{C}(\bar{D}, \mathbb{R}^n)$, there is a $\phi_0 \in \mathscr{C}^1(\bar{D}, \mathbb{R}^n)$ arbitrarily close to any $\phi \in \mathscr{C}(\bar{D}, \mathbb{R}^n)$, and we define $d(\phi, p, D) = d(\phi_0, p, D)$. In order to justify this procedure, it must be shown that the degree is independent of the particular p_0, ϕ_0 chosen, and it is in establishing this that most of the unpleasant technicalities of finite dimensional degree theory arise. Three lemmas are needed. The first, which is quoted from Schwartz (1969, p. 54) enables us to prove the existence of an appropriate p_0. The other two show respectively that under certain conditions $d(\phi, p, D)$ defined by (13.2.1) is invariant under small changes of ϕ in $\|\cdot\|_{\mathscr{C}^1}$, and in p if p does not cross $\phi(\partial D)$.

13.2.3 Sard's Lemma. *If $\phi \in \mathscr{C}^1(\bar{D}, \mathbb{R}^n)$, the Lebesgue measure of $\phi(Z)$ is zero.*

13.2.4 Lemma. *Assume that p, ϕ, D satisfy* (i)–(iv) *above. Then there is an $\varepsilon > 0$ such that $\psi \in \mathscr{C}^1(\bar{D}, \mathbb{R}^n)$ and $\|\phi - \psi\|_{\mathscr{C}^1} < \varepsilon$ implies that $p \notin \psi(\partial D)$, $d(\psi, p, D)$ is defined and $d(\phi, p, D) = d(\psi, p, D)$.*

Proof. The first assertion follows from Lemma 13.2.1. Now let $x_1, \dots, x_k$ be the solutions of $\phi(x) = p$. By (iv) $J_\phi(x_i) \neq 0$ for $1 \leqslant i \leqslant k$, and since $\phi \in \mathscr{C}^1(\bar{D}, \mathbb{R}^n)$, $J_\phi(\cdot)$ is continuous, and for some $\delta > 0$, J_ϕ does not vanish on the closed balls $\bar{S}(x_i, \delta)$ for $1 \leqslant i \leqslant k$. Also, $J_\phi(x)$ regarded as a function of ϕ (and thus mapping $\mathscr{C}^1(\bar{D}, \mathbb{R}^n)$ into $\mathbb{R}$) is continuous, from which it follows that for small enough ε, $J_\psi(x)$ is also non-zero and has the same sign as $J_\phi(x)$ on each of these balls. We shall show by further decreasing δ and ε if necessary, that there is exactly one solution, $\bar{x}_i$ say, if $\psi(x) = p$ in each ball and none elsewhere. This will be enough to prove the result, for $d(\psi, p, D)$ will obviously be defined, and we will have

$$d(\phi, p, D) = \sum_{i=1}^{k} \operatorname{sgn} J_\phi(x_i) = \sum_{i=1}^{k} \operatorname{sgn} J_\psi(\bar{x}_i) = d(\psi, p, D).$$

Define $h : \bar{D} \times [0, 1] \to \mathbb{R}^n$ by setting $h(x, t) = \phi(x) - p + t[\psi(x) - \phi(x)]$, and notice that $h(x, 0) = \phi(x) - p$, $h(x, 1) = \psi(x) - p$. Now use Corollary 4.4.10 of the Implicit Function Theorem in $\bar{S}(x_i, \delta)$ after making the switch of notation $A = h, f = x, g = t, g_0 = 0$. Having chosen δ, simply observe that (4.3.8) and (4.3.9) are satisfied with $t_0 = 1$ if ε is small enough since

$$\|A_1(f, 0) - A_1(f, g)\| = \|t[\psi' - \phi']\|_{\mathscr{C}} \leqslant \varepsilon,$$
$$\|A(f_0, g)\| = t\|\phi(x_i) - \psi(x_i)\| \leqslant \varepsilon.$$

Therefore $\psi(x) = p$ has exactly one solution in $\bar{S}(x_i, \delta)$. Repeat the argument for $i = 1, \dots, k$ decreasing δ, ε further if necessary. Finally note that since there is an $\eta > 0$ such that $\|\phi(x) - p\| \geqslant \eta$ on $\bar{D} \backslash \bigcup \bar{S}(x_i, \delta)$, $\psi(x) = p$ has no solutions other than those already obtained if $\varepsilon < \eta$. □

13.2.5 Lemma. *Suppose that $\phi \in \mathscr{C}^1(\bar{D}, \mathbb{R}^n)$. For $p_1, p_2 \in \mathbb{R}^n$ assume that $p_1, p_2 \notin \phi(\partial D) \cup \phi(Z)$. Then if p_1 and p_2 belong to the same component of $\mathbb{R}^n \backslash \phi(\partial D)$,*

$$d(\phi, p_1, D) = d(\phi, p_2, D). \tag{13.2.2}$$

Proof. The first step is to prove another formula for the degree. We show that if $p \notin \phi(\partial D) \cup \phi(Z)$, for small enough ε (depending on ϕ),

$$d(\phi, p, D) = \int_D j_\varepsilon(\phi(x) - p)\, J_\phi(x)\, \mathrm{d}x, \tag{13.2.3}$$

where the j_ε are mollifiers (Definition 2.6.6) and so have support in $\bar{S}(0, \varepsilon)$. The integral depends only on the behaviour of ϕ in small neighbourhoods of the solutions $x_1, \dots, x_k$, which is consistent with the definition of degree, and the result will follow from the change of variable $y = \phi(x) - p$ after splitting D into small balls on which $J_\phi(x) \neq 0$ together with a region where the integrand vanishes. The details are as follows.

By compactness there is a neighbourhood U of p such that

$$U \cap \{\phi(Z) \cup \phi(\partial D)\} = \emptyset.$$

Now $J_\phi(\cdot)$ is continuous and does not vanish at any solution since $p \notin \phi(Z)$. Hence there are disjoint open balls $S_i = S(x_i, \delta)$ for $i = 1, \ldots, k$ with J_φ non-zero on each S_i, and by choice of δ small enough we can arrange that $\bigcup \phi(S_i) \subset U$. By the Implicit Function Theorem [illegible].4.9, δ may be chosen in such a way that the restrictions of ϕ to each S_i are homeomorphisms onto neighbourhoods of p contained in U. It follows that for some $\varepsilon > 0$, $\|\phi(x) - p\| > \varepsilon$ for $x \in \bar{D} \backslash \bigcup S_i = V$ say, and since j_ε has support in $S(0, \varepsilon)$, $j_\varepsilon(\phi(x) - p) = 0$ for $x \in V$, while J_ϕ has constant sign on each S_i. Using the change of variable $y = \phi(x) - p$ on each S_i, we obtain (13.2.3) as follows:

$$\begin{aligned}\int_D j_\varepsilon(\phi(x) - p)\, J_\phi(x)\, \mathrm{d}x &= \sum_{i=1}^{k} \operatorname{sgn} J_\phi(x_i) \int_{S_i} j_\varepsilon(\phi(x) - p) |J_\phi(x)|\, \mathrm{d}x \\ &= \sum_{i=1}^{k} \operatorname{sgn} J_\phi(x_i) \int_{S_i} j_\varepsilon(y)\, \mathrm{d}y \\ &= \sum_{i=1}^{k} \operatorname{sgn} J_\phi(x_i) \\ &= d(\phi, p, D).\end{aligned}$$

The deduction of (13.2.2) is based on the simple observation that if $v \in \mathscr{C}^1(\bar{D}, \mathbb{R}^n)$ has support in D, by the Divergence Theorem

$$\int_D \operatorname{div} v\, \mathrm{d}x = 0. \tag{13.2.4}$$

We now quote a difficult result from Schwartz (1969, pp. [illegible]3–66) which states that if in addition to the assumed conditions $\phi \in \mathscr{C}^2(\bar{D}, \mathbb{R}^n)$ then the functions

$$[j_\varepsilon(\phi(x) - p_1) - j_\varepsilon(\phi(x) - p_1 + p_2)]\, J_\phi(x)$$
$$[j_\varepsilon(\phi(x) - p_2) - j_\varepsilon(\phi(x) - p_1 + p_2)]\, J_\phi(x)$$

are divergences of functions in $\mathscr{C}^1(\bar{D}, \mathbb{R}^n)$ with supports in D. Hence from (13.2.3) and (13.2.4), for small enough ε,

$$\begin{aligned}d(\phi, p_1, D) &= \int_D j_\varepsilon(\phi(x) - p_1)\, J_\phi(x)\, \mathrm{d}x \\ &= \int_D j_\varepsilon(\phi(x) - p_1 + p_2)\, J_\phi(x)\, \mathrm{d}x \\ &= \int_D j_\varepsilon(\phi(x) - p_2)\, J_\phi(x)\, \mathrm{d}x \\ &= d(\phi, p_2, D).\end{aligned}$$

The last step is the removal of the restriction $\phi \in \mathscr{C}^2(D, \mathbb{R}^n)$. $\mathscr{C}^2(D, \mathbb{R}^n)$ is dense in $\mathscr{C}^1(\bar{D}, \mathbb{R}^n)$. By Lemma 13.2.4 there is a $\psi \in \mathscr{C}^2(D, \mathbb{R}^n)$ so close to ϕ in $\|\cdot\|_{\mathscr{C}^1}$ that $d(\psi, p_1, D)$ and $d(\psi, p_2, D)$ are both defined and equal respectively to $d(\phi, p_1, D), d(\phi, p_2, D)$. (13.2.2) follows from the conclusion of the last paragraph. □

Condition (iv), that is $p \notin \phi(Z)$, can now be removed from the definition of degree. By Sard's Lemma 13.2.3 the set of points which are not images of critical points (that is are not in $\phi(Z)$) is dense in each component of $\mathbb{R}^n \backslash \phi(\partial D)$. For otherwise $\phi(Z)$ would contain an open ball, which has non-zero measure, a possibility ruled out by the lemma. Choose then any p_0 not in $\phi(Z)$, but in the same component of $\mathbb{R}^n \backslash \phi(\partial D)$ as p, and define $d(\phi, p, D) = d(\phi, p_0, D)$. By Lemma 13.2.5 $d(\phi, p, D)$ is the same for every such p_0, and indeed the lemma shows that $d(\phi, p, D)$ is constant with respect to variations in p if p does not cross $\phi(\partial D)$.

To remove the condition $\phi \in \mathscr{C}^1(\bar{D}, \mathbb{R}^n)$, Lemma 13.2.4 is first strengthened.

13.2.6 Lemma. *Lemma 13.2.4 is valid even if $p \in \phi(Z)$.*

Proof. By Sard's Lemma there is a $p_0 \notin \phi(Z)$ such that $\|p - p_0\| < \frac{1}{2}\operatorname{dist}(p, \phi(\partial D))$, and then p, p_0 are in the same component of $\mathbb{R}^n \backslash \phi(\partial D)$—by Lemma 13.2.1. For small enough ε and any $\psi \in \mathscr{C}^1(\bar{D}, \mathbb{R}^n)$ such that $\|\phi - \psi\|_{\mathscr{C}^1} < \varepsilon$, p, p_0 are also in the same component of $\mathbb{R}^n \backslash \psi(\partial D)$. By further decreasing ε if necessary and using Lemma 13.2.4 we can ensure that $d(\phi, p_0, D) = d(\psi, p_0, D)$. The result follows on noting that by Lemma 13.2.5, $d(\phi, p, D) = d(\phi, p_0, D)$ and $d(\psi, p, D) = d(\psi, p_0, D)$. □

13.2.7 Corollary. *Let $\{h_t\}$ be a family of mappings $\bar{D} \to \mathbb{R}^n$ defined for $t \in [0, 1]$. Assume that $h_t \in \mathscr{C}^1(\bar{D}, \mathbb{R}^n)$ for each t, and that h_t depends continuously in the $\mathscr{C}^1(\bar{D}, \mathbb{R}^n)$ sense on t. If $p \notin h_t(\partial D)$ for any t, then $d(h_t, p, D)$ is independent of t.*

Proof. Simply note that $d(h_t, p, D)$ is constant on neighbourhoods of each t by the previous lemma and hence is a continuous function of t. Since it is integer valued, it is constant. □

13.2.8 Lemma. *For given $\phi \in \mathscr{C}(\bar{D}, \mathbb{R}^n)$ assume that $p \notin \phi(\partial D)$. Let $\psi_1, \psi_2 \in \mathscr{C}^1(\bar{D}, \mathbb{R}^n)$. If $\|\phi - \psi_1\|_{\mathscr{C}}, \|\phi - \psi_2\|_{\mathscr{C}} < \frac{1}{2}\operatorname{dist}(p, \phi(\partial D))$, then $d(\psi_1, p, D) = d(\psi_2, p, D)$.*

Proof. Set $h_t(x) = t\psi_1(x) + (1 - t)\psi_2(x)$ for $t \in [0, 1]$ and $x \in \bar{D}$. It is easy to show that $p \neq h_t(x)$ for any $x \in \partial D$ and $t \in [0, 1]$. Also

$$\|h_{t_1} - h_{t_2}\|_{\mathscr{C}^1} \leqslant |t_1 - t_2|[\|\psi_1\|_{\mathscr{C}^1} + \|\psi_2\|_{\mathscr{C}^1}],$$

whence h_t is continuous as a function of t in the $\mathscr{C}^1(\bar{D}, \mathbb{R}^n)$ sense. The result follows from Corollary 13.2.7. □

Condition (iii), that $\phi \in \mathscr{C}^1(\bar{D}, \mathbb{R}^n)$, may finally be relaxed. Since $\mathscr{C}^1(\bar{D}, \mathbb{R}^n)$ is dense in $\mathscr{C}(\bar{D}, \mathbb{R}^n)$, there is a $\phi_0 \in \mathscr{C}^1(\bar{D}, \mathbb{R}^n)$ with $\|\phi - \phi_0\|_{\mathscr{C}} < \frac{1}{2}\operatorname{dist}(p, \phi(\partial D))$. Define $d(\phi, p, D) = d(\phi_0, p, D)$, and note that by the previous lemma $d(\phi, p, D)$ is uniquely determined. The stages leading to the final definition are summarized as follows.

13.2.9 Definition. Let D be a bounded open set in $\mathbb{R}^n$. Assume that $\phi : \bar{D} \to \mathbb{R}^n$ is in $\mathscr{C}(\bar{D}, \mathbb{R}^n)$, and that no solution of $\phi(x) = p$ lies on ∂D. The **degree** $d(\phi, p, D)$ is defined as follows.

(i) If $\phi \in \mathscr{C}^1(\bar{D}, \mathbb{R}^n)$ and $p \notin \phi(Z)$, set

$$d(\phi, p, D) = \begin{cases} 0 & (p \notin \phi(\bar{D})), \\ \displaystyle\sum_{x_i \in \phi^{-1}(p)} \operatorname{sgn} J_\phi(x_i) & (p \in \phi(D)). \end{cases}$$

(ii) If $\phi \in \mathscr{C}^1(\bar{D}, \mathbb{R}^n)$ but $p \in \phi(Z)$, take any p_0 not in $\phi(Z)$, but in the same component of $\mathbb{R}^n \backslash \phi(\partial D)$ as p, and put $d(\phi, p, D) = d(\phi, p_0, D)$.

(iii) If $\phi \notin \mathscr{C}^1(\bar{D}, \mathbb{R}^n)$, take any $\phi_0 \in \mathscr{C}^1(\bar{D}, \mathbb{R}^n)$ such that $\|\phi - \phi_0\|_{\mathscr{C}} < \frac{1}{2}\operatorname{dist}(p, \phi(\partial D))$ and set $d(\phi, p, D) = d(\phi_0, p, D)$.

The definition is now complete. It only remains to check that the degree has as promised properties (i)–(iv) of the introduction.

13.2.10 Definition. Let D be a subset of the Banach space $\mathscr{B}$. Suppose that $\{h_t\}$ is a family of operators mapping $\bar{D} \to \mathscr{B}$ for each $t \in [0, 1]$. Assume that $h_t(x)$ is continuous in t and x—that is as a mapping $[0, 1] \times \bar{D} \to \mathscr{B}$. Then the family is called a **homotopy**, two members of the family being described as **homotopic**.

13.2.11 Theorem. *Let D be a bounded open set in $\mathbb{R}^n$. Assume that $\phi \in \mathscr{C}(\bar{D}, \mathbb{R}^n)$ and that no solution of $\phi(x) = p$ lies on ∂D. Then the following hold.*

(i) *$d(\phi, p, D)$ is an integer-valued function which depends only on p and the values of ϕ on ∂D. $d(\phi, p, D)$ is a constant function of p if p does not cross $\phi(\partial D)$, and of ϕ (in the $\mathscr{C}(\bar{D}, \mathbb{R}^n)$ sense) if p never lies in $\phi(\partial D)$.*

(ii) *Homotopy Invariance. Let $\{h_t\}$ be a homotopy, and suppose that $h_t(x) \neq p$ for any $x \in \partial D$ and $t \in [0, 1]$. Then $d(h_t, p, D)$ is independent of t.*

(iii) *$d(I, p, D) = 1$ if $p \in D$, $d(I, p, D) = 0$ if $p \notin \bar{D}$.*
(iv) *If $d(\phi, p, D) \neq 0$, the equation $\phi(x) = p$ has at least one solution in D.*

Proof. (ii) Note that by Lemma 13.2.8 the degree is constant on some neighbourhood of each t, and proceed as in the proof of Corollary 13.2.7. (i) If ϕ_1 and ϕ_2 have the same values on ∂D, we see by considering the homotopy $h_t = t\phi_1 + (1 - t)\phi_2$ and using (ii) that $d(\phi_1, p, D) = d(\phi_2, p, D)$. The remaining assertions follow from Lemmas 13.2.8 and 13.2.5 respectively. (iii) is obvious. (iv) We show that if there is no solution, then $d(\phi, p, D) = 0$. Choose $\psi \in \mathscr{C}^1(\bar{D}, \mathbb{R}^n)$ such that $\|\phi - \psi\|_{\mathscr{C}} < \frac{1}{2}\,\mathrm{dist}(p, \phi(\bar{D}))$. Then $p \notin \psi(\bar{D})$, and $d(\psi, p, D) = 0$ by Definition 13.2.9(i). From Lemma 13.2.8, $d(\phi, p, D) = d(\psi, p, D)$, and the result follows. □

We conclude the discussion of the theory with one further remark concerning the definition. From (i) of the last theorem, if p is fixed $d(\phi, p, D)$ depends only on the values of ϕ on ∂D. It is therefore natural to ask whether the degree can be defined for a continuous ϕ defined only on ∂D if $p \notin \phi(\partial D)$. Obviously it can if there is a continuous extension of ϕ to the whole of $\bar{D}$ (of course the explicit form of the extension is irrelevant), but if the dimension is greater than one it is by no means obvious whether such an extension exists. In finite dimensions the Tietze Extension Theorem asserts that it does. Remarkably an analogous result holds even in infinite dimensions, and this is quoted here since it will be needed later. For a proof see Schwartz (1969, p. 94).

13.2.12 The Dugundji Extension Theorem. *Assume that X is a closed subset of the Banach space $\mathscr{B}$, and let $S \subset \mathscr{B}$ be convex. Suppose that $A : X \to S$ is continuous. Then if Y is any subset of $\mathscr{B}$ containing X, A has a continuous extension $\tilde{A} : Y \to S$.*

The analysis underlying degree theory is certainly rather heavy. However, this should not obscure the fact that the use of the theory in applications is simple in principle, most of what it is essential to know being contained in Theorem 13.2.11. In summary the position is as follows. The degree is defined for any continuous ϕ such that $\phi(x) = p$ has no solution on ∂D. In practice its computation is scarcely ever based on Definition 13.2.9 but rather on homotopy invariance (Theorem 13.2.11(ii)), a standard technique being to perturb ϕ into the identity and to use the known value of $d(I, p, D)$. Finally, if the degree is not zero, the existence of a solution (although not of course its uniqueness) is assured by part (iv) of the theorem. The following examples are intended to illustrate some of these points.

13.2.13 Example. Let $D \subset \mathbb{R}^n$ be a bounded open set containing the origin, and suppose that $L: \mathbb{R}^n \to \mathbb{R}^n$ is a linear operator. Let L also denote the matrix representation of the operator with respect to some basis. Obviously $d(L, 0, D)$ is defined if and only if L is injective (otherwise a solution will lie on ∂D), but then $\det L \neq 0$ and from Definition 13.2.9(i), $d(L, 0, D) = \operatorname{sgn} \det L$. Now $\det L$ is simply the product of the eigenvalues λ_i, the number of repetitions of any λ_i being its algebraic multiplicity, which is the dimension of $\bigcup N((\lambda_i I - L)^k)$. Since the entries in L are real the eigenvalues are either real or occur in complex conjugate pairs. Hence

$$d(L, 0, D) = \operatorname{sgn} \prod_1^n \lambda_i = (-1)^\beta$$

where β is the sum of the algebraic multiplicities of the negative eigenvalues.

13.2.14 Example. In $\mathbb{R}$ take $D = (-1, 1)$ and $\phi(x) = \frac{1}{2}x^2(x - \frac{1}{2})$. Then ∂D is just the pair of points ± 1, and $\phi(1) = \frac{1}{4}$, $\phi(-1) = \frac{3}{4}$. Thus $d(\phi, p, D)$ is defined for all $p \neq \frac{1}{4}, \frac{3}{4}$. Consider first the calculation of the degree direct from Definition 13.2.9. Which part of this is used depends on whether J_ϕ vanishes at one of the solutions $x = 0, \frac{1}{2}$, that is on whether the derivative of ϕ is zero at one of these points. Obviously this happens if and only if $p = 0$, and if $p \neq 0$ the degree is readily found from Definition 13.2.9(i): $d(\phi, p, D) = 1$ or 0 depending on whether p lies in $(-\frac{3}{4}, \frac{1}{4})$ or outside $[-\frac{1}{3}, \frac{1}{4}]$. If $p = 0$, take any small non-zero p_0. Then (ii) of the definition yields $d(\phi, 0, D) = d(\phi, p_0, D) = 1$.

The calculation using instead homotopy invariance goes as follows. With $h_t(x) = t\phi(x) + (1 - t)x$, $h_1 = \phi$ and $h_0 = I$. Then $h_t(\pm 1) \neq p$ for $0 \leqslant t \leqslant 1$ and $-\frac{3}{4} < p < \frac{1}{4}$. Thus for such p,

$$d(\phi, p, D) = d(h_1, p, D) = d(h_0, p, D) = d(I, p, D) = 1.$$

13.2.15 Example. The Brouwer Fixed Point Theorem. Degree theory provides a simple method for proving this deep theorem. The theorem asserts that if Ω is a bounded closed convex subset of a finite dimensional Banach space $\mathscr{B}$, and $\phi: \Omega \to \Omega$ is continuous, then ϕ has a fixed point in Ω. Observe that $\mathscr{B}$ may be taken real without loss of generality, for $\mathbb{C}^n$ is isomorphic with $\mathbb{R}^{2n}$, and convexity is preserved.

The result is first established for $\Omega = \bar{S}(0, r) = \bar{S}$ say. Assume that ϕ has no fixed point on ∂S (otherwise the proof is complete), and consider the homotopy $h_t = I - t\phi$ for $t \in [0, 1]$. If $t < 1$ and $x \in \partial S$, then $t\|\phi(x)\| < r$, and it follows that $h_t(x) \neq 0$ for any $x \in \partial S$ and $t \in [0, 1]$. Therefore by homotopy invariance (Theorem 13.2.11(ii)),

$$d(I - \phi, 0, S) = d(h_1, 0, S) = d(h_0, 0, S) = d(I, 0, S) = 1.$$

Thus by (iv) of the same theorem, ϕ has a fixed point in $\bar{S}$.

Suppose now that Ω is any bounded closed convex set. Then it is contained in some ball $\bar{S}$ centre the origin. Since Ω is convex, ϕ has a continuous extension, $\tilde{\phi}$ say, to $\bar{S}$ with $\tilde{\phi}: \bar{S} \to \Omega$ (Theorem 13.2.12). By what was proved in the last paragraph $\tilde{\phi}$ has a fixed point, $\bar{x}$ say, in $\bar{S}$. However, $\tilde{\phi}(\bar{S}) \subset \Omega$, so $\bar{x} \in \Omega$. This proves the theorem since $\phi = \tilde{\phi}$ on Ω.

13.3 The Leray–Schauder Degree

For arbitrary continuous operators in infinite dimensions, it is certainly impossible to define a degree with the same properties as in finite dimensions, for if it were the Brouwer Fixed Point Theorem would generalize without change to yield a result known to be false (Problem 4.6). Clearly some further restriction must be placed on the operator. Since the natural generalization of the Brouwer Theorem is the Schauder Fixed Point Theorem, it may be expected that a compactness condition would be sufficient, and in the original theory of Leray–Schauder it was shown that a satisfactory definition of degree was indeed possible for compact perturbations of the identity, that is for operators of the form $I + A$ with A compact. This theory will now be outlined; for a review of later extensions, see Lloyd (1978).

So far as the details of the analysis are concerned, it is clear that the definition of degree cannot readily be based on (13.2.1). The tactics will be rather to use a finite dimensional approximation to the operator and to capitalise on the theory of the previous section.

Throughout D will be a bounded open subset of the real Banach space $\mathscr{B}$, p a vector in $\mathscr{B}$, and $A : \bar{D} \to \mathscr{B}$ a compact operator such that $f + Af \neq p$ for $f \in \partial D$. For technical reasons it is convenient to assume temporarily that $0 \in D$. Set

$$r = \tfrac{1}{2}\,\mathrm{dist}(p, (I + A)(\partial D)). \qquad (13.3.1)$$

13.3.1 Lemma. *Under the above conditions, $r > 0$.*

Proof. If the result is false, there is a sequence (f_n) in ∂D such that $\lim(f_n + Af_n - p) = 0$. Because A is compact there is a subsequence, which will still be denoted by (f_n), such that (Af_n) is convergent. It follows that (f_n) is also convergent with limit f say, and as ∂D is closed, $f \in \partial D$. Therefore $f + Af = p$ contrary to assumption. □

13.3.2 Lemma. *Let A be as above. Then given any ε with $0 < \varepsilon < r$, there is a finite dimensional linear subspace $\mathscr{M}$ of $\mathscr{B}$ which contains p, and a continuous operator $B : \bar{D} \to \mathscr{M}$ such that*

(i) $\|Af - Bf\| \leqslant \varepsilon \qquad (f \in \bar{D})$,
(ii) $\|f + Bf - p\| \geqslant r \qquad (f \in \partial D)$.

Proof. By Lemma 8.2.5 there is a continuous operator B defined on $\bar{D}$ with range in a finite dimensional linear subspace, $\mathscr{M}_0$ say of $\mathscr{B}$, such that (i) is satisfied. Choose $\mathscr{M}$ to be the smallest linear subspace containing both p and $\mathscr{M}_0$. Then $R(B) \subset \mathscr{M}$. Since $\varepsilon < r$, (ii) is a consequence of the following valid for all $f \in \partial D$:

$$\|f + Bf - p\| \geqslant \|f + Af - p\| - \|Bf - Af\| \geqslant 2r - \varepsilon. \qquad \square$$

It will be necessary to consider various restrictions of B; each of these will still be denoted by B since it is always clear from the context which restriction is referred to.

We are now ready to proceed with the definition of the degree of $I + A$. Suppose that B is any operator satisfying the conditions of the lemma, and consider its restriction to $\overline{D \cap \mathscr{M}}$, which is not empty as $0 \in D$. Then B maps the bounded set $D \cap \mathscr{M}$, which is open in $\mathscr{M}$, into $\mathscr{M}$, and $f + Bf \neq p$ for $f \in \partial(D \cap \mathscr{M})$. As $\mathscr{M}$ is finite dimensional, $d(I + B, p, D \cap \mathscr{M})$ may be defined as in the last section. We shall take $d(I + A, p, D)$ to be this integer, but we must first show that this definition is independent of the choice of B. A preparatory result is needed.

13.3.3 Lemma. *Suppose $m, n > 0$ are integers, and let S be a bounded open set in $\mathbb{R}^{n+m}$. Let $(x_1, \ldots, x_n) \to (x_1, \ldots, x_n, 0, \ldots, 0)$ be the natural imbedding of $\mathbb{R}^n$ into $\mathbb{R}^{n+m}$. Assume that $C : \bar{S} \to \mathbb{R}^n$ is continuous, and for $f \in \bar{S}$ let*

$$Cf = (C_1 f, \ldots, C_n f), \qquad \tilde{C}f = (C_1 f, \ldots, C_n f, 0, \ldots, 0),$$

so that $\tilde{C}$ is C followed by the natural imbedding of $\mathbb{R}^n$ into $\mathbb{R}^{n+m}$. If $(I + C)f \neq p$ for $f \in \partial S$, then

$$d(I + C, p, S \cap \mathbb{R}^n) = d(I + \tilde{C}, p, S),$$

where I denotes the identity both in $\mathbb{R}^n$ and $\mathbb{R}^{n+m}$.

Proof. Assume first that $C \in \mathscr{C}^1(\bar{S}, \mathbb{R}^n)$ and that $p \notin C(Z)$. Let J_{n+m} be the Jacobian matrix of $I + \tilde{C}$ and J_n the Jacobian matrix of the restriction of $I + C$ to $\mathbb{R}^n$. Then at each point of $S \cap \mathbb{R}^n$,

$$\det J_{n+m} = \begin{vmatrix} J_n & 0 \\ 0 & I_m \end{vmatrix} = \det J_n,$$

where I_m is the $m \times m$ identity matrix, and the result follows from Definition 13.2.9(a) of degree. For $C \in \mathscr{C}(\bar{S}, \mathbb{R}^n)$, take limits and use Theorem 13.2.11(i).

$\square$

13.3.4 Lemma. *Let $B_1 : \bar{D} \to \mathscr{M}_1$ and $B_2 : \bar{D} \to \mathscr{M}_2$ be any two operators satisfying the conditions of Lemma 13.3.2. Then*

$$d(I + B_1, p, D \cap \mathscr{M}_1) = d(I + B_2, p, D \cap \mathscr{M}_2).$$

Proof. With an appropriate basis any finite dimensional linear subspace of $\mathscr{B}$ can be identified with $\mathbb{R}^k$ for some k, and the choice of basis does not affect the degree. Thus by the previous lemma, for $i = 1, 2$,

$$d(I + B_i, p, D \cap \tilde{\mathscr{M}}) = d(I + B_i, p, D \cap \mathscr{M}_i), \tag{13.3.2}$$

where $\tilde{\mathscr{M}}$ is the smallest linear subspace containing $\mathscr{M}_1$ and $\mathscr{M}_2$. Therefore if it can be proved that the degrees on the left-hand side are equal, the result will follow.

Consider the homotopy $\{H_t\}$ for $0 \leqslant t \leqslant 1$, where $H_t : \overline{D \cap \tilde{\mathscr{M}}} \to \tilde{\mathscr{M}}$ is defined by

$$H_t f = t(f + B_1 f - p) + (1 - t)(f + B_2 f - p).$$

For $f \in \partial(D \cap \tilde{\mathscr{M}})$,

$$\begin{aligned} \|H_t f\| &= \|f + Af - p + t(B_1 f - Af) + (1 - t)(B_2 f - Af)\| \\ &\geqslant \|f + Af - p\| - t\|B_1 f - Af\| - (1 - t)\|B_2 f - Af\| \\ &\geqslant 2r - tr - (1 - t)r \qquad \text{(by Lemma 13.3.2)} \\ &= r. \end{aligned}$$

Hence by homotopy invariance (Theorem 13.2.11(ii)),

$$d(H_1, 0, D \cap \tilde{\mathscr{M}}) = d(H_0, 0, D \cap \tilde{\mathscr{M}}).$$

Since $H_1 f = f + B_1 f - p$, $H_0 f = f + B_2 f - p$, this proves that

$$d(I + B_1, p, D \cap \tilde{\mathscr{M}}) = d(I + B_2, p, D \cap \tilde{\mathscr{M}}),$$

and the result follows from (13.3.2). □

13.3.5 Definition. Let D be an open bounded subset of the real Banach space $\mathscr{B}$. Assume that $A : \bar{D} \to \mathscr{B}$ is compact and that $(I + A)f \neq p$ for $f \in \partial D$.

Suppose first that $0 \in D$. With any $B, \mathscr{M}$ as in Lemma 13.3.2 define

$$d(I + A, p, D) = d(I + B, p, D \cap \mathscr{M}).$$

If $0 \notin D$, first shift the origin to a point in D and then define $d(I + A, p, D)$ as before. The finite integer $d(I + A, p, D)$ is called the (Leray–Schauder) **degree** of $I + A$.

It is an easy consequence of the Dugundji Extension Theorem 13.2.12 that the degree can also be defined for compact $A : \partial D \to \mathscr{B}$, see Problem 13.6.

The definition of degree for compact perturbations of the identity is now complete. The next theorem summarizes those properties which are most useful in applications.

13.3.6 Theorem. *Let D be an open bounded subset of the real Banach space $\mathscr{B}$. The following properties hold.*

(i) *Homotopy Invariance. Suppose that $\{H_t\}$ is a homotopy of operators $\bar{D} \to \mathscr{B}$ for $t \in [0, 1]$, and assume that $H_t - I$ is compact for each t. If $H_t f \neq p$ for any $f \in \partial D$ and $t \in [0, 1]$, then $d(H_t, p, D)$ is independent of t.*

(ii) *$d(I, p, D) = 1$ if $p \in D$, and $d(I, p, D) = 0$ if $p \notin \bar{D}$.*

(iii) *The Leray–Schauder Fixed Point Theorem. Suppose that $A : \bar{D} \to \mathscr{B}$ is compact, and that $f + Af \neq p$ for any $f \in \partial D$. If $d(I + A, p, D) \neq 0$, the equation $f + Af = p$ has at least one solution in D.*

Proof. The proofs are easy, being based on the corresponding finite dimensional result and a limiting procedure, and it will suffice to illustrate the argument if we establish (iii) only.

Choose a sequence (ε_n) of strictly positive numbers tending to zero. Then by Lemma 13.3.2 there is a sequence (B_n) of continuous operators and a sequence $(\mathscr{M}_n)$ of finite dimensional linear subspaces containing p such that B_n maps $\bar{D}$ into $\mathscr{M}_n$, and $\|B_n f - Af\| \leqslant \varepsilon_n$ for $f \in \bar{D}$. Therefore from Definition 13.3.5, for sufficiently large n

$$d(I + B_n, p, D \cap \mathscr{M}_n) = d(I + A, p, D).$$

As $d(I + A, p, D) \neq 0$, we can invoke Theorem 13.2.11(iv) and conclude that there is a sequence (f_n) in D such that $f_n + B_n f_n = p$. But A is compact, so there is a subsequence, denoted still by (f_n), such that (Af_n) is convergent. Thus

$$\|f_n + Af_n - p\| \leqslant \|f_n + B_n f_n - p\| + \|Af_n - B_n f_n\| \leqslant \varepsilon_n.$$

It follows that (f_n) is convergent with limit f say, and as A is continuous, $f + Af = p$. □

So far as the use of this theorem in practice is concerned, as in finite dimensions the most frequently employed technique is to base the calculation of degree on homotopy invariance, and then to prove existence by means of the Leray–Schauder Fixed Point Theorem.

The final result, sometimes also called the Leray–Schauder Fixed Point Theorem, is especially convenient in applications as it contains no explicit mention of degree.

13.3.7 Theorem. *Let D be a bounded open subset of the real Banach space $\mathscr{B}$, and suppose that $A : \bar{D} \to \mathscr{B}$ is compact. Take $p \in D$ and assume that $f + tAf$*

$\neq p$ for any $f \in \partial D$ and $0 \leqslant t \leqslant 1$. Then the equation $f + Af = p$ has at least one solution in D.

Proof. Set $H_t f = (I + tA)f$ for $f \in \bar{D}$. By Theorem 13.3.6(i)

$$d(I, p, D) = d(H_0, p, D) = d(H_1, p, D) = d(I + A, p, D).$$

Since $p \in D$, $d(I, p, D) = 1$ and the result follows from Theorem 13.3.6(iii). □

13.4 A Problem in Radiative Transfer

The problem of radiative transfer in stellar atmospheres is of considerable interest in astronomy and has been studied extensively over the past fifty years. One formulation of the problem leads to a singular linear integral equation which motivated the invention of the Wiener–Hopf Theory. Another approach yields a nonlinear integral equation, often known as the Chandresekhar H-equation, which has advantages from the point of view of numerical computation. The first rigorous treatment of this equation appears to be due to Crum in 1947. Crum's method is based on a complicated complex variable argument, and although this was subsequently simplified by Busbridge, the method remains rather involved. Recently a number of authors have attacked the equation by functional analytic methods. For example the theory of monotone operators has been successfully employed by Leggett (1976), while Stuart (1974) has devised an elegant existence proof using degree theoretic arguments. Stuart's method follows the main steps typically taken in applications of degree theory without involving heavy technicalities, and we shall use it as an illustration of the technique.

The Chandresekhar H-equation is

$$H(x) = 1 + x H(x) \int_0^1 \frac{\Psi(y) H(y)}{x + y} \mathrm{d}y \qquad (0 \leqslant x \leqslant 1). \tag{13.4.1}$$

Here Ψ is a given function which is non-negative and continuous, and

$$\mu = \int_0^1 \Psi(y) \, \mathrm{d}y \leqslant \tfrac{1}{2}. \tag{13.4.2}$$

It is required that the unknown function H should lie in $\mathscr{C}([0, 1])$. A useful fact, readily deduced from (13.4.1) by inspection, is that $H(x) \geqslant 1$ for $x \in [0, 1]$. As it stands (13.4.1) is inconvenient to handle, and the first step is to rewrite it in a more tractable form in which the unknown element is the bounded continuous function $u = 1/H$. The following standard result summarizes some useful facts about the new equation.

13.4.1 Lemma. *Suppose that* $\mu \leq \frac{1}{2}$. *If* $u \in \mathscr{C}([0, 1])$ *is a strictly positive solution of the equation*

$$u(x) = (1 - 2\mu)^{\frac{1}{2}} + \int_0^1 \frac{y\Psi(y)[u(y)]^{-1}}{x + y}\, dy, \tag{13.4.3}$$

then $H = 1/u$ *is a solution of the H-equation* (13.4.1). *Further,* u *satisfies the relations*

$$\int_0^1 \Psi(x)[u(x)]^{-1}\, dx = 1 - (1 - 2\mu)^{\frac{1}{2}}, \tag{13.4.4}$$

$$1 \geq u(x) \geq (1 - 2\mu)^{\frac{1}{2}} + \int_0^1 \frac{y\Psi(y)}{x + y}\, dy \qquad (0 \leq x \leq 1). \tag{13.4.5}$$

Proof. Divide (13.4.3) by u, multiply by Ψ and integrate, to obtain

$$\int_0^1 \Psi(x)\, dx = (1 - 2\mu)^{\frac{1}{2}} \int_0^1 \Psi(x)[u(x)]^{-1}\, dx$$
$$+ \int_0^1 \int_0^1 y\Psi(x)\Psi(y)[(x + y)u(x)u(y)]^{-1}\, dx\, dy.$$

Interchange of x and y in the last integral and addition gives

$$\mu = (1 - 2\mu)^{\frac{1}{2}} \int_0^1 \Psi(x)[u(x)]^{-1}\, dx + \frac{1}{2}\left\{\int_0^1 \Psi(x)[u(x)]^{-1}\, dx\right\}^2.$$

(13.4.4) is obtained by solving this as a quadratic equation. Substitution for $(1 - 2\mu)^{\frac{1}{2}}$ from (13.4.4) in (13.4.3) and the change of variable $u = 1/H$ then yields (13.4.1). Finally, since $y(x + y)^{-1} \leq 1$ for $x, y \in (0, 1]$, from (13.4.3) and (13.4.4),

$$u(x) \leq (1 - 2\mu)^{\frac{1}{2}} + \int_0^1 \Psi(y)[u(y)]^{-1}\, dy = 1.$$

The second inequality in (13.4.5) is an obvious consequence when (13.4.3) is used again. □

The problem is of most mathematical interest when $\mu = \frac{1}{2}$. Indeed if $\mu < \frac{1}{2}$ existence follows by an easy positivity argument—see Problem 13.16. Therefore from here on only the case $\mu = \frac{1}{2}$ will be considered. By the lemma the existence of a solution of the H-equation will follow if it can be proved that the equation

$$u(x) = \int_0^1 \frac{y\Psi(y)}{x + y}[u(y)]^{-1}\, dy \tag{13.4.6}$$

has a strictly positive continuous solution. Let $\mathscr{C}([0, 1])$ be the Banach space of bounded real valued continuous functions with the sup norm. In preparation for the use of a homotopy invariance argument define

$$\Psi_t(x) = (1 - t)\Psi(x),$$

$$K_t f(x) = \int_0^1 \frac{y\Psi_t(y)}{x - y} f(y)\, dy,$$

$$A_t u(x) = t^{\frac{1}{2}} + \int_0^1 \frac{y\Psi_t(y)}{x - y} [u(y)]^{-1}\, dy,$$

for $0 \leqslant t \leqslant 1$ and $0 \leqslant x \leqslant 1$. In this notation (13.4.6) is $u = A_0 u$, and we must show that A_0 has a fixed point in $\mathscr{C}([0, 1])$ which is a strictly positive function. Observe first that for each t, $K_t: \mathscr{C}([0, 1]) \to \mathscr{C}([0, 1])$ is a compact linear operator (see Problem 13.14). The main difficulty is the choice of a suitable domain $\bar{D}$ for A_t. Because of the presence of a term $1/u$ in $A_t u$, $\bar{D}$ must only include strictly positive u, say those u with $u(x) \geqslant a > 0$, but then A_t will be compact. Further, in view of the usual restriction in degree theory, $I - A_t$ must not vanish on ∂D for any t. A crucial observation at this stage, as in many applications of degree theory, is that there is a bound independent of t on all fixed points of A_t. As in the following lemma, this *a priori* bound is obtained by an independent argument. Note that the existence of a fixed point is not assumed.

13.4.2 Lemma. *There is an $m > 0$ such that any strictly positive continuous solution of $u = A_t u$ for any $t \in [0, 1]$ satisfies the inequality $u(x) \geqslant 2m$ $(0 \leqslant x \leqslant 1)$.*

Proof. Since $\mu = \frac{1}{2}$, from (13.4.2),

$$\left\{1 - 2\int_0^1 \Psi_t(x)\, dx\right\}^{\frac{1}{2}} = \left\{1 - 2(1 - t)\int_0^1 \Psi(x)\, dx\right\}^{\frac{1}{2}} = t^{\frac{1}{2}}.$$

Therefore if in (13.4.3) Ψ is replaced by Ψ_t, the equation $u = A_t u$ is obtained. Hence from (13.4.5),

$$u(x) \geqslant t^{\frac{1}{2}} + (1 - t)\int_0^1 \frac{y\Psi(y)}{1 + y}\, dy.$$

An easy calculation shows that for $0 \leqslant t \leqslant 1$ the right hand side of this inequality is a minimum when $t = 0$. □

13.4.3 Theorem. *For $\mu = \frac{1}{2}$, the integral equation (13.4.6) has a strictly positive solution $u \in \mathscr{C}([0, 1])$, and $H = 1/u$ is a continuous solution of the H-*

equation (13.4.1) *which is such that*

$$1 \geqslant 1/H(x) \geqslant \int_0^1 \frac{y\Psi(y)}{x+y}\,\mathrm{d}y \qquad (0 \leqslant x \leqslant 1).$$

Proof. For given a, b with $0 < a < b$ set

$$D = \{u : u \in \mathscr{C}([0,1]),\ a < u(x) < b \text{ for } x \in [0,1]\}.$$

D is evidently a non-empty bounded open subset of $\mathscr{C}([0,1])$, and $A_t : \bar{D} \to \mathscr{C}([0,1])$ is compact. We choose a, b in such a way that $I - A_t$ does not vanish on ∂D for any $t \in [0,1]$, and let

$$a = \min(\tfrac{1}{2}, m), \qquad b = 2 + a^{-1}\|K_0\|.$$

It is evidently enough to show that if $u - A_t u = 0$ for $u \in \bar{D}$, then $u \in D$. To prove this note first that from the *a priori* bound for u (Lemma 13.4.2),

$$u(x) \geqslant 2m > a \qquad (0 \leqslant x \leqslant 1).$$

Secondly,

$$\|u\| = \|A_t u\| \leqslant t^{\frac{1}{2}} + a^{-1}\|K_t\| \leqslant t^{\frac{1}{2}} + a^{-1}(1-t)\|K_0\| < b.$$

Thus $u \in D$ as asserted.

The preparations for the use of degree theory are now complete. Consider the homotopy $H_t = I - A_t$. It has just been proved that this does not vanish on ∂D. Hence by homotopy invariance (Theorem 13.3.6(i)),

$$d(I - A_1, 0, D) = d(I - A_1, 0, D).$$

But $A_1 u = 1$ (the unit function), and

$$d(I - A_1, 0, D) = d(I, 1, D) = 1,$$

by Theorem 13.3.6(ii) as $1 \in D$. Therefore $d(I - A_0, 0, D) = 1$. The result follows from the Leray–Schauder Fixed Point Theorem 13.3.6(iii). □

Problems

13.1 Take $D = (-1, 1)$, $\mathscr{B} = \mathbb{R}$ and $-1 < p < 1$, and consider in turn the functions $|x|^3 \sin(\pi/2x)$, $|x| \sin(\pi/2x)$, both being defined to be zero when $x = 0$. Outline the method for calculating the degree directly from the definition, and compare this with the argument based on homotopy invariance.

13.2 Let $L: \mathbb{R}^n \to \mathbb{R}^n$ be linear, and suppose that $\mu^{-1} \in \rho(L)$. With D the open unit ball show that

$$d(I - \mu L, 0, D) = (-1)^{\beta},$$

where β is the sum of the algebraic multiplicities (defined in Example 13.2.13) of the real eigenvalues of L which have the same sign as μ and are greater in absolute magnitude than $|\mu|^{-1}$.

13.3 Suppose that D_1 and D_2 are disjoint bounded open sets in $\mathbb{R}^n$. Assume that $\phi : \bar{D}_1 \cup \bar{D}_2 \to \mathbb{R}^n$ is continuous and that $\phi(x) \neq p$ for $x \in \partial D_1 \cup \partial D_2$. Prove that

$$d(\phi, p, D_1 \cup D_2) = d(\phi, p, D_1) + d(\phi, p, D_2).$$

13.4 Suppose that D is a bounded open subset of $\mathbb{R}^n$, that $\phi : \bar{D} \to \mathbb{R}^r$ is continuous, and that $\phi(x) \neq p$ on ∂D. Let x_0 be an isolated solution of $\phi(x) = p$. The *index* $i(\phi, x_0)$ is defined to be $d(\phi, p, D_0)$ where D_0 is an open neighbourhood of x_0 containing only the solution x_0 itself. Assume that $\phi(x) = p$ has a finite number of solutions $x_1, \ldots, x_k$ in $\bar{D}$. Prove that

$$d(\phi m\, p, D) = \sum_{j=1}^{k} i(\phi, x_j).$$

13.5 (Poincaré–Bohl). Suppose that D is an open bounded subset of $\mathbb{R}^n$, and assume that for $j = 1, 2$, $\phi_j : \bar{D} \to \mathbb{R}^n$ is continuous and $\phi_j(x) \neq p$ on ∂D. Prove that $d(\phi_1, p, D) = d(\phi_2, p, D)$ if the vectors $\phi_1(x) - p$, $\phi_2(x) - p$ never point in opposite directions for $x \in \partial D$.

13.6 Let D be an open bounded subset of the real Banach space $\mathscr{B}$, and suppose that $A : \partial D \to \mathscr{B}$ is compact. Using Dugundji's Theorem 13.2.12 and Lemma 8.2.4, show that A has a compact extension $\tilde{A}$ to $\bar{D}$ with $R(\tilde{A})$ contained in the convex hull of $R(A)$. Hence define $d(I + A, p, D)$ if $(I + A)f \neq p$ on ∂D, and show that the definition is independent of the choice of extension.

13.7 Prove the Schauder Fixed Point Theorem by a degree theoretic argument. (Recall Lemma 8.2.4, and proceed as in Example 13.2.15).

13.8 Let D be a bounded open subset of the real Banach space $\mathscr{B}$, and assume that $A, B : \bar{D} \to \mathscr{B}$ are compact. If

$$\|Af - Bf\| < \|f - Bf - p\| \qquad (f \in \partial D),$$

show that $d(I - A, p, D) = d(I - B, p, D)$.

13.9 Let $\mathscr{B}$ be a real Banach space. Suppose that $B : \mathscr{B} \to \mathscr{B}$ is linear and compact with $1 \in \rho(B)$, and $A : \mathscr{B} \to \mathscr{B}$ is compact. Assume that A is asymptotically linear in the sense that

$$\lim_{\|f\| \to \infty} \|Af - Bf\|/\|f\| = 0.$$

Deduce from the previous problem and Lemma 14.2.4 that A has a fixed point.

13.10 A powerful global existence theorem for Hammerstein integral equations may be deduced from the last problem. Suppose that $k : [0, 1] \times [0, 1] \to \mathbb{R}$ is continuous and define $K : \mathscr{C}([0, 1]) \to \mathscr{C}([0, 1])$ by

$$Kf(x) = \int_0^1 k(x, y) f(y)\, dy.$$

Suppose that $\psi:[0, 1] \times \mathbb{R} \to \mathbb{R}$ is continuous, and assume that there are a real α, an $\varepsilon > 0$, and continuous g_1, g_2 such that

$$|\psi(x, z) - \alpha z| \leqslant g_1(x)\,|z|^{1-\varepsilon} + g_2(x) \qquad (x \in [0, 1], z \in \mathbb{R}).$$

Prove that if $\alpha^{-1} \in \rho(K)$, the equation

$$f(x) = \int_0^1 k(x, y)\,\psi[y, f(y)]\,\mathrm{d}y$$

has a solution in $\mathscr{C}([0, 1])$.

13.11 Let D be the open unit ball in the real Banach space $\mathscr{B}$, and suppose that $A : \bar{D} \to \mathscr{B}$ is compact. Prove that A has a fixed point in $\bar{D}$ if either of the following hold:

(i) $\|f - Af\|^2 \geqslant \|Af\|^2 - \|f\|^2 \qquad (f \in \partial D)$;
(ii) $\mathscr{B}$ is a real Hilbert space and $(f, Af) \leqslant \|f\|^2$ for $f \in \partial D$.

13.12 Let $\mathscr{B}$ be a real Banach space, and let $A : \mathscr{B} \to \mathscr{B}$ be compact. Suppose that there is an *a priori* bound $m > 0$ such that every solution of $f - tAf = 0$ for $0 \leqslant t \leqslant 1$ satisfies $\|f\| \leqslant m$. Prove that A has a fixed point in $\bar{S}(0, m)$.

13.13 Suppose that $\psi : \mathbb{R} \to \mathbb{R}$ is continuous and non-decreasing, and assume that $\psi(z) \to \pm\infty$ respectively as $z \to \pm\infty$. Let g be a real-valued continuous function. Use 13.12 to prove the existence of a solution in $\mathscr{C}^2([0, 1])$ under homogeneous Dirichlet conditions of

$$f''(x) = \psi(f(x)) + g(x) \qquad (0 \leqslant x \leqslant 1).$$

13.14 Prove that the operator K_t in Section 13.4 is compact.

13.15 Show how the existence proof for the H-equation may be simplified if $\mu < \frac{1}{2}$, an *a priori* bound being unnecessary.

13.16 If $\mu < \frac{1}{2}$ the existence of a solution of the H-equation may also be deduced from Theorem 8.3.9.

Chapter 14

Bifurcation Theory

14.1 Introduction

Among the great range of physical phenomena modelled by nonlinear equations, those leading to eigenfunction problems are of particular difficulty and importance. The following classical example is typical. Suppose an inviscid fluid flows irrotationally along a flat-bottomed canal. Can periodic waves of permanent type exist? The flow can be modelled by an equation of the form $\lambda f = Af$ where A is a nonlinear operator with $A0 = 0$ and λ is a parameter. One possible motion is a uniform stream (corresponding to the trivial solution $f = 0$), but it is of course the non-trivial solutions which are of physical interest. The presence of the trivial solution greatly complicates the mathematical treatment here, and it is a matter of some difficulty to devise arguments sufficiently subtle to deal with problems of this type. It will be the business of this chapter to review some of these arguments.

For A linear, a study of the equation $\lambda f = Af$ has been carried out previously as a part of spectral theory. Although the aim is a generalization of this theory, it is conventional in the nonlinear case to make a change of notation and use instead the parameter $\mu = \lambda^{-1}$. In some applications the operator itself depends on μ, a possibility which is allowed for in the following.

14.1.1 Definition. Let D be a subset of the *real* Banach space $\mathscr{B}$, and let $0 \in D$. For each real μ let A_μ be an operator mapping D into $\mathscr{B}$, and assume that $A_\mu 0 = 0$. The (real) values of μ for which the equation $f = \mu A_\mu f$ has non-zero solutions will be called the **characteristic values** of A_μ, and the corresponding solutions f **eigenfunctions**. As with the linear equation, the term eigenvalue will refer to the inverse of a characteristic value.

14.1.2 Example. For illustration consider the following, in each of which $\phi : \mathbb{R} \to \mathbb{R}$.

Case A. Let ϕ be a linear mapping with $\phi(x) = \alpha x$ and α a non-zero real number. Then the equation $x = \mu\phi(x)$ has a non-zero solution iff $\mu = \alpha^{-1}$,

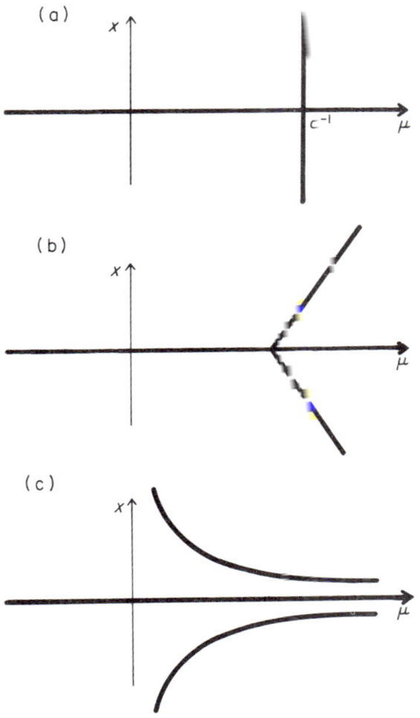

Fig. 14.1 Bifurcation diagrams for (a) $\phi(x) = \alpha x$. (b) $\phi(x) = x/(1 + |x|)$. (c) $\phi(x) = x^3$.

and in this case every real number is a solution. In Fig. 14.1(a) the solutions, represented by a heavy line, are plotted against μ.

Case B. For $\phi(x) = x/(1 + |x|)$, the equation $x = \mu\phi(x)$ has non-zero solutions $(x = \pm(\mu - 1))$ if and only if $\mu > 1$. The analogous diagram is Fig. 14.1(b).

Case C. For $\phi(x) = x^3$, the equation $x = \mu\phi(x)$ has non-zero solutions $(x = \pm\mu^{-\frac{1}{2}})$ iff $\mu > 0$, see Fig. 14.1(c).

Points of interest which emerge from a comparison of these examples are as follows.

(i) In each case there are no eigenfunctions in a fixed ball in $\mathbb{R}$ centre the origin for small μ.

(ii) In the linear case the set of characteristic values is a single point. For the nonlinear operators this set is an interval, $\mu > 1$ in case B and $\mu > 0$ in case C. Thus in striking contrast with the linear case, the "nonlinear spectrum" is not discrete even if the operator is compact.

(iii) For linear ϕ, the equation $x - \mu\phi(x) = a$ has a solution for every real a if and only if μ is not a characteristic value. The corresponding nonlinear equations have a solution for every μ: a Fredholm alternative evidently does not hold.

(iv) In cases A and B, as μ increases a point is reached at which the solutions branch or "bifurcate" from the trivial solution. This behaviour is common, but as can be seen from Fig. 14.1(c) bifurcation need not take place. In each case the behaviour of solutions as functions of μ is neatly summarized by the "bifurcation diagrams" above. Although in higher dimensions the solutions cannot be represented on a single axis, there is often some quantity of significance and physical interest—perhaps the norm of the solution—which may usefully be used as a substitute in an analogous bifurcation diagram.

The above examples together with those of Problem 14.1 give some idea of the diversity possible even in one dimension, and in higher dimensions further complications may occur. The infinite dimensional case is intractable unless some further assumption is made, and the discussion will be restricted here to compact A. Then for linear A the bifurcation diagram (Fig. 14.2) is readily constructed from the spectral theory of compact operators.

If A is nonlinear the dependence on μ is seldom so simple, and in particular

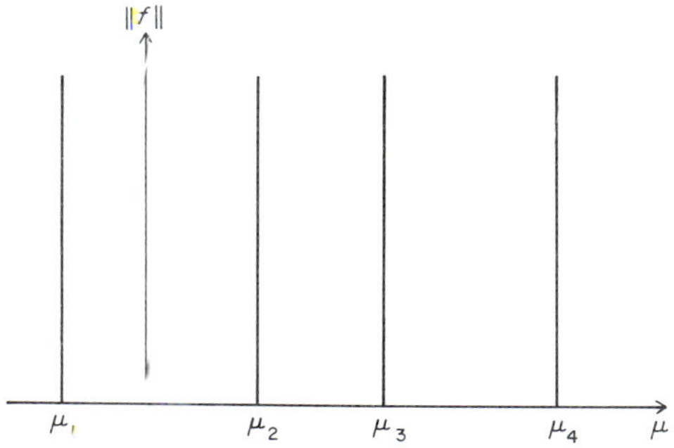

Fig. 14.2 Bifurcation diagram for a linear compact operator. $\mu_1, \mu_2, \ldots$ are its characteristic values.

the behaviour for small and large $\|f\|$ may be quite different. In the sequel the term *local* will be used when only small values of $\|f\|$ are allowed, while *global* will imply that $\|f\|$ is not so restricted.

A formal argument which has often been used for the local problem is based on linearization, the governing equation being replaced by a linear equation whose solutions are used as approximations to the small solutions of the original nonlinear equation. An examination of this procedure in case B above shows that the only characteristic value of the linearized equation is unity, and that this is indeed the only bifurcation point of A. The local bifurcation theory of the next section is based on this method, and we shall see that although the argument sometimes fails, it is valid for a rather wide class of operator. Among the applications to be discussed will be the wave motion problem mentioned above, the theory being used to provide a simple proof of the existence of a periodic train of small nonlinear waves.

In many applications the local theory while useful does not provide a full resolution of the problem. For example in the wave problem a variety of heuristic arguments suggest that waves with maximum angle of inclination up to $\pi/6$ can exist, and to confirm this a global theory is evidently needed. Such a theory usually presents much greater technical difficulties, but some results in this area will be obtained in Section 3 by exploiting the global nature of Leray–Schauder degree theory. The main goal is the powerful Monotone Minorant Theorem of Krasnoselskii; in one of its applications the above prediction for the wave problem is confirmed.

Nonlinear eigenfunction problems are discussed extensively by Krasnoselskii (1964a) and Berger (1977). The *Rocky Mountain Journal of Mathematics* **3**, 2 (1973) contains a number of interesting articles on recent developments, for which Stakgold (1971) and Sattinger (1973) may also be consulted. Several applications to physical situations are reviewed in Keller and Antman (1969), while Dickey (1976) is concerned with problems in elasticity. For a discussion of the important stability question for evolution equations see Sattinger (1973), and Benjamin (1976) for the hydrodynamical implications. Finally it may be noted that for a certain special class of nonlinear operator, a Fredholm type theory does exist, see Fučík *et al.* (1973).

14.2 Local Bifurcation Theory

Throughout D will be an open subset containing the origin of the real Banach space $\mathscr{B}$, and $\{A_\mu\}$ will be a family of operators mapping $\bar{D}$ into $\mathscr{B}$ with $A_\mu f$ continuous in μ and f for μ in some subset of $\mathbb{R}$ and f in $\bar{D}$. Assume further that $A_\mu 0 = 0$ for every μ. The local theory concerns the eigenfunctions of A_μ of small norm.

14.2.1 Definition. Let $\mu_0 \in \mathbb{R}$ be given. Suppose that for every $\varepsilon > 0$ there is a characteristic value μ and corresponding eigenfunction f of A_μ such that $|\mu - \mu_0| < \varepsilon$ and $\|f\| < \varepsilon$. Then μ_0 is called a **bifurcation point** (from zero) of A_μ. Bifurcations from non-zero solutions may also occur but will not be considered here.

If A_μ is linear and independent of μ, its bifurcation points are just its characteristic values. If A_μ is nonlinear and has Fréchet derivative $A'_\mu(0)$ at zero, $A'_\mu(0)$ is an approximation to A_μ near zero. One may therefore expect that the bifurcation points and small eigenfunctions of A_μ and of the linear operator $A'_\mu(0)$ will be related. The validity of this linearization argument is now examined under the assumption that $A_\mu = L + N_\mu$ where

(i) $L : \mathscr{B} \to \mathscr{B}$ is a non-zero compact linear operator independent of μ,
(ii) $N_\mu : \bar{D} \to \mathscr{B}$ is compact for each μ, and $N_\mu f$ is continuous in μ and f with

$$\lim_{h \to 0} \|N_\mu h\| / \|h\| = 0,$$

the convergence being uniform with respect to μ for μ in any finite interval.

If A_μ is independent of μ and compact these conditions are equivalent to Fréchet differentiability at the origin with $A'_\mu(0) = L$, the compactness of L being automatic (Problem 14.5). Here a slightly more general situation is considered, the operator being allowed to depend on μ, although only in a mild manner.

14.2.2 Theorem. *Under the above conditions, μ_0 can be a bifurcation point of A_μ only if it is a characteristic value of L. The set of bifurcation points of A_μ has no finite limit point.*

Proof. Suppose μ_0 is a bifurcation point of A_μ, but is not a characteristic value of L. Since L is compact, a neighbourhood of μ_0^{-1} is contained in $\rho(L)$. Therefore $(I - \mu L)^{-1}$ is uniformly bounded in $\mathscr{L}(\mathscr{B})$ (by Theorem 3.6.3) for μ in this neighbourhood. Now $f = \mu A_\mu f$ may be rewritten as

$$f/\|f\| = \mu(I - \mu L)^{-1} N_\mu f/\|f\|,$$

and by condition (ii), a contradiction is obtained on taking the norm of each side of this equation and choosing sufficiently small f. □

Granted that bifurcation can only take place at characteristic values of L, it is natural to ask whether each of these values is a bifurcation point. We do not need to look far to see that the answer is in the negative.

14.2.3 Example. For vectors $x = (x_1, x_2)$ in $\mathbb{R}^2$ define $\phi: \mathbb{R}^2 \to \mathbb{R}^2$ by

$$\phi(x) = (x_1 - x_2(x_1^2 + x_2^2), x_2 + x_1(x_1^2 + x_2^2)).$$

Then the Fréchet derivative of ϕ at zero is the identity, and thus has the characteristic value unity. If x is an eigenvector of ϕ,

$$x_1 = \mu[x_1 - x_2(x_1^2 + x_2^2)], \qquad x_2 = \mu[x_2 + x_1(x_1^2 + x_2^2)], \qquad (14.2.1)$$

and multiplying these equations by x_1 and x_2 respectively and adding we deduce that

$$x_1^2 + x_2^2 = \mu(x_1^2 + x_2^2).$$

This equation has a non-zero solution only if $\mu = 1$, but then the only solution of (14.2.1) is $x = 0$. Therefore ϕ has *no* characteristic values.

Characteristic values of L need not then be bifurcation points of A_μ. The next theorem asserts that if the multiplicity of a characteristic value is odd, bifurcation does indeed take place. The argument that will be used to prove this is degree theoretic, and is based on the fact that on a neighbourhood of the origin the degrees of $I - \mu A_\mu$ and $I - \mu L$ are the same. In finite dimensions the latter is easy to calculate (Problem 13.2) and the identical formula is valid in infinite dimensions—for a proof see Krasnoselskii (1964a, p. 133). In the following, the algebraic multiplicity is defined as before (Example 13.2.13), and the eigenvalue is said to be *simple* if its algebraic multiplicity is unity.

14.2.4 Lemma. *Suppose $\mu^{-1} \in \mathbb{R}$ is not an eigenvalue of the compact linear operator L. Then if D is any open set containing 0*

$$d(I - \mu L, 0, D) = (-1)^\beta,$$

where β is the sum of the algebraic multiplicities of the real eigenvalues of L which have the same sign as μ and are greater in absolute magnitude than $|\mu|^{-1}$.

14.2.5 Theorem. *Assume that A_μ satisfies conditions* (i) *and* (ii) *above. If μ_0 is a characteristic value of L of odd multiplicity, it is a bifurcation point of A_μ.*

Proof. A contradiction argument is used. Choose $\varepsilon > 0$ so small that μ_0 is the only characteristic value of L in $[\mu_0 - \varepsilon, \mu_0 + \varepsilon]$. If μ_0 is not a bifurcation point, it is possible to arrange by choosing ε smaller if necessary that for every μ in the above interval, zero is the only solution of $f = \mu A_\mu f$ for which $\|f\| \leqslant \varepsilon$. Let S be the open ball in $\mathscr{F}$ centre the origin and radius ε. Then

$I - (\mu_0 + t\varepsilon)A_{\mu_0 + t\varepsilon}$ does not vanish on ∂S for any $t \in [-1, 1]$, and by homotopy invariance

$$d(I - (\mu_0 - \varepsilon)A_{\mu_0 - \varepsilon}, 0, S) = d(I - (\mu_0 + \varepsilon)A_{\mu_0 + \varepsilon}, 0, S). \qquad (14.2.2)$$

It is next shown that the degrees of $I - \mu A_\mu$ and $I - \mu L$ are equal. Consider the homotopy $H_t = I - \mu L - t\mu N_\mu$ for $t \in [0, 1]$. If μ is not a characteristic value of L,

$$\|H_0 f\| = \|f - \mu L f\| \geqslant c\|f\| \qquad (14.2.3)$$

for some $c > 0$ and all $f \in \mathscr{B}$. Also

$$\|H_t f - H_0 f\| = t|\mu|\,\|N_\mu f\|,$$

and in view of the assumed condition on N_μ, we can arrange (by choosing smaller ε if necessary) that $\|N_\mu f\|/\|f\| < c/|\mu|$ for $\|f\| \leqslant \varepsilon$. Thus

$$\|H_t f - H_0 f\| < c\varepsilon \qquad (f \in \bar{S}). \qquad (14.2.4)$$

Therefore, for $f \in \partial S$ and $0 \leqslant t \leqslant 1$,

$$\|H_t f\| \geqslant \|H_0 f\| - \|H_t f - H_0 f\| > 0$$

from (14.2.3) and (14.2.4). Thus the degree of H_t on S is defined, and recalling that $H_0 = I - \mu L$, $H_1 = I - \mu A_\mu$ and using homotopy invariance we deduce that

$$d(I - \mu A_\mu, 0, S) = d(I - \mu L, 0, S). \qquad (14.2.5)$$

Finally, by (14.2.2) and (14.2.5),

$$\begin{aligned} d(I - (\mu_0 + \varepsilon)L, 0, S) &= d(I - (\mu_0 + \varepsilon)A_{\mu_0 + \varepsilon}, 0, S) \\ &= d(I - (\mu_0 - \varepsilon)A_{\mu_0 - \varepsilon}, 0, S) \\ &= d(I - (\mu_0 - \varepsilon)L, 0, S). \end{aligned}$$

However, μ_0 has odd multiplicity, so Lemma 14.2.4 is contradicted. □

It is interesting to note that this theorem may be much strengthened to yield a result essentially global in character, see Rabinowitz (1973) or Berger (1977, p. 276). The following, whose proof is set as exercise, asserts that near a bifurcation point the eigenfunctions of A behave locally like those of L.

14.2.6 Theorem. *Assume that A_μ satisfies conditions* (i) *and* (ii) *above. Let μ_0 be a bifurcation point of A_μ, and denote by X the set of eigenfunctions of A_μ corresponding to characteristic values in $[\mu_0 - \varepsilon, \mu_0 + \varepsilon]$. Let Y be the linear span of the eigenfunctions of L corresponding to μ_0. Then for sufficiently small ε,*

$$\lim_{f \to 0,\, f \in X} \operatorname{dist}(f, Y)/\|f\| = 0.$$

Taken together these theorems provide a rigorous basis for the heuristic linearization argument. A shortcoming is that Theorem 14.2.5 only covers situations in which the multiplicity is odd. This restriction may be lifted under certain circumstances, for example if A is a "gradient operator" (Krasnoselskii, 1964a; p. 332), but in general the higher order terms must be examined; for an introductory discussion see Berger and Berger (1968, p. 123). The theorem is nonetheless useful in many applications. For example if L is an integral operator arising from a boundary value problem for a differential equation, the characteristic values are often simple. Two well known nonlinear eigenfunction problems are next considered. The first is an easy example in elasticity and is intended primarily as an illustration.

14.2.7 Example. The Buckling of a Compressed Rod. Consider a thin elastic rod of length unity hinged at one end and acted on at the other end by a compressive force P, see Fig. 14.3. Let the distance along the rod be s and suppose the density is $\rho(s)$—assumed continuous and strictly positive. Then with $y(s)$ the transverse displacement at s, the governing equations are

$$\frac{d^2y}{ds^2} + \mu\rho y\left[1 - \left(\frac{dy}{ds}\right)^2\right]^{\frac{1}{2}} = 0$$

$$y(0) = y(1) = 0$$

where μ is a constant proportional to P. Physical intuition suggests that for small P the rod will be compressed without deflexion, but that if P is large there may be buckling.

Rewriting the differential equation as an integral equation with the help of the Green's function k, we obtain on setting $\phi = -y''$,

$$\begin{aligned}\phi(s) &= \mu\rho(s)\int_0^1 k(s,t)\,\phi(t)\,dt\,.\left\{1 - \left[\int_0^1 \frac{\partial k}{\partial s}(s,w)\,\phi(w)\,dw\right]^2\right\}^{\frac{1}{2}},\\ &= \mu A\phi(s)\end{aligned}$$

say. This is a nonlinear eigenfunction problem. Take $\mathscr{B}$ to be the real Banach space $\mathscr{C}([0,1])$ with the sup norm, and let D be its open unit ball. Evidently $A : \bar{D} \to \mathscr{C}([0,1])$ is compact, and A has the Fréchet derivative L at zero where

$$L\phi(s) = \rho(s)\int_0^1 k(s,t)\,\phi(t)\,dt.$$

Denote the characteristic values and corresponding eigenfunctions of L by $\mu_1, \mu_2, \ldots$ and $\phi_1, \phi_2, \ldots$ respectively. Since the characteristic values are simple, it is an immediate consequence of Theorems 14.2.2 and 14.2.5 that the set of bifurcation points of A is precisely $\{\mu_n\}$. Further, from Theorem 14.2.6,

For μ near μ_n the small eigenfunctions of A are approximated by appropriate multiples of ϕ_n.

This analysis confirms that bifurcation takes place at the characteristic values of the linearized equation. However, the local theory leaves a number of important questions unanswered. In particular it does not indicate whether any μ lying between two bifurcation points is a characteristic value, nor does it resolve the question (to which we return in Example 14.3.11) of the existence of large buckled states. In the present simple case the full solution can be obtained if the rod is uniform, as the differential equation may be integrated explicitly in terms of elliptic functions, see Keller and Antman (1969, p. 10).

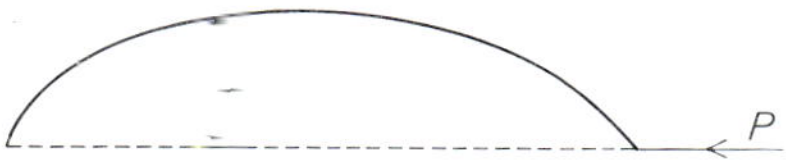

Fig. 14.3

The bifurcation diagram (Fig. 14.4) with the angle α at one end as a measure of deflexion, shows that for any $\mu > \pi^2$ there is at least one buckled state. The question of which state is preferred by the system for large μ can also be settled by further analysis.

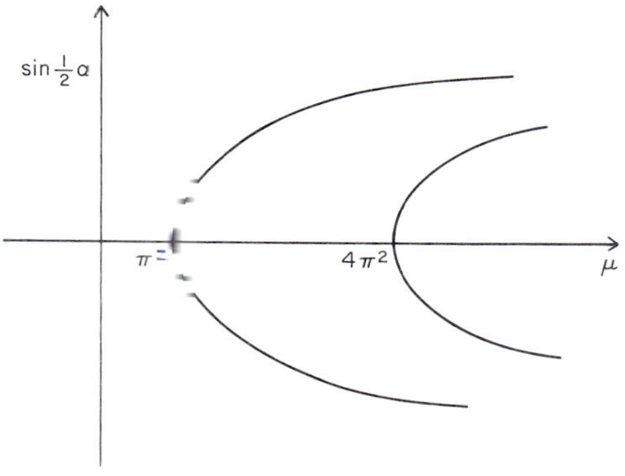

Fig. 14.4 Bifurcation diagram for the buckling of a compressed rod.

14.2.8 Example. One of the most interesting topics in the theory of water waves is the study of large amplitude waves in a canal. A problem in this area which has attracted mathematicians for more than a hundred years is the following. Suppose an inviscid fluid is in irrotational two-dimensional motion in a channel. Does there exist a periodic wave train of permanent type (that is unchanging with respect to an observer moving with the wave speed)? Stokes conjectured as long ago as 1880 that waves with maximum angle of inclination up to $\pi/6$ would exist, but such are the mathematical difficulties that it is only relatively recently that this prediction has been rigorously confirmed, and there remain numerous unsolved problems in this area. The allied solitary wave problem has turned out to be even more intractable.

The first major difficulty in tackling waves which are not infinitesimal, that is with linearized boundary condition, arises in coping with the free boundary. In two dimensions this problem may be resolved by the use of a complex variable technique. However, the basic equation is nonlinear, and further the existence of an undisturbed stream as a solution means that the problem is of eigenfunction type, the parameter being related to the wavelength. In the long history of this problem two main advances stand out. The first is a local result, the existence of small waves, proved independently by Levi–Civita and Nekrasov in the early 1920s; it will be shown next that this result may be obtained very easily by the bifurcation theory methods of this section. The second, which will be discussed in the next section, is the contribution of Krasovskii towards the solution of the much more difficult global existence problem.

The formulation on which the proof of local existence will be based here is that of Nekrasov, who showed that for the case of infinite depth, θ the angle of inclination of the velocity vector at the free boundary satisfies the equation $\theta = \mu A_\mu \theta$ where

$$A_\mu\theta(x) = \int_{-\pi}^{\pi} k(x, y) \sin\theta(y) \left[1 + \mu \int_0^y \sin\theta(u)\, \mathrm{d}u\right]^{-1} \mathrm{d}y.$$

Here μ is a parameter depending on wavelength, particle speed at a trough and g, and the kernel k—which is the Green's function of the Neumann problem for the Laplacian on a disc—is given by

$$k(x, y) = \frac{1}{3\pi} \sum_{n=1}^{\infty} n^{-1} \sin nx \sin ny.$$

The aim is to prove the existence of non-zero continuous solutions. Let $\mathscr{C}([-\pi, \pi])$ be the Banach space of sup normed real-valued continuous functions. For a fixed finite range of values of μ, A_μ is readily shown to be continuous and compact on a sufficiently small ball centre the origin. Its Fréchet derivative at zero is the integral operator L where

$$L\theta(x) = \int_{-\pi}^{\pi} k(x, y)\, \theta(y)\, \mathrm{d}y.$$

L has simple characteristic values $3n\pi$ ($n = 1, 2, \ldots$) and the other conditions on A_μ are easy to verify. We conclude immediately from Theorem 14.2.5

that each of these values is a bifurcation point of A_μ. This proves the existence of a periodic train of waves of small but not infinitesimal amplitude, which is the required local existence result (actually only the first bifurcation point has physical significance, Hyers (1964, p. 324)).

14.3 Global Eigenfunction Theory

The two examples which conclude the last section emphasize a severe limitation of local bifurcation theory, for in both cases the most interesting questions concern large solutions and are thus global in character. Indeed it must be expected that in practice, solutions whose size is limited only by some natural bound derived from the physics of the problem will exist. It is often very difficult to confirm this prediction unless further conditions on the operator hold. In one situation, that is when a partial ordering on the space simplifies matters, considerable progress is possible, and a typical group of results in this area which has proved useful in applications will be considered here.

Throughout $\mathscr{B}$ will be a real Banach space with a partial ordering induced by a cone E, and D will be a bounded open subset of $\mathscr{B}$ with $0 \in D$. Assume that $A : \bar{D} \to \mathscr{B}$ is a given compact operator (note that A is now assumed independent of μ). The aim will be to show that the totality of eigenfunctions of A form a "continuous branch" of some given length, the definition of this concept being motivated by the idea of a continuous curve through the origin in two dimensions.

14.3.1 Definition. Let S be a subset of $\mathscr{B}$ and let $r > 0$ be given. S is said to form a **continuous branch** of length r iff for every $r' < r$, the boundary of every open set containing 0 and contained in the ball centre 0 and radius r' has non-empty intersection with S.

14.3.2 Theorem. *Let D be a bounded open subset of the real Banach space $\mathscr{B}$, and let D contain the origin. Suppose that $A : \bar{D} \to E$ is compact, and that*

$$\operatorname{dist}(0, A(\partial D)) = \inf_{f \in \partial D} \|Af\| > 0. \tag{14.3.1}$$

Then A has an eigenfunction on $\partial D \cap E$ corresponding to a positive characteristic value.

If D is a ball centre the origin this result is readily deduced from the Schauder Fixed Point Theorem (see Problem 14.9). The point of allowing D to be *any* open set is that the existence of continuous branches of eigenfunctions may be deduced from the theorem. The heart of the proof is the next lemma established by a degree theoretic argument.

14.3.3 Definition. Two non-zero vectors $f, g \in \mathscr{B}$ are said to be in the same direction iff there is a $c > 0$ such that $f = cg$.

14.3.4 Lemma. *Take D and $\mathscr{B}$ as in the theorem, and let $B : \bar{D} \to \mathscr{B}$ be compact. Suppose $I - B$ does not vanish on ∂D, and assume that there is a non-zero $p \in \mathscr{B}$ such that $(I - B)f$ and p are never in the same direction for $f \in \partial D$. Then B has an eigenfunction on ∂D corresponding to a positive characteristic value.*

Proof. Consider the homotopy $\{H_t\}$ for $0 \leqslant t \leqslant 1$, where

$$H_t f = t(I - B)f - (1 - t)p.$$

Note first that $H_t f \neq 0$ for any $t \in (0, 1)$ and $f \in \partial D$, for otherwise the vectors $(I - B)f$ and $t^{-1}(1 - t)p$ would, contrary to assumption, be in the same direction for some such f. Also $H_1 = I - B$, and by a condition of the lemma, $I - B$ does not vanish on ∂D. Thus H_t does not vanish on ∂D for any $t \in (0, 1]$. However, since $p \neq 0$, for sufficiently small t, H_t does not vanish in $\bar{D}$ and so has degree zero. Therefore by homotopy invariance, $d(I - B, 0, D) = 0$.

If the assertion of the lemma is false, there do not exist a $t > 0$ and $f \in \partial D$ such that $f = tBf$. It follows that $I - tB$ does not vanish on ∂D for $0 \leqslant t \leqslant 1$. Noting that $0 \in D$ and using homotopy invariance, we deduce that

$$1 = d(I, 0, D) = d(I - B, 0, D).$$

This contradicts the conclusion reached in the previous paragraph. □

The theorem will be proved by applying this lemma to tA for large positive t. The following shows that the last condition of the lemma is satisfied.

14.3.5 Lemma. *There exist $p \in \mathscr{B}$ and $t_0 > 0$ such that for all $t > t_0$ and $f \in \partial D$ the vectors $(I - tA)f$ and p are not in the same direction.*

Proof. Fix any $p \in E$ with $\|p\| = 1$. If the result is false there are a sequence (t_n) of real numbers tending to infinity, a sequence (c_n) of positive numbers, and a sequence (f_n) in ∂D, such that

$$f_n - t_n A f_n = c_n p.$$

It follows on dividing by t_n and taking norms that

$$(t_n^{-1} f_n - Af_n)/\|t_n^{-1} f_n - Af_n\| = p. \tag{14.3.2}$$

Since A is compact, there is a subsequence, still denoted by (f_n), such that (Af_n) is convergent; let $g = \lim Af_n$. From (14.3.1) $g \neq 0$, and as $Af_n \in E$ and E is closed, $g \in E$. Letting $n \to \infty$ in (14.3.2) we deduce that $g/\|g\| = -p$, and since $g \in E$, then $-p \in E$. Therefore, as E is a cone and $p \in E$, $p = 0$. This contradicts the fact that $\|p\| = 1$. □

Proof of Theorem 14.3.2. Suppose that there is no $t > 0$ such that $I - tA$ vanishes on ∂D. Then with $t > t_0$ (t_0 being defined as in Lemma 14.3.5) and $B = tA$, the conditions of Lemma 14.3.4 are satisfied. Thus for some $\mu > 0$ and $f \in \partial D$, $f = \mu Bf = t\mu Af$, and it follows that with $t' = t\mu > 0$, $I - t'A$ vanishes on ∂D contrary to assumption.

Therefore $f = t'Af$ for some $t' > 0$ and $f \in \partial D$. Since $A\bar{D} \subset E$ (a condition of the theorem), then $f \in E$. Thus f is an eigenfunction in $\partial D \cap E$ corresponding to the positive characteristic value t'. □

The conditions of this theorem are extremely restrictive. Indeed the requirement that A should map the whole of $\bar{D}$ into E is not satisfied even by a positive linear operator. This objection is met by the next result in which conditions are imposed on the behaviour of A on $\partial D \cap E$ only.

14.3.6 Theorem. *Suppose D is a bounded open subset of the real Banach space $\mathscr{B}$, and let $0 \in D$. Assume that $A : \partial D \cap E \to E$ is compact, and that*

$$\operatorname{dist}(0, A(\partial D \cap E)) > 0 \tag{14.3.3}$$

Then A has an eigenfunction on $\partial D \cap E$ corresponding to a positive characteristic value.

Proof. The idea of the proof is to show that there is an extension, $\tilde{A}$ say, of A to the whole of $\bar{D}$ with $\tilde{A}$ satisfying the conditions of Theorem 14.3.2. This will be enough to prove the result, for evidently an eigenfunction of $\tilde{A}$ in $\partial D \cap E$ is also an eigenfunction of A.

Since E is convex, the convex hull $\operatorname{co} R(A) \supset R(A) = A(\partial D \cap E)$ is contained in E. By the Dugundji Extension Theorem 3.2.12, A has an extension, still denoted by A, to $\bar{D}$ with range in $\operatorname{co} R(A) \subset E$, and $A : \bar{D} \to \operatorname{co} R(A)$ is compact (Lemma 8.2.4). This extension does not quite do the job. For although $A(\partial D \cap E)$ is safely away from the origin, $A(\partial D)$ is only known to lie in $\operatorname{co} A(\partial D \cap E)$, and in infinite dimensions this is not enough to ensure that $d(0, A(\partial D)) > 0$. Thus condition (14.3.1) may not hold. However, if a continuous operator P can be found which leaves $A(\partial D \cap E)$ unchanged, but maps the part of the cone near the origin away from the origin, then $\tilde{A} = PA$ will have all the required properties. It is now shown that such a P can be obtained by a simple geometrical construction.

Let $a = \text{dist}(0, A(\partial D \cap E))$, and choose any u with $-u \in E$ and $\|u\| = \frac{1}{4}a$. Let S be the intersection of E with the closed ball centre u and radius $\frac{1}{2}a$. Finally, for any $f \in S$ let $\tilde{f}$ be the point where the line through u and f meets the curved boundary of S—see Fig. 14.5—that is

$$\tilde{f} = u + \tfrac{1}{2}a(f - u)/\|f - u\|.$$

Then define $Pf = f$ for $f \in E \backslash S$, and $Pf = \tilde{f}$ for $f \in S$. The operator P projects the interior of S onto its curved boundary and leaves the rest of E unchanged. Hence $d(0, P(E)) > 0$. The continuity of P is easily checked. Thus P has all the required properties, and the theorem follows on applying Theorem 14.3.2 to $\tilde{A} = PA$. □

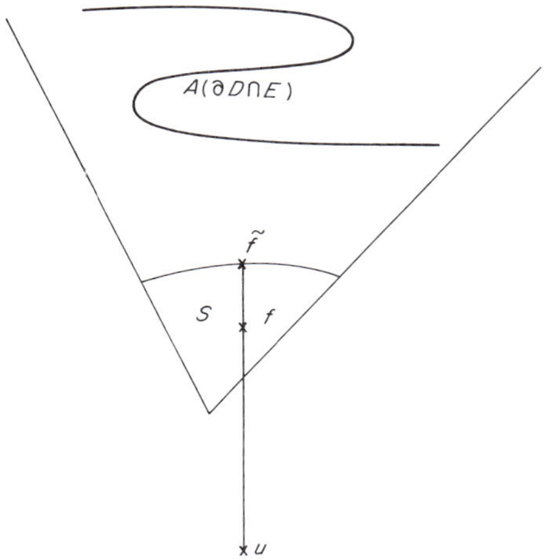

Fig. 14.5

If A is positive and (14.3.3) holds for every open D containing the origin and contained in $S(0, r)$, the eigenfunctions of A form a continuous branch of length r in E. The theorem is now tested on a Hammerstein integral equation.

14.3.7 Example. Consider the equation $f = \mu A f$ where

$$Af(x) = \int_0^1 k(x, y)\, \psi[y, f(y)]\, dy,$$

and k and ψ are continuous real valued functions. If k is strictly positive, there is an $\varepsilon > 0$ such that $k(x, y) \geqslant \varepsilon$ for $0 \leqslant x, y \leqslant 1$. If also $\psi[y, z] \geqslant z$ for $0 \leqslant y \leqslant 1$ and $z \geqslant 0$, A will be positive for E the cone of non-negative functions in some suitable Banach space $\mathscr{B}$. The principal difficulty comes in choosing $\mathscr{B}$ in such a way that (14.3.3) will be satisfied. Unfortunately the simplest possibility $\mathscr{B} = \mathscr{C}([0, 1])$ will not work, essentially because the integral of f may be small even if $\|f\| = 1$ and $f \geqslant 0$. If $\mathscr{B} = \mathscr{L}_1(0, 1)$, for $f \in E$,

$$\begin{aligned}\|Af\| &= \int_0^1 \int_0^1 k(x, y)\,\psi[y, f(y)]\,\mathrm{d}x\,\mathrm{d}y \\ &\geqslant \int_0^1 \int_0^1 \varepsilon f(y)\,\mathrm{d}x\,\mathrm{d}y \\ &= \varepsilon\|f\|,\end{aligned}$$

and (14.3.3) is satisfied. However, a condition on the growth of ψ is needed to ensure the compactness of A. If such a condition holds, Theorem 14.3.6 may be applied, and we can conclude that A has a continuous branch of eigenfunctions of infinite length in E.

In this example if the condition of strict positivity on k is relaxed, the theorem cannot be used, for (14.3.3) may not be satisfied. This condition is too severe for most applications, and in particular does not hold if k is the Green's function for the homogeneous Dirichlet problem. A less restrictive condition replaces (14.3.3) in the next theorem, which is a version of the Monotone Minorant Theorem of Krasnoselskii (1964a, p. 268).

14.3.8 Definition. The operator B will be called **homogenous** iff

$$B(tf) = tBf \qquad (f \in E,\ t \geqslant 0).$$

14.3.9 Theorem. *Let $E_r = E \cap \bar{S}(0, r)$. Suppose A is a positive compact operator defined on E_r. Assume that A has a monotone homogeneous minorant B such that for some $m > 0$ and non-zero $u \in E$, $Bu \geqslant mu$. Then A has a continuous branch of eigenfunctions of length r in E.*

Proof. It must be shown that A has an eigenfunction on $\partial D \cap E$ for any open D containing 0 and contained in $\bar{S}(0, r')$ for $r' < r$ Choose any such D which will be assumed fixed throughout the proof.

Set $A_t f = Af + tu$ for $t > 0$ and $f \in E_r$. Since A is positive, $A_t f \geqslant 0$, and by Problem 8.7(iii), $\operatorname{dist}(0, A_t(\partial D \cap E)) > 0$. Theorem 14.3.6 then shows that for each t there exist $\mu_t > 0$ and $f_t \in \partial D \cap E$ such that

$$\mu_t(Af_t + tu) = \mu_t A_t f_t = f_t. \tag{14.3.4}$$

We next prove that $\mu_t \leqslant m^{-1}$ for all $t > 0$. As B is a minorant of A, $Bf_t \leqslant Af_t$. Hence from (14.3.4),

$$Bf_t + tu \leqslant \mu_t^{-1} f_t. \tag{14.3.5}$$

Since $Af_t \geqslant 0$, (14.3.4) shows that $f_t \geqslant \mu_t t u$, and we deduce from an elementary property of cones (Problem 8.7(iv)) that there is a largest real number s_t for which $f_t \geqslant s_t u$. Now

$$\mu_t^{-1} f_t \geqslant Bf_t \geqslant B(s_t u) \geqslant s_t Bu \geqslant m s_t u,$$

where (14.3.5), the monotonicity and homogeneity of B, and the fact that $Bu \geqslant mu$ have been used in succession. Therefore $f_t \geqslant \mu_t m s_t u$, and it follows from the maximality of s_t that $\mu_t m s_t \leqslant s_t$. Thus $\mu_t \leqslant m^{-1}$ as asserted.

The result is now readily obtained on choosing a sequence (t_n) tending to zero and taking limits. For since $\mu_t \leqslant m^{-1}$ there is a subsequence, still denoted by (t_n), such that (μ_{t_n}) is convergent. Let $\mu = \lim \mu_{t_n}$. As A is compact it is possible to arrange, by choosing a further subsequence if necessary, that (Af_{t_n}) is convergent. Therefore from (14.3.4) (f_{t_n}) is convergent with limit f say, and $f \in \partial D \cap E$ since this set is closed. Hence

$$f = \lim f_{t_n} = \lim \mu_{t_n}(Af_{t_n} + t_n u) = \mu Af. \qquad \square$$

14.3.10 Example. A re-examination of Example 14.3.7 using this theorem is illuminating. Suppose as before that $\psi(y, z) \geqslant z$ for $0 \leqslant y \leqslant 1$ and $z \geqslant 0$, but now assume only that k is non-negative. Then for non-negative f,

$$Af(x) \geqslant \int_0^1 k(x, y) f(y)\, dy = Bf(x)$$

say. If there is a continuous non-negative u and an $m > 0$ such that $Bu \geqslant mu$, the theorem is applicable in $\mathscr{C}([0, 1])$ with E the usual cone of non-negative functions, and it follows that A has a continuous branch of eigenfunctions of infinite length.

The Monotone Minorant Theorem is specially useful in the important case when k is a Green's function. For frequently k will be non-negative and B will have an eigenfunction in E associated with its smallest characteristic value. This eigenfunction will be a valid choice for u. Also as $\mathscr{C}([0, 1])$ is being used no growth condition on ψ is needed.

In the examples from elasticity and water wave theory in the last section, the existence of small solutions was established by local bifurcation theory. These examples will now be examined again with the aim of proving existence of solutions of a physically reasonable range of size.

14.3.11 Example. (Saaty, 1967; p. 298). The buckling of a compressed rod (Example 14.2.7) is governed by the equation $\phi = \mu A\phi$ where

$$A\phi(x) = \rho(s) \int_0^1 k(s, t)\, \phi(t)\, dt \,. \left\{1 - \left[\int_0^1 \frac{\partial k}{\partial s}(s, w)\, \phi(w)\, dw\right]^2\right\}^{\frac{1}{2}}.$$

Choose $\mathscr{B} = \mathscr{C}([0, 1])$ the space of continuous real valued functions with the sup norm, and let E be the cone of non-negative functions. The kernel k is given explicitly by (4.2.5), and a routine calculation shows that

$$\left| \int_0^1 \frac{\partial k}{\partial s}(s, w)\, \phi(w)\, \mathrm{d}w \right| \leqslant \tfrac{1}{2}\|\phi\| \qquad (\phi \in E).$$

Therefore A is defined on $E_r = E \cap \bar{S}(0, r)$ for any $r < 2$, and is evidently positive and compact. A monotone minorant for A is easy to find; it is just the linear operator B where

$$B\phi(s) = (1 - \tfrac{1}{4}r^2)^{\frac{1}{2}}\, \rho(s) \int_0^1 k(s, t)\, \phi(t)\, \mathrm{d}t.$$

Further, with $\phi = y''$ and $\lambda = \mu(1 - \frac{1}{4}r^2)^{\frac{1}{2}}$, the equation $\phi = \mu B\phi$ corresponds to

$$y''(s) + \lambda\rho(s)\, y(s) = 0, \qquad y(0) = y(1) = 0,$$

and since ρ is strictly positive, by the standard Sturm–Liouville theory the system has a non-negative eigenfunction associated with a positive eigenvalue λ. That is there is a non-zero $u \in E$ and a $\mu > 0$ such that $u = \mu Bu$. Thus all the conditions of the Monotone Minorant Theorem 14.3.9 are satisfied, and it follows that A has a continuous branch of eigenfunctions of length 2 in E. This confirms the physical picture of large buckling if the compression is big enough.

14.3.12 Example. A major breakthrough in the periodic wave train problem (Example 14.2.8) was made by Krasovskii (1961), who established the existence of waves with any maximum angle of inclination less than $\pi/6$ in a stream of finite or infinite depth. In the following we shall give a proof of Krasovskii's result for the case of infinite depth. The method, due to Keady (1972), is based on the Monotone Minorant Theorem.

In the Nekrasov formulation the operator depends in a rather complicated manner on the parameter μ and in Krasovskii's proof the problem is rewritten in terms of an operator independent of μ. As before the unknown function θ is the angle of inclination of the velocity vector at the free boundary. With an appropriate scaling the range of the independent variable may be taken to be $[-\pi, \pi]$. However, it will be possible to show that waves symmetrical with respect to a crest have the required properties. The corresponding θ will be an odd function, and it is thus sufficient to determine this function on $[0, \pi]$. The basic equation of the problem may then be expressed in the form $\theta = \mu A\theta$. Here $\mu = gl/2\pi c^2$ where g is the acceleration due to gravity and l and c are the wave length and wave speed respectively. The operator A is defined formally by

$$A\theta(x) = \int_0^\pi k(x, y) \exp[3C\theta(y)] \sin\theta(y)\, dy.$$

As before k is the Green's function

$$k(x, y) = \frac{2}{\pi} \sum_{n=1}^{\infty} n^{-1} \sin nx \sin ny.$$

The harmonic conjugation operator C is defined by taking $C\theta$ to be the boundary value of the real part of an analytic function whose imaginary part has boundary value θ. To make this precise, on $[-\pi, \pi]$ take the odd real valued continuous function θ which vanishes at π, and extend θ to the whole real axis by requiring that the extension should be periodic with period 2π. Let Θ be a harmonic function defined on $(-\infty, \infty) \times (0, \infty)$ such that $\Theta(x, x') \to 0$ as $x' \to \infty$ and $\Theta(x, x') \to \theta(x)$ as $x' \to 0$. Assume that $\Phi(x, x') + i\Theta(x, x')$ is an analytic function of $x + ix'$, and take $C\theta(x) = \lim_{x' \to 0} \Phi(x, x')$. $C\theta$ is then given up to an arbitrary constant which will be fixed by specifying that

$$\int_0^{2\pi} C\theta(x)\, dx = 0.$$

In view of the properties required of the solution, a natural choice of Banach space is $\mathscr{C}_*[0, \pi]$, the set—equipped with the sup norm $\|\cdot\|$—of real-valued continuous functions defined on $[0, \pi]$ and vanishing at the end points. E will be the cone of non-negative functions in $\mathscr{C}_*[0, \pi]$. It will also be necessary to use $\mathscr{L}_p(0, \pi)$ in the proof, and the norm in this space will be denoted by $\|\cdot\|_p$. The goal is to show that given any positive number $\theta_m < \pi/6$, the eigenfunction problem $\theta = \mu A\theta$ has a solution in E with $\|\theta\| = \theta_m$. Most of the difficulties in the proof stem from the awkward nature of the harmonic conjugation operator C. The derivation of the properties of C is a technical exercise in complex variable theory, and the necessary results are quoted—for a proof see Zygmund (1959). The properties of the Green's function are well known. The basic facts are summarized as follows.

14.3.13 Lemma. *The Green's function $k(x, y)$ is continuous and non-negative for all $x \neq y$ and $x, y \in [0, \pi]$. If $1 \leqslant p < \infty$, there is a real number k_p such that for all $x \in [0, \pi]$,*

$$\int_0^\pi |k(x, y)|^p\, dy \leqslant k_p.$$

14.3.14 Lemma. *The harmonic conjugation operator C defined on $\mathscr{C}_*[0, \pi]$ has the following properties.*

(i) *Given any p with $1 < p < \infty$, there is a real number m_p dependent only on p such that $\|C\theta\|_p \leqslant m_p\|\theta\|_p$.*

(ii) *Given any $d > 0$, there is a real number n_d such that for any θ with $a\|\theta\| = \frac{1}{2}\pi - d$,*

$$\int_0^\pi \exp[aC\theta(x)]\,\mathrm{d}x \leqslant n_d.$$

We may now proceed with the verification of the conditions of the Monotone Minorant Theorem. The details are somewhat tedious, although in principle little more than Hölder's inequality (Theorem 2.5.3) is required.

14.3.15 Lemma. *If $d > 0$, A is a positive compact operator on the closed ball $\bar{S}(0, \pi/6 - d)$ in $\mathscr{C}_*[0, \pi]$.*

Proof. The positivity is obvious, and continuity will be established below. The proof of compactness involves similar manipulations and will be omitted.

Throughout the proof $d > 0$ will be fixed. From the properties of the Green's function, $R(A) \subset \mathscr{C}_*[0, \pi]$. Thus continuity will follow if it can be shown that there is a number c such that

$$\|A\theta_1 - A\theta_2\| \leqslant c\|\theta_1 - \theta_2\|. \tag{14.3.6}$$

To prove this set $\phi_j = C\theta_j$ for $j = 1, 2$ and note the identity

$$A\theta_1(x) - A\theta_2(x) = \frac{1}{2}\int_0^\pi k(x, y)\,[\mathrm{e}^{3\phi_1(y)} + \mathrm{e}^{3\phi_2(y)}]\,[\sin\theta_1(y) - \sin\theta_2(y)]\,\mathrm{d}y$$
$$+ \frac{1}{2}\int_0^\pi k(x, y)\,[\mathrm{e}^{3\phi_1(y)} - \mathrm{e}^{3\phi_2(y)}]\,[\sin\theta_1(y) + \sin\theta_2(y)]\,\mathrm{d}y. \tag{14.3.7}$$

For the first integral,

$$\left|\int_0^\pi k(x, y)\,\mathrm{e}^{3\phi_j(y)}\,[\sin\theta_1(y) - \sin\theta_2(y)]\,\mathrm{d}y\right|$$
$$= 2\left|\int_0^\pi k(x, y)\,\mathrm{e}^{3\phi_j(y)}\cos\tfrac{1}{2}[\theta_1(y) + \theta_2(y)]\sin\tfrac{1}{2}[\theta_1(y) - \theta_2(y)]\,\mathrm{d}y\right|$$
$$\leqslant \int_0^\pi k(x, y)\,\mathrm{e}^{3\phi_j(y)}\,|\theta_1(y) - \theta_2(y)|\,\mathrm{d}y$$
$$\leqslant \|\theta_1 - \theta_2\|\left\{\int_0^\pi |k(x, y)|^p\,\mathrm{d}y\right\}^{1/p}\left\{\int_0^\pi \mathrm{e}^{3q\phi_j(y)}\,\mathrm{d}y\right\}^{1/q},$$

by Hölder's inequality for any $1 < p, q < \infty$ with $p^{-1} + q^{-1} = 1$. Since $d > 0$, there is a $q > 1$ such that $3q(\pi/6 - d) < \frac{1}{2}\pi$. For such q, Lemmas

14.3.13 and 14.3.14 respectively show that the first and second curly bracketed terms are bounded. The second integral in (14.3.7) is treated by a similar argument after first noting the elementary inequality

$$|e^x - e^y| \leqslant |x - y| (e^x + e^y).$$

(14.3.6) follows immediately. □

14.3.16 Lemma. *Take $d > 0$ and set $r = \pi/6 - d$. Let E_r be the intersection of E with the ball $\bar{S}(0, r)$ in $\mathscr{C}_*[0, \pi]$. Choose some $p > 1$. For real γ define the operator $B : E \to \mathscr{C}_*[0, \pi]$ by*

$$B\theta(x) = \gamma \left\{ \int_0^\pi k(x, y) |\theta(y)|^{1/p} \, dy \right\}^p.$$

Then for some $\gamma > 0$, B is a monotone homogeneous minorant of A on E_r, and there are an $\alpha > 0$ and a non-zero $u \in E$ such that $Bu = \alpha u$.

Proof. Clearly B is monotone and homogeneous (Definition 14.3.8), and $Bu = \alpha u$ with $\alpha > 0$ if $u(x) = (\sin x)^p$. It only remains to prove that B is a minorant. With $\phi = C\theta$,

$$[\gamma^{-1} B\theta(x)]^{1/p} = \int_0^\pi k(x, y) |\theta(y)|^{1/p} \, dy$$

$$= \int_0^\pi |k(x, y) \, e^{3\phi(y)} \sin \theta(y)|^{1/p} |k(x, y)|^{1/q} \left| \frac{\theta(y)}{\sin \theta(y)} \right|^{1/p} e^{-3p^{-1}\phi(y)} \, dy,$$

where $p^{-1} + q^{-1} = 1$. For any $r, s > 1$ with $p^{-1} + r^{-1} + s^{-1} = 1$, by Hölder's inequality,

$$[\gamma^{-1} B\theta(x)]^{1/p} \leqslant \left\{ \int_0^\pi k(x, y) \, e^{3\phi(y)} \sin \theta(y) \, dy \right\}^{1/p} \left\{ \int_0^\pi |k(x, y)|^{r/q} \, dy \right\}^{1/r} \times$$

$$\times \left\{ \int_0^\pi \left| \frac{\theta(y)}{\sin \theta(y)} \right|^{s/p} e^{-3sp^{-1}\phi(y)} \, dy \right\}^{1/s}.$$

Now choose any s such that $1 < s < p$. Then by Lemma 14.3.14(ii) the last integral is finite. Also it may readily be checked that $rq^{-1} > 1$, and the second integral is thus finite by Lemma 14.3.13. Therefore, for some real c,

$$[\gamma^{-1} B\theta(x)]^{1/p} \leqslant c[A\theta(x)]^{1/p}.$$

The result is obtained on taking $\gamma^{-1} = c^p$. □

Global existence now follows directly from the Monotone Minorant Theorem 14.3.9 which also shows that the solutions form a continuous branch.

14.3.17 Theorem. *Let θ_m be any number such that $0 < \theta_m < \pi/6$. There is a steady symmetrical periodic wave train in deep water for which the maximum angle of inclination of the velocity vector at the surface is θ_m. These solutions form a continuous branch of length $\pi/6$ in $\mathscr{C}_*[0, \pi]$.*

Krasovskii's result represents a quite remarkable advance towards the solution of a problem which has resisted mathematical treatment for more than a century; this result has recently been strengthened by Keady and Norbury (1978) and Toland (1977). Nonetheless the problem is by no means fully resolved and is still an active area of research. Indeed the long standing conjecture that a steady periodic train of waves with maximum angle of slope greater than $\pi/6$ cannot exist has still not been confirmed or disproved —see Toland (1977). Questions concerning uniqueness and stability remain completely open. Apart from the references cited above, the reader may refer to the articles of Hyers (1964), Keady (1972) and Wehausen (1963) which review the extensive literature on the wave problem.

Problems

14.1 Construct bifurcation diagrams for the operator $\phi : \mathbb{R} \to \mathbb{R}$ when

(i) $\phi(x) = x + x^2$.
(ii) $\phi(x) = x + x^3$.
(iii) $\phi(x) = x(1 - x + x^2)$.
(iv) $\phi(x) = \sin x$.

14.2 For the real space $\mathscr{C}([0, 1])$ construct a bifurcation diagram for A_μ where

$$A_\mu f(x) = \int_0^1 [f(x)]^2 \, dx + \mu^{-1}.$$

14.3 Show that the simultaneous equations

$$u'' + \lambda[u + v(u^2 + v^2)] = 0, \qquad v'' + \lambda[v - u(u^2 + v^2)] = 0$$

have no real-valued non-trivial solution satisfying the boundary condition $u(0) = u(1) = v(0) = v(1)$, but note that the linearized system has eigenvalues $\pi^2, 4\pi^2, \ldots$.

14.4 Let D be an open subset of the real Banach space $\mathscr{B}$. Suppose that $A : D \to \mathscr{B}$ is continuous, and assume that there are a non-zero $\phi_0 \in D$ and a neighbourhood S of ϕ_0 such that

(i) A is Fréchet differentiable on S and $A'(\cdot)$ is continuous,
(ii) $\phi_0 = \mu_0 A \phi_0$,
(iii) $\mu_0^{-1} \in \rho(A'(\phi_0))$.

Prove that there is an $\varepsilon > 0$ such that every $\mu \in [\mu_0 - \varepsilon, \mu_0 + \varepsilon]$ is a characteristic value of A. (Use the Implicit Function Theorem 4.4.9).

14.5 Let D be an open subset of the Banach space $\mathscr{B}$, and let $0 \in D$. If $A : D \to \mathscr{B}$ is compact and has Fréchet derivative L at 0, prove that L is compact.

14.6 Let D be a bounded open subset of the real Banach space $\mathscr{B}$, and assume that $A : \bar{D} \to \mathscr{B}$ is compact and that $Af \neq f$ on ∂D. Prove that A has an eigenfunction on ∂D if either of the following hold.

(i) $0 \in D$ and $d(I - A, 0, D) \neq 1$.
(ii) $0 \notin D$ and $d(I - A, 0, D) \neq 0$.

14.7 Consider the boundary value problem

$$f''(x) + \mu\phi[x, f(x)] = 0,$$
$$f(0) = f(1) = 0.$$

Suppose $\phi : [0, 1] \times \mathbb{R} \to \mathbb{R}$ is continuous and assume that $\phi[x, z] = z + \psi[x, z]$ where $z^{-1}\psi[x, z] \to 0$ uniformly as $z \to 0$. Show that the bifurcation points are $\pi^2, 4\pi^2, \ldots$.

14.8 Supply a proof of Theorem 14.2.6.

14.9 If $D = S(0, r)$, show that Theorem 14.3.5 may be proved simply by applying the Schauder Fixed Point Theorem to the operator $\tilde{A}$ where

$$\tilde{A}f = \|f\|A(rf/\|f\|) + (r - \|f\|)g,$$

and g is some vector in E.

14.10 Suppose that A is an $n \times n$ matrix each of whose entries is strictly positive. Show that A has a positive eigenvalue corresponding to an eigenvector with positive components.

14.11 Consider the Hammerstein integral operator A where

$$Af(x) = \int_0^1 k(x, y)\,\psi[y, f(y)]\,\mathrm{d}y,$$

and assume that k is continuous and strictly positive and that $\psi : [0, 1] \times \mathbb{R} \to \mathbb{R}$ is continuous. Suppose there are real numbers $a > 0$ and b, c such that for some p with $1 \leqslant p < \infty$,

$$az^p \leqslant \psi[y, z] \leqslant bz^p + c \qquad (0 \leqslant y \leqslant 1, z \geqslant 0).$$

Deduce from Theorem 14.3.6 that A has a continuous branch of infinite length of non-negative eigenfunctions in $\mathscr{L}_p(0, 1)$.

References

Agmon, S. (1965). "Lectures on Elliptic Boundary Value Problems". Van Nostrand, New York.

Aleksandrjan, R. A., Berezanskii, Ju. M., Il'in, V. A. and Kostjučenko, A. G. (1976). Some questions in spectral theory for partial differential equations, *Amer. Math. Soc. Transl.* (2) **105**, 1–53.

Amann, H. (1976). Fixed point theorems and nonlinear eigenvalue problems, *SIAM Rev.* **18**, 620–709.

Anselone, P. M. (ed.) (1964). "Nonlinear Integral Equations". University of Wisconsin Press, Madison.

Anselone, P. M. (1971). "Collectively Compact Operator Approximation Theory". Prentice Hall, Englewood Cliffs, New Jersey.

Atkinson, K. E. (1976). "A Survey of Numerical Methods for the Solution of Fredholm Integral Equations of the Second Kind". SIAM, Philadelphia.

Aubin, J. P. (1972). "Approximation of Elliptic Boundary Value Problems". Wiley, New York.

Aziz, A. K. (ed.) (1972). "The Mathematical Foundations of the Finite Element Method with Applications to Partial Differential Equations". Academic Press, New York and London.

Babuška, I. and Aziz, A. K. (1976). On the angle condition in the finite element method, *SIAM J. Numer. Anal.* **13**, 214–226.

Baker, C. T. H. (1977). "The Numerical Treatment of Integral Equations". Clarendon Press, Oxford.

Barnhill, R. and Whiteman, J. R. (1973). Error analysis of finite element methods with triangles for elliptic boundary value problems. *In* "The Mathematics of Finite Elements and Applications" (Whiteman, J. R., ed.), pp. 83–112. Academic Press, London.

Bartle, R. G. (1966). "The Elements of Integration". Wiley, New York.

Bellman, R. and Kalaba, R. (1965). "Quasilinearization and Nonlinear Boundary Value Problems". Elsevier, New York.

Benjamin, T. B. (1976). Applications of Leray–Schauder degree theory to problems of hydrodynamic stability, *Math. Proc. Cambridge Philos. Soc.* **79**, 373–392.

Berger, M. and Berger, M. (1968). "Perspectives in Nonlinearity". Benjamin, New York.

Berger, M. (1977). "Nonlinearity and Functional Analysis". Academic Press, New York.

Bernkopf, M. (1966). The development of function spaces with particular reference to their origins in integral equation theory, *Arch. History Exact Sci.* **3**, 1–136.

Bourbaki, N. (1969). "Éléments d'Histoire des Mathématiques". Hermann, Paris.

Bramble, J. H. and Zlámal, M. (1970). Triangular elements in the finite element method, *Math. Comput.* **24**, 809–820.

Burkill, J. C. (1951). "The Lebesgue Integral". Cambridge University Press.

Chandra, J. and Davis, P. (1974). A monotone method for quasilinear boundary value problems. *Arch. Rational Mech. Anal.* **54**, 257–266.

Courant, R. (1950). "Dirichlet's Principle, Conformal Mapping, and Minimal Surfaces". Wiley, New York.

De Barra, G. (1974). "Introduction to Measure Theory". Van Nostrand, New York.

Dennis, J. E. (1971). Towards a unified convergence theory for Newton-like methods. *In* "Nonlinear Functional Analysis and Applications" (Rall, L. B., ed.), pp. 425–472. Academic Press, New York.

Dickey, R. W. (1976). "Bifurcation Problems in Nonlinear Elasticity". Pitman, London.

Dunford, N. and Schwartz, J. (1958). "Linear Operators", Part I; (1963), Part II. Wiley, New York.

Flett, T. M. (1979). "Differential Analysis". Cambridge University Press.

Folland, G. B. (1976). "Introduction to Partial Differential Equations". Princeton University Press, Princeton, New Jersey.

Friedman, A. (1969). "Partial Differential Equations". Holt, Rinehart and Winston, New York.

Friedman, A. (1970). "Foundations of Modern Analysis". Holt, Rinehart and Winston, New York.

Fučík, S., Nečas, J., Souček, J. and Souček, V. (1973). "Spectral Analysis of Nonlinear Operators" (Lecture Notes in Math. 343). Springer-Verlag, Berlin.

Garabedian, P. R. (1964). "Partial Differential Equations". Wiley, New York.

Gilbarg, D. and Trudinger, N. S. (1977). "Elliptic Partial Differential Equations of Second Order". Springer-Verlag, Berlin.

Groetsch, C. W. (1977). "Generalized Inverses of Linear Operators". Marcel Dekker, New York.

Halmos, P. R. (1948). "Finite Dimensional Vector Spaces". Princeton University Press, Princeton, New Jersey.

Halmos, P. R. (1967). "A Hilbert Space Problem Book". Van Nostrand, Princeton, New Jersey.

Hewitt, E. (1960). The rôle of compactness in analysis. *Amer. Math. Monthly* **67**, 499–516.

Higgins, J. R. (1977). "Completeness and Basis Properties of Sets of Special Functions". Cambridge University Press.

Hilgers, J. W. (1976). On the equivalence of regularization and certain reproducing kernel Hilbert space approaches for solving first kind problems. *SIAM J. Numer. Anal.* **13**, 172–184.

Holtzman, J. M. (1970). "Nonlinear System Theory". Prentice Hall, Englewood Cliffs, New Jersey.

Hutson, V., Kendall, P. C. and Malin, S. (1972). Computation of the solution of geomagnetic induction problems: a general method with applications. *Geophys. J. R. astr. Soc.* **28**, 489–498.

Hyers, D. H. (1964). Some nonlinear equations of hydrodynamics. *In* "Nonlinear Integral Equations" (Anselone, P. M., ed.), pp. 319–344. University of Wisconsin Press, Madison.

Kato, T. (1966). "Perturbation Theory for Linear Operators". Springer-Verlag, Berlin.

Keady, G. (1972). Large-amplitude water waves and Krasovskii's existence proof. (Report No. **37**). Fluid Mechanics Research Institute, University of Essex.

Keady, G. and Norbury, J. (1978). On the existence theory for irrotational water waves. *Math. Proc. Cambridge Philos. Soc.* **83**, 137–157.

Keller, J. B. and Antman, S. (eds.) (1969). "Bifurcation Theory and Nonlinear Eigenvalue Problems". Benjamin, New York.

Krasnoselskii, M. A. (1958). Some problems of nonlinear analysis. *Amer. Math. Soc. Transl.* (2) **10**, 345–408.

Krasnoselskii, M. A. (1964a). "Topological Methods in the Theory of Nonlinear Integral Equations". Pergamon Press, London.

Krasnoselskii, M. A. (1964b). "Positive Solutions of Operator Equations". Noordhoff, Groningen.

Krasnoselskii, M. A. *et al.* (1972). "Approximate Solution of Operator Equations". Noordhoff, Groningen.

Krasovskii, Yu. P. (1961). On the theory of steady state waves of large amplitude. *USSR Computational Mathematics and Mathematical Physics* **1**, 966–1018.

Ladas, G. and Lakshmikantham, V. (1972). "Differential Equations in Abstract Spaces". Academic Press, New York.

Ladyzenskaya, O. A. (1969). "The Mathematical Theory of Viscous Incompressible Flow", 2nd edn. Gordon and Breach, New York.

Leggett, R. W. (1976). A new approach to the H-equation of Chandrasekhar. *SIAM J. Math. Anal.* **7**, 542–550.

Lloyd, N. G. (1978). "Degree Theory". Cambridge University Press.

Love, E. R. (1974). Inequalities for the capacity of an electrified conducting annular disc. *Proc. Roy. Soc. Edinburgh Sect. A.* **74**, 257–270.

Lusternik, L. A. and Sobolev, V. J. (1974). "Elements of Functional Analysis", 2nd edn. Hindustan Publishing Corpn., Delhi.

Mitchell, A. R. and Wait, R. (1977). "The Finite Element Method in Partial Differential Equations". Wiley, New York.

Monna, A. (1973). "Functional Analysis in Historical Perspective". Oosthoek, Scheltema and Holkema, Utrecht.

Monna, A. (1975). "Dirichlet's Principle". Oosthoek, Scheltema and Holkema, Utrecht.

Mooney, J. and Roach, G. (1976). Iterative bounds for the stable solutions of convex nonlinear boundary value problems. *Proc. Roy. Soc. Edinburgh Sect. A.* **76**, 81–94.

Naimark, M. A. (1968). "Linear Differential Operators". Ungar, New York.

Nashed, Z. (1974). Approximate regularized solutions to improperly posed linear integral and operator equations. *In* "Constructive and Computational Methods for Differential and Integral Equations" (Lecture Notes in Math. **430**), pp. 289–332. Springer-Verlag, Berlin.

Oden, J. T. and Reddy, J. N. (1976). "An Introduction to the Mathematical Theory of Finite Elements". Wiley, New York.

Ortega, J. and Rheinboldt, W. (1970). "Iterative Solutions of Nonlinear Equations in Several Variables". Academic Press, New York.
Prenter, P. M. (1975). "Splines and Variational Methods". Wiley, New York.
Rabinowitz, P. H. (1973). Some aspects of nonlinear eigenvalue problems. *Rocky Mountain J. Math.* **3**, 161–202.
Rall, L. B. (1969). "Computational Solution of Nonlinear Operator Equations". Wiley, New York.
Rall, L. B. (ed.) (1971). "Nonlinear Functional Analysis and Applications". Academic Press, New York.
Reed, M. and Simon, B. (1972). "Methods of Modern Mathematical Physics", Vol. I; (1975), Vol. II. Academic Press, New York.
Riesz, F. (1913). "Les Systèmes d'Équations Linéaires à une Infinité d'Inconnues". Gauthier-Villars, Paris.
Riesz, F. and Sz.-Nagy, B. (1955). "Functional Analysis". Ungar, New York.
Saaty, T. L. (1967). "Modern Nonlinear Equations". McGraw-Hill, New York.
Sattinger, D. H. (1973). "Topics in Stability and Bifurcation Theory" (Lecture Notes in Math. **309**). Springer-Verlag, Berlin.
Schechter, M. (1977). "Modern Methods in Partial Differential Equations". McGraw-Hill, New York.
Schwartz, J. T. (1969). "Nonlinear Functional Analysis". Gordon and Breach, New York.
Serrin, J. (1976). The solvability of boundary value problems. *Proc. Symp. P. M.* **18**, 507–524. Amer. Math. Soc., Providence, Rhode Island.
Shinbrot, M. (1969). Fixed point theorems. *In* "Mathematics in the Modern World", pp. 145–150. Freeman, San Francisco.
Showalter, R. E. (1977). "Hilbert Space Methods for Partial Differential Equations". Pitman, London.
Simmons, G. F. (1963). "Introduction to Topology and Modern Analysis". McGraw-Hill, New York.
Smart, D. R. (1974). "Fixed Point Theorems". Cambridge University Press.
Sneddon, I. (1972). "The Use of Integral Transforms". McGraw-Hill, New York.
Stakgold, I. (1968). "Boundary Value Problems of Mathematical Physics". Macmillan, New York.
Stakgold, I. (1971). Branching of solutions of nonlinear equations. *SIAM Rev.* **13**, 289–332.
Steen, L. A. (1973). Highlights in the history of spectral theory. *Amer. Math. Monthly* **80**, 359–381.
Strang, G. and Fix, G. (1973). "An Analysis of the Finite Element Method". Prentice-Hall, Englewood Cliffs, New Jersey.
Stuart, C. A. (1974). Existence theorems for a class of nonlinear integral equations. *Math. Z.* **137**, 49–66.
Stuart, C. A. (1975). Integral equations with decreasing nonlinearities. *J. Differential Equations* **18**, 202–217.
Taylor, A. E. (1958). "Introduction to Functional Analysis". Wiley, New York.
Temam, R. (1970). "Analyse Numerique: Résolution Approchée d'Équations aux Dérivées Partielles". Presses Universitaires, Paris.
Titchmarsh, E. C. (1962). "Eigenfunction Expansions". Clarendon, Oxford.
Todd, M. J. (1976). "The Computation of Fixed Points and Applications". Springer-Verlag, Berlin.
Toland, J. F. (1977). On the existence of a wave of greatest height and Stokes' con-

jecture (Report No. **87**). Fluid Mechanics Research Institute, University of Essex.
Treves, F. (1975). "Basic Linear Partial Differential Equations". Academic Press, New York.
Tricomi, F. G. (1957). "Integral Equations". Wiley, New York.
Vandergraft, J. S. (1967). Newton's method for convex operators in partially ordered spaces. *SIAM J. Numer. Anal.* **4**, 406–432.
Vilenkin, N. Ya. *et al.* (1972). "Functional Analysis". Noordhoff, Groningen.
Wehausen, J. V. (1963). Recent developments in free-surface flows. (Report No. **NA-63-5**). Institute of Engineering Research, University of California, Berkeley.
Weinberger, H. F. (1974). "Variational Methods of Eigenvalue Approximation". SIAM, Philadelphia.
Weinstein, A. and Stenger, W. (1972). "Methods of Intermediate Problems for Eigenvalue Theory and Ramifications". Academic Press, New York.
Whiteman, J. R. (ed.) (1973). "The Mathematics of Finite Elements and Applications", Vol. I; (1977), Vol. II. Academic Press, New York.
Whittaker, E. T. and Watson, G. N. (1927). "A Course of Modern Analysis", 4th edn. Cambridge University Press.
Zabreyko, P. P. *et al.* (1975). "Integral Equations—a Reference Text". Noordhoff, Groningen.
Zygmund, A. (1959). "Trigonometric Series". Cambridge University Press.

List of Symbols

Spaces

$\mathcal{B}, \mathcal{C}$	Banach spaces always, 19
$\mathcal{B}^*, \mathcal{C}^*$	Dual spaces, 50
$\mathbb{C}^n$	n-dimensional complex space
$\mathscr{C}(\Omega), \mathscr{C}(\Omega, \mathbb{C}^n)$	Bounded continuous functions, 15
$\mathscr{C}^k(\Omega), \mathscr{C}^k(\Omega, \mathbb{C}^n)$	Functions with k bounded continuous derivatives, 15
$\mathscr{C}_0^k(\Omega), \mathscr{C}_0^k(\Omega, \mathbb{C}^n)$	15
$\mathcal{H}$	Hilbert space always, 29
$\mathcal{H}^m, \mathcal{H}_0^m$	Sobolev spaces, 290, 291
ℓ	Vector space of sequences, 5
ℓ_p	Space of sequences with norm $\|\cdot\|_p$, 12
$\mathscr{L}_p(\Omega)$	Space of functions with norm $\|\cdot\|_p$, 54
$\mathscr{L}_p^{\text{loc}}(\Omega)$	Functions in $\mathscr{L}_p(S)$ for each compact $S \subset \Omega$, 55
$\mathscr{L}(\mathcal{B}, \mathcal{C}), \mathscr{L}(\mathcal{B})$	Spaces of bounded linear operators $\mathcal{B} \to \mathcal{C}, \mathcal{B} \to \mathcal{B}$, 79
$\mathcal{M}$	Linear subspace, 5
$\mathcal{M}_k$	314
$\mathbb{R}^n$	n-dimensional real space
$\mathcal{V}, \mathcal{W}$	Normed vector spaces, 8

Roman Alphabet

A	Nonlinear operator
A'	Fréchet derivative of A, 123
$A(S)$	Image of S under A, 64
$A^{-1}(S)$	Pre-image of S under A, 64
$A: S \to \mathcal{W}$	A maps S into $\mathcal{W}$, 64
B	Nonlinear operator
$B[\cdot, \cdot]$	Bilinear form, 175

$\mathbb{C}$	Complex number system
co	Convex hull, 7
$D(A)$	Domain of A, 64
$d(\cdot, \cdot)$	Metric, 8
$d(I + A, p, D)$	Degree, 340
dist (f, S)	Distance of f from the set S, 8
E	Cone, 212
ess sup	Essential supremum, 50
f, g, h	Points in a space
f^*, g^*, h^*	Points in dual
$\hat{f}$	Fourier transform, 58
$\tilde{f}_k$	Ritz approximation, 314
$G, \tilde{G}$	Green's operators, 331
$G(L), G'(L)$	Graph, inverse graph of L, 99
h_t, H_t	Homotopies, 335
I	Identity operator, 69
Im	Imaginary part of
j_ε	Mollifier, 59
$J_\phi(x)$	Jacobian determinant of ϕ at x
K	Integral operator
L	Linear operator, 66
L^{-1}	Inverse of L, 69
L^*	Adjoint of L, 163, 170
$\tilde{L}$	Extension of L, 65
$\bar{L}$	Closure of L, 100
L', L_0	263, 265
l	Formal differential operator, 97, 286, 305
l^*	Formal adjoint of l, 261, 287
l_p	Principal part of l, 286, 305
M	Linear operator
m_+, m_-	167

$N(L)$	Null space of L, 69
$N_{\pm}$	Deficiency spaces, 254
$n_{\pm}$	Dimensions of $N_{\pm}$ respectively, 254
P_{λ}	Spectral projection, 241
$\mathbb{R}$	Real number system
$\overline{\mathbb{R}}$, $\overline{\mathbb{R}}^{+}$	Extended, non-negative extended real number system, 41
Re	Real part of
$R(A)$	Range of A, 64
$R(\lambda; L)$	Resolvent $(\lambda I - L)^{-1}$ of L, 91
$r_{\sigma}(L)$	Spectral radius of L, 93
S	Set
$\bar{S}$, ∂S	Closure, boundary of S, 10–11
$[S]$	Linear span of S, 6
$S^{\perp}$	Orthogonal complement of S, 28, 155
$S(f, r)$, $\bar{S}(f, r)$	Open, closed balls with centre f and radius r, 8
$\mathscr{S}$	Class of sets
$\mathscr{S}_{\sigma}$	σ-algebra generated by $\mathscr{S}$, 40
T	Compact operator, 179
$(X, \mathscr{S}, \mu)$	Measure space, 42
Z	Set of critical points, 331

Greek Alphabet

μ	Measure, 41
$\rho(L)$	Resolvent set of L, 91
$\sigma(L)$	Spectrum of L, 91
$\sigma_{p}(L)$	Point spectrum of L, 91
ϕ'	Fréchet derivative of ϕ, 123
χ_{s}	Characteristic function of S, 38
Ω	Subset of $\mathbb{R}^{n}$

General

$\exists$	There exists
$\in$	Belongs to
$\notin$	Does not belong to
$\varnothing$	The empty set
$\subset$	For sets: Is a (not necessarily proper) subset of
	For operators: Is a restriction of, 65
$\cap, \cup$	Intersection, union
$S \backslash U$	Complement of U in S.
$\{x : P(x)\}$	Set of all x for which the statement $P(x)$ holds
$X \times Y$	Product set, 6
$\oplus$	Direct sum, 6
$+$	Vector sum, 6
$\perp$	Orthogonality, 2[illegible], 155
$\|\cdot\|$	Norm of vector, 8; norm of an operator, 71
$\|\cdot\|_p$	Norm in ℓ_p, 11; norm in $\mathscr{L}_p$, 54
$\|\cdot\|_m$	Norm in $\mathscr{H}^m$, 290
$\|\cdot\|_E$	Energy norm, 313
$\vert\cdot\vert_m$	319
$\vert\vert\vert\cdot\vert\vert\vert_p$	74, 75
$(\cdot, \cdot)$	Inner product, 26
$(\cdot, \cdot)_m$	Inner product in $\mathscr{H}^m$, 290
$(\cdot, \cdot)_E$	Energy inner product, 313
$[\cdot, \cdot]_a^b$	261, 266
$[\cdot, \cdot]$	Element of vector space $\mathscr{V} \times \mathscr{W}$, 7
	Order interval, 213
∇^2	The Laplacian
$\rightarrow$	Convergence of vectors, 10
	Uniform convergence of operators, 80
$\rightharpoonup$	Weak convergence of vectors, 156, 160
$\xrightarrow{s}$	Strong convergence of operators, 83
$\stackrel{w}{=}$	Weak equality, 288, 306

Index

Italics indicate principal definitions and statements of theorems.